Recent Advances in Microbiology

Recent Advances in Microbiology

Edited by Dean Watson

SYRAWOOD
PUBLISHING HOUSE
New York

Published by Syrawood Publishing House,
750 Third Avenue, 9th Floor,
New York, NY 10017, USA
www.syrawoodpublishinghouse.com

Recent Advances in Microbiology
Edited by Dean Watson

Cataloging-in-Publication Data

Recent advances in microbiology / edited by Dean Watson.
    p. cm.
Includes bibliographical references and index.
ISBN 978-1-68286-664-1
1. Microbiology. 2. Microorganisms. 3. Biology. I. Watson, Dean.
QR41.2 .R43 2019
579--dc23

TABLE OF CONTENTS

PREFACE

The aim of this book is to present researches that have transformed the discipline of microbiology and aided its advancement. Microbiology is the study of all types of microorganisms: unicellular, multicellular or a cellular. It is a vast subject that encapsulates a range of other fields of study such as parasitology, mycology, bacteriology, etc. The concepts and principles of microbiology are applied to a number of areas such as in the development of efficient methods of industrial fermentation or in the use of microorganisms as molecular carriers of DNA material, etc. Chapters compiled herein cover in detail some of the existing theories and innovative concepts revolving around this field of study. Different approaches, evaluations, methodologies and advanced studies on microbiology have also been included in this book. Those in search of information to further their knowledge will be greatly assisted by this book.

This book is the end result of constructive efforts and intensive research done by experts in this field. The aim of this book is to enlighten the readers with recent information in this area of research. The information provided in this profound book would serve as a valuable reference to students and researchers in this field.

At the end, I would like to thank all the authors for devoting their precious time and providing their valuable contribution to this book. I would also like to express my gratitude to my fellow colleagues who encouraged me throughout the process.

Editor

The rumen microbial metaproteome as revealed by SDS-PAGE

Timothy J. Snelling and R. John Wallace[*]

Abstract

Background: Ruminal digestion is carried out by large numbers of bacteria, archaea, protozoa and fungi. Understanding the microbiota is important because ruminal fermentation dictates the efficiency of feed utilisation by the animal and is also responsible for major emissions of the greenhouse gas, methane. Recent metagenomic and metatranscriptomic studies have helped to elucidate many features of the composition and activity of the microbiota. The metaproteome provides complementary information to these other –omics technologies. The aim of this study was to explore the metaproteome of bovine and ovine ruminal digesta using 2D SDS-PAGE.

Results: Digesta samples were taken via ruminal fistulae and by gastric intubation, or at slaughter, and stored in glycerol at −80 °C. A protein extraction protocol was developed to maximise yield and representativeness of the protein content. The proteome of ruminal digesta taken from dairy cows fed a high concentrate diet was dominated by a few very highly expressed proteins, which were identified by LC-MS/MS to be structural proteins, such as actin and α- and β-tubulins, derived from ciliate protozoa. Removal of protozoa from digesta before extraction of proteins revealed the prokaryotic metaproteome, which was dominated by enzymes involved in glycolysis, such as glyceraldehyde-3-phosphate dehydrogenase, phosphoenolpyruvate carboxykinase, phosphoglycerate kinase and triosephosphate isomerase. The enzymes were predominantly from the Firmicutes and Bacteroidetes phyla. Enzymes from methanogenic archaea were also abundant, consistent with the importance of methane formation in the rumen. Gels from samples from dairy cows fed a high proportion of grass silage were consistently obscured by co-staining of humic compounds. Samples from beef cattle and fattening lambs receiving a predominantly concentrate diet produced clearer gels, but the pattern of spots was inconsistent between samples, making comparisons difficult.

Conclusion: This work demonstrated for the first time that 2D-PAGE reveals key structural proteins and enzymes in the rumen microbial community, despite its high complexity, and that taxonomic information can be deduced from the analysis. However, technical issues associated with feed material contamination, which affects the reproducibility of electrophoresis of different samples, limits its value.

Keywords: Cattle, Proteomics, Rumen, Sheep

Background

The rumen is the primary digestive organ in ruminants such as cattle, sheep, buffaloes and deer. It contains a vast number of anaerobic eukaryotic and prokaryotic microorganisms, which break down ingested feed materials to short chain fatty acids that are absorbed, to be used by the host animal for energy [1, 2]. The cellulolytic microbes are particularly important, because the host animal lacks the necessary enzymes to break down cellulose, which is abundant in forage diets [3]. The microbial cells formed during fermentation constitute the majority source of amino acids flowing to the gastric stomach [4]. Thus, ruminal fermentation is of vital importance to the nutrition of the animal. There are environmental problems associated with modern ruminant livestock production, principally the excretion of nitrogenous wastes and the emission of methane [5, 6]. Understanding the composition and activity of the rumen microbial community is therefore crucial if we are to improve productivity and to lessen the environmental impact associated with ruminant livestock. Furthermore, the rumen has a major influence on the health of the animal, so understanding the composition and function

---
* Correspondence: john.wallace@abdn.ac.uk
Rowett Institute, University of Aberdeen, Foresterhill, Aberdeen AB16 5BD, UK

of the ruminal microbiota will also help to improve the health and welfare of the livestock [1, 7–9].

Traditional microbiological methods have largely given way to powerful non-cultivation methodologies, such as metagenomics and metatranscriptomics, in the study of complex microbial communities from environmental samples. Shotgun metagenomic sequencing has enabled a much deeper understanding of the composition of microbial communities and their gene contents, including those of the rumen [10–12]. Entire genomes of new, uncultured species have been assembled [10]. Novel enzymes have been extracted by so-called gene mining strategies [10, 13–15]. Most important for the livestock industry, we are beginning to understand the relations between microbial species and gene abundances and production characteristics, including methane emissions [16–18]. Metatranscriptomics describes the transcription of the genes to mRNA, which gives an impression of the activity of the microbial community rather than just its genetic complement [19]. It might be argued, however, that it is the combined output of transcription and translation, the metaproteome, that might tell us most about actual metabolic activity in the ecosystem.

Metaproteomic analysis aims to characterise the entire protein content of an environmental sample at a given point in time [20] Two main technical approaches are available in proteomics. The first is the well established 2D SDS-PAGE technology originated by O'Farrell [21]. Individual proteins are separated by isoelectric point in the first dimension and size in the second. The proteome is then visible when the gel is stained. Individual spots can be identified by peptide analysis following trypsinisation. Protein identification depends heavily on databases that generally do not include ruminal species. Activated sludge systems [20], anaerobic waste water [22], soil and sediments [23, 24], rhizosphere [25] and human faecal samples [26] have already been analysed by this method, but not until now the rumen. The second method utilises state-of-the-art mass spectrometric analysis of peptides derived by partial hydrolysis of protein mixtures without protein separation, so-called shotgun peptide sequencing. Many believe that the shotgun method, with the much larger volume of data generated, will supplant the gel-based method. Once again, the method relies upon peptide databases in which the great majority of ruminal species are not represented. The first analysis of the ruminal metaproteome using the latter methodology was published in 2015 [27].

The rumen shares some characteristics of the communities that have previously characterised by metaproteomics, in terms of microbial diversity and relative abundance of microorganisms and food materials in others, but it provides a unique challenge in the combination of these properties. The metaproteome will provide

an alternative insight into the function of the rumen microbial community compared to the nucleic acid metaomes, arguably one that might prove more useful as part of the campaign to lower methane emissions and to better understand the role of key enzymes involved in feed utilisation efficiency in ruminants. The aims of this study were to investigate the effectiveness of SDS PAGE methods for generating metaproteomic information from ruminal digesta and to evaluate the information that can be obtained by this method.

Methods
Animals and digesta sampling methods
Samples of ruminal digesta were obtained from dairy cows in Sweden and Finland, reindeer in Finland, and beef cattle and lambs in Scotland. Sampling was from live animals in the cases of dairy cows, reindeer and beef cattle and *post mortem* in the case of lambs. The dairy cows were kept according to licences granted by national regulatory authorities and the experimental protocols were scrutinised by local welfare committees.

Red dairy cows from Sweden received a diet containing a mixture of grass silage (632 g/kg DM) and barley (218 g/kg DM), with rapeseed expeller added as a protein supplement (100 g/kg DM). The diets were fed ad libitum as a total mixed ration.

Cows and reindeer from Finland were fed the same total mixed ration based on grass silage and concentrates (60:40 forage:concentrate ratio on a DM basis) at a restricted level of intake to meet maintenance energy and protein requirements.

Beef cattle kept at Easter Bush Farm, Midlothian, Scotland received 60% forage with 40% concentrate diet in which the main ingredient was barley (20%).

Lambs received a diet comprising 70% forage and 30% complete feed concentrate containing 22.6% barley, 4% wheat, 2% soya, minerals and supplements.

Ruminal digesta samples (4 ml) were taken manually from cows and reindeer in Finland and Sweden via a ruminal cannula, diluted in 8 ml PBS buffer containing 20% glycerol and transported to the laboratory on dry ice where they were stored at −80 °C. Fresh digesta was obtained from the beef cattle and lambs immediately after slaughter and stored in an insulated container prior to protein extraction.

Sample processing
The digesta obtained from the lambs and cannulated reindeer and cows contained large amounts of dietary fibre. After thawing the sample, the coarse fibres were separated by gentle centrifugation ($200 \times g$) and the supernatant retained. The remaining fibres were washed in a sodium phosphate and detergent buffer (50 mM pH 6.5, 0.1% (v/v) Tween 80) and the centrifugation

process was repeated three more times. The pooled supernatant was centrifuged at 12,000 × g at 4 °C for 20 min to collect the enriched microbial fraction [28]. The dairy cow and beef cattle rumen fluid samples taken by gastric tube contained very little coarse fibre and were processed as received. After thawing, the samples were left to settle for 15 min to reduce any residual course fibre and feed particles. This step also reduced the number of protozoa, which would otherwise dominate the microbial proteome. The remaining fraction was aspirated and immediately centrifuged at 12,000 × g at 4 °C for 20 min. In all cases, the pellet was then resuspended in 1.5 ml lysis buffer based on Rabilloud [28] containing 7 M urea, 2 M thiourea, 4% CHAPS, 1% dithiothreitol and a protease inhibitor cocktail (Sigma Aldrich). Protein extraction was carried out using six rounds of 30 s bead beating with 3 min cooling on ice based on the DNA extraction protocol by Yu and Morrison [29].

Proteins were precipitated by adding 6 M trichloroacetic acid/80 mM DTT solution at a ratio of 1:3 protein extract, vortexing and incubating overnight at 4 °C. The tubes were centrifuged at 16,000 × g for 20 min at 4 °C and the supernatants were discarded. The pellets were washed twice in a –20 °C solution of 20% DMSO in acetone and twice in –20 °C acetone based on the protocol by Song et al. [30]. The pellet was resuspended in the Rabilloud buffer described previously and the concentration of total protein extract was assayed using the RC/DC method (Bio-Rad) against a BSA standard.

Visual inspection of the dissolved protein extract revealed some brown discolouration of the solution indicating the presence of dissolved organic compounds. This was most notable in the samples from the silage/concentrate fed reindeer and cow samples. Where this occurred, additional extraction methods to remove these organic compounds were attempted including; wash dilution, column filtration and phenol extraction. After centrifugation, the enriched microbial pellet was resuspended in a sodium phosphate wash buffer (50 mM pH 6.5) containing 0.1% v/v Tween 80. The solution was shaken gently for 5 min and the microbial pellet retrieved by centrifugation at 12,000 × g. This process was repeated three to four times, or until the supernatant was clear. Column filtration was carried out using an Amicon® Ultra-4 centrifugal filter (Millipore) 10 kDa molecular weight cut-off according to the manufacturer's protocol. Phenol extraction was carried out subsequent to the TCA and acetone precipitation stage and based on the protocol by Wu et al. [25].

Metaproteome analysis

Protein separation was carried out on individual samples using 1D and 2D SDS-PAGE using precast gels according to the manufacturer's protocol (Bio-Rad). 100–300 μg protein in 325 μl Rabilloud buffer were applied to immobilised pH gradient (IPG) strips pH from 4–7 as standard after previous work with pH 3–10 IPG strips determined that the isoelectric points of most proteins were contained within this range. Gels were stained using Coomassie Blue. Visibly abundant protein spots were selected on the 2D gels for protein identification by LC-MS/MS. These were cut manually and the gel plugs destained and digested with trypsin. The resulting peptide solution was processed using an Ultimate nano LC system, with Famos autosampler and Switchos microcolumn (Thermo Scientific). Ionisation and mass spectra measurement was carried out using an AB Sciex Q-Trap triple quadrupole mass spectrometer. Total ion current data was submitted to the MASCOT server (Matrix Science) to identify the most likely protein hit from the NCBI nr database. The search criteria were: allowance of 0 or 1 missed cleavages; tolerance of ± 1.5 Da; fragment mass tolerance of ± 1.5 Da, trypsin as digestion enzyme; carbamidomethyl fixed modification of cysteine; methionine oxidation as a variable modification; and charged state as 2+ and 3+. Potential function was inferred from the annotation of the protein best hit provided by the NCBI nr database entry. This was carried out using MASCOT (Matrix Science Inc.), which uses a probabilistic scoring system for protein identification/inference adapted from the MOlecular Weight SEarch (MOWSE) algorithm.

Results and discussion
Effects of different sample types on electrophoresis

The ruminal metaproteome from a variety of species and sample types was characterised by SDS-PAGE using digesta samples obtained from a variety of ruminant species, including cattle, sheep and reindeer, fed different diets. The digesta samples were taken via cannula from the silage/concentrate fed cows and reindeer and the fresh samples taken from the forage/concentrate fed lambs contained a large proportion of coarse dietary fibre. For these samples, a method was adapted to remove the fibre and enrich the bacteria fraction using differential centrifugation and a wash dilution procedure [31]. In comparison, the samples from the forage/barley concentrate fed dairy and beef cattle taken by gastric tube contained very little in the way of plant fibre and were processed as received. The difference in sample quality was accepted to be a result of the methods and was not considered as a factor associated either with the individual animal or species. Different sampling methods - stomach tube, rumen cannula and slaughter – were used in this study, but no two methods were compared directly, therefore it is not possible to assess if sampling method had any influence on gel quality.

Samples were firstly subjected to 1D SDS-PAGE in order to anticipate possible problems associated with impurities in the samples (Fig. 1). The lack of clear banding in 1D SDS-PAGE was due partly by the complexity of the metaproteome and partly by the presence of contaminants, possibly humic compounds, co-precipitating with the proteins. This assumption was based on similar studies reporting co-staining with Coomassie blue of these compounds on SDS-PAGE derived from similar environmental samples [32]. Therefore, no attempt was made to identify proteins using LC-MS/MS from these gels. The lack of resolution of proteins was also evident in the 2D SDS-PAGE results of the dairy cow and reindeer digesta samples (Fig. 2). Here, contaminants co-stained by the Coomassie Blue consistently obscured proteins in all of the gels prepared from the protein extracts [33]. A typical feature of the 2D gel was a dark streak in the acidic region of the isoelectric focussing [24]. Even where it was possible to identify faint spots on these gels, there was not sufficient protein to identify using LC-MS/MS.

Repeated 2D SDS-PAGE gels from different extracts from the same sample produced highly reproducible results (see Additional file 1). In contrast, spot patterns were poorly reproduced between 2D gels of different samples, preventing any meaningful comparison of protein abundance between gels. Scans of the gels were loaded onto PDQuest analysis software (Bio-Rad) to align the gels and detect spots in an attempt to assess relative values of protein abundance. Setting the levels for fainter spots was hampered by high background staining and the complexity of the pattern. In most cases the 2D gels were dominated by a few highly abundant proteins with fainter spots at the limit of detection by the software. The estimation of protein loading on the gel was also affected by contaminants affecting the result of the absorbance readings taken as part of the RC/DC assay used to measure protein concentration (BioRad) [22].

The additional methods, including the wash dilution, column filtration and phenolic extraction steps carried out to eliminate or reduce humic compound contamination were all unsuccessful, and did not produce any clearly resolved 2D SDS-PAGE results. The use of DMSO during acetone precipitation was previously reported to reduce contamination in plant root proteins [30] and although the effect was inconclusive in the case of rumen digesta, it was used in all the extractions performed here.

Identification of proteins on 2-D gels

Coomassie Blue was used to stain gels in all cases following the principle that any resolved protein bands or spots should yield sufficient material (10–100 ng) to identify using mass spectroscopy [34]. MASCOT MS/MS ion search results of the selected spots from the 2D SDS-PAGE gels are shown in Fig. 2b-d. Proteins were selected on the basis of highest MASCOT score and percent coverage. A number of MASCOT hits that gave human keratin or trypsin as a result were disregarded and considered as either contaminants for the former or the enzyme used for protein digestion for the latter. A taxonomic summary of the unique proteins found is shown in Fig. 3 and a complete list of the unique highly abundant proteins (by GI sequence identification number) from the successfully resolved gels is given in Table 1 (prokaryotic proteins) and 2 (eukaryotic proteins). The number of peptides mapping to the protein hit and the percent coverage gave a degree of confidence to the protein identity. Proteins identified from a single peptide and coverage less than 5% were excluded from the results. The theoretical (MW_t) size was also compared to the position of the spot on the gel (MW_e). However, in some cases the MW_t and MW_e value differed due to a partial protein reference sequence or the protein separating into subunits on the gel.

Protozoal proteins

The 2D SDS-PAGE of the rumen fluid taken from high concentrate fed dairy cows was dominated by a few very abundant proteins (Fig. 2b). LC-MS/MS identified these as actin, alpha and beta tubulin and axonemal isoforms of dynein light chains. All these proteins can be found in the cilia of rumen protozoa and taxonomic identification confirmed the rumen ciliate species *Entodinium caudatum* as the most likely source. While this group of microorganisms is not as abundant as the bacteria, they can make up a large proportion of the microbial biomass [1].

Fig. 1 1D SDS-PAGE of ruminal digesta protein extract. Standard ranges 15–250 kDa. Lanes from left to right: Five reindeer and five cows from Finland (MTT)

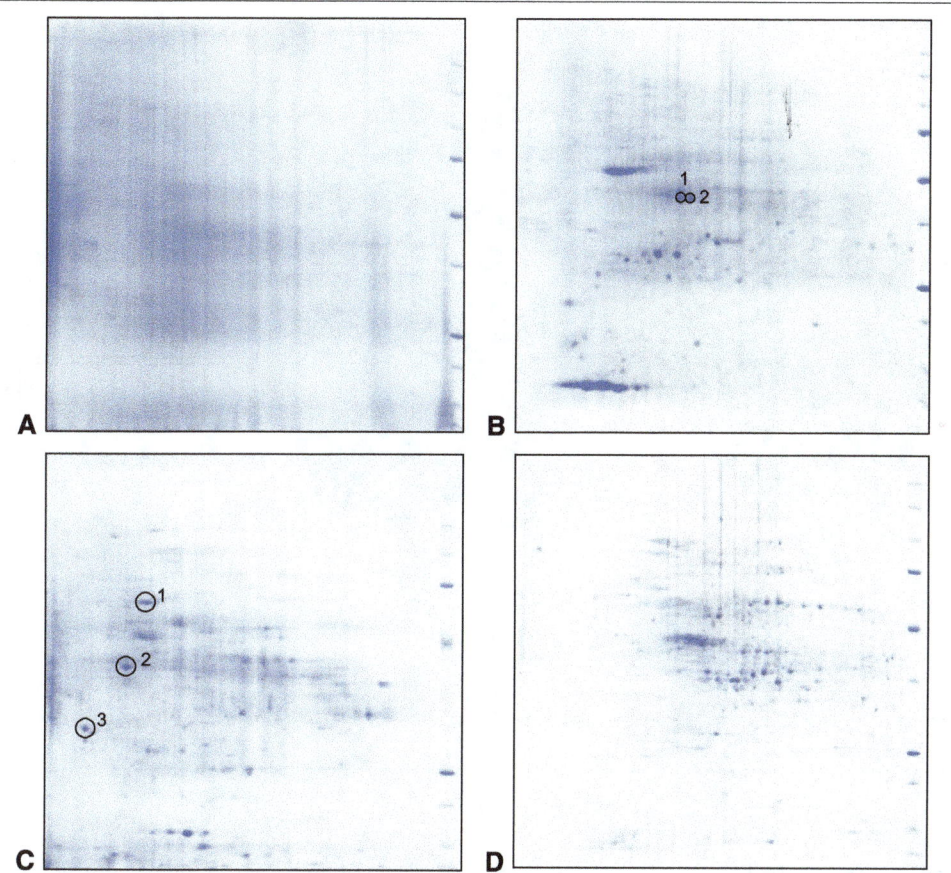

Fig. 2 2D SDS-PAGE of proteins extracted from ruminal digesta. Digesta samples were obtained from different host species using different sampling methods, mixed with PBS/glycerol buffer and stored at −80 °C. In all gels size standards range from 10–250 kDa from bottom to top, isoelectric points (pI) range from pH 4–7 from left to right. **a**. Reindeer from Finland (MTT) fed silage forage based diet. Samples were obtained manually via ruminal cannulae. Protein extraction was carried out after bacterial enrichment by differential centrifugation and wash dilution stages. The gel shows severe protein degradation and spots are obscured by co-staining of humic compounds. **b**. Dairy cows from Sweden (SLU) fed a high protein diet. Samples were taken by intubation via ruminal cannula. Protein was extracted with no sample pre-processing. Spots identified: 1. Actin, *Entodinium caudatum*, GI: 3377675. Based on eight peptide matches, 36% coverage, theoretical size 41.7 kDa. 2. Actin, *E. caudatum*, GI: 3386579. Based on eight peptide matches, 34% coverage, theoretical size 41.7 kDa. **c**. Beef cattle from Scotland on a fattening high concentrate diet. Samples were taken by nasogastric intubation. Large particles were separated and removed by settling for 5 min before continuing to the protein extraction stages. Spots identified: 1. Methyl-CoM reductase McrA, *Methanobrevibacter smithii*, GI: 518094697. Based on five peptide matches, 12% coverage, theoretical size 61.1 kDa. 2. Methyl-CoM reductase beta subunit McrB, *M. ruminantium* M1, GI:288561184. Based on three peptide matches, 8% coverage, theoretical size 47.2 kDa. 3. 5,10-methylenetetrahydromethanopterin reductase, *M. ruminantium* M1 GI:288559826. Based on five peptide matches, 22% coverage, theoretical size 33.1 kDa. Additional proteins identified are described in Tables 1 and 2. **d**. *Post-mortem* digesta from lambs from Scotland fed on a finishing concentrate diet. Protein extraction was carried out on fresh samples after bacterial enrichment by differential centrifugation and wash dilution stages. Proteins identified are described in Tables 1 and 2

To date, there are no completed genome sequences for rumen ciliate species; consequently a reference proteome was not available. However, some 52 coding cDNA sequences obtained from functional screening and structural protein analysis of the rumen ciliate genus *Entodinium* were contained in the NCBInr protein database [35]. Five of the 15 eukaryotic structural proteins mapped to this small reference dataset with four others mapped to related ciliate genera: *Euplotes*, *Ichthyophthirius*, *Spathidium*, *Epiphyllum* and *Amphileptus* (Table 2). In the NCBI nr database used in the present study, reference sequences for similar structural proteins such as actin are abundant, originating from a wide range of organisms. The narrow range of taxa that the proteins mapped to here was an indication of the specificity of the amino acid sequence identity to rumen species.

Bacterial proteins To increase the abundance of the prokaryotic microbial proteins relative to the ciliate structural proteins, it was necessary to fractionate the samples by settling for 15 min. SDS-PAGE results with

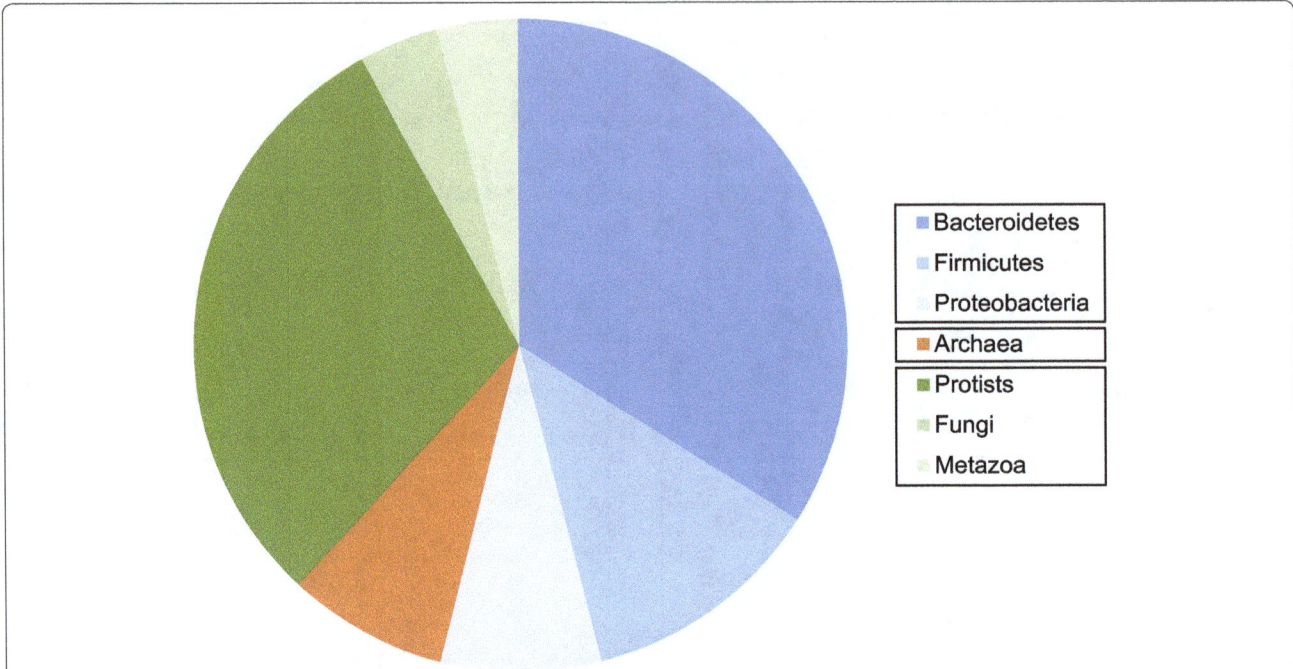

Fig. 3 Taxonomy summary of rumen proteins. The taxon pie chart is based on the 50 unique proteins identified from the 2D gels shown in Fig. 2b, c and d and listed in Table 1. The chart shows the relative richness of unique proteins from bacteria (blue), archaea (orange) and eukaryotic (green) groups

this step included are shown in Fig. 2c. This gel lacked the dominant spot pattern apparent in Fig. 2b and highly abundant proteins identified by LC-MS/MS from a broad range of organisms. Based on the best hits results provided by the MASCOT search of total ion current data, the majority of proteins originated from bacteria with others from the host, plants, fungi, archaea and protozoa.

Similar results were seen in the *post-mortem* samples from the lambs on the high concentrate diet (Fig. 2d). The dominant spot pattern indicating abundant ciliate structural proteins identified in gel 2B was not seen on the gel and no ciliate proteins were detected by LC-MS/MS. In this sample, the bacteria enrichment steps for these samples to remove larger contaminating particles used repeated dilution and centrifugation [31] and may well have resulted in the complete removal of ciliates.

Many of the prokaryotic proteins were central metabolic enzymes such as glyceraldehyde-3-phosphate dehydrogenase, phosphoenolpyruvate carboxykinase, phosphoglycerate kinase and triosephosphate isomerase, involved in central carbohydrate metabolism pathways and present in almost any cellular organism. However, in a similar manner to the ciliate structural proteins, the amino acid sequence identity of these enzymes was associated with of rumen prokaryote species that had been characterised previously in genomic studies [36] (Table 1). Bacterial

phyla included Bacteroidetes, Firmicutes and Proteobacteria. The Firmicutes proteins were from a diverse range of species, some associated directly with the rumen or the human gut and some with anaerobic sewage environments, the latter being a result of the paucity of protein sequences of rumen microorganisms in the reference database. All the Bacteroidetes proteins were from species of *Prevotella* and accounted for over half of the total prokaryotic proteins. The dominance of *Prevotella* proteins reflected the abundance of this genus in the rumen [37, 38].

Archaeal proteins

The ruminal archaea are much lower in abundance than bacteria, on average approximately 5% of the bacterial population based on relative abundance of 16S rRNA subunit [17, 39, 40]. Two archaeal proteins were discovered here as dominant spots. 5,10-methylenetetrahydromethanopterin reductase (*mer*) from *Methanobrevibacter ruminantium* and methyl coenzyme M reductase beta subunit (*mcrB*) from *M. ruminantium* and *Methanobrevibacter smithii* are both important components of the pathway converting CO_2 and H_2 into methane [19]. The detection of these enzymes from a relatively small proportion of the microbial community highlights the importance of methane metabolism in the rumen. Ruminants

Table 1 Abundant prokaryotic proteins from ruminal digesta from different species visualised by 2D SDS-PAGE and identified using LC-MS/MS

Taxon	Gel	Protein	Organism	GI Number	MW t (kDa)	No. of Peptides	Coverage (%)
Firmicutes	Beef	pyruvate, phosphate dikinase	Eubacterium sp	510873014	97	9	12
	Beef	putative uncharacterized protein	Faecalibacterium sp. CAG:74	548227312	25.5	2	9
	Lamb	cysteine synthase	Syntrophobotulus glycolicus DSM 8271	325290835	34.1	2	7
	Lamb	beta hydroxybutyryl-CoA dehydrogenase	Butyrivibrio fibrisolvens	21623533	31.5	2	10
	Lamb	Glu/Leu/Phe/Val dehydrogenase	Acidaminococcus fermentans DSM 20731	284049088	44.8	5	19
	Lamb	fructose-1,6-bisphosphate aldolase	Mitsuokella multacida DSM 20544	255657994	35.7	4	17
Bacteroidetes	Lamb	glutamate dehydrogenase	Prevotella bergensis DSM 17361	261879552	49.8	6	17
	Lamb	glutamate-ammonia ligase	Prevotella bryantii B14	300726017	82.5	5	10
	Lamb	glucose-6-phosphate isomerase	Prevotella bryantii B14	300727003	49.7	2	6
	Lamb	glyceraldehyde-3-phosphate dehydrogenase	Prevotella copri DSM 18205	281419707	37.1	5	14
	Lamb	phosphoglycerate kinase	Prevotella copri DSM 18205	281419885	44.3	10	32
	Lamb	transketolase	Prevotella copri DSM 18205	281420193	73.2	11	20
	Lamb	NifU-related domain containing protein	Prevotella copri DSM 18205	281422452	25.4	6	33
	Lamb	triose-phosphate isomerase	Prevotella dentalis DSM 3688	340347512	27.8	5	41
	Lamb	glyceraldehyde-3-phosphate dehydrogenase	Prevotella multisaccharivorax DSM 17128	336398137	37.3	4	17
	Beef	chaperonin GroEL	Prevotella ruminicola 23	294673078	58.5	9	23
	Beef	glutamate dehydrogenase	Prevotella ruminicola 23	294673915	49.3	8	20
	Beef	phosphoglycerate kinase	Prevotella ruminicola 23	294674314	44.5	9	27
	Beef	triose-phosphate isomerase	Prevotella ruminicola 23	294674529	27.8	7	44
	Beef	phosphoenolpyruvate carboxykinase (ATP)	Prevotella ruminicola 23	294674921	59.8	12	26
	Beef	phosphoribosylaminoimidazolecarboxamide formyltransferase/IMP cyclohydrolase	Prevotella sp.	493902114	21.8	2	12
	Beef	glyceraldehyde-3-phosphate dehydrogenase	Prevotella sp.	495177024	36.7	8	34
	Beef	glyceraldehyde-3-phosphate dehydrogenase	Prevotella sp. CAG:604	547555062	36.6	8	35
	Beef						
Proteobacteria	Beef	tellurium resistance protein TerY	Thiorhodovibrio sp. 970	496438280	22.8	2	10
	Beef	GDP-D-mannose dehydratase	Burkholderia cenocepacia	493522632	39.4	2	7
	Beef	60 kDa chaperonin	Glaciecola polaris	494167481	57.6	4	8
	Beef	elongation factor Tu	Baumannia cicadellinicola	94676917	43.6	4	10
	Beef						
Archaea	Beef	5,10-methylenetetrahydromethanopterin reductase	Methanobrevibacter ruminantium M1	288559826	33.1	5	22
	Beef	M reductase beta subunit McrB	Methanobrevibacter ruminantium M1	288561184	47.2	3	8
	Beef	methyl-coenzyme M reductase McrB	Methanobrevibacter smithii	490132552	47	4	9
	Beef	methyl-coenzyme M reductase McrA	Methanobrevibacter smithii	518094697	61.1	5	12

produce abundant quantities of methane, up to 500 L/d in a dairy cow [6]. Methane is a greenhouse gas (GHG) with a global warming potential 28-fold that of carbon dioxide [41]. Methane production from ruminants accounts for the majority of the 37% of total GHG from agriculture in the UK [42]. Ruminal methanogenesis derives from fermentation by bacteria, protozoa and fungi, which produce short-chain fatty acids and H_2; the latter which, with CO_2, is the main substrate for methane formation by methanogenic archaea [6]. Understanding this complex process is vital if we are to develop methods to lower methane emissions from ruminants and thereby lessen the environmental impact of livestock agriculture. The present work indicates that metaproteomics may be a useful tool in achieving that aim.

Perspective
Deusch and Seifert [27] made the first description of the ruminal metaproteome by shotgun peptide sequencing. 2-D gels clearly provide a visual image that the shotgun method does not. However, the power of the shotgun technique compares very favourably compared to 2D SDS-PAGE, in the sense that thousands of peptides were analysed in a non-selective way, whereas only a few hundred were analysed by SDS-PAGE here. Furthermore, the electrophoresis itself is subject to major problems that

Table 2 Abundant eukaryotic proteins from ruminal digesta from different species visualised by 2D SDS-PAGE and identified using LC-MS/MS

Taxon	Gel	Protein	Organism	GI Number	MW (kDA)	No. of Peptides	Coverage (%)
Protists	Cow	dynein light chain	*Entodinium caudatum*	3386573	12.5	5	72
	Cow	actin	*Entodinium caudatum*	3386579	41.7	8	34
	Cow	heat shock 70 protein	*Entodinium caudatum*	4324942	72.2	7	13
	Cow	actin	*Entodinium caudatum*	3377675	41.8	5	15
	Cow	alpha-tubulin	*Entodinium sp.*	860930	42.8/50	6	28
	Beef	beta-tubulin	*Euplotes octocarinatus*	586077	50	3	9
	Beef	alpha-tubulin	*Spathidium sp.*	861142	42.7	3	9
	Beef	hypothetical protein	*Ichthyophthirius multifiliis*	471220775	48	3	10
	Cow	alpha tubulin	*Epiphyllum shenzhenense*	376000791	43.4	6	21
	Beef	beta-tubulin	*Cryptosporidium parvum*	1944528	50.3	3	10
	Beef	beta-tubulin	*Andalucia incarcerata*	13649518	43.5	5	17
	Beef	beta-tubulin	*Bigelowiella natans*	3790451	43.7	6	18
	Beef	alpha-tubulin	*Streblomastix strix*	28779301	42.9	3	11
	Beef	alpha-tubulin	*Amphileptus marinus*	376000769	40	6	25
	Beef	hypothetical protein	*Trypanosoma cruzi marinkellei*	407410997	49	2	6
Fungi	Cow	beta-tubulin	*Phytophthora citrophthora*	374707001	40	3	14
	Beef	beta-tubulin	*Phytophthora polonica*	156988212	43.4	7	22
Metazoa	Beef	Odorant-binding protein (*obp*)	*Bos taurus*	129022	18.5	4	28
	Beef	Chain A, Complex Of Bovine Odorant Binding Protein (*obp*)	*Bos taurus*	2392495	18.4	7	48

affect the comparison of different samples. Yet the ability to identify visible proteins on a gel has some merit, we believe. Both methods enable an analysis of the phylogenetic origin of a peptide/protein. It is extremely important to point out that the weakness of the databases with regard to the poor representation of true ruminal organisms is a handicap to both methods. As far as we are aware, there is no corresponding proteomic initiative to match the Hungate 1000 genomic project [43], which will enable precise assignment of gene sequences to phylogenetic taxa.

Conclusions

Despite the taxonomic diversity of the rumen, a relatively small number of protein spots dominated the metaproteome. Co-precipitation of grass-derived contaminants had a critical effect on the outcome of the SDS-PAGE protein separation and visualisation, such that within-sample replication was excellent and between-sample replication was poor, lowering the value of SDS-PAGE as a tool to predict rumen function. Although the volume of data retrievable from 2D-SDS-PAGE was a couple of orders of magnitude less than shotgun peptide sequencing analysis [27], the conclusions on protein complement were qualitatively similar. 2D-SDS-PAGE, when successful, has the advantage of creating an image that can be compared with others visually. Enzymes from methanogenic archaea were among the most readily identifiable proteins, indicating a possible role for metaproteomics in exploring low-emitting phenotypes of ruminants, which in turn may enable mitigation of greenhouse gas emissions from farm livestock production by selective breeding.

Abbreviations
1D- and 2D-SDS-PAGE: One- and two-dimensional gel electrophoresis; DM: Dry matter; DMSO: Dimethyl sulfoxide; DTT: Dithiothreitol; GHG: Greenhouse gas; LC-MS/MS: Liquid chromatography followed by tandem mass spectrometry; PBS: Phosphate-buffered saline

Acknowledgements
Ruminal digesta samples were provided by Pekka Huhtanen of the Swedish University of Agricultural Sciences, Umeå, Sweden, Kevin Shingfield of MTT, Jokionen, Finland, and Christine McCartney and Thulile Sgwane of the Rowett Institute, University of Aberdeen. The analytical work, including 1D- and 2D- SDS-PAGE and MS/MS ion searching, was carried out by the proteomics facility of the Rowett Institute of Nutrition and Health, University of Aberdeen. Thanks go to Fergus Nicol, Martin Reid and Louise Cantlay for providing this service.

Funding
This work was supported by the RuminOmics project and funded by the European Commission (Grant Agreement No. 289319). The Rowett Institute is funded by the Rural and Environment Science and Analytical Services Division (RESAS) of the Scottish Government. The funding bodies had no role in the design of the study or collection, analysis, or interpretation of data or in writing the manuscript.

Authors' contributions

RJW initiated the project. TJS carried out all the practical and bioinformatics analysis. Both had an equal input in writing the manuscript and approved the final manuscript.

Competing interests

The authors declare that they have no competing interests.

References

1. Hungate RE. The rumen and its microbes. New York and London: Academic; 1966.
2. Morgavi DP, Forano E, Martin C, Newbold CJ. Microbial ecosystem and methanogenesis in ruminants. Animal. 2010;4:1024–36.
3. Hobson PN, Stewart CS. The rumen microbial ecosystem. 3rd ed. London: Chapman and Hall; 1997.
4. Leng RA, Nolan JV. Nitrogen metabolism in the rumen. J Dairy Sci. 1984;67: 1072–89.
5. Pfeffer E, Hristov AN. Interactions between cattle and the environment: a general introduction. In: Pfeffer E, Hristov AN, editors. Nitrogen and phosphorus nutrition of cattle. Wallingford: CABI Publishing; 2007. p. 1–12.
6. Moss AR, Jouany JP, Newbold CJ. Methane production by ruminants: its contribution to global warming. Ann Zootech. 2000;49:231–53.
7. Wang Y, Majak W, McAllister TA. Frothy bloat in ruminants: cause, occurrence, and mitigation strategies. Anim Feed Sci Technol. 2012;172: 103–14.
8. Plaizier JC, Khafipour E, Li S, Gozho GN, Krause DO. Subacute ruminal acidosis (SARA), endotoxins and health consequences. Anim Feed Sci Technol. 2012;172:9–21.
9. Gressley TF, Hall MB, Armentano LE. Ruminant nutrition symposium: productivity, digestion, and health responses to hindgut acidosis in ruminants. J Anim Sci. 2011;89:1120–30.
10. Hess M, Sczyrba A, Egan R, Kim TW, Chokhawala H, Schroth G, et al. Metagenomic discovery of biomass-degrading genes and genomes from cow rumen. Science. 2011;331:463–7.
11. Brulc JM, Antonopoulos DA, Miller MEB, Wilson MK, Yannarell AC, Dinsdale EA, et al. Gene-centric metagenomics of the fiber-adherent bovine rumen microbiome reveals forage specific glycoside hydrolases. Proc Natl Acad Sci U S A. 2009;106:1948–53.
12. McAllister TA, Meale SJ, Valle E, Guan LL, Zhou M, Kelly WJ, et al. Use of genomics and transcriptomics to identify strategies to lower ruminal methanogenesis. J Anim Sci. 2015;93:1431–49.
13. Ferrer M, Golyshina OV, Chernikova TN, Khachane AN, Reyes-Duarte D, Dos Santos VAPM, et al. Novel hydrolase diversity retrieved from a metagenome library of bovine rumen microflora. Environ Microbiol. 2005;7:1996–2010.
14. Privé F, Newbold CJ, Kaderbhai NN, Girdwood SG, Golyshina OV, Golyshin PN, et al. Isolation and characterization of novel lipases/ esterases from a bovine rumen metagenome. Appl Microbiol Biotechnol. 2015;99:5475–85.
15. Liu K, Wang J, Bu D, Zhao S, McSweeney C, Yu P, Li D. Isolation and biochemical characterization of two lipases from a metagenomic library of China Holstein cow rumen. Biochem Biophys Res Commun. 2009;385:605–11.
16. Zhou M, Hernandez-Sanabria E, Guan LL. Assessment of the microbial ecology of ruminal methanogens in cattle with different feed efficiencies. Appl Environ Microbiol. 2009;75:6524–33.
17. Wallace RJ, Rooke JA, Duthie CA, Hyslop JJ, Ross DW, McKain N, et al. Archaeal abundance in post-mortem ruminal digesta may help predict methane emissions from beef cattle. Sci Rep. 2014;4:5892.
18. Martin C, Morgavi DP, Doreau M. Methane mitigation in ruminants: from microbe to the farm scale. Animal. 2010;4:351–65.
19. Shi W, Moon CD, Leahy SC, Kang D, Froula J, Kittelmann S, et al. Methane yield phenotypes linked to differential gene expression in the lambs rumen microbiome. Genome Res. 2014;24:1517–25.
20. Wilmes P, Bond PL. The application of two-dimensional polyacrylamide gel electrophoresis and downstream analyses to a mixed community of prokaryotic microorganisms. Environ Microbiol. 2004;6:911–20.
21. O'Farrell PH. High resolution two-dimensional electrophoresis of proteins. J Biol Chem. 1975;250:4007–21.
22. Abram F, Gunnigle E, O'Flaherty V. Optimisation of protein extraction and 2-DE for metaproteomics of microbial communities from anaerobic wastewater treatment biofilms. Electrophoresis. 2009;30:4149–51.
23. Chourey K, Jansson J, VerBerkmoes N, Shah M, et al. Direct cellular lysis/protein extraction protocol for soil metaproteomics. J Proteome Res. 2010;9:6615–22.
24. Benndorf D, Vogt C, Jehmlich N, Schmidt Y, et al. Improving protein extraction and separation methods for investigating the metaproteome of anaerobic benzene communities within sediments. Biodegradation. 2009;20: 737–50.
25. Wu L, Wang H, Zhang Z, Lin R, et al. Comparative metaproteomic analysis on consecutively Rehmannia glutinosa-monocultured rhizosphere soil. PLoS One. 2011;6:e20611.
26. Klaassens ES, de Vos WM, Vaughan EE. Metaproteomics approach to study the functionality of the microbiota in the human infant gastrointestinal tract. Appl Environ Microbiol. 2007;73:1388–92.
27. Deusch S, Seifert J. Catching the tip of the iceberg – evaluation of sample preparation protocols for the metaproteomic studies of the rumen microbiota. Proteomics. 2015;15:1–6.
28. Rabilloud T, Adessi C, Giraudel A, Lunardi J. Improvement of the solubilization of proteins in two-dimensional electrophoresis with immobilized pH gradients. Electrophoresis. 1997;18:307–16.
29. Yu ZT, Morrison M. Improved extraction of PCR-quality community DNA from digesta and fecal samples. Biotechniques. 2004;36:808–12.
30. Song Y, Zhang H, Wang G, Shen Z. DMSO, an organic cleanup solvent for TCA/acetone-precipitated proteins, improves 2-DE protein analysis of rice roots. Plant Mol Biol Rep. 2012;30:1204–9.
31. Apajalahti JHA, Sarkilahti LK, Maki BRE, Heikkinen JP, Nurminen PH, et al. Effective recovery of bacterial DNA and percent-guanine-plus-cytosine-based analysis of community structure in the gastrointestinal tract of broiler chickens. Appl Environ Microbiol. 1998;64:4084–8.
32. Schneider T, Keiblinger KM, Schmid E, Sterflinger-Gleixner K, Ellersdorfer G, et al. Who is who in litter decomposition? Metaproteomics reveals major microbial players and their biological functions. ISME J. 2012;6:1749–62.
33. Heyer R, Kohrs F, Benndorf D, Rapp E, Kausmann R, et al. Metaproteome analysis of the microbial communities in agricultural biogas plants. N Biotechnol. 2013;30:614–22.
34. Miller I, Crawford J, Gianazza E. Protein stains for proteomic applications: Which, when, why? Proteomics. 2006;6:5385–408.
35. McEwan NR, Eschenlauer SC, Calza RE, Wallace RJ, Newbold CJ. The 3′ untranslated region of messages in the rumen protozoan Entodinium caudatum. Protist. 2000;151:139–46.
36. Purushe J, Fouts DE, Morrison M, White BA, Mackie RI, et al. Comparative genome analysis of Prevotella ruminicola and Prevotella bryantii: insights into their environmental niche. Microb Ecol. 2010;60:721–9.
37. Edwards JE, McEwan NR, Travis AJ, Wallace RJ. 16S rDNA library-based analysis of ruminal bacterial diversity. Anton v Leeuwen. 2004;86:263–81.
38. Kim M, Morrison M, Yu Z. Status of the phylogenetic diversity census of ruminal microbiomes. FEMS Microbiol Ecol. 2011;76:49–63.
39. Pinares-Patino CS, Seedorf H, Kirk MR, Ganesh S, McEwan JC, Janssen PH. Simultaneous amplicon sequencing to explore co-occurrence patterns of bacterial, archaeal and eukaryotic microorganisms in rumen microbial communities. PLoS One. 2013;8:e47879.
40. Janssen PH, Kirs M. Structure of the archaeal community of the rumen. Appl Environ Microbiol. 2008;74:3619–25.
41. Intergovernmental Panel on Climate Change. Guidelines for national greenhouse gas inventories vol. 4 Agriculture, Forestry and Other Land Use. 2006.
42. Cottle DJ, Nolan JV, Wiedemann SG. Ruminant enteric methane mitigation: a review. Anim Prod Sci. 2011;51:491–514.
43. Kelly WJ. http://www.rmgnetwork.org/hungate1000.html.

2

Genome-wide comparative analysis of putative Pth11-related G protein-coupled receptors in fungi belonging to Pezizomycotina

Xihui Xu[1], Guopeng Li[2], Lu Li[1], Zhenzhu Su[3] and Chen Chen[1]*

Abstract

Background: G-protein coupled receptors (GPCRs) are the largest family of transmembrane receptors in fungi, where they play important roles in signal transduction. Among them, the Pth11-related GPCRs form a large and divergent protein family, and are only found in fungi in Pezizomycotina. However, the evolutionary process and potential functions of Pth11-related GPCRs remain largely unknown.

Results: Twenty genomes of fungi in Pezizomycotina covering different nutritional strategies were mined for putative Pth11-related GPCRs. Phytopathogens encode much more putative Pth11-related GPCRs than symbionts, saprophytes, or entomopathogens. Based on the phylogenetic tree, these GPCRs can be divided into nine clades, with each clade containing fungi in different taxonomic orders. Instead of fungi from the same order, those fungi with similar nutritional strategies were inclined to share orthologs of putative Pth11-related GPCRs. Most of the CFEM domain-containing Pth11-related GPCRs, which were only included in two clades, were detected in phytopathogens. Furthermore, many putative Pth11-related GPCR genes of phytopathogens were upregulated during invasive plant infection, but downregulated under biotic stress. The expressions of putative Pth11-related GPCR genes of saprophytes and entomopathogens could be affected by nutrient conditions, especially the carbon source. The gene expressions revealed that Pth11-related GPCRs could respond to biotic/abiotic stress and invasive plant infection with different expression patterns.

Conclusion: Our results indicated that the Pth11-related GPCRs existed before the diversification of Pezizomycotina and have been gained and/or lost several times during the evolutionary process. Tandem duplications and trophic variations have been important factors in this evolution.

Keywords: Fungi, G-protein coupled receptors, Gene family evolution, Gene expression pattern, Pezizomycotina, Phytopathogens, Phylogenetics analysis, Pth11-related GPCRs

Background

Fungi live in a complex environment, where they receive and integrate abiotic and biotic stimuli, then respond in the manner most appropriate for survival. For example, fungal endophytes in the rhizosphere recognize and colonize specific host plants from which they obtain nutrients [1, 2]. However, the cell wall and membrane, acting as a barrier, separate the interior of the cell from the outside environment [3, 4]. Consequently, communication of cells, both with their environment and with each other, is crucial for the survival of fungi.

Membrane proteins play several essential roles in a cell, including receiving extracellular signals and triggering intracellular responses to them, and the maintenance of interactions between cells [5–7]. The fungal cell membrane is equipped with many protein receptors. These receptors sense both abiotic and biotic stimuli from the surrounding environment, and facilitate the response to these stimuli,

---

* Correspondence: chenchen@njau.edu.cn
[1]College of Life Sciences, Nanjing Agricultural University, Nanjing 210095, China
Full list of author information is available at the end of the article

which may include altering fungal development, morphogenesis, and metabolism [8, 9]. Cell-surface G-protein coupled receptors (GPCRs) are the largest family of transmembrane receptors, and are characterized by seven transmembrane domains anchored in the plasma membrane with an intracellular carboxyl- and extracellular amino-terminus [10, 11]. GPCRs sense a diverse array of stimuli including light, sugars, amino acids, and pheromones [12, 13]. In fungi, many signaling pathways are regulated by GPCRs, such as the mitogen-activated protein kinase and cAMP-dependent protein kinase cascades. These pathways regulate growth, morphogenesis, metabolism, mating, virulence, and stress responses [14, 15].

Many GPCR receptors have been identified in fungi, including pheromone receptors, cAMP receptor-like receptors, carbon-sensing receptors, Stm1-related receptors, and regulator of G protein signaling (RGS) proteins [16]. After first being identified in *Saccharomyces cerevisiae* [17], Ste2- and Ste3-like pheromone receptors have been found in many ascomycete fungi, while basidiomycete pheromone receptors are only of the Ste3-like type [16, 18]. *Neurospora crassa* GPR-1 was the first cAMP receptor-like GPCR characterized in ascomycete fungi [19] and the number of this type of GPCR varies among fungal species [16]. *S. cerevisiae* Gpr1p and *N. crassa* GPR-4 are carbon-sensing receptors [20, 21], and homologues of Gpr1p and GPR-4 are universally present in fungi [16]. *S. pombe* Stm1 is involved in the recognition of nitrogen starvation signals [22], and Stm1-related receptors are widely distributed in fungi [16]. RGS proteins are GTPase-activating proteins, which provide negative control of Gα protein signaling [23]. GprK were found to contain an RGS domain in *Aspergillus* sp. [24], and GprK homologues are present in ascomycetes [16].

A novel class of receptors, the Pth11-related group, was identified by Kulkarni et al. [25]. This group is typified by *Magnaporthe oryzae* Pth11, a cell-surface integral membrane protein implicated in pathogenesis [25, 26]. These Pth11-related proteins share many characteristics diagnostic of GPCRs, including seven transmembrane regions. For Pth11-related GPCRs, conserved residues (termed as Pth11-domain) occur within the membrane-spanning regions, which is consistent with other GPCRs that sequence conservation is typically limited to the transmembrane sequences [25]. It has been showed that the Pth11-domain is remarkably different from the conserved sequences of other GPCR classes, such as domains conserved in cAMP-, STM1-, and mPR-related GPCRs [25]. The conserved Pth11-domain distinguishes Pth11-related proteins from others and defines a new class of GPCR-like proteins. Except for Pth11-domain, Pth11 also has an amino-terminal extracellular cysteine-rich CFEM domain (pfam05730). However, only a subset of Pth11-related proteins from *M. grisea* and *N. crassa*

contained the CFEM domain, and these CFEM domain-containing proteins occur together in one clade on the phylogeny tree [25], indicating that the sequences are closely related. The gene duplication may leads to the arisen of these CFEM domain-containing proteins [25].

The Pth11-related GPCRs form a large and diverse protein family [16, 25]. Interestingly, Pth11-related GPCRs were only found in fungi belonging to Pezizomycotina (a subphylum within Ascomycota), while none were found in other subphyla of Ascomycota or Basidiomycota [16, 25]. These results reveal that Pth11-related GPCRs are very ancient in origin, and may have evolved to serve functions specific to this subphylum of fungi. However, the previous studies only focused on some phytopathogens such as *M. oryzae* and *Fusarium graminearum* while few of them covered symbionts or entomopathogens. Besides, the evolution of Pth11-related GPCRs and their potential functions are largely unknown, especially at the subphylum level. Recently, an increasing number of genome sequences have become available for fungi in Pezizomycotina, making it possible to mine and compare Pth11-related GPCR sequences. Here we explore the genomes of 20 model organisms in Pezizomycotina with different nutritional strategies and identify Pth11-related GPCRs in the deduced proteomes. The phylogenetic analysis and chromosomal distribution has shed light on the evolution of Pth11-related GPCRs. We also mined expression trends for Pth11-related GPCR genes during growth and invasion, as well as under biotic and abiotic stress, using existing mRNA profiles or microarray datasets.

Results

Identification of putative Pth11-related GPCRs in Pezizomycotina

In total, 20 genomes of fungi in Pezizomycotina were searched for putative Pth11-related GPCRs using a homology (hmmscan and BLAST)-based strategy. These species include members of the Magnaporthales, Ophiostomatales, Sordariales, Glomerellales, Hypocreales, and Eurotiales (Fig. 1), and cover phytopathogens (*M. oryzae*, *Gaeumannomyces graminis*, *Magnaporthe poae*, *Verticillium dahlia*, *Colletotrichum higginsianum*, *F. graminearum*, and *Plectosphaerella cucumerina*), symbionts (*Harpophora oryzae*, and *Epichloe festucae*), saprophytes (*Ophiostoma piceae*, *N. crassa*, *Chaetomium globosum*, *Myceliophthora thermophile*, *Podospora anserina*, *Sodiomyces alkalinus*, *Trichoderma reesei*, *Aspergillus niger*, and *Penicillium digitatum*), and entomopathogens (*Grosmannia clavigera* and *Metarhizium acridum*) (Fig. 1). All the identified proteins were evaluated for the typical topology of seven transmembrane regions, which resulted in 296 putative Pth11-related GPCRs being identified in the 20 proteomes (Fig. 1). The different numbers of predicted Pth11-related GPCRs

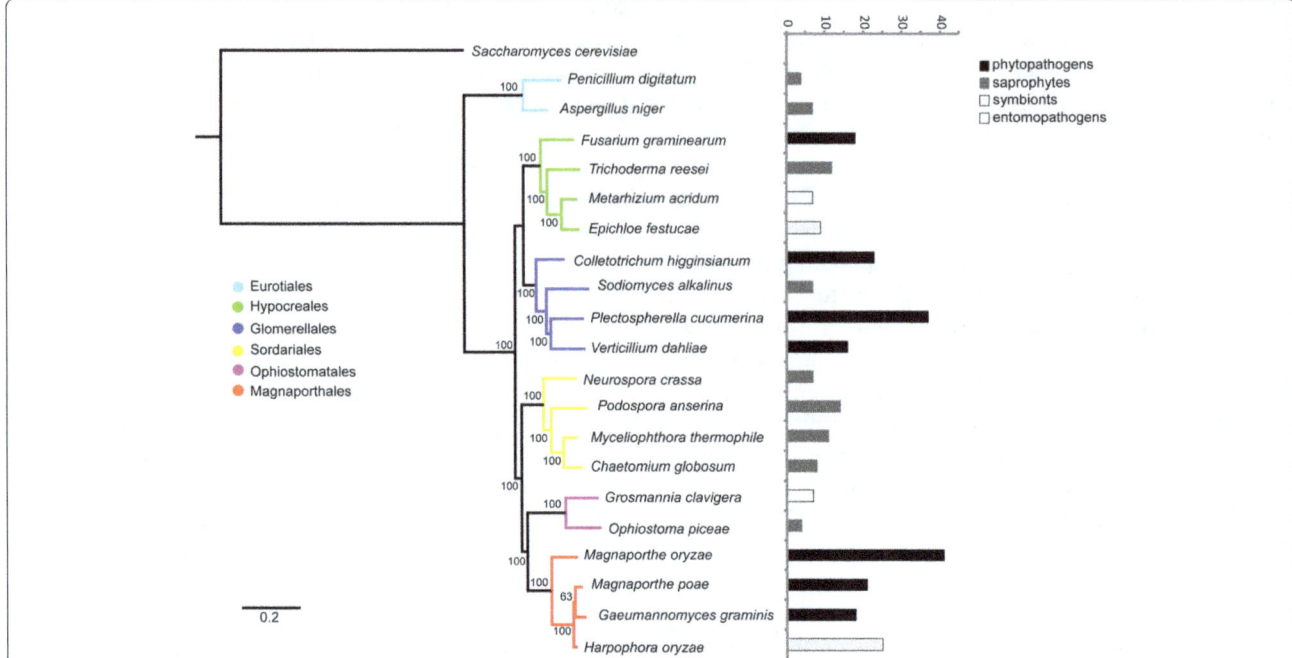

Fig. 1 Phylogenetic relationships and number of putative Pth11-related GPCRs among 20 selected fungal species. *S. cerevisiae* was used as outgroup. Maximum likelihood (ML) phylogenetic tree shows the evolutionary relationships of 2 Eurotiales species (cyan), 4 Hypocreales species (green), 4 Glomerellales species (blue), 4 Sordariales species (yellow), 2 Ophiostomatales species (purple), and 4 Magnaporthales species (red). The ML bootstrap values are sequentially indicated above the branches. Numbers of Pth11-related GPCRs of the same 20 fungal species are shown on the right of the phylogenetic tree, including 7 phytopathogens, 9 saprophytes, 2 entomopathogens, and 2 symbionts

between the present research and previous studies may result from the newer genome database version and strengthened hmmscan and BLAST cut-off to avoid the false positive.

By multiple sequence alignments of all 296 putative Pth11-related GPCRs, ten blocks of conserved sequences in Pth11-domain were detected (Fig. 2a). Each block contained a conserved motif (Fig. 2b). For examples, motif 1, 2, and 3 were conserved in LXXXR, DD, and GXH/D patterns respectively. The conserved residues of Pth11-domain detected from the 20 proteomes were consistent with previously study in which proteins used were limited to *M. grisea* and *N. crassa* [25], indicating the high probability of authenticity of these putative Pth11-related GPCRs.

Chromosomal distribution of putative Pth11-related GPCR genes

To determine the chromosomal distribution of putative Pth11-related GPCRs, chromosome maps were constructed for *M. oryzae*, *C. higginsianum*, and *F. graminearum* (Fig. 3 and Additional file 1). These three species all have complete chromosomal (or scaffold) information available, encode more putative Pth11-related GPCRs than others, and belong to different orders. In *M. oryzae*, the putative Pth11-related GPCR genes are distributed among all seven chromosomes (Fig. 3a). Both chromosome 2 and

3 encoded the highest number (8 genes) of putative Pth11-related GPCR genes, followed by chromosome 6 and 4, encoding 6 and 5 genes respectively. Tandem duplications were found in the chromosome 2 and 6. In *C. higginsianum*, scaffold 10 was devoid of putative Pth11-related GPCR genes, whereas scaffold 2 encoded the maximum of 5 genes (Fig. 3b). Moreover, tandem duplication was found in scaffold 6. No tandem duplication of putative Pth11-related GPCR genes was found in *F. graminearum* (Additional file 1).

Phylogenetic and hierarchical clustering analysis

To elucidate the evolutionary relationships among putative Pth11-related GPCRs across fungi in Pezizomycotina, a phylogenetic analysis was performed using the conserved regions of putative Pth11-related GPCR sequences (Fig. 4). Generally, the putative Pth11-related GPCRs from each taxonomic order (i.e., Magnaporthales, Ophiostomatales, Sordariales, Glomerellales, Eurotiales, and Hypocreales) were scattered throughout the phylogenetic tree rather than clustered together. According to the phylogenetic tree, all putative members of the Pth11-related GPCR family can be divided into nine major clades (Fig. 4). Almost all clades were comprised of putative Pth11-related GPCRs from different orders, and no order-specific clade was found. We also performed a hierarchical clustering analysis of the 20 species based on the counts of putative

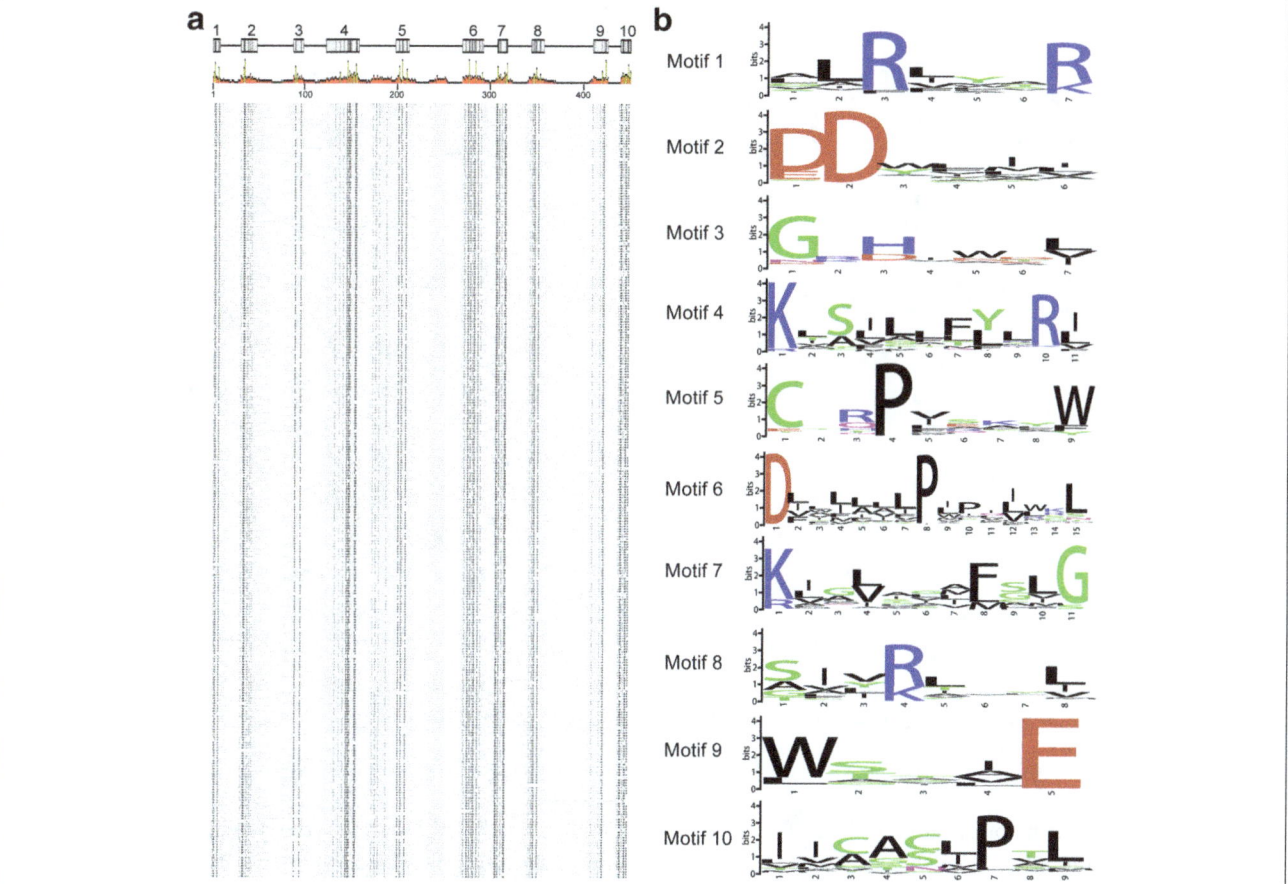

Fig. 2 Multiple sequence alignments and conserved motifs in Pth11-domain. **a** Alignment of 296 putative Pth11-related GPCRs. Each line refers to one protein sequence. The high conserved residues were in black (*yellow in scale bar*) while the low conserved ones were in grey (*red in scale bar*). The ten conserved blocks were indicated above the scale bar. **b** Sequences of the ten motifs. The ten motifs were detected based on the sequence alignment and each motif respectively associated with a corresponding block

Pth11-related GPCRs in each clade (Fig. 5). This revealed that species with similar nutritional strategies from the same order clustered together, including Magnaporthales (*H. oryzae* arose from phytopathogens and can be considered a phytopathogen [27]), Sordariales, Eurotiales, and Glomerellales. Meanwhile, species from the same order, but with different nutritional strategies, did not cluster together. To be specific, the phytopathogen *F. graminearum* (Hypocreales) was clustered with phytopathogens instead of with other fungi in Hypocreales while the saprophyte *S. alkalinus* (Glomerellales) was clustered with saprophytes rather than other phytopathogens in Glomerellales. Entomopathogens shared a cluster with *E. festucae*. This may be because *E. festucae* was derived from insect-parasitic ancestors [28].

CFEM domains in putative Pth11-related GPCRs

Only a subset of the identified putative Pth11-related GPCRs (46, 15%) contained cysteine-rich CFEM domain (Fig. 4 and Additional file 2). The *P. cucumerina* genome encodes the highest number of CFEM-containing Pth11-

related GPCRs (8) followed by *M. oryzae* and *C. higginsianum* (6). All CFEM-containing Pth11-related GPCRs are included in two clades, i.e., clade 2 (25), and clade 3 (5), except for *P. anserina* Pa_5_7120 in clade 1, indicating that the sequences are very closely related. Similarly, Kulkarni et al. [25] showed that CFEM-containing Pth11-related GPCRs in *M. oryzae* and *N. crassa* occurred in one clade while Gruber et al. [29] showed that this type of GPCR in *T. reesei*, *T. atroviride*, and *T. virens* also clustered together. It is worth noticing that most of CFEM-containing Pth11-related GPCRs were detected in phytopathogens (67%).

Expression patterns of putative Pth11-related GPCR genes in phytopathogens

In order to gain an insight into the possible function of Pth11-related GPCRs, we analyzed the expression patterns of putative Pth11-related GPCR genes under various conditions, including during biotic stress, invasive plant infection, and growth under different nutritional conditions (Fig. 6). Gene expression data for putative Pth11-related

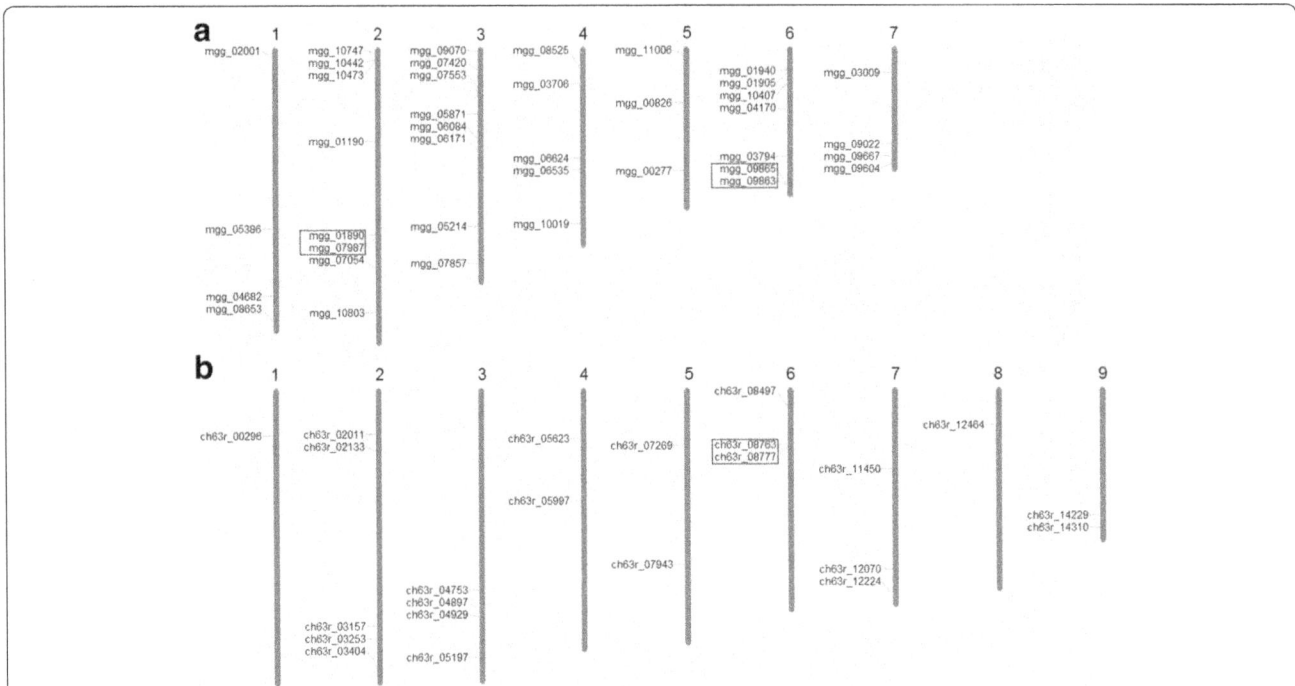

Fig. 3 Chromosomal distribution of putative *M. oryzae* (**a**) and *C. higginsianum* (**b**) Pth11-related GPCR genes. The chromosome numbers are shown at the top of the chromosomes (*M. oryzae*) or scaffolds (*C. higginsianum*), and tandemly duplicated genes are shown in boxes. No putative Pth11-related GPCR genes were detected on scaffold 10 of *C. higginsianum*

GPCRs was mined from publically available datasets, including experiments of GSE65311 [30] and GSE49597 [31] for biotic stress, GSE37886 [32], GSE21908 [33], and GSE33683 [34] for invasive plant infection, and GSE43006 [35], GSE53040 [36], GSE42692 [37], and GSE46155 [38] for nutritional stress. RNA-seq data for *H. oryzae* were also used to study the invasive infection of plants by *H. oryzae* [27].

Expressions of most putative Pth11-related GPCR genes were downregulated when *M. oryzae* was treated with bacteria that inhibit *M. oryzae* growth, including EA105 (a pseudomonad naturally isolated from rice soil), CHA0 (a *Pseudomonas fluorescens* biocontrol strain) and CHA77 (the non-cyanide-producing mutant of CHA0) (Fig. 6a). Transcriptional expression was analyzed when *F. graminearum* was exposed to bacterial MAMPs (microbe-associated molecular patterns), such as flagellin (FLG), lipooligosaccharides (LOS), and peptidoglycans (PGN) (Fig. 6d). Three time-points were used (1, 2, and 4 h after treatment with MAMPs). Similarly to the expression patterns in *M. oryzae*, most putative *F. graminearum* Pth11-related GPCR genes were downregulated during the biotic stress, especially after 2 h treatment (Fig. 6d). In contrast, upregulation of putative Pth11-related GPCR gene expression was detected during infection of both rice and barley by *M. oryzae*, including MGG_01884, MGG_09022, MGG_06171, MGG_06535, and MGG_01940 (Fig. 6a). The expressions of putative *F. graminearum*

Pth11-related GPCRs during the infection time course 1, 2, and 4 days after inoculation of plants (wheat and barley) were compared to control (Fig. 6d). Consistent with *M. oryzae* again, most putative *F. graminearum* Pth11-related GPCR genes were upregulated during invasive plant infection. Besides, putative Pth11-related GPCR genes could also respond to invasive plant infection for *H. oryzae* (Fig. 6b) and *C. higginsianum* (Fig. 6c) by induced regulations with suppressions of a few of them. These results indicated that Pth11-related GPCRs can respond to both biotic stress and invasive plant infection, and in clearly different manners. The two different expression patterns were also suggested by the two separate clusters in the hierarchical clustering analysis (Fig. 6).

Expression patterns of putative Pth11-related GPCR genes in saprophytes and entomopathogens

We examined putative Pth11-related GPCR gene expression in *N. crassa* treated with five antifungal compounds [three thioxanthone derivatives (TX129, TX34, TX87), XP13 (a prenylated analogue of 3,4-dihydroxyxanth-9-one), and D1 (2,4-dihydroxy-3-methylacetophenone)] and during growth under different nutritional conditions [media with either sucrose, xylan, pectin, orange peel powder (OPP), or avicel as a sole carbon source] (Fig. 6g). Almost all of the putative Pth11-related GPCRs were suppressed by the five antifungal compounds. Downregulation of putative Pth11-related GPCRs was also detected

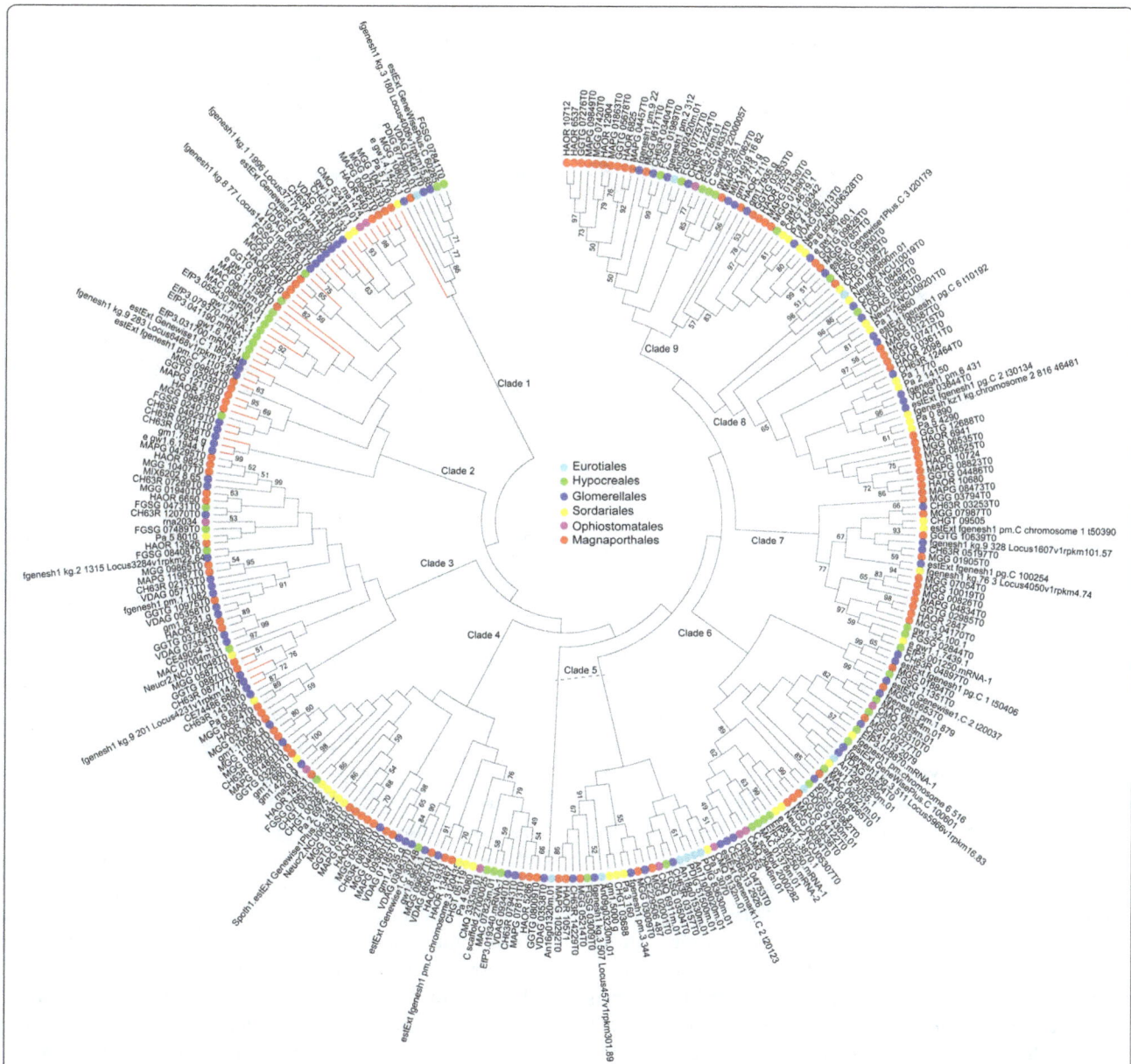

Fig. 4 Maximum likelihood phylogenetic tree of all predicted Pth11-related GPCRs among 20 selected fungal species. Species belonging to Eurotiales, Hypocreales, Glomerellales, Sordariales, Ophiostomatales, and Magnaporthales are indicated by cyan, green, blue, yellow, purple, and red circles, respectively. The thick red lines denote Pth11-related GPCRs containing a CFEM domain. Bootstrap values greater than 50% are shown at branches

when *N. crassa* grew under given nutrient conditions. However, NCU02903 and NCU05307 were significantly induced by the 5 nutrient conditions. These two distinct expression patterns were supported by the fact that the GPCRs involved fell into two different clusters (Fig. 6g). Our results not only indicated that putative Pth11-related GPCRs of *N. crassa* can respond to different nutrient conditions and antifungal compounds, but also showed the clearly opposite expression manners responding to them. We also compared the gene expression of putative Pth11-related GPCR in *T. reesei* growing on glucose, cellulose, or lactose as carbon sources (Fig. 6e). Induced expression of

most putative Pth11-related GPCRs was detected, indicating that a carbon source could affect the expression of putative *T. reesei* Pth11-related GPCR genes (Fig. 6e). Similar results were found for *G. clavigera*, which showed that a carbon source, including mannose, olive oil, oleic acid, and terpene, could alter the expression of putative *G. clavigera* putative genes (Fig. 6f).

Discussion

It has been shown that Pth11 is involved in pathogenesis and is required for the plant pathogen *M. oryzae* to cause disease [25, 26]. Genomes of phytopathogens such

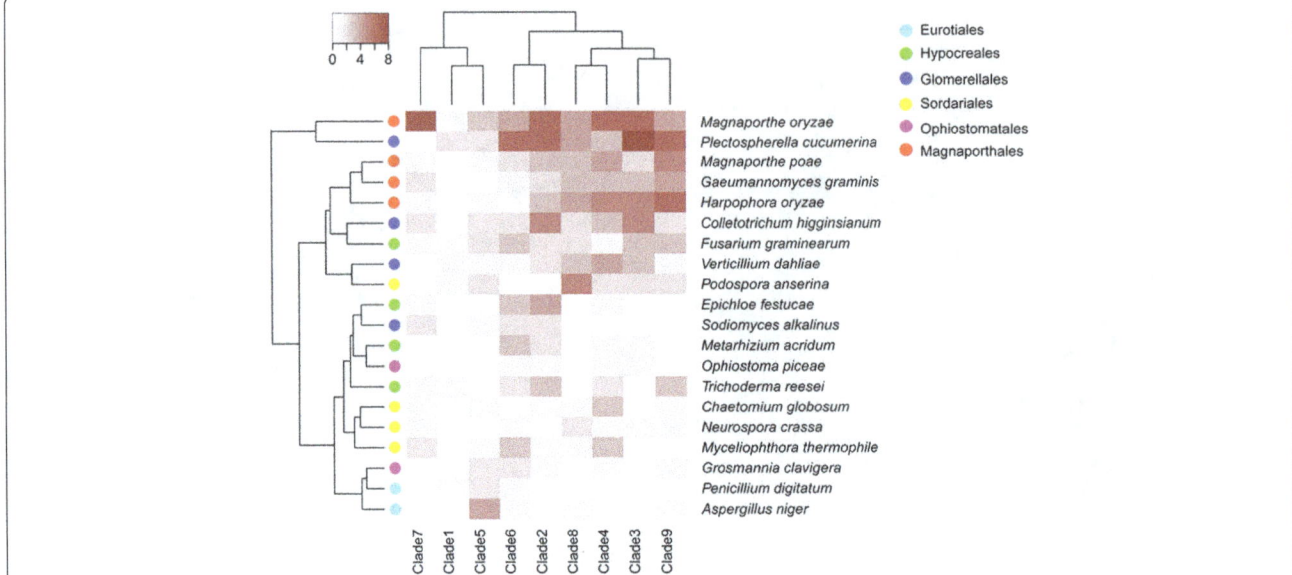

Fig. 5 Hierarchical clustering analysis of 20 selected fungal species based on the counts of putative Pth11-related GPCRs in each clade. The clades are defined based on the phylogenetic tree in Fig. 4. Cyan, Eurotiales; Green, Hypocreales; Blue, Glomerellales; Yellow, Sordariales; Purple, Ophiostomatales; Red, Magnaporthales

as *P. cucumerina*, *V. dahlia* and *C. higginsianum* in Glomerellales and *M. oryzae*, *G.graminis*, and *M. poae* in Magnaporthales consistently have the largest number of putative Pth11-related GPCRs (from 16 to 41 genes). Compared with phytopathogens, much fewer putative Pth11-related GPCRs were detected in both saprophytes and entomopathogens. The phytopathogen *F. graminearum* encodes 17 putative pth11-related GPCRs, many more than the other three species (with 7 to 12 each) in Hypocreales. Both *H. oryzae* and *E. festucae* are symbionts. However, *H. oryzae* encodes much more pth11-related GPCRs than *E. festucae*. One possible explanation is that *E. festucae* is derived from insect-parasitic ancestors [28], while *H. oryzae* arose from phytopathogenic ancestors [27]. These results showed that the arsenal of Pth11-related GPCRs might be related to nutritional strategies, especially for phytopathogens. Besides, the hierarchical clustering analysis revealed that species with similar nutritional strategies from the same order clustered together while species from the same order, but with different nutritional strategies were detected in different clusters. These results revealed that instead of fungi from the same order, those fungi with similar nutritional strategies were inclined to share orthologs of putative Pth11-related GPCRs. Moreover, Pth11 has an extracellular amino-terminal CFEM domain [25, 39]. Although only a subset of putative Pth11-related GPCRs contained this fungal-specific cysteine-rich CFEM domain, most of CFEM-containing Pth11-related GPCRs were detected in phytopathogens and were very closely related. This phenomenon also indicated that trophic variations have

been important factors in the evolution of Pth11-related GPCRs. Furthermore, the topology of the phylogenetic tree indicated that each order's putative Pth11-related GPCRs were derived from GPCRs of their common ancestors and that Pth11-related GPCRs gained and/or lost several times during the evolutionary process. Overall, these results make it fairly safe to infer that Pth11-related GPCRs existed before the divergence of Pezizomycotina, and later evolved independently in a species-specific manner. And during the evolution of Pth11-related GPCRs, the different nutritional strategies of these fungi could be an important evolutionary stress.

The evolution of Pth11-related GPCRs involved in trophic variations of fungi in Pezizomycotina could also be revealed by their possible functions. We analyzed the expression pattern of putative Pth11-related GPCR genes under various conditions, including during biotic stress, invasive plant infection, and growth under different nutritional conditions. Expressions of most putative Pth11-related GPCR genes from both *M. oryzae* and *F. graminearum* were downregulated during the biotic stress while upregulation were detected during invasive plant infection by both of them. The two clearly different expression patterns revealed that Pth11-related GPCRs can respond to both biotic stress and invasive plant infection for phytopathogens. For saprophytes, almost all the putative Pth11-related GPCRs from *N. crassa* were suppressed by the antifungal compounds but some of them were induced when *N. crassa* were subjected to different nutrient conditions. Similar results were also found by Cabrera et al. [9], who revealed that many

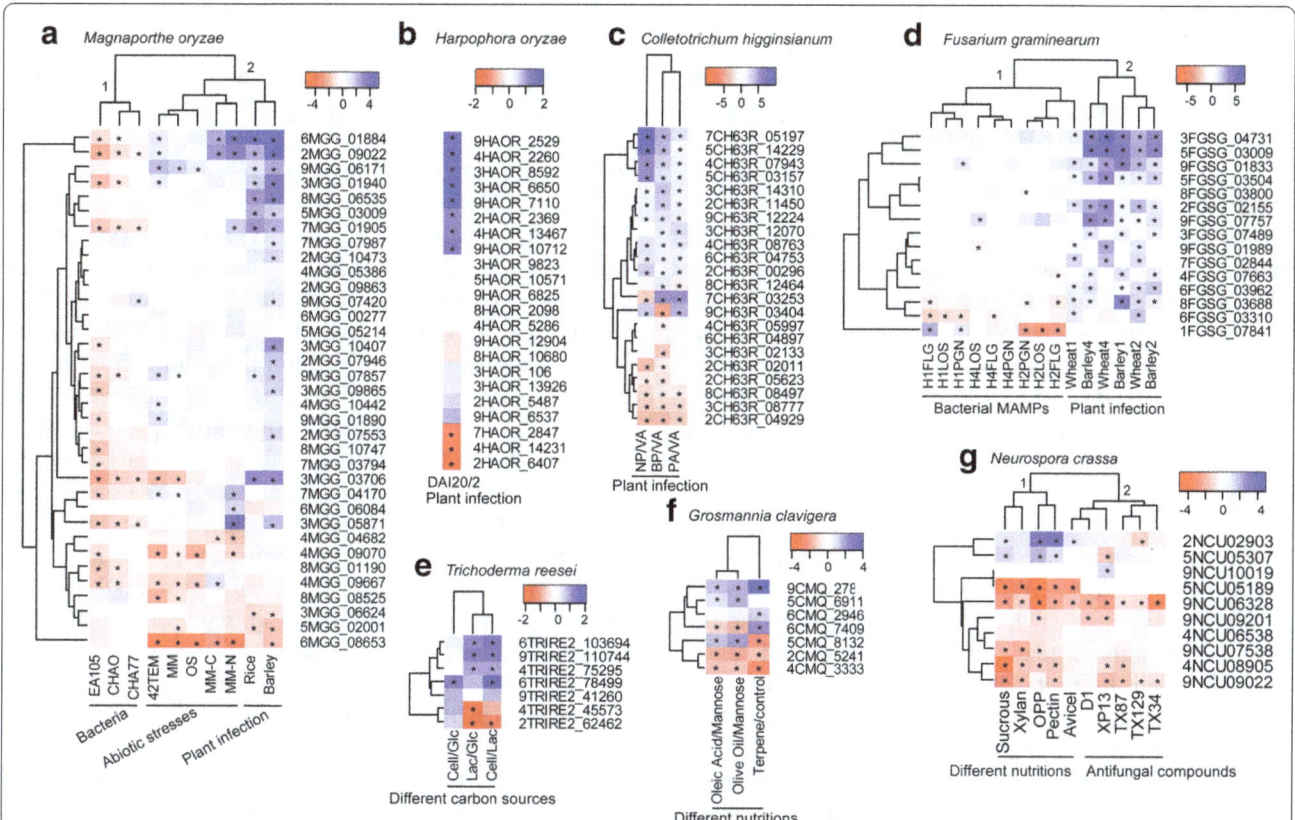

Fig. 6 Heatmaps of gene expression of putative Pth11-related GPCRs. **a** Gene expression in *M. oryzae* under various treatments. EA105, CHAO, and CHA77 refer to bacteria, including a pseudomonad naturally isolated from rice soil, a *P. fluorescens* biocontrol strain, and the non-cyanide-producing mutant of CHAO, respectively. 42TEM: heat shock (42 °C for 45 min); MM, minimal media; OS, oxidative stress (treated with methyl viologen); MM-C, carbon limitation; MM-N, nitrogen limitation; Rice: rice at 72 h post-inoculation (hpi); Barley: barley at 72 hpi. **b** Comparative gene expression of *H. oryzae* between DAI20 and DAI2 which refer to genes expressed by *H. oryzae* infecting rice roots at 20 and 2 days after inoculation (DAI), respectively. **c** Comparative gene expression of *C. higginsianum* in four stages during the infection process of *Arabidopsis*. The four stages are: VA (in vitroappressoria), PA (in planta appressoria), BP (biotrophic phase), and NP (necrotrophic phase). **d** Gene expression of *F. graminearum* under various treatments. *F. graminearum* was treated with bacterial MAMPs including flagellin (FLG), lipooligosacharides (LOS), and peptidoglycans (PGN). Time points are 1, 2, and 4 h after treatment with MAMPs. Barley and wheat indicated gene expression during the infection time course (1, 2, and 4 days after inoculation). **e** Comparative gene expression of *T. reesei* growing on glucose (Glc), cellulose (Cell), or lactose (Lac) as a sole carbon source. **f** Comparative gene expression of *G. clavigera* growing on mannose, olive oil, oleic acid, or terpene as a sole carbon source. **g** Gene expression of *N. crassa* under various conditions. *N. crassa* was subjected to five different nutrient conditions [pectin, orange peel powder (OPP), xylan, avicel, or sucrose as a sole carbon source] and five antifungal compounds [three thioxanthone derivatives (TX129, TX34, TX87), XP13 (a prenylated analogue of 3,4-dihydroxyxanth-9-one,) and D1 (2,4-dihydroxy-3-methylacetophenone)]. Asterisks denote differential expressions greater than twofold change

Pth11-related GPCRs are related to chemical sensitivity or nutritional phenotypes by analyzing the phenotypes of mutants. We also found that carbon source could affect the expression of putative Pth11-related GPCR genes of *T. reesei* and *G. clavigera*. The results revealed the common functions of Pth11-related GPCRs in respond to nutritional conditions for saprophytes and entomopathogens.

Expression patterns of putative Pth11-related GPCR genes in *M. oryzae* were also detected under abiotic stresses including growth on minimal medium, carbon and nitrogen starvation, heat shock (42 °C for 45 min), and oxidative stress (treated with methyl viologen). Interestingly, the hierarchical clustering analysis showed that the abiotic stress and invasive plant infection clustered together, indicating the similarity in expression patterns of putative Pth11-related GPCR genes between abiotic stress and invasive plant infection. These similar expression patterns may have resulted from that *M. oryzae* typically encountering nutrient-limited environments at the invasive growth stage [33].

The key role of tandem duplication in the evolution of other gene families has been reported, including the P450 family [40]. It has also been shown that gene duplication often plays a central role in both fungal and plant diversification, and is a key process generating the raw material necessary for adaptive evolution [41, 42]. Tandem duplications were detected in the chromosomes of both *M. oryzae* and *C. higginsianum*, indicating that tandem duplication may contribute to the evolution of the Pth11-related GPCR family. However, no tandem duplication of putative Pth11-related GPCR genes was found in *F. graminearum* (Additional file 1), indicating the complex evolutionary history of Pth11-related GPCRs.

Conclusions
This study provided a thorough examination of 20 genomes in Pezizomycotina for putative Pth11-related GPCRs. The 20 selected fungi cover different nutritional strategies, including phytopathogens, symbionts, saprophytes, and entomopathogens. To elucidate the evolution of putative Pth11-related GPCRs, we performed a chromosome distribution and phylogenetic analysis of them. This analysis indicated that putative Pth11-related GPCRs existed before the divergence of Pezizomycotina, and that the GPCRs in each species were derived from GPCRs of their common ancestors. During this evolutionary process, putative Pth11-related GPCRs could have been gained and lost several times, possibly involving tandem duplication. Our results showed that putative Pth11-related GPCRs could respond to bacterial challenges, antifungal compounds, different nutritional conditions, and invasive plant infection, and different expression patterns were used to in response to these stimuli. Based on the common functions of putative Pth11-related GPCRs in respond to nutritional conditions and the results of fungi with similar nutritional strategies were inclined to share orthologs of putative Pth11-related GPCRs, we suggested that the different nutritional strategies of fungi could have been an important evolutionary stress during the evolution of Pth11-related GPCRs. It is worth mentioning that the proteins identified as putative Pth11-related GPCRs in this study have only been characterized in silico. Compared with only three types of G protein in most fungi, a large number of putative Pth11-related GPCRs were detected. Therefore, determining the intracellular interactions of Pth11-related GPCRs and their signaling output will help us understand how fungi adapt to different challenges and nutritional conditions.

Methods
Identification of putative Pth11-related GPCRs
Twenty genome sequences and deduced proteomes of Pezizomycotina were used from the following fungi: *M. oryzae, G. graminis, M. poae* [43], *H. oryzae* [44], *G. clavigera* [45], *C. globosum* [46], *N. crassa* [47], *V. dahlia* [48], *C. higginsianum* [49], *F. graminearum* [50], *E. festucae* [51], *M. acridum* [52], *T. reesei* [53], *P. digitatum* [54], *A. niger* [55], *S. alkalinus* [56], *P. cucumerina* [57], *P. anserine* [58], *M. thermophile* [59], and *O. piceae* [60]. Genome sequence of *S. cerevisiae* [61] was used as outgroup. These species cover Magnaporthales, Ophiostomatales, Sordariales, Glomerellales, and Hypocreales,

and comprise phytopathogens (*M. oryzae, G. graminis, M. poae, V. dahlia, C. higginsianum, F. graminearum* and *P. cucumerina*), symbionts (*H. oryzae*, and *E. festucae*), saprophytes (*O. piceae, N. crassa, C. globosum, M. thermophile, P. anserina, S. alkalinus, T. reesei, A. niger,* and *P. digitatum*), and entomopathogens (*G. clavigera* and *M. acridum*). A pipeline was used to identify the putative Pth11-related GPCRs in the 21 selected proteomes. Firstly, as the Pth11-domain distinguishes Pth11-related proteins from other class of GPCR-like proteins [25], the hmmscan program from the HMMER v3.1 package [62] was used to search for the Pth11-domain across all the 21 proteomes with an e-value cutoff of 1e-20. Then the obtained Pth11-domain containing proteins were used in a BLASTP search against Pth11-related GPCRs of *M. oryzae* [25]. An e-value limit of 1e-09 was applied, and all proteins that had at least 30% identity and 80% overlap over the length of the proteins were retained [25]. Finally, the obtained proteins were evaluated for the typical seven-transmembrane domain by TMHMM, HMMTOP, and Phobius [63–65], and proteins with seven or more transmembrane domains predicted by at least two algorithms were retained as putative Pth11-related GPCRs and used for further analysis. By this analysis pipeline, the authenticity of these predicted Pth11-related GPCRs was highly improved.

Chromosomal organization of putative Pth11-related GPCRs
Chromosome location images were generated using MapInspect software to localize putative Pht11-related GPCRs of *M. oryzae, C. higginsianum*, and *F. graminearum*. Any putative Pth11-related GPCRs separated by no more than 15 genes were identified as tandemly duplicated genes.

Protein alignments and phylogenetic analysis
Protein sequences were aligned using ClustalW v2 [66]. To select conserved regions, the alignments were analyzed with Gblocks v0.91b [67] using the default parameters. The best amino acid substitution model was chosen using ProtTest v3.2 [68], and LG + G + F model was selected as the best-fit model for the datasets. The phylogenetic tree of putative Pth11-related GPCRs was constructed in MEGA v7 [69] using maximum likelihood (ML) methods with the best-fit model followed by bootstrap analysis (1000 bootstrap replications). To infer phylogenetic relationships among the 20 species, 100 clusters of 1:1 orthologs were chosen based on our previous study [27]. The proteins of the 100 orthologs were aligned and then concatenated. Phylogenetic analyses were performed using the ML criterion implemented in RAxML [70] through the RAxML-HPC BlackBox web server at the Cyber Infrastructure for Phylogenetic Research with LG + G + I model. The sequence logo was created using

WebLogo v2.8.2 [71] based on a multiple sequence alignment of 296 putative Pth11-related GPCRs.

CFEM domain search
The CFEM domain architectures of putative Pth11-related GPCRs were predicted using two search methods, including Pfam [72], and Conserved Domain Search (CDSearch) [73] applying the default settings for each.

Expression analysis
Expression data were mined for putative Pth11-related GPCR genes from several datasets, including RNA sequencing and microarray data, which were downloaded from the Gene Expression Omnibus. The datasets with accession numbers GSE65311, GSE37886 (*F. graminearum*), GSE21908, GSE49597 (*M. oryzae*), GSE33683 (*C. higginsianum*), GSE43006 (*G. clavigera*), GSE53040, GSE42692 (*N. crassa*), and GSE46155 (*T. reesei*) were analyzed. For *H. oryzae*, RNA-seq data from a previous study [27] were used. These data pertained to different environmental conditions, including biotic and abiotic stress conditions and various stages of colony development. The data were visualized using heatmaps generated with the heatmap.2 package in R, which is based on log2 fold changes after normalization.

Additional files

Additional file 1: Chromosomal distribution of putative *F. graminearum* Pth11-related GPCR genes. Chromosome numbers are shown at the top of the chromosomes.

Additional file 2: Predicted Pth11-related GPCRs among thirteen fungi belonging to Pezizomycotina.

Abbreviations
BLAST: Basic local alignment search tool; GPCR: G-protein coupled receptors; GPCRDB: GPCR database; MAMP: Microbe-associated molecular patterns

Acknowledgements
Not applicable.

Funding
The research was funded by the National Key Research and Development Program of China (2016YFD0800803), the National Natural Science Foundation of China (31501689 and 31400328), the Science Foundation of Jiangsu Province, China (BK20150670 and BK20140697), and the Fundamental Research Funds for the Central Universities (KJQN201641).

Authors' contributions
XX and CC conceived and designed the experiments; XX and GL performed the experiments; XX, GL, LL, and ZS analyzed the data; XX and CC wrote the paper. All authors read and approved the final manuscript.

Competing interests
The authors declare that they have no competing interests.

Author details
[1]College of Life Sciences, Nanjing Agricultural University, Nanjing 210095, China. [2]Agricultural Product Processing Research Institute, Chinese Academy of Tropical Agricultural Sciences, Zhanjiang 524001, China. [3]State Key Laboratory of Rice Biology, Institute of Biotechnology, Zhejiang University, Hangzhou 310058, China.

References
1. Rodriguez RJ, White JF, Arnold AE, Redman RS. Fungal endophytes: diversity and functional roles. New Phytol. 2009;182:314–30.
2. Xu XH, Wang C, Li SX, Su ZZ, Zhou HN, Mao LJ, et al. Friend or foe: differential responses of rice to invasion by mutualistic or pathogenic fungi revealed by RNAseq and metabolite profiling. Sci Rep. 2015;5:13624.
3. Robertson JD. The ultrastructure of cell membranes and their derivatives. Biochem Soc Symp. 1959;16:3–43.
4. Zimmermann W, Rosselet A. Function of the outer membrane of Escherichia Coli as a permeability barrier to beta-lactam antibiotics. Antimicrob Agents Chemother. 1977;12:368–72.
5. von Heijne G. Membrane-protein topology. Nat Rev Mol Cell Biol. 2006;7:909–18.
6. Engel A, Gaub HE. Structure and mechanics of membrane proteins. Annu Rev Biochem. 2008;77:127–48.
7. Müller DJ, Wu N, Palczewski K. Vertebrate membrane proteins: structure, function, and insights from biophysical approaches. Pharmacol Rev. 2008;60:43–78.
8. Elion EA. Pheromone response, mating and cell biology. Curr Opin Microbiol. 2000;3:573–81.
9. Cabrera IE, Pacentine IV, Lim A, Guerrero N, Krystofova S, Li L, et al. Global analysis of predicted G protein-coupled receptor genes in the filamentous fungus, Neurospora crassa. G3 (Bethesda). 2015;5:2729–43.
10. Shukla AK, Singh G, Ghosh E. Emerging structural insights into biased GPCR signaling. Trends Biochem Sci. 2014;39:594–602.
11. Chini B, Parenti M, Poyner DR, Wheatley M. G-protein-coupled receptors: from structural insights to functional mechanisms. Biochem Soc Trans Biochem Soc Trans. 2013;41:135–6.
12. Maller JL. Fishing at the cell surface. Science. 2003;300:594–5.
13. Hamm HE. The many faces of G protein signaling. J Biol Chem. 1998;273:669–72.
14. Lengeler KB, Davidson RC, D'souza C, Harashima T, Shen WC, Wang P, et al. Signal transduction cascades regulating fungal development and virulence. Microbiol Mol Biol Rev. 2000;64:746–85.
15. Rispail N, Soanes DM, Ant C, Czajkowski R, Grünler A, Huguet R, et al. Comparative genomics of MAP kinase and calcium-calcineurin signalling components in plant and human pathogenic fungi. Fungal Genet Biol. 2009;46:287–98.
16. Li L, Wright SJ, Krystofova S, Park G, Borkovich KA. Heterotrimeric G protein signaling in filamentous fungi. Annu Rev Microbiol. 2007;61:423–52.
17. Dohlman HG, Thorner JW. Regulation of G protein-initiated signal transduction in yeast: paradigms and principles. Annu Rev Biochem. 2001;70:703–54.
18. Casselton LA, Olesnicky NS. Molecular genetics of mating recognition in basidiomycete fungi. Microbiol Mol Biol Rev. 1998;62:55–70.
19. Krystofova S, Borkovich KA. The predicted G-protein-coupled receptor GPR-1 is required for female sexual development in the multicellular fungus Neurospora crassa. Eukaryot Cell. 2006;5:1503–16.
20. Lorenz MC, Pan X, Harashima T, Cardenas ME, Xue Y, Hirsch JP, et al. The G protein-coupled receptor Gpr1 is a nutrient sensor that regulates pseudohyphal differentiation in Saccharomyces Cerevisiae. Genetics. 2000;154:609–22.
21. Li L, Borkovich KA. GPR-4 is a predicted G-protein-coupled receptor required for carbon source-dependent asexual growth and development in Neurospora crassa. Eukaryot Cell. 2006;5:1287–300.

22. Chung KS, Won M, Lee SB, Jang YJ, Hoe KL, Kim DU, et al. Isolation of a novel Gene from Schizosaccharomyces pombe: Stm1 encoding a seven-transmembrane loop protein that may couple with the Heterotrimeric Gα2 protein, Gpa2. J Biol Chem. 2001;276:40190–201.

23. Chen JG, Willard FS, Huang J, Liang J, Chasse SA, Jones AM, et al. A seven-transmembrane RGS protein that modulates plant cell proliferation. Science. 2003;301:1728–31.

24. Lafon A, Han KH, Seo JA, Yu JH, d'Enfert C. G-protein and cAMP-mediated signaling in aspergilli: a genomic perspective. Fungal Genet Biol. 2006;43:490–502.

25. Kulkarni RD, Thon MR, Pan H, Dean RA. Novel G-protein-coupled receptor-like proteins in the plant pathogenic fungus Magnaporthe grisea. Genome Biol. 2005;6:R24.

26. DeZwaan TM. Magnaporthe grisea Pth11p is a novel plasma membrane protein that mediates Appressorium differentiation in response to inductive substrate cues. Plant Cell. 1999;11:2013–30.

27. Xu XH, Su ZZ, Wang C, Kubicek CP, Feng XX, Mao LJ, et al. The rice endophyte Harpophora oryzae genome reveals evolution from a pathogen to a mutualistic endophyte. Sci Rep. 2014;4:5783.

28. Spatafora JW, Sung GH, Sung JM, Hywel-Jones NL, White JF. Phylogenetic evidence for an animal pathogen origin of ergot and the grass endophytes. Mol Ecol. 2007;16:1701–11.

29. Gruber S, Omann M, Zeilinger S. Comparative analysis of the repertoire of G protein-coupled receptors of three species of the fungal genus Trichoderma. BMC Microbiol. 2013;13:108.

30. Ipcho S, Sundelin T, Erbs G, Kistler HC, Newman M-A, Olsson S. Fungal innate immunity induced by bacterial microbe-associated molecular patterns (MAMPs). G3 (Bethesda). 2016;6:1585–95.

31. Spence CA, Raman V, Donofrio NM, Bais HP. Global gene expression in rice blast pathogen Magnaporthe oryzae treated with a natural rice soil isolate. Planta. 2014;239:171–85.

32. Harris LJ, Balcerzak M, Johnston A, Schneiderman D, Ouellet T. Host-preferential Fusarium graminearum gene expression during infection of wheat, barley, and maize. Fungal Biol. 2016;120:111–23.

33. Mathioni SM, Belo A, Rizzo CJ, Dean RA, Donofrio NM. Transcriptome profiling of the rice blast fungus during invasive plant infection and in vitro stresses. BMC Genomics. 2011;12:20.

34. O'Connell RJ, Thon MR, Hacquard S, Amyotte SG, Kleemann J, Torres MF, et al. Lifestyle transitions in plant pathogenic Colletotrichum fungi deciphered by genome and transcriptome analyses. Nat Genet. 2012;44:1060–5.

35. Wang Y, Lim L, Madilao L, Lah L, Bohlmann J, Breuil C. Gene discovery for enzymes involved in limonene modification or utilization by the mountain pine beetle-associated pathogen Grosmannia clavigera. Appl Environ Microbiol. 2014;80:4566–76.

36. Pedro Gonçalves A, Silva N, Oliveira C, Kowbel DJ, Glass NL, Kijjoa A, et al. Transcription profiling of the Neurospora crassa response to a group of synthetic (thio)xanthones and a natural acetophenone. Genomics Data. 2015;4:26–32.

37. Benz PJ, Chau BH, Zheng D, Bauer S, Glass NL, Somerville CR. A comparative systems analysis of polysaccharide-elicited responses in Neurospora crassa reveals carbon source-specific cellular adaptations. Mol Microbiol. 2014;91:275–99.

38. Bischof R, Fourtis L, Limbeck A, Gamauf C, Seiboth B, Kubicek CP. Comparative analysis of the Trichoderma Reesei transcriptome during growth on the cellulase inducing substrates wheat straw and lactose. Biotechnol Biofuels. 2013;6:127.

39. Kulkarni RD, Kelkar HS, Dean RA. An eight-cysteine-containing CFEM domain unique to a group of fungal membrane proteins. Trends Biochem Sci. 2003;28:118–21.

40. Doddapaneni H, Chakraborty R, Yadav JS. Genome-wide structural and evolutionary analysis of the P450 monooxygenase genes (P450ome) in the white rot fungus Phanerochaete chrysosporium: evidence for gene duplications and extensive gene clustering. BMC Genomics. 2005;6:92.

41. Flagel LE, Wendel JF. Gene duplication and evolutionary novelty in plants. New Phytol. 2009;183:557–64.

42. Wapinski I, Pfeffer A, Friedman N, Regev A. Natural history and evolutionary principles of gene duplication in fungi. Nature. 2007;449:54–61.

43. The Magnaporthe comparative genome database. https://www.broadinstitute.org/scientificcommunity/science/projects/fungal-genome-initiative/magnaporthe-comparative-genomics-proj. Accessed 16 Mar 2016.

44. The Harpophora oryzae genome database. http://www.ncbi.nlm.nih.gov/nuccore/667821394. Accessed 16 Mar 2016.

45. The Grosmannia clavigera genome database. http://genome.jgi.doe.gov/Grocl1/Grocl1.home.html. Accessed 16 Mar 2016.

46. The Chaetomium globosum genome database. http://genome.jgi.doe.gov/Chagl_1/Chagl_1.home.html. Accessed 16 Mar 2016.

47. The Neurospora crassa genome database. http://genome.jgi.doe.gov/Neucr_trp3_1/Neucr_trp3_1.home.html. Accessed 16 Mar 2016.

48. The Verticillium dahlia genome database. http://genome.jgi.doe.gov/Verda1/Verda1.home.html. Accessed 16 Mar 2016.

49. The Colletotrichum higginsianum genome database. http://genome.jgi.doe.gov/Colhi1/Colhi1.home.html. Accessed 16 Mar 2016.

50. The Fusarium graminearum genome database. http://genome.jgi.doe.gov/Fusgr1/Fusgr1.home.html. Accessed 16 Mar 2016.

51. The Epichloe festucae genome database. http://csbio-l.csr.uky.edu/ef894. Accessed 16 Mar 2016.

52. The Metarhizium acridum genome database. http://genome.jgi.doe.gov/Metac1/Metac1.home.html. Accessed 16 Mar 2016.

53. The Trichoderma reesei genome database. http://genome.jgi.doe.gov/Trire2/Trire2.home.html. Accessed 16 Mar 2016.

54. The Penicillium digitatum genome database. http://genome.jgi.doe.gov/Pendi1/Pendi1.home.html. Accessed 10 June 2017.

55. The Aspergillus niger genome database. http://genome.jgi.doe.gov/Aspni_DSM_1/Aspni_DSM_1.home.html. Accessed 10 June 2017.

56. The Sodiomyces alkalinus genome database. http://genome.jgi.doe.gov/Sodal1/Sodal1.home.html. Accessed 10 June 2017.

57. The Plectospherella cucumerina genome database. http://genome.jgi.doe.gov/Plecu1/Plecu1.home.html. Accessed 10 June 2017.

58. The Podospora anserine genome database. http://genome.jgi.doe.gov/Podan2/Podan2.home.html. Accessed 10 June 2017.

59. The Myceliophthora thermophile genome database. http://genome.jgi.doe.gov/Spoth2/Spoth2.home.html. Accessed 10 June 2017.

60. The Ophiostoma piceae genome database. http://genome.jgi.doe.gov/Ophpic1/Ophpic1.home.html. Accessed 10 June 2017.

61. The Saccharomyces cerevisiae database. http://genome.jgi.doe.gov/Sacce1/Sacce1.home.html. Accessed 10 June 2017.

62. Finn RD, Clements J, Eddy SR. HMMER web server: interactive sequence similarity searching. Nucleic Acids Res. 2011;39(Web Server issue):W29–37.

63. Tusnády GE, Simon I. The HMMTOP transmembrane topology prediction server. Bioinformatics. 2001;17:849–50.

64. Krogh A, Larsson B, von Heijne G, Sonnhammer EL. Predicting transmembrane protein topology with a hidden Markov model: application to complete genomes. J Mol Biol. 2001;305:567–80.

65. Käll L, Krogh A, Sonnhammer EL. A combined transmembrane topology and signal peptide prediction method. J Mol Biol. 2004;338:1027–36.

66. Larkin MA, Blackshields G, Brown NP, Chenna R, Mcgettigan PA, McWilliam H, et al. Clustal W and Clustal X version 2.0. Bioinformatics. 2007;23:2947–8.

67. Castresana J. Selection of conserved blocks from multiple alignments for their use in Phylogenetic analysis. Mol Biol Evol. 2000;17:540–52.

68. Abascal F, Zardoya R, Posada D. ProtTest: selection of best-fit models of protein evolution. Bioinformatics. 2005;21:2104–5.

69. Kumar S, Stecher G, Tamura K. MEGA7: molecular evolutionary genetics analysis version 7.0 for bigger datasets. Mol Biol Evol. 2016;33:1870–4.

70. Stamatakis A. RAxML-VI-HPC: maximum likelihood-based phylogenetic analyses with thousands of taxa and mixed models. Bioinformatics. 2006;22:2688–90.

71. Crooks GE, Hon G, Chandonia JM, Brenner SE. WebLogo: A sequence logo generator. Genome Res. 2004;14:1188–90.

72. Finn RD, Bateman A, Clements J, Coggill P, Eberhardt RY, Eddy SR, et al. Pfam: the protein families database. Nucleic Acids Res. 2014;42:D222–30.

73. Marchler-Bauer A, Derbyshire MK, Gonzales NR, Lu S, Chitsaz F, Geer LY, et al. CDD: NCBI's conserved domain database. Nucleic Acids Res. 2015;43:D222–6.

3

Effects of *Pseudomonas aeruginosa* and *Streptococcus mitis* mixed infection on TLR4-mediated immune response in acute pneumonia mouse model

Chao Song[1,2], Hongdong Li[1,2], Yunhui Zhang[1,2] and Jialin Yu[1,2]*

Abstract

Background: Our previous research on the diversity of microbiota in the endotracheal tubes (ETTs) of neonates in the neonatal intensive care unit found that *Pseudomonas aeruginosa* (*P. aeruginosa*) and *Streptococcus mitis* (*S. mitis*) were the dominant bacteria on the ETT surface and the existence of *S. mitis* could promote biofilm formation and pathogenicity of *P. aeruginosa*. Toll-like receptor 4 (TLR4), which has been widely detected on the surface of airway epithelial cells, is the important component of the innate immune system. Therefore, we hypothesized that the co-existence of these two bacteria might impact the host immune system through TLR4 signaling.

Results: *S. mitis* rarely caused inflammation, whereas *P. aeruginosa* caused the most severe inflammation accompanied by increases in the number of inflammatory cells, interleukin (IL)-6 and tumor necrosis factor (TNF)-α expression, and total cell counts in BALF ($p < 0.05$). In the PAO1 + *S. mitis* group, moderate inflammation, reduced IL-6 and TNF-α protein levels, and decreased total cell counts were observed. Additionally, levels of these indicators were decreased lower in TLR4-deficient mice than in wild-type mice ($p < 0.05$).

Conclusions: Our results demonstrated that infection with *S. mitis* together with *P. aeruginosa* could alleviate lung inflammation in acute lung infection mouse models possibly via the TLR4 signaling pathway.

Keywords: Acute lung infection, *Pseudomonas Aeruginosa*, *Streptococcus Mitis*, TLR4, Il-6, Tnf-α

Background

Ventilator-associated pneumonia (VAP) is a common device-related infection in neonatal intensive care units (NICUs), and it is associated with high morbidity and mortality as well as increased antibiotic resistance and high economic costs [1, 2]. Our previous research demonstrated that *Pseudomonas aeruginosa* (*P. aeruginosa*) and *Streptococcus mitis* (*S. mitis*) were the most dominant microbes on the surface of neonatal endotracheal tubes (ETTs) [3, 4]. In addition, the co-existence of *S. mitis* from the oral microbiome and the opportunistic

pathogen *P. aeruginosa* on the same ETT may play a crucial role in biofilm formation. Khosravi et al. found that *S. mitis* produced and released a tenovin 6-like molecule that induced growth inhibition and coccoid conversion of *Helicobacter pylori* cells [5]. Moreover, recent discoveries regarding the pathogenesis of VAP has found an interesting phenomenon in which autoinducer-2 produced by *S. mitis* can act as an important molecule to promote *P. aeruginosa* biofilm formation and to increase proinflammatory cytokine secretion in endotracheal intubation rat models [6]. Given that *S. mitis* is the most abundant bacteria of the normal human oral flora and rarely causes diseases, we questioned whether *S. mitis* could also modulate the pathogenicity of *P. aeruginosa* in acute lung infection. If so, how does this co-existence impact the immune system of the host and what is the possible mechanism in acute lung infection?

---

* Correspondence: yujialin486@126.com
[1]Department of Neonatology, Children's Hospital of Chongqing Medical University, Chongqing, China
[2]Ministry of Education Key Laboratory of Child Development and Disorders – Chongqing Key Laboratory of Pediatrics, China International Science and Technology Cooperation Base of Child Development and Critical Disorders, Chongqing, China

Toll-like receptors (TLRs), one of the most important receptor families of the innate immune system, can recognize pathogen-associated molecular patterns (PAMPs), including microbial products, and play an important role in the immune response [7, 8]. At present, 11 TLRs have been identified in humans [9]. Recent studies have confirmed that TLR2, 4, 5, and 9 play pivotal roles in the response to bacterial infections [10]. TLR4 is a receptor for lipopolysaccharide (LPS), an important cell wall component of gram-negative bacteria. Through ligand binding, TLR4 recruits signaling adaptors and initiates signaling cascades, which results in the activation of nuclear factor (NF)-κB and the release of inflammatory cytokines, such as interleukin (IL)-6 [11] and tumor necrosis factor (TNF)-α [12]. Moreover, airway epithelial cells are believed to contribute to the inflammatory response in the lung, and TLR4 has been widely detected on the surface of airway epithelial cells as well as cells of the myeloid lineage, such as macrophages and neutrophils [13, 14]. Some studies have reported that TLR4-deficient mice show increased lung inflammation and higher bacterial load, and TLR4 signaling may have a critical function in the fine tuning of inflammation during chronic mycobacterial infection [15]. Therefore, we hypothesized that TLR4 signaling might participate in the response to acute lung infection.

Based on our previous findings that *P. aeruginosa* and *S. mitis* are the dominant bacteria in the mixture of organisms causing lung infections, these two bacteria were selected for the present study. The aim of our study was to explore the relationship between *P. aeruginosa* and *S. mitis* in lung infection and ascertain their roles in the immune response. After acute lung infection mouse models were established, lung bacteriological and histopathological examinations were performed, and total cell counts and levels of related cytokines in bronchoalveolar lavage fluid (BALF) were determined.

Methods

Bacteria and growth conditions

P. aeruginosa strain PAO1 (ATCC27853) and *S. mitis* (ATCC49456) were used in our study. PAO1 was kindly provided by Professor Li Shen (Institute of Molecular Cell and Biology, New Orleans, LA, USA). *S. mitis* was purchased from American Type Culture Collection. The bacterial strains were both cultured overnight in brain-heart infusion (BHI) broth. The *S. mitis* strain was grown at a neutral pH at 37 °C in a 5% CO_2 atmosphere. The PAO1 strain was grown at 37 °C on an orbital shaker at 200 rpm. Overnight-grown cultures of the strains were standardized to 0.2 (OD_{600}), and then diluted to a working concentration of OD_{600} = 0.1, if necessary.

Animals

Forty C57BL/6 mice (6–8 weeks, 18–20 g) were obtained from the Experimental Animal Center of Chongqing Medical University and housed in the Laboratory Animal Center at the Children's Hospital of Chongqing Medical University. Twenty TLR4-deficient mice (C57BL/10ScNJ) were obtained from the Key Laboratory of Diagnostic Medicine Designated by the Ministry of Education at Chongqing Medical University. TLR4 gene expression was determined by polymerase chain reaction (PCR) before the experiments to confirm the reliability of the TLR4-deficient mice. The experimental protocol was approved by the Animal Care and Use Committee of Chongqing Medical University.

Acute lung infection mouse models

Twenty wild-type mice were randomly divided into four groups (n = 5/group) treated with: PAO1 (OD_{600} = 0.1, 1 × 10^8 CFU/mL, 50 μL), *S. mitis* (OD_{600} = 0.1, 1 × 10^8 CFU/mL, 50 μL), PAO1 + *S. mitis* (OD_{600} = 0.2, 2 × 10^8 CFU/mL, 25 μL + 25 μL), or phosphate-buffered saline (PBS) (control group, 50 μL). Twenty TLR4-deficient mice were divided into the same treatment groups. Additionally, another 20 wild-type mice were subjected to the same treatments and used to estimate the bacterial burden, in order to confirm successful establishment of the acute lung infection mouse models.

The mouse models of acute lung infection were established as previously described with some modifications [16]. Briefly, mice were anesthetized with an intraperitoneal injection of chloral hydrate (0.02 mL/g) before tracheotomy. Then they were infected with an intratracheal instillation of 50 μL of bacterial suspension and kept upright in standing posture for 10 s to ensure the bacteria were fully delivered to the lung tissues. Finally, the trachea and skin were sutured. The mice were kept for 48 h before sacrificed.

Lung bacteriological and histological examinations

The whole lung of each mouse (n = 5 mice/group) was homogenized in 1 mL PBS and then serially diluted. A 50-μL sample from each tissue homogenate specimen was cultured quantitatively on Columbia sheep blood agar plates overnight at 37 °C in a 5% CO_2 atmosphere, and then the colonies were counted to estimate the number of colony-forming units (CFU). The right lungs of the other five mice in each group were fixed in 10% formalin buffer for 48 h, embedded in paraffin, dehydrated, cut into 5-μm slices, and stained with hematoxylin-eosin (H&E) for histopathological examination [17]. Images were obtained using light microscopy (Nikon eclipse 55i, Japan).

Total cell counts and cytokine analyses in BALF

The left lungs were lavaged and collected with 1 mL PBS. The fluid was instilled and withdrawn three times, and then the total cell counts in the BALF were determined using a cell counter (Countstar, Beijing, China). BALF samples were centrifuged at 3000 g and 4 °C for 5 min. The concentrations of pro-inflammatory cytokines TNF-α and IL-6 in the BALF supernatant were determined using mouse cytokine enzyme-linked immunosorbent assay (ELISA) kits (Beijing 4A Biotech Co., Ltd., Beijing, China) according to the manufacturer's instructions.

Statistical analysis

Statistical analyses were carried out using SPSS 22.0 (SPSS, Inc., Chicago, IL, USA). Analysis of variance (ANOVA) was used to identify significant differences among all groups. $p < 0.05$ was considered statistically significant.

Results

Bacterial CFU in lung tissues

To confirm the successful establishment of the acute lung infection mouse models, the bacterial burdens in harvested lung tissues were estimated. As shown in Table 1, the numbers of CFU in the PAO1 and PAO1 + S.mitis groups were significantly higher than that in the PBS control group ($p = 0.01$ and $p = 0.01$, respectively). More interestingly, although more bacteria were injected initially in the PAO1 + S. mitis group, after 48 h, the numbers of CFU did not differ significantly between the PAO1 and PAO1 + S. mitis groups ($p > 0.05$). Additionally, the number of CFU in the S. mitis group did not differ from that in the PBS control group ($p > 0.05$).

Histological observation of lung tissues from acute lung infection mouse models

Wild-type mice had severe lung damage after *P. aeruginosa* challenge (Fig. 1b). Numerous foci of necrosis and inflammatory infiltrates were discovered, with increased numbers of alveolar macrophages infiltrating the alveolar septa, as compared with the PBS group. However, little change was observed upon infection of wild-type mice with *S. mitis*, with almost no macrophage infiltration in the lungs (Fig. 1c). Notably, in the PAO1 + S. mitis group, moderate lung inflammation was observed, with recruitment of inflammatory cells in the peribronchial

wall and surrounding the vessels (Fig. 1d). Similar to wild-type mice, TLR4-deficient mice infected with *S. mitis*, PAO1 + S. mitis, or PAO1 showed slight, moderate, and severe lung inflammation, respectively (Fig. 2).

Cell counts in BALF

Among infected wild-type mice, the S. mitis group showed little change in the total cell count in BALF, as compared to the PBS group ($2.17 \pm 1.90 (\times 10^6)$ vs $3.2 \pm 1.9 (\times 10^5)$, $p > 0.05$). However, the total cell counts were slightly increased in the PAO1 and PAO1 + S. mitis groups (Fig. 3a). Similar results were obtained in TLR4-deficient mice (Fig. 3b). Interestingly, TLR4-deficient mice infected with PAO1 + S. mitis had significantly lower total cell counts in BALF compared with TLR4-deficient mice infected with PAO1 only ($p < 0.001$), which is consistent with the results of histological examination of lung tissues. In comparing wild-type and TLR4-deficient mice (Fig. 3c), we found that the BALF of TLR4-deficient mice showed significantly lower total cell counts in each comparative group (all $p < 0.05$).

IL-6 and TNF-α protein expression

TLR4 recognizes specific bacterial molecules, and their binding initiates host immune responses such as expression of proinflammatory cytokines TNF-α and IL-6. To better understand the immune response in our mouse models, the concentrations of TNF-α and IL-6 were measured. Significantly increased protein levels of IL-6 and TNF-α ($p < 0.001$ and $p < 0.001$, respectively) were observed in wild-type mice challenged with PAO1 compared with levels in the PBS group (Fig. 4). By contrast, S. mitis infection hardly induced the secretion of these two cytokines ($p > 0.05$). While increased IL-6 and TNF-α expression was observed the PAO1 + S. mitis group relative to the PBS group, these cytokine levels were still lower than those in the PAO1 group ($p = 0.001$ and $p = 0.002$, respectively). Interestingly, after infection of TLR4-deficient with the different bacteria (Fig. 4), expression of both IL-6 and TNF-α was almost completely inhibited, with no statistically significant differences in expression levels among the four treatment groups ($p > 0.05$). In comparison to IL-6 and TNF-α expression levels in wild-type mice, the concentrations of IL-6 and TNF-α were lower in TLR4-deficient mice infected with PAO1 with or without S. mitis ($p < 0.001$ and $p = 0.005$, respectively; Fig. 4).

Discussion

As is known, VAP is a principal cause of morbidity, mortality, and economic burden in ICUs. Increasing antimicrobial resistance has drawn attention to the failure of antibiotic treatment. Recent progress has mainly focused on the pathogenicity and antimicrobial resistance of a

Table 1 Bacterial counts in lung tissues of wild-type mice

	PBS	PAO1	S. mitis	PAO1 + S. mitis
CFU	0 (0–0)	1000 (500–3466)●▲	0 (0–173)	200 (166–800)●▲

●: Significantly different compared to the PBS control group ($p < 0.05$)
▲: Significantly different compared to the S. mitis group ($p < 0.05$)

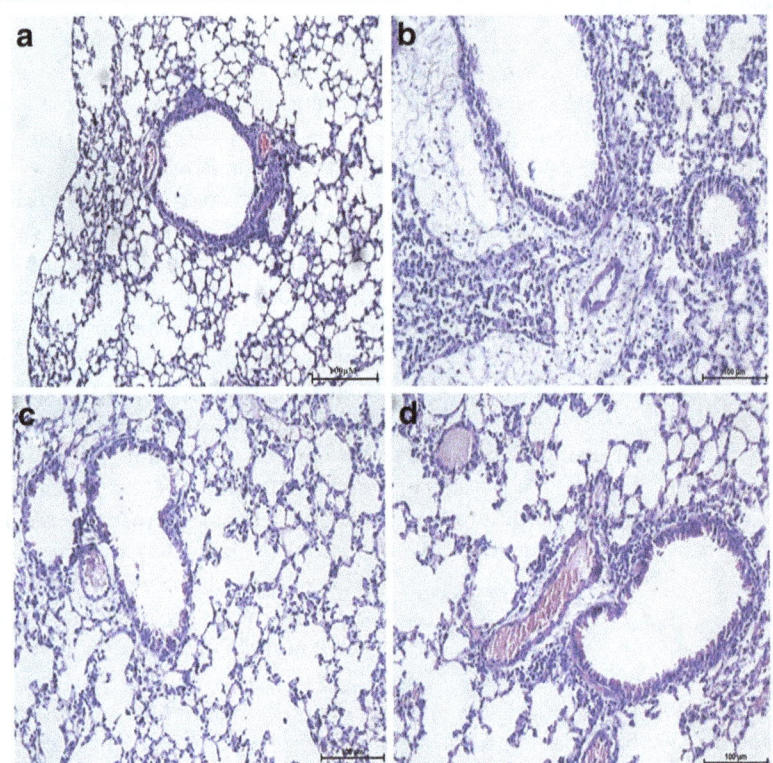

Fig. 1 HE-stained lung tissues from wild-type acute lung infection mouse models. After the wild-type mice were infected with PAO1 (**b**), *S. mitis* (**c**), PAO1 + *S. mitis* (**d**), or PBS (**a**) for 48 h, mice were sacrificed and right lungs were stained with HE to observe histological changes. Original magnification, ×200

single strain, specifically *P. aeruginosa*, on the ETTs. However, few studies have investigated the interaction of multiple bacteria in the pathogenesis of VAP. Based on our previous research, we further investigated the presence of *S. mitis* and *P. aeruginosa* on the ETTs. Here, for the first time to our knowledge, we found that *S. mitis* can counteract the inflammatory potential of *P. aeruginosa* possibly through TLR4 signaling in acute lung infection.

S. mitis is the most abundant bacteria of the normal human oral flora and also a predominant colonizer of the mucosal site, usually inhabiting the human oral cavity as early as 1–3 days after birth. Moreover, except for endocarditis [18], *S. mitis* rarely causes diseases. In our previous research, we found that wild-type mice as well as TLR4-deficient mice infected with an *S. mitis* strain for 48 h showed little change in pulmonary lesions, supporting the common notion that *S. mitis* is a normal commensal bacteria in the human oropharynx [19–21].

In the present study, we observed more severe inflammation in lung tissues accompanied with infiltration of more inflammatory cells in the PAO1 + *S. mitis* group compared with the *S. mitis* group and PBS control group, but less severe lung inflammation in comparison with that in the PAO1 group. These findings suggest

that in acute lung infection, *S. mitis* helps to alleviate the inflammatory response so as to reduce the local tissue damage. Several studies have found similar effects, in which the bacteria mainly act as "protectors" during the process of infection. Recent findings in *Bifidobacteria* and *Lactobacilli* have confirmed that they are effective at reducing allergic symptoms and inhibiting the allergic airway response in murine models of acute airway inflammation [22, 23]. More important, this inhibition is possibly achieved by regulating TLR signaling [24]. Therefore, we have reason to believe that, unlike the traditional concept in which *S. mitis* is considered normal commensal bacteria, our results indicate that *S. mitis* might play a potential protective role in the respiratory tract during acute lung infection. Also considering our results in TLR4-deficient mice, we conclude that *S. mitis* may help alleviate lung inflammation, possibly through modulation of TLR pathway signaling.

Additionally, the total cell counts in BALF are important indicators of the proliferation and differentiation of inflammatory cells. After infection of wild-type mice, total cell counts in the *S. mitis* group were hardly increased, whereas those in the PAO1 group were significantly increased compared with the PBS group. These findings are consistent with the results of lung

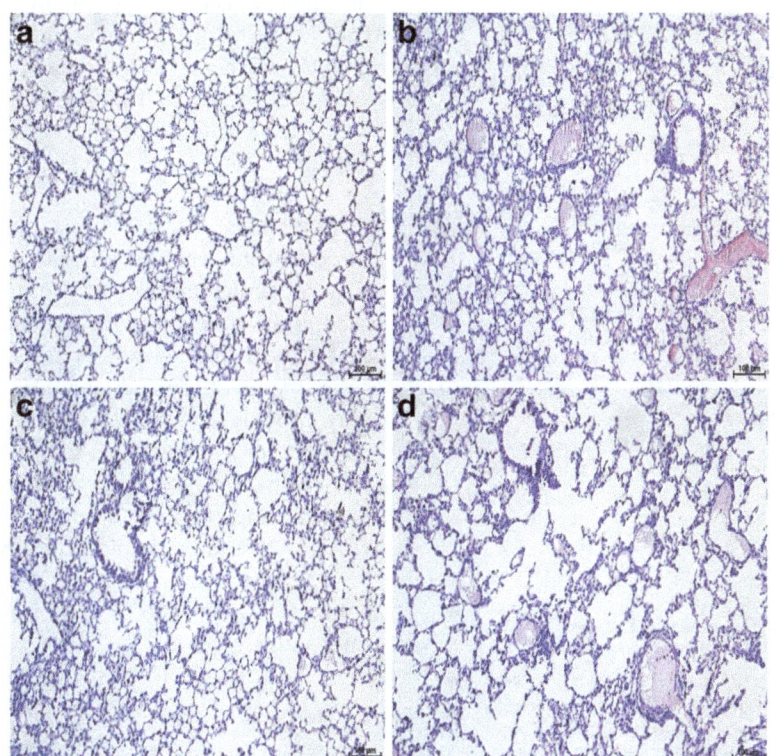

Fig. 2 HE-stained lung tissues from TLR4-deficient acute lung infection mouse models. After the TLR4-deficient mice were infected with PAO1 (**b**), *S. mitis* (**c**), PAO1 + *S. mitis* (**d**), or PBS (**a**) for 48 h, mice were sacrificed and right lungs were stained with HE to observe histological changes. Original magnification, ×100

histopathological examination. However, the total cell counts in the PAO1+ *S. mitis* group were slightly lower than those in the PAO1 group. We speculate that *S. mitis* might activate some other signaling pathways in the process of promoting inflammatory cells. Some studies have found that *S. mitis* reduces proliferation of T cells specific to an unrelated antigen (TT), which suggests an inhibitory effect of *S. mitis*. This inhibition can affect either the antigen-presenting cells (APCs) directly by preventing activation and/or antigen presentation or on the memory Th cells directly by interfering with the

APC–T cell interaction [5]. We speculate that this might be one possible explanation for our observations. After infection of TLR4-deficient mice, the total cell counts were significantly lower than those in wild-type mice, which indicates that TLR4 expressed on cell surface is needed for inflammatory cell activation [25].

Several studies have demonstrated that TNF-α activates the expression of endothelial adhesion molecules to facilitate migration of neutrophils into inflamed lungs [26]. Similarly, in our acute lung infection mouse models, we found that wild-type mice infected with

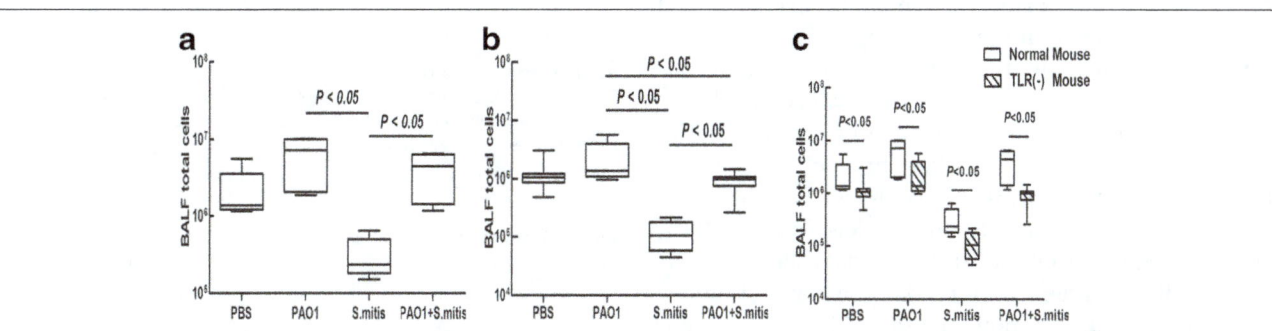

Fig. 3 Total cell counts in BALF. The total cell counts in BALF of wild-type mice (**a**) and TLR4-deficient mice (**b**) and the differences in total cell counts between wild-type and TLR4-deficient mice (**c**) are shown

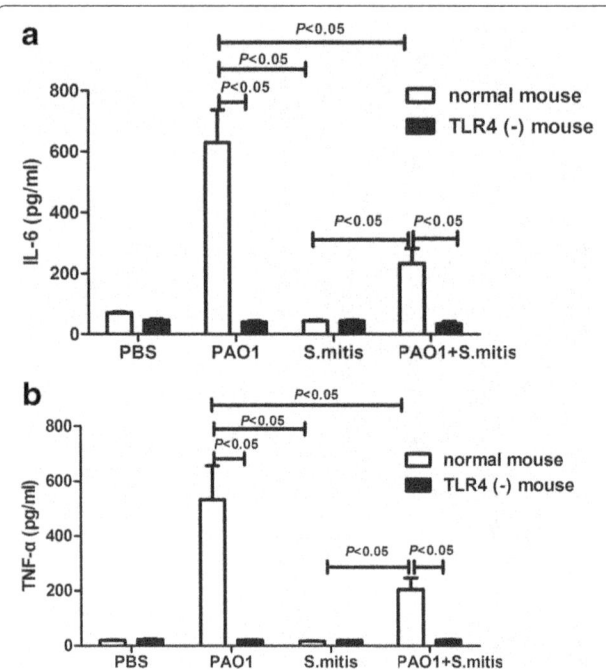

Fig. 4 IL-6 and TNF-α concentrations in BALF. The IL-6 protein levels in BALF of wild-type mice and TLR4-deficient mice and the differences in IL-6 protein levels between wild-type and TLR4-deficient mice (**a**) are shown. The TNF-α protein levels in BALF of wild-type mice and TLR4-deficient mice and the differences between in TNF-α protein levels wild-type and TLR4-deficient mice (**b**) are shown

PAO1 showed significantly increased TNF-α protein expression as well as increased IL-6 expression. These findings are consistent with the literature, which supportes that IL-6 modulates almost every aspect of the innate immune system, including the accumulation of neutrophils at sites of infection or trauma through the control of granulopoiesis [27, 28]. Previous studies have confirmed that TLR4 plays a crucial role in the recognition of Gram-negative bacteria, but has no obvious effects on Gram-positive bacteria [25, 29]. However, our present study showed that *P. aeruginosa* together with *S. mitis* could affect activity of the TLR4 pathway. Although we examined the TLR4 pathway specifically, the roles of other TLR pathways cannot be excluded and should be considered in future investigations.

In contrast with our study, Wang et al. found an opposite result, which confirmed that *S. mitis* mainly secrets autoinducer-2 to promote biofilm formation and to increase proinflammatory cytokine production, resulting in severe inflammation [6]. This difference might be attributed to the different infection models. In their study, chronic pneumonia rat models were established by dual-biofilm catheter intubation for 7 days. In contrast, in our study, acute lung infection mouse models were established by planktonic bacteria infection for only 48 h. However, further investigations are needed to elaborate

the mechanisms. Additionally, we speculate that the conflicting results are related to the protective effects of *S. mitis* in acute infection, which resulted in reduced cytokine secretion by the host immune system and weakened bacterial clearance, finally resulting in persistent and chronic infection.

Conclusions

In conclusion, the results of this study will contribute to a better understanding of the co-existence of *P. aeruginosa* and *S. mitis*, and for the first time, we demonstrate that *S. mitis* plays a potentially protective role in the respiratory tract during acute infection. These findings may provide a new perspective in the treatment of *P. aeruginosa* infection in the NICU. The mechanisms underlying the protective effect of *S. mitis* remain unknown, and further studies are needed to discover whether this effect is mediated by TLR signaling directly or via some small molecules or whether *S. mitis* simply competes with PAO1 for nutrients, limiting PAO1 replication.

Abbreviations
BALF: Bronchoalveolar lavage fluid; CFU: Colony-forming units; ETTs: Endotracheal tubes; H&E: Hematoxylin-eosin; LPS: Lipopolysaccharide; *P. aeruginosa*: *Pseudomonas aeruginosa*; PAMPs: Pathogen-associated molecular patterns; PBS: Phosphate-buffered saline; *S. mitis*: *Streptococcus mitis*; TLR: Toll-like receptor; TT: Unrelated antigen; VAP: Ventilator-associated pneumonia

Acknowledgments
We are grateful to Professor Yibing Yin (Key Laboratory of Diagnostic Medicine Designated by the Ministry of Education, Chongqing Medical University, Chongqing, China) for excellent technical support.

Funding
This research was supported by the National Natural Science Foundation of China (Nos. 81,370,744, 81,571,483), which provided money to buy reagents. Doctoral Degree Funding from the Chinese Ministry of Education (No. 20135503110009), and the State Key Clinic Discipline Project (No. 2011–873), which provided money for the labor costs, and the Clinical Research Foundation of Children's Hospital of Chongqing Medical University, which provide the money for article modification.

Authors' contributions
Conceived and designed the experiments: CS, JY. Performed the experiments: CS, HL, YZ. Analyzed the data: CS, YZ. Wrote the paper: CS. All authors read and approved the final manuscript.

Authors' information
Not applicable.

Competing interests
The authors declare that they have no competing interests.

References

1. Ismail A, El-Hage-Sleiman AK, Majdalani M, Hanna-Wakim R, Kanj S, Sharara-Chami R. Device-associated infections in the pediatric intensive care unit at the American University of Beirut Medical Center. J Infect Dev Ctries. 2016; 10(6):554–62.

2. Hocevar SN, Edwards JR, Horan TC, Morrell GC, Iwamoto M, Lessa FC. Device-associated infections among neonatal intensive care unit patients: incidence and associated pathogens reported to the National Healthcare Safety Network, 2006-2008. Infect Control Hosp Epidemiol. 2012;33(12): 1200–6.

3. Hongdong Li CS, Liu D, Ai Q, Yu J. Molecular analysis of biofilms on the surface of neonatal endotracheal tubes based on 16S rRNA PCR-DGGE and species-specific PCR. Int J Clin Exp Med. 2015;8(7):11075–84.

4. YJ SC, Ai Q, Liu D, Lu W, Lu Q, Peng NN. Diversity analysis of biofilm bacteria on tracheal tubes removed from intubated neonates. Zhonghua Er Ke Za Zhi. 2013;51(8):602–6.

5. Khosravi Y, Dieye Y, Loke MF, Goh KL, Vadivelu J. Streptococcus Mitis induces conversion of helicobacter pylori to coccoid cells during co-culture in vitro. PLoS One. 2014;9(11):e112214.

6. Wang Z, Xiang Q, Yang T, Li L, Yang J, Li H, He Y, Zhang Y, Lu Q, Yu J. Autoinducer-2 of Streptococcus Mitis as a target molecule to inhibit pathogenic multi-species Biofilm formation in vitro and in an Endotracheal intubation rat model. Front Microbiol. 2016;7(88):1–11.

7. Knapp S. Update on the role of Toll-like receptors during bacterial infections and sepsis. Wien Med Wochenschr. 2010;160(5–6):107–11.

8. Wieland CW, van Lieshout MH, Hoogendijk AJ, van der Poll T. Host defence during Klebsiella pneumonia relies on haematopoietic-expressed Toll-like receptors 4 and 2. Eur Respir J. 2011;37(4):848–57.

9. Frazao JB, Errante PR, Condino-Neto A. Toll-like receptors' pathway disturbances are associated with increased susceptibility to infections in humans. Arch Immunol Ther Exp. 2013;61(6):427–43.

10. Brown J, Wang H, Hajishengallis GN, Martin M. TLR-signaling networks: an integration of adaptor molecules, kinases, and cross-talk. J Dent Res. 2011; 90(4):417–27.

11. Greenhill CJ, Rose-John S, Lissilaa R, Ferlin W, Ernst M, Hertzog PJ, Mansell A, Jenkins BJ. IL-6 trans-signaling modulates TLR4-dependent inflammatory responses via STAT3. J Immunol. 2011;186(2):1199–208.

12. Liu Y, Yin H, Zhao M, Lu Q. TLR2 and TLR4 in autoimmune diseases: a comprehensive review. Clinical reviews in allergy & immunology. 2014;47(2): 136–47.

13. Perros F, Lambrecht BN, Hammad H. TLR4 signalling in pulmonary stromal cells is critical for inflammation and immunity in the airways. Respir Res. 2011;12:125.

14. Schoeniger A, Fuhrmann H, Schumann J. LPS- or Pseudomonas Aeruginosa-mediated activation of the macrophage TLR4 signaling cascade depends on membrane lipid composition. PeerJ. 2016;4:e1663.

15. Fremond CM, Nicolle DM, Torres DS, Quesniaux VF. Control of Mycobacterium Bovis BCG infection with increased inflammation in TLR4-deficient mice. Microbes Infect. 2003;5(12):1070–81.

16. Caron E, Desseyn JL, Sergent L, Bartke N, Husson MO, Duhamel A, Gottrand F. Impact of fish oils on the outcomes of a mouse model of acute Pseudomonas Aeruginosa pulmonary infection. Br J Nutr. 2015;113(2):191–9.

17. Ding FM, Zhu SL, Shen C, Ji XL, Zhou X. Regulatory T cell activity is partly inhibited in a mouse model of chronic Pseudomonas Aeruginosa lung infection. Exp Lung Res. 2015;41(1):44–55.

18. Khan ST, Ahmad J, Ahamed M, Musarrat J, Al-Khedhairy AA. Zinc oxide and titanium dioxide nanoparticles induce oxidative stress, inhibit growth, and attenuate biofilm formation activity of Streptococcus Mitis. J Biol Inorg Chem. 2016;21(3):295–303.

19. Mitchell J. Streptococcus Mitis: walking the line between commensalism and pathogenesis. Mol Oral Microbiol. 2011;26(2):89–98.

20. Jennifer L, Kirchherra GHB, Michael F. Colea, Yoshiaki Kawamurac, Dorothy, A. Richmondd MJS, and Katherine a. Wirth: physiological and serological variation in S. mitis biovar 1 from the human oral cavity during the first year of life. Arch Oral Biol. 2007;52(1):90–9.

21. Mager DL, Ximenez-Fyvie LA, Haffajee AD, Socransky SS. Distribution of selected bacterial species on intraoral surfaces. J Clin Periodontol. 2003; 30(7):644–54.

22. Sagar S, Morgan M, Chen S, Vos AP, Garssen J, van Bergenhenegouwen J, Boon L, Georgiou NA, Kraneveld AD, Folkerts G. Bifidobacterium breve and lactobacillus rhamnosus treatment is as effective as budesonide at reducing inflammation in a murine model for chronic asthma. Respir Res. 2014;16(15): 1–17.

23. Hougee S, Vriesema AJ, Wijering SC, Knippels LM, Folkerts G, Nijkamp FP, Knol J, Garssen J. Oral treatment with probiotics reduces allergic symptoms in ovalbumin-sensitized mice: a bacterial strain comparative study. Int Arch Allergy Immunol. 2010;151(2):107–17.

24. Kalliomaki M, Antoine JM, Herz U, Rijkers GT, Wells JM, Mercenier A. Guidance for substantiating the evidence for beneficial effects of probiotics: prevention and management of allergic diseases by probiotics. J Nutr. 2010; 140(3):713S–21S.

25. van Helden SF, van den Dries K, Oud MM, Raymakers RA, Netea MG, van Leeuwen FN, Figdor CG. TLR4-mediated podosome loss discriminates gram-negative from gram-positive bacteria in their capacity to induce dendritic cell migration and maturation. J Immunol. 2010;184(3):1280–91.

26. Pandit AA, Choudhary S, Ramnee K, Singh B, Sethi RS. Imidacloprid induced histomorphological changes and expression of TLR-4 and TNFalpha in lung. Pestic Biochem Physiol. 2016;131:9–17.

27. Hunter CA, Jones SA. IL-6 as a keystone cytokine in health and disease. Nat Immunol. 2015;16(5):448–57.

28. Tanaka T, Narazaki M, Kishimoto T. IL-6 in inflammation, immunity, and disease. Cold Spring Harb Perspect Biol. 2014;6(10):1–17.

29. Neyen C, Lemaitre B. Sensing gram-negative bacteria: a phylogenetic perspective. Curr Opin Immunol. 2016;38:8–17.

4

Antioxidant properties, antimicrobial and anti-adhesive activities of DCS1 lipopeptides from *Bacillus methylotrophicus* DCS1

Nawel Jemil[1]*(iD), Hanen Ben Ayed[1], Angeles Manresa[2], Moncef Nasri[1] and Noomen Hmidet[1]

Abstract

Background: The present work aims to investigate the antioxidant and antimicrobial activities as well as the potential of DCS1 lipopeptides produced by *Bacillus methylotrophicus* DCS1 strain at inhibition and disruption of biofilm formation.

Results: The produced biosurfactants were characterized as lipopeptides molecules by using thin layer chromatography (TLC) and Fourier transform infrared spectroscopy (FT-IR). The DCS1 lipopeptides were assayed for their antioxidant activity through five different tests. The scavenging effect on DPPH radicals at a concentration of 1 mg mL^{-1} was 80.6%. The reducing power reached a maximum value of 3.0 (OD$_{700\ nm}$) at 2 mg mL^{-1}. Moreover, the DCS1 lipopeptides exhibited a strong inhibition of β-carotene bleaching by linoleic acid assay with 80.8% at 1 mg mL^{-1} and showed good chelating ability and lipid peroxidation inhibition. The in vitro antimicrobial activity of DCS1 lipopeptides showed that they display significant antibacterial and antifungal activities. The anti-adhesive activity of DCS1 lipopeptides was evaluated against several pathogenic microorganisms. The lipopeptides showed excellent anti-adhesive activity, even at low concentrations, in a polystyrene surface pre-treatment against all the microorganisms tested. Further, they can disrupt performed biofilms.

Conclusion: This study shows the potentiality of DCS1 lipopeptides as natural antioxidants, antimicrobial and/or anti-adhesive agent for several biomedical and industrial applications.

Keywords: *Bacillus methylotrophicus* DCS1, Lipopeptides like substances, Antioxidant, Antimicrobial, Anti-adhesive activities

Background

Bacteria are the main group of biosurfactants producing microorganisms [1]. Several studies have reported the potential of *Bacillus* species as biosurfactant producers such as lipopeptide type biosurfactants [2]. These are amphiphilic cyclic peptides that are linked to a fatty acid hydrocarbon chain and they belong to the surfactin, iturin and fengycin families. They are synthesized by non-ribosomal peptide synthetases without involving messenger RNA [3]. Lipopeptides were found to have specific biological activities, such as antioxidant [4, 5], antimicrobial [6–8], and antitumor activities [9]. The advantages related to the use of microbial biosurfactants over their chemical counterparts include their lower toxicity and higher biodegradability, as well as their effectiveness at different environmental conditions such as extreme pH, temperature and high ionic strength, in addition to their biocompatibility [10]. These advantages allow using them in pharmaceutical, cosmetic and food additives industries [11]. In the food industry, biosurfactants offer several functions such as emulsifying/foaming agents, stabilizers, antioxidant agents, and anti-adhesives [12, 13].

* Correspondence: nawel.1501@yahoo.com
[1]Laboratoire de Génie Enzymatique et de Microbiologie, Université de Sfax, Ecole Nationale d'Ingénieurs de Sfax, B.P, 1173-3038 Sfax, Tunisia
Full list of author information is available at the end of the article

Biosurfactants have also been proved to be great inhibitors of microbial adhesion and biofilm formation. Biofilms are communities of microbes adhering to biotic or abiotic surfaces. Broadly, microbial biofilms are a daily challenge faced by the food industry and society [14]. Adhesion is the first stage of biofilm formation and the best time for the action of anti-adhesive compounds [15]. In fact, life in a biofilm probably represents the predominate mode of growth for microbes in most environments. The role of biosurfactants in microbial anti-adhesion and desorption has been widely described, and adsorption of biosurfactants to solid surfaces can be an effective strategy to reduce microbial adhesion and combating colonization by pathogenic microorganisms [16, 17].

This report represents an investigation of the antioxidant, antimicrobial and anti-adhesive activities of lipopeptides produced by *Bacillus methylotrophicus* DCS1 strain.

Methods
Microorganism
Biosurfactant-producing strain used in this study was isolated from hydrocarbon contaminated soil in Sfax City, Tunisia. It was selected on the basis of the high hemolytic activity and decreasing surface tension of the culture medium from 72 mN m^{-1} to 31 mN m^{-1}. This strain was identified as *Bacillus methylotrophicus* DCS1 based on its biochemical and 16S rDNA gene sequence analysis [18]. The bacterial strain was maintained at 4 °C and also preserved in glycerol at –80 °C.

Production media
Bacillus methylotrophicus DCS1 was inoculated into a 250 mL Erlenmeyer flask containing 25 mL Luria-Bertani (LB) broth medium (g L^{-1}): peptone, 10.0; yeast extract, 5.0; and NaCl, 5.0; pH 7.0. Culture was incubated at 37 °C with shaking at 200 rpm for 18 h. A 3% (v/v) of inoculum [OD$_{600 \text{ nm}}$ = 7.6] was transferred into a 2 L Erlenmeyer flask containing 250 mL of Landy medium [19] which contains: glucose 20 g L^{-1}, L-glutamic acid 5 g L^{-1}, yeast extract 1 g L^{-1}, K$_2$HPO$_4$ 1 g L^{-1}, MgSO$_4$ 0.5 g L^{-1}, KCl 0.5 g L^{-1}, CuSO$_4$ 0.0016 g L^{-1}, Fe$_2$ (SO$_4$)$_3$ 0.0004 g L^{-1}, MnSO$_4$ 0.0012 g L^{-1}. The initial pH of the medium was adjusted to 7.0 and culture was incubated for 72 h at 30 °C with shaking at 150 rpm.

Biosurfactant recovery
The culture medium was centrifuged at 8000 rpm and 4 °C for 20 min to remove bacterial cells. Supernatant free-cell was subjected to acid precipitation with 6 N HCl until pH 2.0, then left overnight at 4 °C with agitation. The precipitate was collected by centrifugation at 8000 rpm, for 20 min at 4 °C, suspended in distilled water and the pH was adjusted to 8.0 with NaOH

(1.0 N) [20]. The crude biosurfactants obtained was lyophilized.

Thin layer chromatography
The extracted biosurfactants were analyzed on silica plates 60 F (Merck, Macherel-Nagel) with a solvent system consisting of (chloroform: methanol: water) solution in the ratio (65:25:4). The compounds on the plate were detected after spraying with the ninhydrin reagent and iodine vapor which respectively detected the peptide and lipid parts in the biosurfactants [21].

FT-IR spectra of the dried biosurfactants
Fourier transform infrared spectroscopy (FT-IR) was used to determine the functional groups and the chemical bonds present in the biosurfactants extract and then to determine its chemical nature. The FT-IR analysis was realized by using Analect Instruments fx-6160 FT-IR spectrometer. 1 mg of lyophilized sample was ground in about 100 mg of spectral grade KBr and pressed for 30 s to obtain translucent pellets. The spectral measurements were carried out in the absorbance mode. The FTIR spectrum of the prepared biosurfactants was recorded between 400 and 4000 cm^{-1}.

Antioxidant activities
DPPH radical-scavenging assay
The 2, 2-diphenyl-1-picrylhydrazyl (DPPH) free radical-scavenging potential of DCS1 lipopeptides at different concentrations (0.1–2 mg mL^{-1}) was determined according to the report of Bersuder et al. [22]. The DPPH radical scavenging activity was calculated as follows:

$$\text{DPPH radical-scavenging activity } (\%) = \frac{A_{control} + A_{blanc} - A_{sample}}{A_{control}} \times 100$$

where $A_{control}$ is the absorbance of the control reaction (containing all reagents except lipopeptides), A_{blank} is the absorbance of lipopeptides (containing all reagents except the DPPH solution) and A_{sample} is the absorbance of lipopeptides with the DPPH solution.

Lower absorbance of the reaction mixture indicated higher DPPH radical-scavenging activity. BHA (2.0 mM) was used as positive standard. The test was carried out in triplicate and the results are mean values.

The effective concentration of DCS1 lipopeptides required to scavenge DPPH radical by 50% (IC$_{50}$ value) was obtained by linear regression analysis of dose-response curve plotting between percentage of DPPH radical-scavenging activity and concentrations.

Ferric-reducing activity

The ability of DCS1 lipopeptides at different concentrations (0.1 to 3 mg mL^{-1}) to reduce iron III was determined according to the method of Yildirim et al. [23]. All data values presented are the mean of triplicate analyses.

Ferrous ion-chelating activity

The Fe^{2+} chelating activity of DCS1 lipopeptides at different concentrations (0.5–5 mg mL^{-1}) was estimated by the method of Carter [24]. The percentage of inhibition of ferrozine–Fe^{2+} complex formation was calculated using the following formula:

$$\text{Ferrous ion-chelating activity} \ (\%) = \frac{A_{\text{control}} + A_{\text{blanc}} - A_{\text{sample}}}{A_{\text{control}}} \times 100$$

where A_{control} is the absorbance of the control (without sample), A_{blank} is the absorbance of the blank (without ferrozine) and A_{sample} is the absorbance of the sample. Ethylene diamine tetra acetic acid (EDTA) was used as positive control and the test was carried out in triplicate.

β-carotene bleaching assay

The capacity of DCS1 lipopeptides to prevent bleaching of β-carotene at different concentrations (0.025–1 mg mL^{-1}) was evaluated as described by Koleva et al. [25]. The antioxidant activity was calculated in terms of β-carotene bleaching inhibition using the following formula:

$$\text{β-carotene bleaching inhibition} = \left[1 - \left(\frac{A_0 - A_t}{A'_0 - A'_t}\right)\right] \times 100$$

where A_0 and A'_0 are the absorbances of the test sample and the control, respectively, measured at time zero; and A_t and A'_t are the absorbance of the sample and the control, respectively, measured after incubation. BHA was used as a positive standard. Tests were performed in triplicate and values presented are the mean of triplicate analyses.

The half maximal inhibitory concentration (IC$_{50}$) is the concentration where 50% inhibition of the β-carotene bleaching radical is obtained. IC$_{50}$ value was obtained by linear regression analysis of dose-response curve plotting between percentage of inhibition and concentrations.

Inhibition of linoleic acid peroxidation

In vitro lipid peroxidation inhibitory activity of lipopeptides like substances DCS1 was determined by evaluating their ability to inhibit oxidation of linoleic acid in an emulsified model system. DCS1 lipopeptides at 0.1 mg mL^{-1} concentration were dissolved in 2.5 mL of distilled water and added to 2.5 mL of ethanol and 32.5 µL linoleic acid. The final volume was then adjusted to 6.25 mL with distilled water. The mixture was incubated in a 10 mL tube with silicone rubber caps at 45 °C for 9 days in dark. Vitamin C, a natural antioxidant agent, was used as a positive control, and distilled water as negative control. Lipid peroxidation was estimated as evidenced by the formation of thiobarbituric acid reactive substances (TBARS) such as malondialdehyde (MDA). TBARS were tested in samples by the method described by Yagi [26]. MDA and other TBARS were measured by their reactivity with TBA in an acidic condition to generate pink colored chromospheres which was read at 530 nm. A sample (375 µL) was homogenized with 150 µL of TBS (50 mM Tris containing 150 mM NaCl, pH 7.4) and 375 µL of TCA 20% (w/w) in order to precipitate proteins, and then centrifuged (1000 g, 10 min, 4 °C). A 400 µL of the supernatant was mixed with 80 µL of HCl (0.6 M) and 320 µL of Tris–TBA (Tris 26 mM; TBA 120 mM), and the mixture was heated for 10 min at 80 °C. The absorbance of the resulting solution was read at 530 nm. Inhibition of linoleic acid peroxidation was estimated by the following formula:

$$\text{Inhibition of linoleic acid peroxidation} = \left[1 - \frac{A_{\text{sample}}}{A_{\text{control}}}\right] \times 100$$

where A_{sample} was the absorbance of the sample and A_{control} was the absorbance of the control (replacing the sample by water).

Antimicrobial activities

Microbial strains

Antibacterial activity of DCS1 lipopeptides was tested against three Gram-positive bacteria: *Staphylococcus aureus* (ATCC 25923), *Bacillus cereus* (ATCC 11778), *Micrococcus luteus* (ATCC 4698) and five Gram-negative bacteria: *Klebsiella pneumoniae* (ATCC 13883), *Escherichia coli* (ATCC 25922), *Salmonella typhimurium* (ATCC 19430), *Salmonella enterica* (ATCC 27853) and *Enterobacterium* sp.

Antifungal activity was tested using *Aspergillus niger*, *Aspergillus flavus*, *Fusarium oxysporium*, *Pythium ultiumum*, *Fusarium solani* and *Rhizoctonia bataticola*.

Agar diffusion method

DCS1 lipopeptides were tested in vitro for antimicrobial activities by conventional agar diffusion method against human pathogens [27].

Using sterile pipette, culture suspension (200 µL) of the tested microorganisms (10^6 colony-forming units CFU mL^{-1}) of bacteria cells (estimated by absorbance at 600 nm) and 10^8 spores mL^{-1} of fungal strains (measured by Malassez blade) were spread uniformly on Luria-Bertani agar and malt extract agar media, respectively. Then, wells were made using a sterile well borer and were filled with 100 µL of lipopeptides sample (2 mg mL^{-1} concentration).

The zone of growth inhibition was measured in millimetres after incubation for 24 h at 37 °C for bacteria and for 72 h at 30 °C for fungal strains. All the results were represented as the average of three independent experiments.

Effect of proteolytic enzymes, heat, and pH on crude lipopeptides antibacterial activity

The susceptibility of the crude lipopeptides to proteolytic enzymes, heat and pH treatments was assessed as described elsewhere [28]. To evaluate its stability to proteolytic enzymes, the crude lipopeptides sample (2 mg mL^{-1} concentration) was incubated at 37 °C for 60 min with 1 mg mL^{-1} final concentration of trypsin, chymotrypsin and pepsine. To analyze thermal stability, samples were incubated at temperatures ranging 40, 60, 80 and 100 °C for 20 min. The effect of pH on DCS1 lipopeptides activity was studied by assaying antibacterial activity against *K. pneumoniae*, the pH was adjusted at values from 3.0 to 10.0. Samples were neutralized to pH 7.0 before measurement of the antibacterial activity, the following buffer systems were used: 100 mM glycine–HCl buffer, pH 3.0–4.0; 100 mM acetate buffer, pH 4.0–6.0; 100 mM Tris–HCl buffer, pH 7.0–8.0 and 100 mM glycine–NaOH buffer, pH 9.0–10.0. After the treatments, residual antibacterial activity against *K. pneumoniae* was determined as follows:

$$\text{Residual antibacterial activity } (\%) = \frac{\text{Zone of growth inhibition after treatment}}{\text{Zone of growth inhibition without treatment}} \times 100$$

Anti-adhesion treatment with lipopeptides extract DCS1

The anti- and post-adhesion treatments were studied with lipopeptides extract obtained as follows: acid-precipitated lipopeptides (1 g) was subjected to extraction with 25 ml tetrahydrofuran (THF) solvent four times and the mixture was stirred and centrifuged at 8000 rpm, for 15 min at 4 °C. The organic phases recuperated were combined and evaporated to dryness in a rotary vacuum evaporator (Büchi, Switzerland) at 40 °C, and then lipopeptides extract was suspended in distilled water and lyophilized.

For surface pre-treatment, the wells of a sterile micro-titer plate were loaded with 200 µL of DCS1 lipopeptides extract at different concentrations ranging from 0.016 to 2 mg mL^{-1}, dissolved in PBS (pH 7.2). Micro-titer plates were incubated for 6 h at room temperature (25 °C) and then washed twice with PBS.

For biofilm formation, *Salmonella typhimurium* ATCC 14028, *Klebsiella pneumoniae* ATCC 13883, *Staphylococcus*

aureus ATCC 25923, *Bacillus cereus* ATCC 11778 and *Candida albicans* ATCC 10231 were cultured overnight in Luria-Bertani medium (LB). Cultures were diluted 1/100 in the medium proposed by O'Toole [29] (g L^{-1}): glucose, 2; casamino acids, 5; KH$_2$PO$_4$, 3; K$_2$HPO$_4$, 7; (NH$_4$)$_2$SO$_4$, 2, MgSO$_4$ 7H$_2$O, 0.12. Then, 200 µL of each dilution were added to the micro-titer plate wells and incubated for 20 h at 37 °C. After incubation, wells were washed three times with distilled water to remove non-adherent cells, fixed for 15 min with methanol and stained with 125 µL crystal violet (CV) (0.1%) for 20 min, then washed with water and dried. For quantifying the microbial adhesion, the stain in the wells was diluted with 200 µL acetic acid in water (33%) and the absorbance was determined at 595 nm [13]. Percentages of microbial adhesion inhibition were calculated using the formula:

$$\text{Microbial adhesion inhibition} = [1-(A_c/A_0)] \times 100$$

where A_c represents the absorbance of the well with lipopeptides at concentration c and A_0 represents the absorbance of the positive control wells (in absence of lipopeptides). Negative control wells contained only lipopeptides dissolved in PBS. Assays were carried out three times.

Mature biofilm treatment with lipopeptides extract DCS1

The wells of a polystyrene micro-titer plates were filled with 200 µL of bacterial suspension prepared as described above, and then the plates were incubated for 20 h at 37 °C. After incubation, the unattached microbial cells were removed by washing the wells three times with distilled water. Thereafter, 200 µL of DCS1 lipopeptides at different concentrations ranging from 0.016 to 2 mg mL^{-1}, were added to each well and the plates were incubated for 6 h at room temperature (25 °C). The quantification was realized as in the pre-treatment. All the results were represented as the average of three independent experiments.

All data presented are the average of at least three measurements which deviated by not more than 5%.

Results and discussion
Preliminary chemical characterization

TLC analysis separates compounds in a mixture and can be used to determine the number of components in solutions. The sample of crude biosurfactants revealed yellow spots with iodine vapour, suggesting the presence of polar lipids. Treatment with ninhydrin revealed pink spots suggesting the presence of protein portions. The presence of both protein units and lipid moieties on the same spot suggested that the sample is a lipopeptide type. Similar results for other lipopeptide type biosurfactant were also reported elsewhere [30].

The IR spectrum of the crude biosurfactants DCS1 was analyzed to gain insight into its chemical nature (Fig. 1). A broad absorbance with wave numbers ranging approximately from 3700 cm^{-1} to 3000 cm^{-1} having its maximum at 3297 cm^{-1} was detected. Absorbance in this region is a result of -CH and -NH stretching vibrations, and it is a characteristic of carbon-containing compounds with amino groups [31]. Other sharp absorbance peaks are seen at 2959, 2928 and 2855 cm^{-1}, indicating the presence of –C-CH$_3$ banding or long alkyl chains [32]. The peak with highest absorbance in the spectrum was observed at 1651 cm^{-1}. Absorbance in this region signifies the presence of peptide groups in the molecules [33]. Another high intensity peak at 1537 cm^{-1}, corresponded to the deformed N–H band. The weak band at 1398 cm^{-1} in the absorption range 1370–1470 cm^{-1} could result from deformation and bending vibrations of –C-CH$_2$ and –C-CH$_3$ groups in aliphatic chains [34]. Peaks at 1234 and 1111 cm^{-1} are probably due to the presence of C-O-C vibrations in esters [32, 33].

The observed peaks are those commonly found in the IR spectra of lipopeptide biosurfactants produced by several species [21, 35, 36].

Antioxidant activities

DPPH radical-scavenging assay

The DPPH radical-scavenging assay has been generally used to study the ability of compounds to act as free radical scavengers or hydrogen donors [37]. When DPPH encounters a hydrogen-donating substance, the radical would be scavenged, as visualized by changing its color from purple to yellow, and the absorbance is reduced [38]. The stable DPPH radical displays a

maximum absorbance at 517 nm in ethanol. As displayed in Fig. 2a, the DCS1 lipopeptides exhibited effective antioxidant activity against DPPH in a dose dependent manner. In fact, at 1 mg mL^{-1}, the DCS1 lipopeptides showed a potential scavenging effect of 80.6%, which is three-times higher than that obtained at 0.1 mg mL^{-1} (25.9%). Our results are in accordance with previous works of Yalçin and Çavuşoğlu [39] who reported that the DPPH scavenging activity increased with increasing the concentration of biosurfactant synthesized by *Bacillus subtilis* RW-I.

However, DCS1 lipopeptides exhibited lower radical-scavenging activity than did BHA used as reference at the same concentrations. The IC$_{50}$ value was about 357 µg mL^{-1}. Scavenging effect can be assigned to the presence of the hydrocarbon fatty acid chain and some active residues in the peptide ring, which acted as a good hydrogen atom or an electron donor and could react with free radicals of DPPH.

Ferric reducing antioxidant power

The reducing power of a compound may serve as a significant indicator of its potential antioxidant activity [40]. In this test, the presence of reducers causes the reduction of the Fe^{3+}/ferricyanide complex to the ferrous form. The reductive activity of DCS1 lipopeptides as well as the synthetic antioxidant BHA as a function of their concentrations is illustrated in Fig. 2b. The reducing capacity of DCS1 lipopeptides increased with increasing their concentration. It indicates a dose-dependent response and reached a maximum of 3.0 (OD$_{700 nm}$) at a concentration of 2.0 mg mL^{-1}. The obtained values are lower than those of BHA at concentrations below 2.0 mg mL^{-1}.

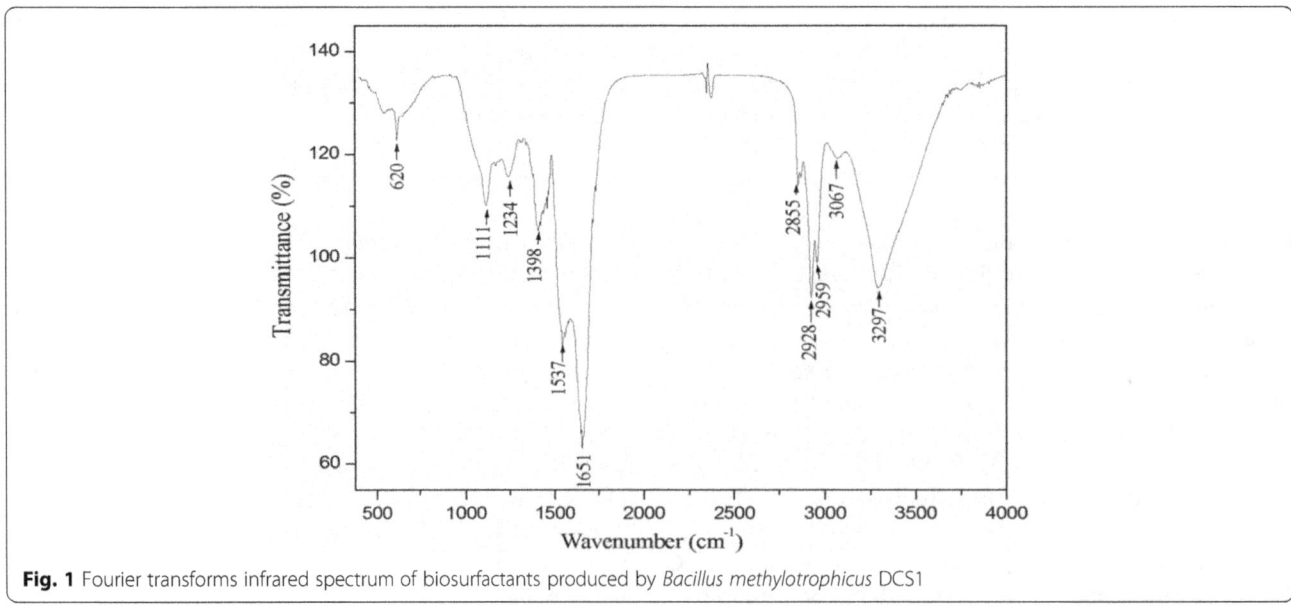

Fig. 1 Fourier transforms infrared spectrum of biosurfactants produced by *Bacillus methylotrophicus* DCS1

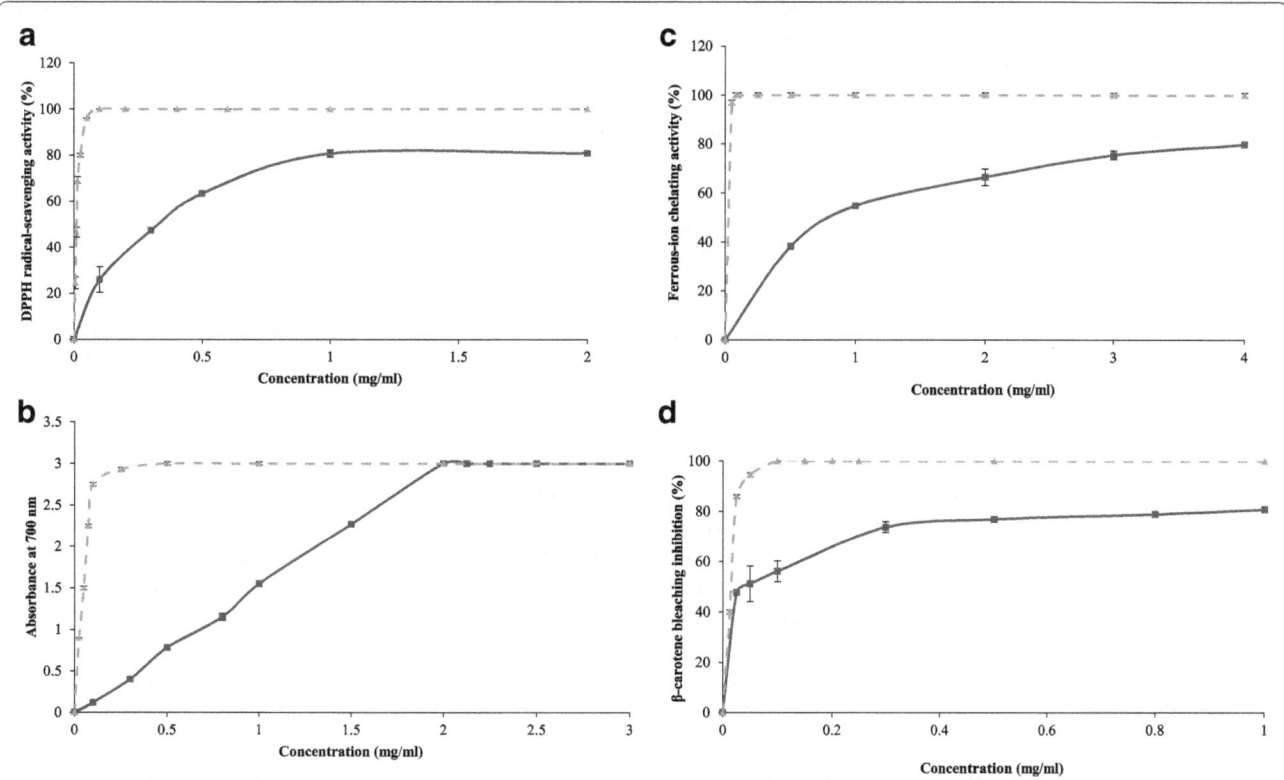

Fig. 2 Scavenging effect on DPPH free radical (**a**), ferric-reducing antioxidant power (**b**), ferrous-ion chelating activity (**c**) and β-carotene bleaching assay (**d**) of DCS1 lipopeptides at different concentrations. BHA and EDTA (2.0 mM) were used as positive control. Values presented are the mean of triplicate analysis. ━■━ Lipopeptides DCS1 - ∗ - BHA - ∗ - EDTA

The reductive ability of DCS1 lipopeptides is higher than those of lipopeptides produced by *Bacillus mojavensis* A21 which reached 2.0 ($OD_{700\ nm}$) at a concentration of 10 mg mL^{-1} [5] and surfactin lipopeptide produced by *B. subtilis* RW-I, which reached 2.0 ($OD_{700\ nm}$) at a concentration of 2.5 mg mL^{-1} [39]. Yalçin and avuşo lu [39] reported that the reductive ability could be related to the presence of hydroxyl groups in the lipopeptides molecules.

Ferrous ion-chelating activity

Ferrous ion (Fe^{2+}) is the most potent pro-oxidant among metal ions. This ion can interact with hydrogen peroxide in a Fenton reaction to produce the reactive oxygen species and hydroxyl free radical (OH), leading to the initiation and/or acceleration of lipid oxidation [41]. Chelating agents may inhibit lipid oxidation by stabilizing transition metals. Measurement of color reduction allows, therefore, estimating the metal chelating activity of the co-existing chelator [42].

The Fe^{2+} chelating capacity of DCS1 lipopeptides against Fe^{2+} was determined by measuring the iron ferrozine complex. As shown in Fig. 2c, DCS1 lipopeptides exhibited strong ferrous-chelating activity and chelated almost 79.8% of ferrous ions at 4 mg mL^{-1}. However, DCS1 lipopeptides exhibited lower metal chelating

activity than EDTA, a well-known metal ion chelator. The obtained results suggest that some lipopeptide isoforms could be potential antioxidant through metal chelating ability.

β-carotene bleaching assay

The antioxidant assay using the discoloration of β-carotene is widely used to measure the antioxidant activity of bioactive compounds. The β-carotene bleaching inhibition was tested at different concentrations and compared with BHA. As shown in Fig. 2d, the DCS1 lipopeptides inhibited significantly the discoloration of β-carotene and its antioxidant activity increased with increasing sample concentration. The inhibitor concentration IC_{50} was estimated to be 42 μg mL^{-1}. However BHA has better β-carotene bleaching inhibition than that of DCS1 lipopeptides at all concentrations tested. DCS1 lipopeptides are more effectives than A21 lipopeptides, which showed a β-carotene bleaching inhibition of 72% at 10 mg mL^{-1} concentration and the IC_{50} was estimated to be 3.7 mg mL^{-1} [5].

Inhibition of linoleic acid peroxidation

Antioxidant activity of DCS1 lipopeptides, at a concentration of 0.1 mg mL^{-1}, against the peroxidation of linoleic acid during 3, 6 and 9 days of storage at 45 °C, was evaluated and compared to that of vitamin C, used as a natural antioxidant (Fig. 3). After 3 days of incubation period, DCS1 lipopeptides displayed a lipid peroxidation inhibition of about 60.22%, and reached about 76.8% after 9 days. The vitamin C showed higher protective effect than biosurfactants. The inhibitory effect could be explained by the presence of acyl chain of fatty acids by improving the interaction between lipopeptide like substances and linoleic acid [5]. The crude lipopeptides synthesized by *B. methylotrophicus* DCS1 strain can be an important potential source of antioxidants.

Antimicrobial activity of DCS1 lipopeptides

The antimicrobial activity of DCS1 lipopeptides, against various microorganisms, was estimated by agar well diffusion method and the results are summarized in Table 1. Lipopeptides were effective against several tested bacteria with different degrees. The highest antibacterial activity was observed against *K. pneumoniae*, with a maximum zone diameter inhibition of 30 mm at a concentration of 2 mg mL^{-1}, while the lowest activity was detected against *Enterobacterium* sp. However, DCS1 lipopeptides did not exhibit antimicrobial activity against *M. luteus* at 2 mg mL^{-1}. Several lipopeptide biosurfactants obtained from *Bacillus amyloliquefaciens* M1 [8] and *Bacillus mojavensis* A21 [7] showed strong antibacterial activity.

Regarding the antifungal activity, DCS1 lipopeptides showed a significant activity against various fungal strains. Higher activity was observed against *A. niger* and *A. flavus*, less activity was against *P. ultimum*, *F. solani* and *R. bataticola*, no effect was observed on *F. oxysporium*. Donio et al. [43] reported that *Bacillus* sp. BS3's biosurfactants possess an important antifungal activity against different pathogens. Sheppard et al. [44] indicated that

lipopeptide biosurfactants interact with the cell membrane of the target cells and exhibit significant antimicrobial properties.

Effect of proteolytic enzymes, heat and pH on lipopeptides antibacterial activity

The resistance of lipopeptides against proteases and extreme conditions, including temperature and pH was a prerequisite for their potential therapeutic and pharmaceutical applications. DCS1 lipopeptides retained their antibacterial activity against *K. pneumoniae* after incubation for 1 h with proteolytic enzymes pepsin, trypsin and chymotrypsin at a concentration of 1 mg mL^{-1} (Table 2). These results exclude the eventual existence of bacteriocin-like substances in the crude lipopeptides and they indicate that the antimicrobial compounds could be cyclic peptides containing unusual amino acids [45].

Regarding thermostability, DCS1 lipopeptides were resistant to heating for 20 min at temperatures up to 100 ° C retaining 100% their initial activity (Table 2). The maintenance of antimicrobial activity after treatment with proteolytic enzymes and after incubation at a wide range of temperature resembles the characteristics of cyclic lipopeptides of *Bacillus* sp. [46].

Antimicrobial activity was also tested at different pH values. The antimicrobial substance was active at pH 8.0 and pH 10.0, retaining 100% of the activity, while at pH 5.0 and 3.0 the activity was decreased (84.4% and 74% of its initial activity at pH 5.0 and pH 3.0, respectively). The decrease in antibacterial activity could be due to partial precipitation of lipopeptides.

Anti-adhesive activity

Biosurfactants are known to decrease the adhesion of pathogenic microorganisms to solid surfaces [47]. Thus, the anti-adhesive activity of DCS1 lipopeptides was

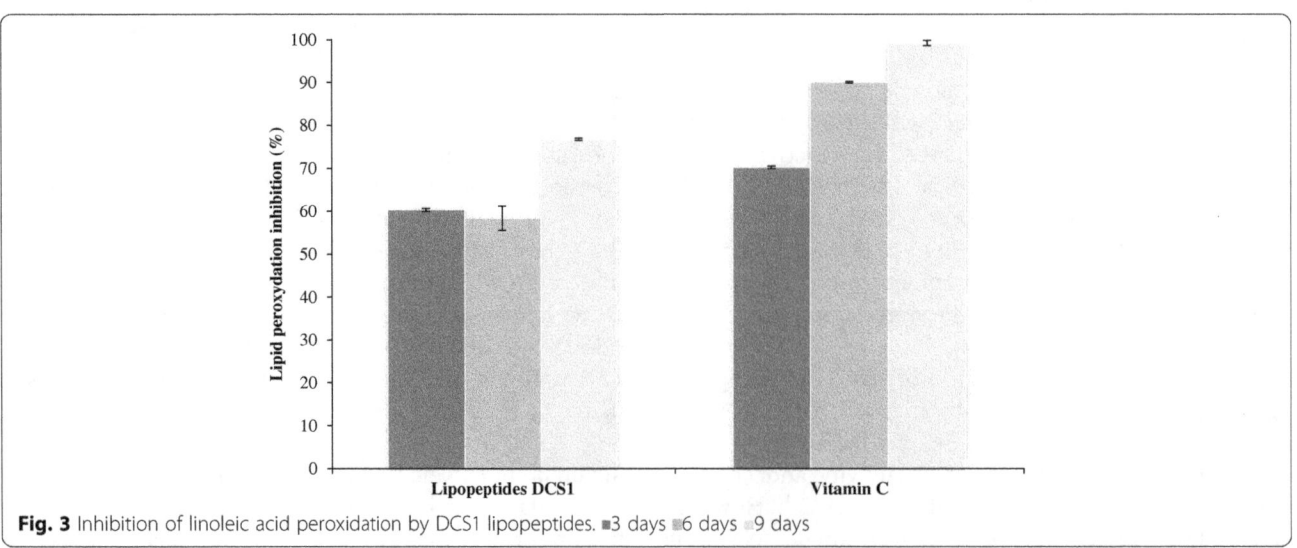

Fig. 3 Inhibition of linoleic acid peroxidation by DCS1 lipopeptides. ■3 days ■6 days ■9 days

Table 1 Antimicrobial activity spectrum of DCS1 lipopeptides (2 mg mL^{-1})

Indicator organisms	Inhibition zone diameter (mm)
Gram (+)	
S. aureus (ATCC 25923)	14 ± 0.4
B. cereus (ATCC 11778)	15 ± 0.1
M. luteus (ATCC 4698)	–
Gram (−)	
K. pneumoniae (ATCC 13883)	30 ± 0.1
E. coli (ATCC 25922)	16.5 ± 0.6
S. typhimyrium (ATCC 19430)	15 ± 0.2
S. enterica (ATCC 27853)	15 ± 0.1
Enterobacterium sp.	7.66 ± 0.6
Fungi	
A. niger	+++
A. flavus	+++
F. oxysporium	–
P. ultimum	++
F. solani	++
R. bataticola	++

Determinations were performed in triplicate and data correspond to mean values ± standard deviations

investigated against five strains. As can be seen in Fig. 4a, the pre-treatment of polystyrene surfaces with DCS1 lipopeptides significantly inhibited the adhesion of all tested bacteria, even at low concentrations. The biofilm formation inhibition was concentration-dependent, and the anti-adhesive effect remains nearly constant above a

Table 2 Influence of proteolytic enzymes, temperature and pH on DCS1 lipopeptides antibacterial activity against K. pneumoniae

Treatment	Residual activity (%)
Enzyme (1 mg mL^{-1})	
Pepsin	100
Trypsin	100
Chymotrypsin	100
Temperature (°C)	
40	100
60	100
80	100
100	100
pH	
3	74
5	84
8	100
10	100

Residual activity compared with the antimicrobial activity before the treatment. Data are means of three independent experiments

concentration of 1 mg mL^{-1} lipopeptides. The highest anti-adhesive effect was observed against C. albicans with an inhibition percentage of about 89.3%. Our results are in accordance with those of Janek et al. [15] who reported that the pre-treatment of a polystyrene surface with 0.5 mg mL^{-1} pseudofactin II inhibited bacterial adhesion by 36–90% and that of C. albicans by 92%. In another study, Coronel-león et al. [13] reported that the highest anti-adhesive activity of the biosurfactant lichenysin produced by Bacillus licheniformis AL1.1 was observed against C. albicans (74.35%) at a concentration of 4 mg mL^{-1}.

High anti-adhesion effect was also observed for S. aureus, B. cereus and S. typhimurium (77.3, 77.1 and 75.7%, respectively, at 1 mg mL^{-1}) with DCS1 lipopeptides, while the effect on K. pneumoniae was low (50%). Gudiña et al. [47] demonstrated high anti-adhesion activity against many pathogens, among them S. aureus (72.0%), Staphylococcus epidermidis (62.1%) and Sreptococcus agalactiae (60%), but at high concentration of biosurfactant produced by Lactobacillus paracasei A20 (25 mg mL^{-1}). Earlier reports [48] described the inhibition of pathogenic bacteria adhesion (Escherichia coli CFT073 and S. aureus ATCC 29213) to polystyrene surfaces by two lipopeptide biosurfactants synthesized by Bacillus subtilis and Bacillus licheniformis. Furthermore, Pontibacter korlensis SBK-47 strain produces pontifactin, a new lipopeptide biosurfactant that exhibits anti-adhesive activity against many pathogenic bacteria ranging from 87% to 99% inhibition at a maximum concentration of 2 mg mL^{-1} [36].

DCS1 lipopeptides exhibit high efficiency as the calculated effective dose (ED$_{50}$) with 50% adhesion inhibition was very low with all microorganisms tested: 0.36 mg mL^{-1} for S. typhimurium, 2 mg mL^{-1} for K. pneumoniae, 0.4 mg mL^{-1} for B. cereus and 0.015 mg mL^{-1} for S. aureus.

The inhibition of biofilm formation is due to the capacity of lipopeptides to modify the physico-chemical properties of the surface to reduce adhesion and biofilm formation. Some lipopeptides are considered anionic due to the negative charges of amino acids, also surfaces of most bacterial cells are negatively charged. Therefore, in the pre-treatment with lipopeptides, the potential effect of biofilm formation inhibition could be the result of the electrostatic repulsion forces between the negative charge of polystyrene surface recovered with lipopeptides molecules and negative charge of the microbial membrane. Bacterial adhesion depends on bacterial charge, surface type and on the biosurfactant charge which interferes in their influences.

Disruptive activity on pre-formed biofilm

Disruption of biofilm consists in removing the attached microorganisms from the surface after biofilm formation.

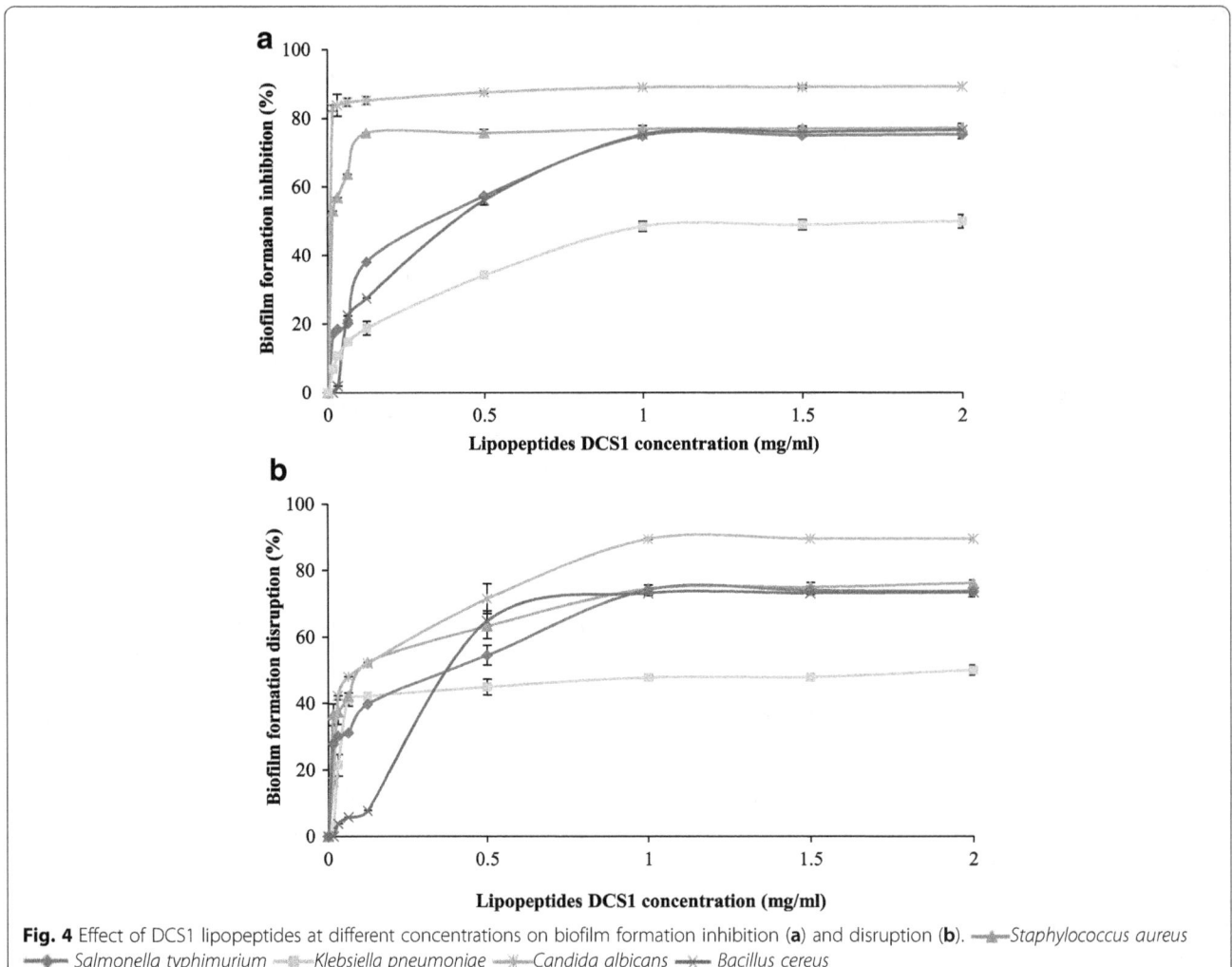

Fig. 4 Effect of DCS1 lipopeptides at different concentrations on biofilm formation inhibition (**a**) and disruption (**b**). ▲ *Staphylococcus aureus* ◆ *Salmonella typhimurium* ▩ *Klebsiella pneumoniae* ✳ *Candida albicans* ✴ *Bacillus cereus*

As illustrated in Fig. 4b, the efficacy of biofilm disruption increased with increasing sample concentration, and the percentages of disruption remain nearly constant above a concentration of 1 mg mL^{-1} lipopeptides. The maximum disruption produced by DCS1 lipopeptides was observed for *C. albicans* with 89.50%, followed by *S. aureus* (77%), *S. typhimurium* (75%), *B. cereus* (74.8%) and *K. pneumoniae* (50.3%). Our results are in agreement with those of Dalili et al. [12] who reported that Coryxin, a cyclic lipopeptide, produced by *Corynebacterium xerosis* NS5 displayed inhibitory and disruptive activities against biofilm formation by a variety of bacteria.

The effectiveness of lipopeptides in post-treatment of biofilm formation using various microorganisms was similar to that in pre-treatment. Our findings are in contrast with those of Janek et al. [15] who stated that the effectiveness in biofilm disruption in a post-treatment is lower than in biofilm inhibition in a pre-treatment with an activity ranging from 26 to 70% with pseudofactin II (0.5 mg mL^{-1}).

The effective dose (ED$_{50}$) was very low in post-treatment with all pathogens tested: 0.51 mg mL^{-1} for *S. typhimurium*, 2 mg mL^{-1} for *K. pneumoniae*, 0.11 mg mL^{-1} for *S. aureus*, 0.37 mg mL^{-1} for *B. cereus* and 0.096 mg mL^{-1} for *C. albicans*.

The disruption of biofilm could be explained by the adsorption of lipopeptides at the interface between the attached biofilm-forming bacteria and the solid surface, thus favoring bacterial detachment. According to McLandsborough et al. [49], surfactants could penetrate and adsorb at the interface between the solid surface and the biofilm owing to their high surface activity and reduce the interfacial tension.

When comparing both processes, pre-treatment and post-treatment of polystyrene surfaces, it is clear that the action of DCS1 lipopeptides was more effective against *C. albicans*. The use of the yeast *C. albicans* is of particular interest because it is recognized as an important pathogen in nosocomial infections [50]. Also DCS1 lipopeptides could be considered as a good alternative for controlling the growth of biofilms of *S. typhimyrium*, *S. aureus* and *B. cereus* which are responsible for opportunistic food-borne illness.

Conclusion

The current study demonstrated that lipopeptides synthesized by *B. methylotrophicus* DCS1 strain exerted considerable antioxidant action involving several antioxidant mechanisms, including metal ion chelating, hydrogen or electron donation and radical scavenging during peroxidation. Further, the DCS1 lipopeptides showed antimicrobial activities against several microorganisms tested. In addition, they showed anti-adhesive activity against biofilm formation as well as their potential to disrupt pre-formed biofilm. The results obtained suggest the possible use of DCS1 lipopeptides as a potential antioxidant and antimicrobial as well as anti-adhesive agent to reduce microbial adhesion and biofilm formation in biomedical field and food industry.

Abbreviations

BHA: Butylhydroxyanisol; CFU: Colony-forming units; CV: Crystal violet; DPPH: 2, 2-diphenyl-1-picrylhydrazyl; ED_{50}: Effective dose; EDTA: Ethylene diamine tetra acetic acid; FT-IR: Fourier transform infrared spectroscopy; IC_{50}: Half maximal inhibitory concentration; LB: Luria-Bertani media; MDA: Malondialdehyde; OD: Optical density; PBS: Phosphate-buffered saline; rDNA: Ribosomal Deoxyribonucleic acid; RNA: Ribonucleic acid; TBARS: Thiobarbituric acid reactive substances; THF: Tetrahydro-furan; TLC: Thin layer chromatography

Funding

This work was funded by the Ministry of Higher Education and Scientific Research, Tunisia. This work was partially financed by the Ministerio de Economía y Competitividad. Spain, project CTQ2014–59632-R, and by the IV Pla de Recerca de Catalunya, Spain (Generalitat de Catalunya), grant 2014SGR-534.

Authors' contributions

NJ conceived and performed all in vitro experiments. HB-A analyzed and interpreted the data. AM analyzed and interpreted the results of anti-adhesion assay. NJ wrote the manuscript. AM, HB-A, NH and MN contributed in revising the manuscript. All authors read and approved the final manuscript.

Competing interests

The authors declare that they have no competing interests.

Author details

[1]Laboratoire de Génie Enzymatique et de Microbiologie, Université de Sfax, Ecole Nationale d'Ingénieurs de Sfax, B.P. 1173-3038 Sfax, Tunisia. [2]Section of Microbiology, Department of Biology, Health and Environment, Faculty of Pharmacy, University of Barcelona, Joan XXIII s/n, 08028 Barcelona, Spain.

References

1. Femi-Ola TO, Oluwole OA, Olowomofe TO, Yakubu H. Isolation and screening of biosurfactant-producing bacteria from soil contaminated with domestic waste water. BJES. 2015;3:58–63.
2. Joshi SJ, Suthar H, Yadav AK, Hingurao K, Nerurkar A. Occurrence of biosurfactant producing *Bacillus* spp. in diverse habitats. ISRN Biotechnol 2013;2013 Article ID 652340, 6 pp.
3. Leclère V, Béchet M, Adam A, Guez JS, Wathelet B, Ongena M, et al. Mycosubtilin overproduction by *Bacillus subtilis* BBG100 enhances the organism's antagonistic and biocontrol activities. Appl Environ Microbiol. 2005;71:4577–84.
4. Tabbene O, Gharbi D, Slimene IB, Elkahoui S, Alfeddy MN, Cosette P, et al. Antioxidative and DNA protective effects of bacillomycin D-like lipopeptides produced by B38 strain. Appl Biochem Biotechnol. 2012;168:2245–56.
5. Ben Ayed H, Bardaa S, Moalla D, Jridi M, Maalej H, Sahnoun Z, et al. Wound healing and in vitro antioxidant activities of lipopeptides mixture produced by *Bacillus mojavensis* A21. Process Biochem. 2015;50:1023–30.
6. Benitez LB, Velho RV, da Motta ADS, Segalin J, Brandelli A. Antimicrobial factor from *Bacillus amyloliquefaciens* inhibits *Paenibacillus* larvae, the causative agent of American foulbrood. Arch Microbiol. 2012;194:177–85.
7. Ben Ayed H, Hmidet N, Béchet M, Chollet M, Chataigné G, Leclère V, et al. Identification and biochemical characteristics of lipopeptides from *Bacillus mojavensis* A21. Process Biochem. 2014;49:1699–707.
8. Xu HM, Rong YJ, Zhao MX, Song B, Chi ZM. Antibacterial activity of the lipopeptides produced by *Bacillus amyloliquefaciens* M1 against multidrug-resistant *Vibrio* spp. isolated from diseased marine animals. Appl Microbiol Biotechnol. 2014;98:127–36.
9. Damalaon R, Findlay B, Ogunsina M, Arthur G, Schweizer F. Ultrashort cationic lipopeptides and lipopeptoids: evaluation and mechanistic insights against epithelial cancer cells. Peptides. 2016;84:58–67.
10. Zheng C, Wang M, Wang Y, Huang Z. Optimization of biosurfactant-mediated oil extraction from oil sludge. Bioresour Technol. 2012;110:338–42.
11. Md F. Biosurfactant: production and application. J Pet Environ Biotechnol. 2012;3:124.
12. Dalili D, Amini M, Framarzi MA, Fazeli MR, Khoshayand MR, Samadi N. Isolation and structural characterization of Coryxin, a novel cyclic lipopeptide from *Corynebacterium xerosis* NS5 having emulsifying and anti-biofilm activity. Colloids Surf B Biointerfaces. 2015;135:425–32.
13. Coronel-León J, Marqués AM, Bastida J, Manresa A. Optimizing the production of the biosurfactant lichenysin and its application in biofilm control. J Appl Microbiol. 2015;150:99–111.
14. Sadekuzzaman M, Yang S, Mizan MFR, Ha SD. Current and recent advanced strategies for combating biofilms. Compr Rev Food Sci Food Saf. 2015;14: 491–509.
15. Janek T, Łukaszewicz M, Krasowka A. Antiadhesive activity of the biosurfactant pseudofactin II secreted by the Arctic bacterium *Pseudomonas fluorescens* BD5. BMC Microbiol. 2012;12:24.
16. Nitschke M, Ferraz C, Pastore GM. Selection of microorganisms for biosurfactant production using agroindustrial wastes. Braz J Microbiol. 2004;35:336–41.
17. Falagas ME, Makris GC. Probiotic bacteria and biosurfactants for nosocomial infection control: a hypothesis. J Hosp Infect. 2009;71:301–6.
18. Jemil N, Ben Ayed H, Hmidet N, Nasri M. Characterization and properties of biosurfactants produced by a newly isolated strain *Bacillus methylotrophicus* DCS1 and their applications in enhancing solubility of hydrocarbon. World J Microbiol Biotechnol. 2016;32:175.
19. Landy M, Warren GH, Rosenman SB, Colio LG. Bacillomycin: an antibiotic from *Bacillus subtilis* active against pathogenic fungi. Proc Soc Exp Biol Med. 1948;67:539–41.
20. Abouseoud M, Yataghene A, Amrane A, Maachi R. Biosurfactant production by free and alginate entrapped cells of *Pseudomonas fluorescens*. J Ind Microbiol Biotechnol. 2008;35:1303–8.
21. Ibrahim ML, Ijah UJJ, Manga SB, Bilbis LS, Umar S. Production and partial characterization of biosurfactant produced by crude oil degrading bacteria. Int Biodeterior Biodegrad. 2013;81:28–34.
22. Bersuder P, Hole M, Smith G. Antioxidants from a heated histidine–glucose model system. I. Investigation of the antioxidant role of histidine and isolation of antioxidants by high performance liquid chromatography. J Am Oil Chem Soc. 1998;75:181–7.
23. Yildirim A, Mavi A, Kara AA. Determination of antioxidant and antimicrobial activities of *Rumex crispus* L. extracts. J Agric Food Chem. 2001;49:4083–9.
24. Carter P. Spectrophotometric determination of serum iron at the submicrogram level with a new reagent (ferrozine). Anal Biochem. 1971;40:450–8.

25. Koleva II, van Beek TA, Linssen JP, de Groot A, Evstatieva LN. Screening of plant extracts for antioxidant activity: a comparative study on three testing methods. Phytochem Anal. 2002;13:8–17.
26. Yagi K. A simple fluorometric assay for lipoperoxide in blood plasma. Biochem Med. 1976;15:212–6.
27. Nanda A, Saravanan M. Biosynthesis of silver nanoparticles from *Staphylococcus aureus* and its antimicrobial activity against MRSA and MRSE. Nanomedicine. 2009;5:452–6.
28. Cladera-Olivera F, Caron GR, Brandelli A. Bacteriocin-like substance production by *Bacillus licheniformis* strain P40. Lett Appl Microbiol. 2004;38:251–6.
29. O'Toole GA. Microtiter dish biofilm formation assay. J Vis Exp. 2011;47:2437.
30. Alajlani M, Shiekh A, Hasnain S, Brantner A. Purification of bioactive lipopeptides produced by *Bacillus subtilis* strain BIA. Chromatographia. 2016;79:1527–32.
31. Kong J, Yu S. Fourier transform infrared spectroscopic analysis of protein secondary structures. Acta Biochim Biophys Sinica. 2007;39:549–59.
32. Thaniyavarn J, Roongsawang N, Kameyama T, Haruki M, Imanaka T, Morikawa M, et al. Production and characterization of biosurfactants from *Bacillus licheniformis* F2.2. Biosci Biotechnol Biochem. 2003;67:1239–44.
33. Das P, Mukherjee S, Sen R. Antimicrobial potential of a lipopeptide biosurfactant derived from a marine *Bacillus circulans*. J Appl Microbiol. 2008;104:1675–84.
34. Dehghan-Noudeh G, Housaindokht M, Bazzaz BS. Isolation, characterization, and investigation of surface and hemolytic activities of a lipopeptide biosurfactant produced by *Bacillus subtilis* ATTC 6633. J Microbiol. 2005;43:272–6.
35. Pereira JFB, Gudina EJ, Costa R, Vitorino R, Teixeira JA, Coutinho JAP, et al. Optimization and characterization of biosurfactant production by *Bacillus subtilis* isolated towards microbial enhanced oil recovery applications. Fuel. 2013;111:259–68.
36. Balan SS, Kumar CG, Jayalakshmi S. Pontifactin, a new lipopeptide biosurfactant produced by a marine *Pontibacter korlensis* strain SBK-47: purification, characterization and its biological evaluation. Process Biochem. 2016;51:2198–207.
37. Uddin SN, Ali ME, Yesmin MN. Antioxidant and antibacterial activities of *Sena tora Roxb*. American J Plant Physiol. 2008;3:096–100.
38. Shimada K, Fujikawa K, Yahara K, Nakamura T. Antioxidative properties of xanthan on the autoxidation of soybean oil in cyclodextrin emulsion. J Agric Food Chem. 1992;40:945–8.
39. Yalçin E, Çavuşoğlu K. Structural analysis and antioxidant activity of a biosurfactant obtained from *Bacillus subtilis* RW1. Turk J Biochem. 2010;35:243–7.
40. Gulcin I, Buyukokuroglu ME, Oktay M, Kufrevioglu OI. On the in vitro antioxidant properties of melatonin. J Pineal Res. 2002;33:167–71.
41. Stohs SJ, Bagchi D. Oxidative mechanisms in the toxicity of metal ions. Free Radic Biol Med. 1995;18:321–36.
42. Costa LS, Fidelis GP, Cordeiro SL, Oliveira RM, Sabry DA, Câmara RBG, et al. Biological activities of sulfated polysaccharides from tropical seaweeds. Biomed Pharmacother. 2010;64:21–8.
43. Donio MBS, Ronica SFA, Thanga Viji V, Velmurugan S, Adlin Jenifer J, Michaelbabu M, et al. Isolation and characterization of halophilic *Bacillus* sp. BS3 able to produce pharmacologically important biosurfactants. Asian Pac J Trop Med. 2013;6:876–83.
44. Sheppard JD, Jumarie C, Cooper DG, Laprade R. Ionic channels induced by surfactin in planar lipid bilayer membranes. Biochim Biophys Acta. 1991;1064:13–23.
45. Bizani D, Brandelli A. Characterization of a bacteriocin produced by a newly isolated *Bacillus* sp. strain 8A. J Appl Microbiol. 2002;93:512–9.
46. Stein T. *Bacillus subtilis* antibiotics: structures, syntheses and specific functions. Mol Microbiol. 2005;56:845–57.
47. Gudiña EJ, Rocha V, Teixeira JA, Rodrigues LR. Antimicrobial and anti-adhesive properties of a biosurfactant isolated from *Lactobacillus paracasei* ssp. *paracasei* A20. Lett Appl Microbiol. 2010;50:419–24.
48. Rivardo F, Turner RJ, Allegrone G, Ceri H, Martinotti MG. Anti-adhesion activity of two biosurfactants produced by *Bacillus* spp. prevents biofilm formation of human bacterial pathogens. Appl Microbiol Biotechnol. 2009;83:541–53.
49. McLandsborough L, Rodriguez A, Pérez-Conesa D, Weiss J. Biofilms: at the interface between biophysics and microbiology. Food Biophys. 2006;1:94–114.
50. Boucherit-Atmani Z, Siddiki SML, Boucherit K, Sari-Belkharoubi L, Kunkel D. *candida albicans* Biofilms formed into catheters and probes and their resistance to amphotericin B. J Mycol Med. 2011;21:182–7.

5

Physiological roles of sigma factor SigD in *Corynebacterium glutamicum*

Hironori Taniguchi[1,2], Tobias Busche[2], Thomas Patschkowski[2,3], Karsten Niehaus[2,3], Miroslav Pátek[4], Jörn Kalinowski[2] and Volker F. Wendisch[1,2*] (iD)

Abstract

Background: Sigma factors are one of the components of RNA polymerase holoenzymes, and an essential factor of transcription initiation in bacteria. *Corynebacterium glutamicum* possesses seven genes coding for sigma factors, most of which have been studied to some detail; however, the role of SigD in transcriptional regulation in *C. glutamicum* has been mostly unknown.

Results: In this work, pleiotropic effects of *sigD* overexpression at the level of phenotype, transcripts, proteins and metabolites were investigated. Overexpression of *sigD* decreased the growth rate of *C. glutamicum* cultures, and induced several physiological effects such as reduced culture foaming, turbid supernatant and cell aggregation. Upon overexpression of *sigD*, the level of Cmt1 (corynomycolyl transferase) in the supernatant was notably enhanced, and carbohydrate-containing compounds were excreted to the supernatant. The real-time PCR analysis revealed that *sigD* overexpression increased the expression of genes related to corynomycolic acid synthesis (*fadD2*, *pks*), genes encoding corynomycolyl transferases (*cop1*, *cmt1*, *cmt2*, *cmt3*), L, D-transpeptidase (*lppS*), a subunit of the major cell wall channel (*porH*), and the envelope lipid regulation factor (*elrF*). Furthermore, overexpression of *sigD* resulted in trehalose dicorynomycolate accumulation in the cell envelope.

Conclusions: This study demonstrated that SigD regulates the synthesis of corynomycolate and related compounds, and expanded the knowledge of regulatory functions of sigma factors in *C. glutamicum*.

Keywords: *Corynebacterium glutamicum*, Sigma factor, SigD, Mycomembrane, Trehalose dicorynomycolate

Background

Sigma factors are a component of bacterial RNA polymerase holoenzymes essential for promoter recognition and transcription initiation [1]. Most bacteria encode multiple sigma factors, and each sigma factor containing RNA polymerase holoenzyme initiates transcription from the cognate promoter sequences [2–5]. By replacing a sigma factor in RNA polymerase holoenzyme, bacteria activate transcription of a different gene set under different conditions, and cope with environmental changes [6]. Therefore, sigma factors play an important role in transcriptional regulation in a global manner. The knowledge of regulations by sigma factors is helpful to elucidate the regulatory network of the organism.

Corynebacterium glutamicum was first isolated as an organism secreting high amounts of L-glutamate [7]. Nowadays, this bacterium is used for production of L-amino acids in million tons per year, especially L-glutamate and L-lysine [8]. *C. glutamicum* ATCC 13032 has seven sigma factor genes in its chromosome, *sigA*, *sigB*, *sigC*, *sigD*, *sigE*, *sigH* and *sigM* [9, 10]. The physiological functions of SigA, SigB, SigC, SigE, SigH and SigM have been studied to some extent [11]; however, the regulation and physiological roles of SigD in *C. glutamicum* have not yet been revealed. *sigD* gene is well conserved among corynebacteria, and 17 out of 19 examined *Corynebacterium* species possess *sigD* genes [11]. Therefore, it is assumed that SigD plays a substantial role in transcriptional regulation and subsequent adaptation of *C. glutamicum* to changing environments.

C. glutamicum belongs to the CMN (<u>C</u>orynebacterium, <u>M</u>ycobacterium, <u>N</u>ocardia) group, which is characterized by the unique molecular constituents of their cell

* Correspondence: volker.wendisch@uni-bielefeld.de
[1]Genetics of Prokaryotes, Faculty of Biology, Bielefeld University, Bielefeld, Germany
[2]Center for Biotechnology, Bielefeld University, Bielefeld, Germany
Full list of author information is available at the end of the article

envelopes such as the mycomembrane [12]. The myco-membrane is composed of a monolayer of corynomyco-late (α-alkyl, β-hydroxy fatty acid), which is covalently linked to arabinogalactan or forms other lipids such as trehalose monocorynomycolate (TMCM) and trehalose dicorynomycolate (TDCM) [13].

In this work, we evaluated the effects of deletion and overexpression of *sigD* on the cell phenotype, and revealed the influence of *sigD* overexpression on transcripts, proteins and metabolites. The achieved results disclosed the important roles of SigD in mycomembrane synthesis and maintaining cell wall integrity.

Methods

Bacterial strains, plasmids and oligonucleotides

The strains, plasmids and oligonucleotides used in this work are listed in Additional file 1: Table S1. A plasmid for *sigD* overexpression was constructed based on pVWEx1, which is an IPTG inducible expression vector for *E. coli* and *C. glutamicum* [14]. A plasmid for gene disruption was constructed based on pK18mobsacB [15]. For plasmid construction, DNA fragments were amplified from the genomic DNA of *C. glutamicum* ATCC 13032 by PCR with the oligonucleotide pairs shown in Additional file 1: Table S1. These fragments were inserted into the digested plasmid by ligation or Gibson assembly [16, 17]. *E. coli* DH5α was used for cloning. *E. coli* competent cells were transformed by the heat shock method [16] or by the electroporation method [18]. DNA sequences of all cloned DNA fragments were confirmed to be correct by sequencing. *C. glutamicum* competent cells were transformed by electroporation at 2.5 kV, 200 Ω, and 25 μF [8, 19]. Gene disruption via two-step homologous recombination and the following selections were carried out as previously described [8]. Disruption was verified by PCR with the respective oligonucleotide pairs.

Medium and conditions for growth experiments

Unless otherwise specified, *C. glutamicum* cells were precultured overnight in lysogeny broth (LB) medium [16] supplemented with 56 mM of glucose, washed once with chemically defined CGXII medium [8] without carbon source, and then inoculated into CGXII medium with 222 mM of glucose at an initial OD_{600} of 1. The cultivation was performed at 30 °C, 120 rpm. OD_{600} was measured with UV-1202 spectrophotometer (Shimadzu, Duisburg, Germany) with suitable dilutions. When necessary, 25 μg/mL of kanamycin and appropriate concentrations of IPTG were added as indicated in the text. For the growth experiment in BioLector® cultivation system (m2pLabs, Baesweiler, Germany), cells were cultivated in 1 mL of CGXII medium with 222 mM of glucose using FlowerPlate® (m2pLabs, Baesweiler,

Germany) at 30 °C, 1100 rpm. Cell growth was monitored online every 10 min, and the maximum growth rate (h^{-1}) was determined from the growth rates μ (h^{-1}) which were calculated with regression analysis from backscattering light intensity (wavelength of 620 nm) at 20 consecutive measuring points.

Photometric determination of supernatant turbidity and observation of cell aggregation by microscopy

To quantify turbidity of supernatants, cell cultures were centrifuged for 30 min with 15,000 x *g* at room temperature. The absorption of supernatants was measured at a wavelength of 600 nm. For microscopic imaging, the culture of each strain in the stationary phase was diluted in CGXII medium without carbon source to an OD_{600} of 1, and observed by microscopy with 100 x oil immersion objective lens and 10 x ocular lens. Quantification of cell aggregate size was performed using ImageJ (https://imagej.nih.gov/ij/).

Protein analysis of the supernatant

Supernatants were taken from the stationary phase cultures. Four volumes of acetone were added to one volume of supernatant, and stored at −20 °C overnight. After centrifugation at 4 °C, 20,000 x *g* for 15 min, precipitates were resuspended in 20 mM Tris-HCl (pH 7). SDS-PAGE was performed using Tris-glycine discontinuous buffer, and visualized by staining with Coomassie Brilliant Blue R250 as previously described [16]. Quantification of the intensity of each band was performed using ImageJ. Protein bands with different intensities observed in the control strain and the *sigD* overexpressing strain were excised from SDS-PAGE gels and transferred to a new tube which had been washed with trifluoroacetic acid: acetonitrile: H_2O (0.1:60:40 *v/v*) in advance. The digestion of protein in the excised band was performed with trypsin overnight as previously described [20]. Protein sequences were identified using an ultrafleXtreme MALDI-TOF/TOF mass spectrometry (Bruker, Bremen, Germany) and Mascot search engine (Matrix Science, London, UK) as previously described with some modifications for *C. glutamicum* ATCC 13032 [21].

Quantification of carbohydrate in acetone precipitates

Supernatants were precipitated with acetone as described above. Precipitates were resuspended with 100 μL of 20 mM Tris-HCl (pH 7), placed at room temperature for 30 min, and then the insoluble fraction was recovered by centrifugation. Insoluble fractions were hydrolyzed as previously described [22]. Briefly, the precipitates were resuspended with 75 μL of 72% (*w/w*) sulfuric acid and incubated at room temperature for 3 h. The slurry was diluted to 1 mL with water, heated at 100 °C for 4 h, and cooled down on ice. Colorimetric quantification for carbohydrates was performed by the phenol sulfuric acid method as previously described with some modifications [23, 24]. Briefly, 200 μL of hydrolysate was mixed with 600 μL of concentrated sulfuric acid rapidly, and 120 μL of 5% phenol (*w/v*) in water was added immediately. The mixture was incubated for 5 min at 90 °C, and cooled to room temperature for 5 min. The absorption at 490 nm was measured, and compared to the absorption of the control samples with different concentrations of arabinose.

RNA extraction

Cells were first precultured in LB medium, and inoculated in CGXII medium with 222 mM of glucose for adaptation. Then, the appropriate amount of cell culture was inoculated into fresh CGXII medium with 222 mM of glucose at an initial OD_{600} of 1. The cells were harvested at an OD_{600} between 6 and 8. Cell cultures (1 mL) were centrifuged for 30 s at 20,000 x *g*, and immediately frozen with liquid nitrogen after removing supernatant. RNA isolation was performed using the RNeasy mini kit along with the RNase-free DNase set (Qiagen, Hilden, Germany) as previously described [25]. The absence of contaminating genomic DNA in RNA samples was confirmed by PCR with multiple oligonucleotide pairs specific to genomic DNA.

RNA isolation, library preparation and RNA-seq

For transcriptome sequencing, RNA samples isolated individually from biological triplicates were mixed for each stain. Quality check of the isolated RNA, library preparation, RNA-seq and data analysis were performed as previously described [26]. Briefly, RNA quality was checked by Trinean Xpose (Gentbrugge, Belgium) and Agilent RNA Nano 6000 kit with Agilent 2100 Bioanalyzer (Agilent Technologies, Böblingen, Germany). Ribo-Zero rRNA Removal Kit (Bacteria) from Illumina (San Diego, CA, USA) was used to remove the ribosomal RNA molecules from the isolated total RNA. Removal of rRNA was checked by Agilent RNA Pico 6000 kit on Agilent 2100 Bioanalyzer (Agilent Technologies, Böblingen, Germany). TruSeq Stranded mRNA Library Prep Kit from Illumina (San Diego, CA, USA) was used to prepare cDNA libraries. cDNAs were sequenced paired end on an Illumina MiSeq system (San Diego, CA, USA) using 50 bp read length.

Read mapping, data visualization and analysis of gene expression

Trimmed reads (26 nt) were mapped to the *C. glutamicum* ATCC 13032 reference genome sequence [9] with SARUMAN [27], allowing for up to one error per read. The forward and reverse read, if both present and with a maximum distance of 1 kb, were combined to one read that contains the reference sequence as insert. Paired mappings with a distance >1 kb were discarded, and

paired reads with either only the forward or only the reverse read mapping were retained as single mapping reads, as previously described [28]. ReadXplorer 2.2.0 was used for visualization of short read alignments [29]. For differential gene expression analysis, ReadXplorer and Bioconductor package DESeq implemented in ReadXplorer were used [29, 30]. Genes with the mean value of signal intensity less than 30 were discarded. A-value and M-value of each gene were calculated based on the intensity value of the strain of interest and the control strain.

Real-time PCR analysis

For real-time PCR analysis, relative abundance of mRNA of each gene was quantified using the same amount of total RNA. RNA sample extracted from biological triplicates were quantified individually. The experiment and analysis were performed as previously described with respective oligonucleotide pairs in Additional file 1: Table S1 [31].

Detection of trehalose dicorynomycolate by thin layer chromatography

Lipid extraction and thin layer chromatography (TLC) were performed based on the method previously described [32, 33]. Briefly, the crude lipid was extracted from cell pellet with $CHCl_3/CH_3OH$ (1:1 *v/v*) once, and with $CHCl_3/CH_3OH$ (2:1 *v/v*) for three times. All the extracts were pooled for each strain, and mixed with $CHCl_3/CH_3OH/H_2O$ (8:4:2 *v/v*) resulting the aqueous layer and the organic layer. The lower organic layer was collected and evaporated to dryness. The dried lipid was weighed and resuspended with $CHCl_3/CH_3OH$ (4:1 *v/v*) to the same concentration for each sample. TLC was performed with ALUGRAM SIL G/UV254 (Macherey-Nagel, Germany) with $CHCl_3/CH_3OH/H_2O$ (30:8:1 *v/v*) as a developing solvent. Total lipid (600 μg) was developed for each strain, and the bands were visualized by spraying TLC plate with sulfuric acid and heating to 110 °C. The mobility of TDCM or TMCM was determined by the retardation factor (Rf) value from the previous studies [32, 33]. Quantification of the intensity of each band was performed with ImageJ.

Results
Deletion and overexpression of *sigD* influenced the maximum growth rate

The *sigD* deletion mutant (Δ*sigD*) and the *sigD* overexpressing strain were cultivated in CGXII medium containing 222 mM of glucose as carbon source. *sigD* was overexpressed from the plasmid pVWEx1-*sigD* using an IPTG inducible promoter and different IPTG concentrations (0, 10, 50, 250 or 1000 μM). The Δ*sigD* mutant grew slightly slower than the wild type (WT) strain (Fig. 1a). The maximum growth rate of the *sigD* overexpressing strain WT(pVWEx1-*sigD*) decreased in an IPTG-

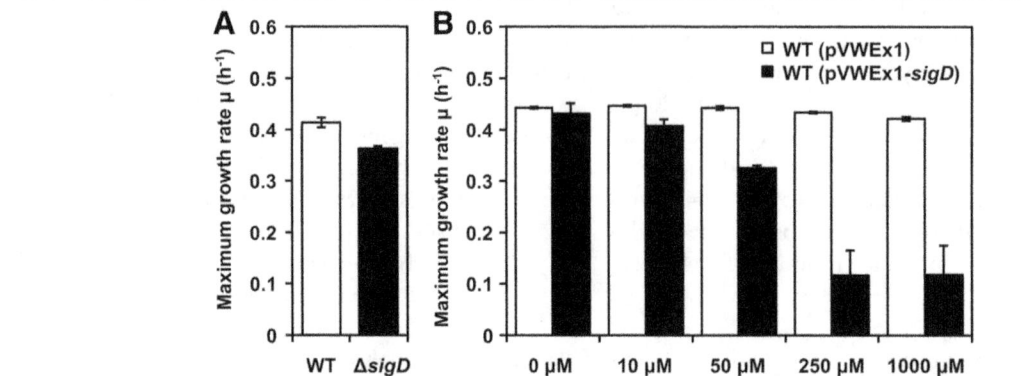

Fig. 1 Maximum growth rates of the Δ*sigD* strain and the *sigD* overexpressing strain with different IPTG concentrations. The maximum specific growth rate (h^{-1}) was shown for (**a**) *C. glutamicum* WT and Δ*sigD*, and (**b**) WT(pVWEx1) and WT(pVWEx1-*sigD*) with different IPTG concentrations (0, 10, 50, 250, 1000 μM). Error bars represent standard deviations from biological triplicates

dependent manner, which was not observed for WT(pVWEx1) (Fig. 1b). Higher concentrations of IPTG (250 and 1000 μM) severely inhibited the growth of WT(pVWEx1-*sigD*). The final biomass concentrations after 24 h of cultivation were comparable for WT, Δ*sigD*, WT(pVWEx1) and WT(pVWEx1-*sigD*) with 0, 10 or 50 μM of IPTG (data not shown). The slower growth of the Δ*sigD* strain indicated that *sigD* is beneficial for growth in minimal CGXII medium although it is not essential for growth under optimum conditions. As excessive *sigD* overexpression at the high IPTG concentrations was found to be harmful to cells, induction with 50 μM of IPTG was used for further experiments.

sigD overexpression influenced cell culture characteristics

WT(pVWEx1-*sigD*) with 50 μM of IPTG showed distinct characteristics compared to WT(pVWEx1) as the cultures of WT(pVWEx1-*sigD*) foamed significantly less than those of WT(pVWEx1) (Fig. 2a). Cell cultures with different IPTG concentrations (0, 10, 50 μM) using FlowerPlates and a BioLector system revealed that the supernatants of WT(pVWEx1-*sigD*) cultures showed higher turbidity than those of the control strain, and this turbidity increased in an IPTG-dependent manner (Fig. 2b). In addition to these distinct characteristics of the cell culture, cell aggregation was observed under the microscope in WT(pVWEx1-*sigD*) with IPTG (Fig. 2c–e). Taken together, overexpression of *sigD* induced pleiotropic changes of the phenotype of *C. glutamicum*.

sigD overexpression changed the pattern of secreted proteins

Overexpression of *sigD* influenced the culture characteristic of foaming. Therefore, the profile of proteins in the culture supernatant was compared between WT(pVWEx1) and WT(pVWEx1-*sigD*). Proteins in the supernatants were analyzed by 1D SDS-PAGE after concentrating by

acetone precipitation. We observed different band patterns of proteins secreted by WT(pVWEx1) and WT(pVWEx1-*sigD*) (Fig. 3a). The proteins in the bands with different intensity were further characterized by tryptic digestion and MALDI-TOF/TOF mass spectrometry (Fig. 3b, c). In this way, corynomycolyl transferase Cmt1 was identified in the band 3, which showed a higher intensity when *sigD* was overexpressed. The band 2 was identified as L, D-transpeptidase LppS, and the band 4 was identified as mixture of two proteins, corynomycolyl transferase Cmt2 and putative secreted protein Cg2052. On the other hand, Psp3, which was found in the band 1 and annotated as putative secreted protein, was less abundant in WT(pVWEx1-*sigD*). These results demonstrated that overexpression of *sigD* altered the secreted protein profile. Since Cmt1 abundance in the supernatant increased most upon *sigD* overexpression (Fig. 3a, c), *cmt1* was overexpressed in the WT strain. However, less foaming was not observed under these conditions (data not shown). Therefore, we concluded that less foaming of the culture did not occur only due to Cmt1 protein abundance but to complex or other unknown reasons.

sigD overexpression induced the secretion of carbohydrate-containing compounds

During acetone precipitation of the supernatant, we observed an insoluble fraction only in the sample of the *sigD* overexpressing strain. This insoluble fraction still existed after an overnight protease K treatment. For the strain *C. glutamicum* CGL2005, it was reported that the ethanol-precipitated fraction of extracellular components consisted primarily of carbohydrates [34]. Therefore, the carbohydrate content in this fraction was determined by the phenol sulfuric acid method [23], which detects all classes of carbohydrates [35]. Arabinose, which is one of the components of peptidoglycan in *C. glutamicum*, was used as control. Colorimetric determination after

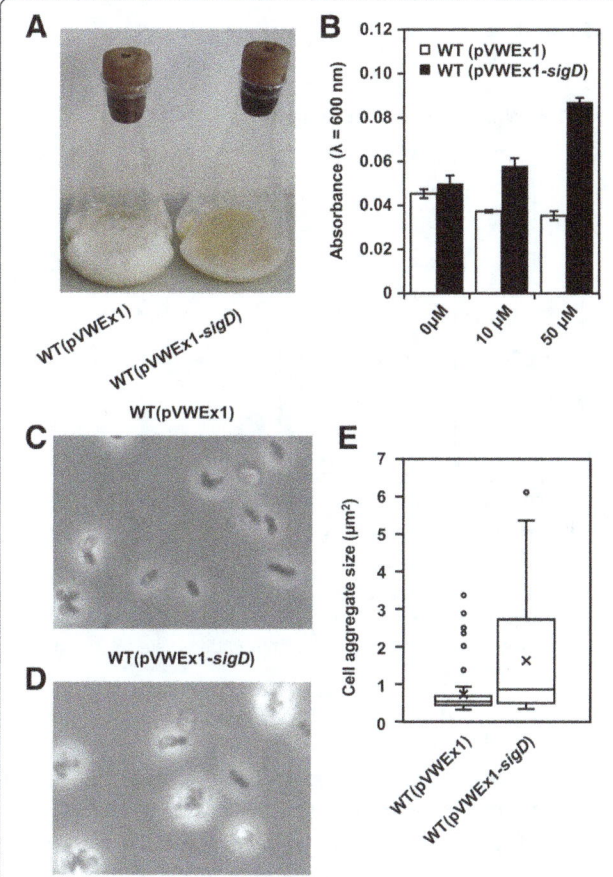

Fig. 2 Influence of *sigD* overexpression on cell cultures and cell morphology. **a** Cell cultures with 50 μM of IPTG after 36 h of cultivation are shown. **b** Supernatant turbidity with different concentrations of IPTG (0 μM, 10 μM, 50 μM) after 36 h is shown. Error bars represent standard deviations from biological triplicates. Microscopic images of the WT strain (**c**) and the *sigD* overexpressing strain (**d**) are shown. Cells in the stationary phase were observed under the microscope with a magnification of 1000. **e** Distribution of the size of cell aggregates is shown. The size of cell aggregates was analyzed by ImageJ, and the distribution was visualized by the box-and-whisker plot. Lower whisker, lower quantile, median, upper quantile and upper whisker are shown. The cross point indicates mean, and outliners were plotted as individual points

hydrolysis confirmed that the insoluble fraction of WT(pVWEx1-*sigD*) supernatants contained carbohydrates (arabinose equivalent of 1.3 mM). On the other hand, the same treatment of WT, Δ*sigD* or WT(pVWEx1) supernatants resulted in carbohydrate contents below the detection limit (arabinose equivalent <0.1 mM) (Table 1). These results showed that *sigD* overexpression induced the secretion of polysaccharides or carbohydrate-containing compounds into the supernatant.

SigD regulated transcription of several genes related to cell envelope integrity

RNA-seq and real-time PCR analysis were performed to understand the effects caused by *sigD* overexpression or

deletion at the transcriptional level. The relative abundance of mRNA of each gene was compared in WT(pVWEx1-*sigD*) without IPTG or with 50 μM of IPTG to analyze the effects of *sigD* overexpression. On the other hand, the mRNA abundance was compared between the Δ*sigD* and the WT strain to study the effects of *sigD* deletion. First, RNA - seq analysis was applied to screen the genes whose expression levels changed upon *sigD* overexpression or deletion. RNA-seq analysis implied that expression of 29 genes increased upon *sigD* overexpression (M-value >1) (Additional file 2: Table S2). Of these 29 genes, several genes are annotated as genes related to cell wall integrity; 6 genes (*cop1*, *cmt1*, *cmt2*, *cmt3*, *elrF* and *fadD2*) as corynomycolyl related proteins, a gene annotated as L, D-transpeptidase (*lppS*) and three genes (cg0420, cg0532, cg1181) as glycosyltransferases. In addition, expression of two further genes encoding proteins with mycomembrane-related functions, *porH* (cg3009) and *pks* (cg3178), increased upon *sigD* overexpression (M-value >$\log_2(1.5)$) and slightly decreased in the Δ*sigD* strain (M-value <$-\log_2(1.5)$). Based on these RNA-seq results, the expression of those genes was further analyzed by real-time PCR (Fig. 4). All of the tested genes were shown to be upregulated significantly as consequence of *sigD* overexpression. Increased transcript levels of *cmt1*, *cmt2* and *lppS* were consistent with the increased protein levels of Cmt1, Cmt2 and LppS in culture supernatants upon *sigD* overexpression (Fig. 3b, c). These results showed that the function of SigD is related to the regulation of cell envelope integrity such as mycomembrane synthesis and cell wall synthesis.

Overexpression of *sigD* increased the amounts of trehalose dicorynomycolate

The results of the real-time PCR analysis confirmed that the expression of several genes annotated as corynomycolyl transferase (*cop1*, *cmt1*, *cmt2*, *cmt3*) and genes related to corynomycolic acid production (*pks* and *fadD2*) increased due to *sigD* overexpression. Overexpression of *cop1* is known to increase the trehalose dicorynomycolate (TDCM) content in the cells [33]. To confirm the effect of *sigD* overexpression on the TDCM content, the crude lipid was extracted from cells, and its composition was analyzed by TLC (Fig. 5). The intensity of the band corresponding to TDCM increased 39% upon *sigD* overexpression. On the contrary, the intensity of the band corresponding to trehalose monocorynomycolate (TMCM) did not show a notable difference between the two strains. This result indicates that upregulation of multiple genes by *sigD* overexpression increased the flux toward corynomycolic acid synthesis which resulted in an alteration of TMCM/TDCM ratio as well as accumulation of other carbohydrate containing compounds.

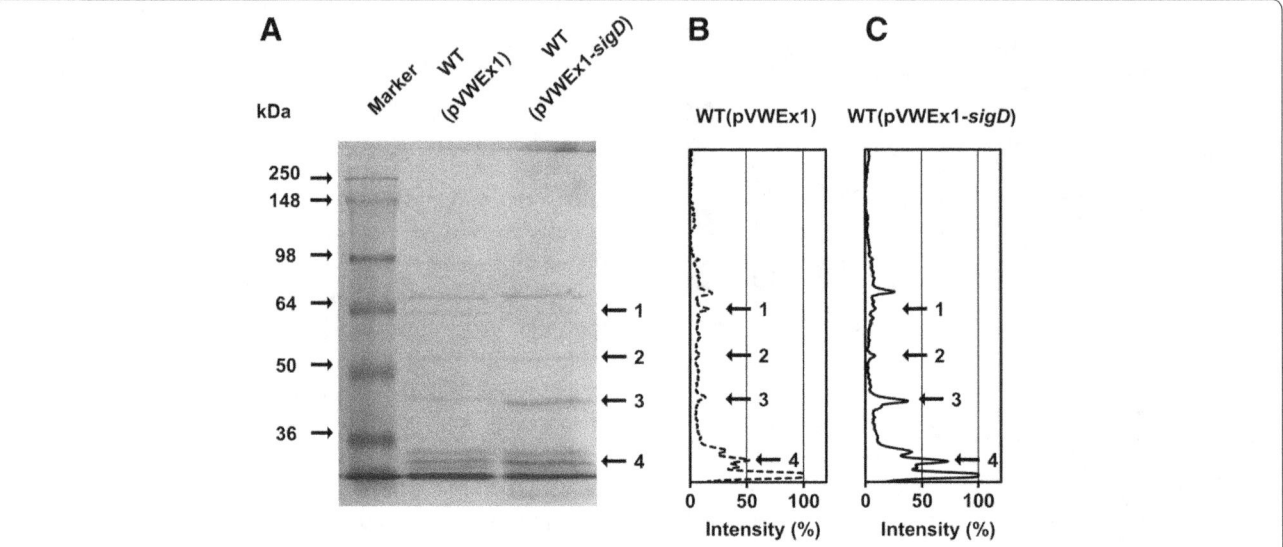

Fig. 3 1D-SDS PAGE of proteins in the supernatants. **a** Secreted proteins were analyzed by 12% SDS-PAGE. The molecular sizes of proteins in the marker are shown in kDa. WT(pVWEx1) and WT(pVWEx1-sigD) protein samples were obtained by acetone precipitation of the supernatants. Proteins from 200 µL of the supernatant was loaded on each lane. The intensity of secreted protein bands was quantified for WT(pVWEx1) (**b**) and for WT(pVWEx1-sigD) (**c**). The highest intensity of the band was normalized to 100%. The protein bands labeled with numbers were subjected to MALDI-TOF/TOF MS. 1: Psp3 (Cg2061), 2: LppS (Cg2720), 3: Cmt1 (Cg0413), 4: Cmt2 (Cg3186) and Cg2052

Discussion

Understanding the regulatory mechanisms of bacteria is important in many fields varying from biotechnology to public health. *C. glutamicum* has been used for the production of amino acids for several decades, however, the transcriptional regulation by sigma factors has not been fully elucidated. As for SigD, selection of high oxygen requiring mutants in a transposon library accidentally revealed that deletion of *sigD* in *C. glutamicum* loses the ability to grow under low oxygen concentrations [36], however, the knowledge is still limited. In the previous works, we demonstrated that overexpression of one of global regulators can artificially perturb the cellular regulation and influence metabolites as well as transcripts [37, 38]. This approach of sigma factor gene overexpression was found to be a useful approach for investigation of regulatory mechanisms and activation of specific biosynthesis pathways in *C. glutamicum*. In this work, deletion and overexpression of *sigD* revealed that sigma factor SigD plays a role in regulating maintenance of the cell wall integrity in *C. glutamicum* (Fig. 6).

Table 1 Carbohydrate content in acetone precipitated culture supernatants

	WT	ΔsigD	WT(pVWEx1)	WT(pVWEx1-sigD)
Arabinose equivalent carbohydrate (mM)	<0.1	<0.1	<0.1	1.29 ± 0.28

Carbohydrate content was measured by the phenol sulfuric acid method. Arabinose samples with known concentrations were used as standards. Carbohydrate content was calculated to the concentration (mM) of arabinose equivalent carbohydrate. The detection limit is 0.1 mM. Standard deviations were calculated from three biological replicates

The real time PCR analysis revealed that overexpression of *sigD* induced the expression of multiple genes related to mycomembrane synthesis (*cop1*, *cmt1*, *cmt2*, *cmt3*, *elrF*, *fadD2*, *porH* and *pks*) which exist at the different loci in the *C. glutamicum* genome. For mycomembrane biosynthesis, one molecule of fatty acid is carboxylated via the carboxylation complex composed of AccD2, AccD3, AccBC and AccE [39], and a second fatty acid is activated to a fatty acyl-CoA by FadD2 [40]. These two molecules are condensed and attached to trehalose by Pks [41, 42]. This product is reduced to TMCM (trehalose monocorynomycolate) by CmrA [43], and TMCM is exported from cytoplasm [44, 45]. Then, corynomycolyl transferases transfer the corynomycolate group of TMCM onto arabinogalactan, TMCM itself or proteins such as PorH [32, 33, 46, 47]. *C. glutamicum* possesses six corynomycolyl transferase genes, *cop1*, *cmt1*, *cmt2*, *cmt3*, *cmt4* and *cmt5* [32], and Cop1, Cmt1 and Cmt2 catalyze TDCM synthesis from TMCM [32, 46]. Cop1 is also reported to transfer corynomycolate from TMCM to arabinogalactan in the cell wall in *C. glutamicum* CGL2005 [33]. Interestingly, *sigD* overexpression induced the expression of four out of six corynomycolyl transferase genes at the same time, and enhanced the secretion of Cmt1 and Cmt2 to the supernatant. PorH forms the major cell wall channel penetrating the mycomembrane together with the other protein PorA [48], and corynomycolation of PorH and PorA catalyzed by Cmt1 was shown to be necessary for the pore forming activity [47, 49]. Furthermore, ElrF was identified as the envelope lipids regulation factor which regulates lipid composition

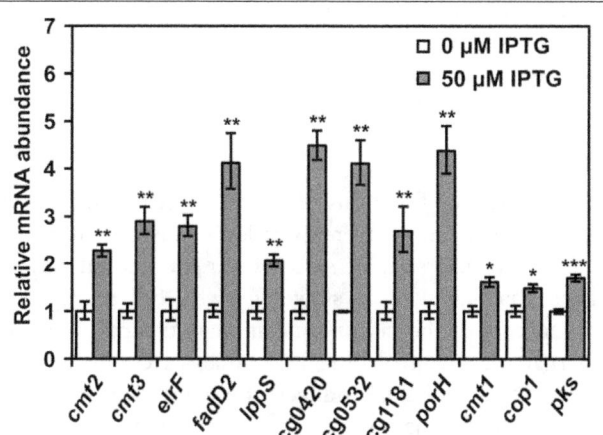

Fig. 4 Relative mRNA abundance of genes upregulated during *sigD* overexpression. Relative abundance of mRNA of each gene was quantified by real-time PCR. The white and gray columns show the abundance in WT(pVWEx1-*sigD*) without IPTG (0 μM IPTG), and in WT(pVWEx1-*sigD*) with 50 μM of IPTG (50 μM IPTG), respectively. Error bars represent standard deviations calculated from biological triplicates. The *p*-value of mRNA abundance was calculated by Student's t-test (two-tail, unpaired) between 0 μM and 50 μM of IPTG, and is shown by *, ** and *** for <0.05, <0.01 and <0.001, respectively

SigD of *C. glutamicum* is classified as ECF40 type sigma factor by the ECFfinder program [3], as is SigD of *M. tuberculosis*. SigD in *M. tuberculosis* was shown to be essential for virulence, and inactivation of *sigD* decreased expression of some mycolyl transferase genes as well as other genes related to lipid metabolism and cell wall processes [51, 52]. For example, Calamita et al. reported that the expression of *fbpA* encoding antigen 85A, which is a homolog for corynomycolyl transferase, decreased in the *sigD* deletion strain [52]. Raman et al. showed that the expression of *fbpC* encoding antigen 85C, which is also a homolog for corynomycolyl transferase, decreased two-fold in the *sigD* deletion strain [51]. Even though a different organism has a different regulatory architecture, *C. glutamicum* and *M. tuberculosis* may share the similar regulatory network by SigD.

SigD gene overexpression in *C. glutamicum* led to excretion of carbohydrate-containing compounds and cell aggregation. In *C. glutamicum* CGL2005, various types of polysaccharides were detected extracellularly [34, 53]. In addition, *C. glutamicum* CCTCC M201005 was found in soil as a producer of a bioflocculant consisting of galacturonic acid as the main structural unit [54]. In *M. tuberculosis* and *M. smegmatis*, polysaccharides containing arabinose are suggested to be involved in the aggregation of cells via interaction with Antigen 85s, which are homologs for corynomycolyl transferases [55]. Furthermore, arabinose was shown to promote the cell aggregation of *M. smegmatis* [56]. Therefore, excretion of polysaccharides or carbohydrate-containing compounds caused by *sigD* overexpression may induce cell aggregation also in *C. glutamicum*.

of corynomycolic acids and phospholipids in cell envelope [50]. In this study, increased content of TDCM in the crude lipid extract indicated that *sigD* overexpression influences not only the transcription of those genes but also the metabolic flux toward mycomembrane synthesis (Fig. 6). These results indicate that SigD controls the integrity of the cell envelope, especially of the mycomembrane in *C. glutamicum*.

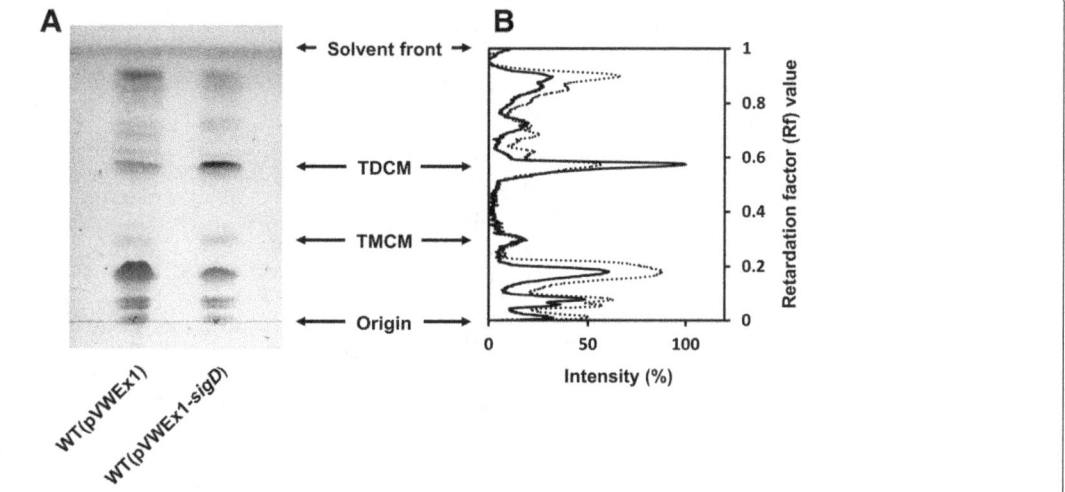

Fig. 5 TLC analysis of lipid crude extracts. **a** Lipid crude extracts were analyzed by thin layer chromatography. **b** The intensity of bands was quantified for WT(pVWEx1) and WT(pVWEx1-*sigD*). The highest intensity of the band was normalized to 100%. The dotted and black lines indicate the intensity profile for WT(pVWEx1) and WT(pVWEx1-*sigD*), respectively. The mobility of TDCM (trehalose dicorynomycolate) and TMCM (trehalose monocorynomycolate) were determined by their Rf values taken from the previous studies [32, 33]. CHCl$_3$/CH$_3$OH/H$_2$O (30:8:1 *v/v*) was used as development solvent

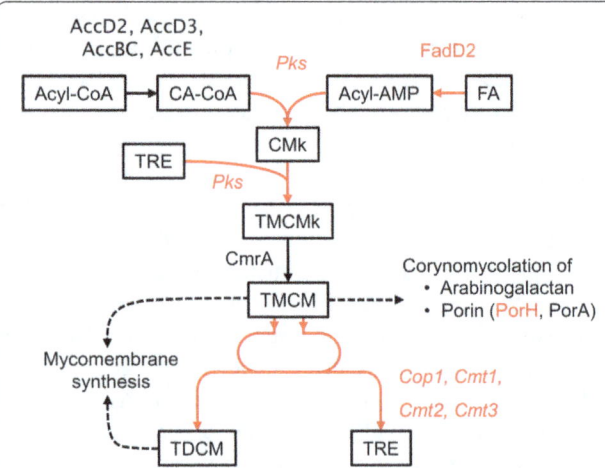

Fig. 6 Summary of effects induced by *sigD* overexpression in *C. glutamicum*. Biosynthesis pathway of trehalose dicorynomycolate (*TDCM*) and trehalose monocorynomycolate (*TMCM*) starting from acyl-CoA (*A-CoA*) and fatty acid (*FA*) is shown. The names of enzymes are shown in red letters and the catalyzing reactions are shown in red lines, only when the expression of corresponding genes were confirmed to be upregulated under *sigD* overexpression by transcriptome analysis. *CA-CoA*: carboxylated acyl-CoA, *TRE*: trehalose, *CMk*: Keto corynomycolic acid, *TMCMk*: TMCM keto form

Sigma factors regulate transcription in a global manner and their effects are pleiotropic. In this study, the link between *sigD* overexpression and the regulation of cell wall integrity was confirmed, however, other effects remain to be elucidated, if exist. Furthermore, overexpression of *sigD* turns on the cascade regulation which includes direct and indirect outcomes inside the cells. Based on our results, the physiological role of *C. glutamicum* SigD in cell wall integrity seems apparent, however, the elucidation of the SigD regulon control and definition of the class of SigD-dependent promoters will require further molecular studies.

The non-pathogenic *C. glutamicum* serves as a good model organism for understanding the cell wall biosynthesis and resistance to antibiotics in Corynebacterineae, which include human pathogenic bacteria such as *M. tuberculosis* and *C. diphtheriae* [42]. Therefore, the findings described in this work can be helpful to understand the cell wall biosynthesis of Corynebacterineae. Furthermore, TDCM was shown to induce priming and activation of macrophages in vivo and in vitro in a similar manner as TDM from *M. tuberculosis* [57]. Chemical synthesis of TDM is not easy because it requires multiple steps, and those compounds are extracted from organisms. Considering the common biotechnological use of *C. glutamicum*, the production of TDCM with *C. glutamicum* seems to be an attractive idea. On the other hand, a corynomycolate-less strain is known to excrete more L-glutamate and L-lysine [58] as well as to take up glycerol and acetate more efficiently [33]. Reorganization of the

mycomembrane by controlling *sigD* expression can be therefore helpful to understand the permeability barrier of *C. glutamicum* cells, and construct strains with higher or lower permeability barriers.

Conclusion

In this work, the functions of *C. glutamicum* sigma factor SigD were studied by overexpression or disruption of *sigD* gene. Overexpression of *sigD* led to the several physiological changes such as slower growth, cell aggregation, less foaming of the culture and increased turbidity of the supernatant. The real-time PCR analysis confirmed that overexpression of *sigD* induced the expression of several genes related to maintenance of cell envelop integrity and mycomembrane biosynthesis. Furthermore, overexpression of *sigD* increased the content of trehalose dicorynomycolate in the lipid extract.

Additional files

Additional file 1: Table S1. Bacterial strains, plasmids and oligonucleotides used in this work. Bacterial strains, plasmids and oligonucleotides used in this work are listed.

Additional file 2: Table S2. RNA-seq analysis of genes differentially transcribed upon sigD overexpression or sigD disruption. The name of genes which expression levels increased under *sigD* overexpression (M-value > 1.0) are listed.

Abbreviations
IPTG: Isopropyl β-D-1-thiogalactopyranoside; TDCM: Trehalose dicorynomycolate; TDM: Trehalose dimycolate; TMCM: Trehalose monocorynomycolate

Acknowledgements
Not applicable.

Funding
HT acknowledges support by DAAD (Deutscher Akademischer Austauschdienst) for providing a full PhD fellowship and MP acknowledges support by grant 17-06991S from Czech Science Foundation. Those foundations had no role in the design of the study in collection, analysis, and interpretation of data and in writing the manuscript.

Authors' contributions
HT, TB, TP, KN, MP, JK and VFW designed the study. HT, TB, TP performed the experiments and analyzed the data. HT wrote the manuscript. TB, TP, KN, MP and JK reviewed the manuscript. VFW finalized the manuscript. All authors read and approved the final manuscript.

Author's information
HT currently belongs to Synthetic Bioengineering lab, Dept.of Biotechnology, Graduate School of Engineering, Osaka University (Yamadaoka 2-1, Suita, Osaka, 565-0871, Japan).

Competing interests

The authors declare that they have no competing interests.

Author details

[1]Genetics of Prokaryotes, Faculty of Biology, Bielefeld University, Bielefeld, Germany. [2]Center for Biotechnology, Bielefeld University, Bielefeld, Germany. [3]Proteome and Metabolome Research, Faculty of Biology, Bielefeld University, Bielefeld, Germany. [4]Institute of Microbiology, Academy of Sciences of the Czech Republic, Prague, Czech Republic.

References

1. Feklístov A, Sharon BD, Darst SA, Gross CA. Bacterial sigma factors: a historical, structural, and genomic perspective. Annu Rev Microbiol. 2014;68:357–76.
2. Rodrigue S, Provvedi R, Jacques P-E, Gaudreau L, Manganelli R. The sigma factors of Mycobacterium tuberculosis. FEMS Microbiol Rev. 2006;30:926–41.
3. Staroń A, Sofia HJ, Dietrich S, Ulrich LE, Liesegang H, Mascher T. The third pillar of bacterial signal transduction: classification of the extracytoplasmic function (ECF) σ factor protein family. Mol Microbiol. 2009;74:557–81.
4. Nicolas P, Mäder U, Dervyn E, Rochat T, Leduc A, Pigeonneau N. Condition-dependent transcriptome reveals high-level regulatory architecture in Bacillus subtilis. Science. 2012;335:1103–6.
5. Cho B-K, Kim D, Knight EM, Zengler K, Palsson BO. Genome-scale reconstruction of the sigma factor network in Escherichia coli: topology and functional states. BMC Biol. 2014;12:4.
6. Österberg S, Peso-Santos T de., and Shingler V. Regulation of alternative sigma factor use. Annu Rev Microbiol 2011;65:37–55.
7. Kinoshita S, Udaka S, Shimono M. Studies on the amino acid fermentation. J Gen Appl Microbiol. 1957;3:193–205.
8. Eggeling L, Bott M. Handbook of Corynebacterium glutamicum. FL, USA: CRC Press; 2005.
9. Kalinowski J, Bathe B, Bartels D, Bischoff N, Bott M, Burkovski A. The complete Corynebacterium glutamicum ATCC 13032 genome sequence and its impact on the production of L-aspartate-derived amino acids and vitamins. J Biotechnol. 2003;104:5–25.
10. Ikeda M, Nakagawa S. The Corynebacterium glutamicum genome: features and impacts on biotechnological processes. Appl Microbiol Biotechnol. 2003;62:99–109.
11. Pátek M, Nešvera J. Sigma factors and promoters in Corynebacterium glutamicum. J Biotechnol. 2011;154:101–13.
12. Barksdale L. Corynebacterium diphtheriae and its relatives. Bacteriol Rev. 1970;34:378–422.
13. Lanéelle M-A, Tropis M, Daffé M. Current knowledge on mycolic acids in Corynebacterium glutamicum and their relevance for biotechnological processes. Appl Microbiol Biotechnol. 2013;97:9923–30.
14. Peters-Wendisch PG, Schiel B, Wendisch VF, Katsoulidis E, Möckel B, Sahm H. Pyruvate carboxylase is a major bottleneck for glutamate and lysine production by Corynebacterium glutamicum. J Mol Microbiol Biotechnol. 2001;3:295–300.
15. Schäfer A, Tauch A, Jäger W, Kalinowski J, Thierbach G, Pühler A. Small mobilizable multi-purpose cloning vectors derived from the Escherichia coli plasmids pK18 and pK19: selection of defined deletions in the chromosome of Corynebacterium glutamicum. Gene. 1994;145:69–73.
16. Sambrook J. Molecular cloning: A laboratory manual. third ed. Cold Spring Harbor: Cold Spring Harbor Laboratory Press; 2001.
17. Gibson DG, Young L, Chuang R-Y, Venter JC, Hutchison CA, Smith HO. Enzymatic assembly of DNA molecules up to several hundred kilobases. Nat Methods. 2009;6:343–5.
18. Nováková J, Izsáková A, Grivalský T, Ottmann C, Farkašovský M. Improved method for high-efficiency electrotransformation of Escherichia coli with the large BAC plasmids. Folia Microbiol (Praha). 2014;59:53–61.
19. van der Rest ME, Lange C, Molenaar D. A heat shock following electroporation induces highly efficient transformation of Corynebacterium glutamicum with xenogeneic plasmid DNA. Appl Microbiol Biotechnol. 1999;52:541–5.
20. Shevchenko A, Tomas H, Havlis J, Olsen JV, Mann M. In-gel digestion for mass spectrometric characterization of proteins and proteomes. Nat Protoc. 2006;1:2856–60.
21. Musa YR, Bäsell K, Schatschneider S, Vorhölter F-J, Becher D, Niehaus K. Dynamic protein phosphorylation during the growth of Xanthomonas campestris pv. campestris B100 revealed by a gel-based proteomics approach. J Biotechnol. 2013;167:111–22.
22. Dallies N, François J, Paquet V. A new method for quantitative determination of polysaccharides in the yeast cell wall. Application to the cell wall defective mutants of Saccharomyces cerevisiae. Yeast. 1998;14:1297–306.
23. DuBois M, Gilles KA, Hamilton JK, Rebers PA, Smith F. Colorimetric method for determination of sugars and related substances. Anal Chem. 1956;28:350–6.
24. Masuko T, Minami A, Iwasaki N, Majima T, Nishimura S-I, Lee YC. Carbohydrate analysis by a phenol–sulfuric acid method in microplate format. Anal Biochem. 2005;339:69–72.
25. Hüser AT, Becker A, Brune I, Dondrup M, Kalinowski J, Plassmeier J. Development of a Corynebacterium glutamicum DNA microarray and validation by genome-wide expression profiling during growth with propionate as carbon source. J Biotechnol. 2003;106:269–86.
26. Busche T, Winkler A, Wedderhoff I, Rückert C, Kalinowski J, Ortiz de Orué Lucana D. Deciphering the Transcriptional Response Mediated by the Redox-Sensing System HbpS-SenS-SenR from Streptomycetes. PLoS One. 2016;11:e0159873.
27. Blom J, Jakobi T, Doppmeier D, Jaenicke S, Kalinowski J, Stoye J, et al. Exact and complete short-read alignment to microbial genomes using Graphics Processing Unit programming. Bioinforma Oxf Engl. 2011;27:1351–8.
28. Pfeifer-Sancar K, Mentz A, Rückert C, Kalinowski J. Comprehensive analysis of the Corynebacterium glutamicum transcriptome using an improved RNAseq technique. BMC Genomics. 2013;14:888.
29. Hilker R, Stadermann KB, Schwengers O, Anisiforov E, Jaenicke S, Weisshaar B, et al. ReadXplorer 2-detailed read mapping analysis and visualization from one single source. Bioinforma Oxf Engl. 2016;32:3702–8.
30. Anders S, Huber W. Differential expression analysis for sequence count data. Genome Biol. 2010;11:106.
31. Busche T, Silar R, Pičmanová M, Pátek M, Kalinowski J. Transcriptional regulation of the operon encoding stress-responsive ECF sigma factor SigH and its anti-sigma factor RshA, and control of its regulatory network in Corynebacterium glutamicum. BMC Genomics. 2012;13:445.
32. Brand S, Niehaus K, Pühler A, Kalinowski J. Identification and functional analysis of six mycolyltransferase genes of Corynebacterium glutamicum ATCC 13032: the genes cop1, cmt1, and cmt2 can replace each other in the synthesis of trehalose dicorynomycolate, a component of the mycolic acid layer of the cell envelope. Arch Microbiol. 2003;180:33–44.
33. Puech V, Bayan N, Salim K, Leblon G, Daffé M. Characterization of the in vivo acceptors of the mycoloyl residues transferred by the corynebacterial PS1 and the related mycobacterial antigens 85. Mol Microbiol. 2000;35:1026–41.
34. Puech V, Chami M, Lemassu A, Lanéelle M-A, Schiffler B, Gounon P. Structure of the cell envelope of corynebacteria: importance of the non-covalently bound lipids in the formation of the cell wall permeability barrier and fracture plane. Microbiology. 2001;147:1365–82.
35. Nielsen SS. Food Analysis Laboratory Manual. Springer US: NY, USA; 2010.
36. Ikeda M, Baba M, Tsukumoto N, Komatsu T, Mitsuhashi S, Takeno S. Elucidation of genes relevant to the microaerobic growth of Corynebacterium glutamicum. Biosci Biotechnol Biochem. 2009;73:2806–8.
37. Taniguchi H, Wendisch VF. Exploring the role of sigma factor gene expression on production by Corynebacterium glutamicum: sigma factor H and FMN as example. Front Microbiol. 2015;6:740.
38. Taniguchi H, Henke NA, Heider SAE, Wendisch VF. Overexpression of the primary sigma factor gene sigA improved carotenoid production by Corynebacterium glutamicum: Application to production of β-carotene and the non-native linear C50 carotenoid bisanhydrobacterioruberin. Metab Eng Commun. 2017;4:1–11.
39. Gande R, Dover LG, Krumbach K, Besra GS, Sahm H, Oikawa T. The two carboxylases of Corynebacterium glutamicum essential for fatty acid and mycolic acid synthesis. J Bacteriol. 2007;189:5257–64.
40. Portevin D, de Sousa-D'Auria C, Montrozier H, Houssin C, Stella A, Lanéelle M-A. The acyl-AMP ligase FadD32 and AccD4-containing acyl-CoA carboxylase are required for the synthesis of mycolic acids and essential for mycobacterial growth: identification of the carboxylation product and determination of the acyl-CoA carboxylase components. J Biol Chem. 2005;280:8862–74.

41. Gavalda S, Bardou F, Laval F, Bon C, Malaga W, Chalut C. The polyketide synthase Pks13 catalyzes a novel mechanism of lipid transfer in mycobacteria. Chem Biol. 2014;21:1660–9.
42. Portevin D, De Sousa-D'Auria C, Houssin C, Grimaldi C, Chami M, Daffé M, et al. A polyketide synthase catalyzes the last condensation step of mycolic acid biosynthesis in mycobacteria and related organisms. Proc Natl Acad Sci U S A. 2004;101:314–9.
43. Lea-Smith DJ, Pyke JS, Tull D, McConville MJ, Coppel RL, Crellin PK. The reductase that catalyzes mycolic motif synthesis is required for efficient attachment of mycolic acids to arabinogalactan. J Biol Chem. 2007;282:11000–8.
44. Varela C, Rittmann D, Singh A, Krumbach K, Bhatt K, Eggeling L. MmpL genes are associated with mycolic acid metabolism in mycobacteria and corynebacteria. Chem Biol. 2012;19:498–506.
45. Yamaryo-Botte Y, Rainczuk AK, Lea-Smith DJ, Brammananth R, van der Peet PL, Meikle P. Acetylation of trehalose mycolates is required for efficient mmpl-mediated membrane transport in *Corynebacterineae*. ACS Chem Biol. 2015;10:734–46.
46. De Sousa-D'Auria C, Kacem R, Puech V, Tropis M, Leblon G, Houssin C. New insights into the biogenesis of the cell envelope of corynebacteria: identification and functional characterization of five new mycoloyltransferase genes in *Corynebacterium glutamicum*. FEMS Microbiol Lett. 2003;224:35–44.
47. Huc E, de Sousa-D'Auria C, de la Sierra-Gallay IL, Salmeron C, van Tilbeurgh H, Bayan N. Identification of a mycoloyl transferase selectively involved in O-acylation of polypeptides in *Corynebacteriales*. J Bacteriol. 2013;195:4121–8.
48. Burkovski A. Cell envelope of corynebacteria: structure and influence on pathogenicity. ISRN Microbiol. 2013;2013:935736.
49. Barth E, Barceló MA, Kläckta C, Benz R. Reconstitution experiments and gene deletions reveal the existence of two-component major cell wall channels in the genus *Corynebacterium*. J Bacteriol. 2010;192:786–800.
50. Meniche X, Labarre C, de Sousa-d'Auria C, Huc E, Laval F, Tropis M, et al. Identification of a stress-induced factor of *Corynebacterineae* that is involved in the regulation of the outer membrane lipid composition. J Bacteriol. 2009;191:7323–32.
51. Raman S, Hazra R, Dascher CC, Husson RN. Transcription regulation by the *Mycobacterium tuberculosis* alternative sigma factor SigD and its role in virulence. J Bacteriol. 2004;186:6605–16.
52. Calamita H, Ko C, Tyagi S, Yoshimatsu T, Morrison NE, Bishai WR. The *Mycobacterium tuberculosis* SigD sigma factor controls the expression of ribosome-associated gene products in stationary phase and is required for full virulence. Cell Microbiol. 2004;7:233–44.
53. Kacem R, De Sousa-D'Auria C, Tropis M, Chami M, Gounon P, Leblon G. Importance of mycoloyltransferases on the physiology of *Corynebacterium glutamicum*. Microbiol Read Engl. 2004;150:73–84.
54. He N, Li Y, Chen J, Lun S-Y. Identification of a novel bioflocculant from a newly isolated *Corynebacterium glutamicum*. Biochem Eng J. 2002;11:137–48.
55. Anton V, Rougé P, Daffé M. Identification of the sugars involved in mycobacterial cell aggregation. FEMS Microbiol Lett. 1996;144:167–70.
56. Jayawardana KW, Wijesundera SA, Yan M. Aggregation-based detection of *M. smegmatis* using D-arabinose-functionalized fluorescent silica nanoparticles. Chem Commun. 2015;51:15964–6.
57. Chami M, Andréau K, Lemassu A, Petit J-F, Houssin C, Puech V. Priming and activation of mouse macrophages by trehalose 6,6-dicorynomycolate vesicles from *Corynebacterium glutamicum*. FEMS Immunol Med Microbiol. 2002;32:141–7.
58. Gebhardt H, Meniche X, Tropis M, Krämer R, Daffé M, Morbach S. The key role of the mycolic acid content in the functionality of the cell wall permeability barrier in *Corynebacterineae*. Microbiology. 2007;153:1424–34.

6

A shift in the virulence potential of *Corynebacterium pseudotuberculosis* biovar *ovis* after passage in a murine host demonstrated through comparative proteomics

Wanderson M. Silva[1,4,5], Fernanda A. Dorella[1], Siomar C. Soares[1], Gustavo H. M. F. Souza[3], Thiago L. P. Castro[1], Núbia Seyffert[1], Henrique Figueiredo[6], Anderson Miyoshi[1], Yves Le Loir[4,5], Artur Silva[2] and Vasco Azevedo[1*]

Abstract

Background: *Corynebacterium pseudotuberculosis* biovar *ovis*, a facultative intracellular pathogen, is the etiologic agent of caseous lymphadenitis in small ruminants. During the infection process, *C. pseudotuberculosis* changes its gene expression to resist different types of stresses and to evade the immune system of the host. However, factors contributing to the infectious process of this pathogen are still poorly documented. To better understand the *C. pseudotuberculosis* infection process and to identify potential factors which could be involved in its virulence, experimental infection was carried out in a murine model using the strain 1002_*ovis* and followed by a comparative proteomic analysis of the strain before and after passage.

Results: The experimental infection assays revealed that strain 1002_*ovis* exhibits low virulence potential. However, the strain recovered from the spleen of infected mice and used in a new infection challenge showed a dramatic change in its virulence potential. Label-free proteomic analysis of the culture supernatants of strain 1002_*ovis* before and after passage in mice revealed that 118 proteins were differentially expressed. The proteome exclusive to the recovered strain contained important virulence factors such as CP40 proteinase and phospholipase D exotoxin, the major virulence factor of *C. pseudotuberculosis*. Also, the proteome from recovered condition revealed different classes of proteins involved in detoxification processes, pathogenesis and export pathways, indicating the presence of distinct mechanisms that could contribute in the infectious process of this pathogen.

Conclusions: This study shows that *C. pseudotuberculosis* modifies its proteomic profile in the laboratory versus infection conditions and adapts to the host context during the infection process. The screening proteomic performed us enable identify known virulence factors, as well as potential proteins that could be related to virulence this pathogen. These results enhance our understanding of the factors that might influence in the virulence of *C. pseudotuberculosis*.

Keywords: *Corynebacterium pseudotuberculosis*, Bacterial label-free proteomic, Caseous lymphadenitis, Bacterial virulence, Serial passage, Extracellular proteins

* Correspondence: vasco@icb.ufmg.br
[1]Departamento de Biologia Geral, Instituto de Ciências Biológicas, Universidade Federal de Minas Gerais, Belo Horizonte, Minas Gerais, Brazil
Full list of author information is available at the end of the article

Background

Corynebacterium pseudotuberculosis biovar *ovis* is a Gram-positive facultative intracellular pathogen. It is the etiologic agent of Caseous Lymphadenitis (CLA) in small ruminants, a disease characterized by abscess formation in lymph nodes and internal organs [1]. Cases of human infection caused by *C. pseudotuberculosis* have been reported and are associated with occupational exposure [1]. CLA is globally distributed and causes significant economic losses in goats, and sheep herds [2]. The pathogenic process of *C. pseudotuberculosis* in the host comprises two phases: (i) initial colonization and replication in lymph nodes that drain the site of infection, which is associated with pyogranuloma formation, and (ii) a secondary cycle of replication and dissemination via the lymphatic or circulatory systems. This dissemination is promoted by the action of phospholipase D (PLD) exotoxin, the major virulence factor of *C. pseudotuberculosis*, which allows this pathogen to contaminate visceral organs and lymph nodes, where it ultimately induces lesion formation [3–5].

Exported proteins reportedly favor the infection process in pathogenic bacteria; this class of proteins is involved in adhesion and invasion of host cells, nutrient acquisition, toxicity, and in the evasion of the host immune system [6]. Different strategies like the transposon mutagenesis have been adopted to identify *C. pseudotuberculosis* biovar *ovis* exported proteins [7]. Additionally, comparative proteomics has been applied to characterize the extracellular proteome of *C. pseudotuberculosis* biovar *ovis*, as well as, the extracellular immunoproteome (strains C231_*ovis* and 1002_*ovis*) [8–11]. In these studies, some proteins of the strain 1002_*ovis*, suspected to be virulence factors, were not detected suggesting this strain presents a low virulence. The surface proteome of *C. pseudotuberculosis* biovar *ovis* was also characterized using bacterial strains isolated from the lymph nodes of naturally infected sheep. This proteomic analysis allowed the identification of proteins that could favor the survival of this pathogen during the chronic phase of CLA [12].

The experimental passage of bacterial pathogens through in vitro or in an in vivo model is a strategy that has been applied to evaluate the virulence potential of several pathogens. By generating a confrontation between the pathogen and the dynamic network of host factors, including the immune system components, it helps to identify bacterial factors involved in virulence [12–19]. In this study, the strain 1002_*ovis* was experimentally inoculated in mice [20, 21] to identify factors which could contribute to virulence in *C. pseudotuberculosis* biovar *ovis*. Comparative proteomics of the culture supernatant from this strain collected before and after the experimental passage in mice was carried out to identify factors that might contribute to virulence of 1002_*ovis*.

Methods

Bacterial strains and growth conditions

The *C. pseudotuberculosis* biovar *ovis* strain 1002 (1002_*ovis*) was isolated from a goat in Brazil; this strain was cultivated under standard conditions in brain–heart infusion broth (BHI-HiMedia Laboratories Pvt. Ltd., India) at 37 °C. When necessary, 1.5% of agar was added to the medium for a solid culture. For extracellular proteomic analyses, 1002_*ovis* was grown in a chemically defined medium (CDM) [(Na$_2$HPO$_4$.7H$_2$O (12.93 g/L), KH$_2$PO4 (2.55 g/L), NH$_4$Cl (1 g/L), MgSO$_4$.7H$_2$O (0.20 g/L), CaCl$_2$ (0.02 g/L) and 0.05% (v/v) Tween 80], 4% (v/v) MEM Vitamins Solution (Invitrogen, Gaithersburg, MD, USA), 1% (v/v) MEM Amino Acids Solution (Invitrogen), 1% (v/v) MEM Non-Essential Amino Acids Solution (Invitrogen), and 1.2% (w/v) glucose at 37 °C [22].

Experimental infection of strain 1002_*ovis* in a murine model (in vivo assay)

The standardization of the parameters for infection was performed according to Moraes et al. [20] and Ribeiro et al. [21]. Female BALB/c mice between six and eight weeks old were used in all experiments. They were provided by the Animal Care Facility of the Biological Sciences Institute from the Federal University of Minas Gerais and were handled by the guidelines of the UFMG Ethics Committee on Animal Testing (Permit Number: CETEA 103/2011). For the bacterial passage assay using the murine model, two groups of three mice each was infected via intraperitoneal injection with 10^6 colony forming units (CFU) of strain 1002_*ovis*. Thirty-six hours after infection, all animals were sacrificed. Their spleens were aseptically removed to recover the bacterial strain, as described below: the spleen removed from each animal was then, individually macerated in sterile saline solution (0.9% NaCl$_2$), seeded onto BHI agar plates and incubated for 48 h at 37 °C. Subsequently, one recovered bacterial colony was cultured in BHI broth. The recovered bacteria were then referred to as Recovered (Rc). For the bacterial virulence assay, we used the freshly recovered bacteria and bacteria that did not contact the murine host as a control, which is referred to as Control (Ct). Groups of five mice were infected with Rc and Ct, via intraperitoneal injection of a suspension containing 10^6 CFU or 10^5 CFU. The animals' survival rates were calculated and represented in GraphPad Prism v.5.0 (GraphPad Software, San Diego, CA, USA) using the Kaplan-Meier survival function. The results of 1002_*ovis* CFU count in the organs were calculated using the two-way ANOVA test.

Preparation of proteins from culture filtrates for proteome analysis

For proteomic analysis, the Ct and Rc (three independently recovered colonies) that was obtained from infected mice spleens as described above were grown in CDM at $OD_{600} = 0.8$. The cultures were then centrifuged for 20 min at $2700 \times g$. The supernatants were then filtered using 0.22-µm filters, 30% (w/v) ammonium sulfate was added to the samples, and the pH of the mixtures was adjusted to 4.0. Next, 20 mL N-butanol was added to each sample. The samples were centrifuged for 10 min at 1350 xg and 4 °C. The interfacial precipitate was collected and resuspended in 1 mL of 20 mM Tris–HCl pH 7.2 [23]. Finally the concentration protein was determined by Bradford method [24].

2D-PAGE electrophoresis and Mass Spectrometry

The 2-DE procedure and in-gel protein digestion were performed as described previously [9, 10]. Approximately 300 µg of the protein extract from of each condition was dissolved in rehydration buffer (Urea 7 M, thiourea 2 M, CHAPS 2%, Tris–HCl 40 mM, bromophenol blue 0.002%, DTT 75 mM, IPG Buffer 1%). Samples were applied to 18 cm pH 3–10 N.L strips (GE Healthcare, Pittsburgh, USA). Isoelectric focusing (IEF) was performed using the apparatus IPGphor 2 (GE Healthcare) under the following voltages: 100 V 1 h, 500 V 2 h, 1000 V 2 h, 10,000 V 3 h, 10,000 V 6 h, 500 V 4 h. The IPG strips were placed on 12% acrylamide/bis acrylamide gels in an Ettan DaltSix II system (GE Healthcare). The gels were stained with Coomassie Blue G-250 staining solution, and 2-DE gels were scanned using an Image Scanner (GE Healthcare). The Image Master 2D Platinum 7 (GE Healthcare) software was used to analyze the generated images and all spots were matched and analyzed by gel-to-gel comparison. The quantification of the spots was calculated according percentage volume (% Vol) and spots with reproducible changes in abundance were considered to be differentially expressed. Protein spots were excised from the gels, and in-gel digestion was carried out using trypsin enzyme (Promega, Sequencing Grade Modified Trypsin, Madison, WI, USA). The peptides were then desalted and concentrated using ZIP TIP C18 tips (Eppendorf).

The samples were subsequently analyzed for MS and MS/MS modes, using an MALDI-TOF/TOF mass spectrometer Autoflex IIITM (Bruker Daltonics, Billerica USA). The equipment was controlled in a positive/reflector way using the Flex-ControlTM software (Brucker Daltonics). External calibration was performed using peptide standards samples (angiotensin II, angiotensin I, substance P, bombesin, ACTH clip 1–17, ACTH clip 18–39, somatostatin 28, bradykinin Fragment 1–7, Renin Substrate tetra decapeptide porcine) (Bruker Daltonics). The peptides were added to the alpha-cyano-4-hydroxycinnamic acid matrix, applied on an Anchor-ChipTM 600 plate (Brucker Daltonics) and analyzed by Autoflex III. The search parameters were as follows: enzyme; trypsin; fixed modification, carbamidomethylation (Cys); variable modifications, oxidation (Met);

mass values, monoisotopic; maximum missed cleavages, 1; and peptide mass tolerance of 0.005% Da (50 ppm). The results obtained by MS/MS were used to identify proteins utilizing the MASCOT_ (http://www.matrixscience.com) program and compared with the genomic data of the Actinobacteria class deposited in the NCBI nr database.

2D nanoUPLC-HDMSE data acquisition and Data Processing

The protein extracts from three biological replicates of each condition were concentrated using spin columns with a 10 kDa threshold (Millipore, Billerica, MA, USA) to perform the label-free proteomic analysis. The protein was denatured (0.1% RapiGEST SF at 60 °C for 15 min) (Waters, Milford, CA, USA), reduced (10 mM DTT), alkylated (10 mM iodoacetamide) and enzymatically digested with trypsin (Promega). The digestion process was stopped by adding 10 µL of 5% TFA (Fluka, Buchs, Germany), and glycogen phosphorylase (Sigma, Aldrich, P00489) was added to the digested samples after digest at 20 fmol.uL^{-1} as an internal standard for normalization. Each replicate was injected using a two-dimensional reversed phase (2D RPxRP) nanoUPLC-MS (Nano Ultra Performance Liquid Chromatography Mass Spectrometry) approach with 171 multiplexed high definition mass spectrometry (HDMSE) label-free quantitation [25]. Qualitative and quantitative experiments were performed using both a 1 h reversed phase gradient from 7% to 40% (v/v) acetonitrile (0.1% v/v formic acid) at 500 nL.min^{-1} and a nanoACQUITY UPLC 2D RPxRP Technology system [26]. A nanoACQUITY 174 UPLC HSS T3 1.8 µm, 75 µm × 15 cm column (pH 3) was used with an RP XBridge BEH130 C18 5 µm 300 µm x 50 mm nanoflow column (pH 10). Typical on-column sample loads were 250 ng of the total protein digests for each of the 5 fractions (250 ng/fraction/load). All analyses were performed using nano electrospray ionization in the positive ion mode nanoESI (+) and a NanoLockSpray (Waters, Manchester, UK) ionization source. The mass spectrometer was calibrated using an MS/MS spectrum of [Glu1]-Fibrinopeptide B human (Glu-Fib) solution (100 fmol.uL-1) delivered through the NanoLockSpray source reference sprayer. Multiplexed data-independent (DIA) scanning with additional specificity and selectivity for non-linear 'T-wave' ion mobility (HDMSE) experiments were performed using a Synapt G2-S HDMS mass spectrometer (Waters, Manchester, UK).

Following the identification of proteins, the quantitative data were packaged using dedicated algorithms [27] and searching against a database with default parameters to account for ions [28]. The databases used were reversed on-the-fly during the database queries and appended to the original database to assess the false positive rate during identification. For proper spectra processing and database searching conditions, the ProteinLynxGlobalServer v.2.5.2 (PLGS) with IdentityE and ExpressionE informatics v.2.5.2 (Waters, Manchester, UK) was used. UniProtKB (release 2013_01) with manually reviewed annotations was used, and the search conditions were based on taxonomy (*Corynebacterium pseudotuberculosis*). One missed cleavage by trypsin was allowed be up to 1 and various modifications as carbamidomethyl (C), Acetyl N terminal, phosphoryl (STY) and oxidation (M) were allowed [29]. The proteins collected were organized by the PLGS ExpressionE tool algorithm into a statistically significant list that corresponded to higher or lower regulation ratios between the different groups. For protein quantitation, we used the PLGS v2.5.2 software with the IdentifyE algorithm using the Hi3 methodology. The search threshold to accept each spectrum was the default value for a false discovery rate 4%. The quantitation values were averaged over all samples, and the standard deviations of $p < 0.05$, which were determined using the ExpressionE software, refer to the differences between biological replicates.

Bioinformatic analysis

The proteins identified in 1002_*ovis* under both conditions were analyzed using the following prediction tools: SecretomeP 2.0 server, to predict proteins exported from non-classical systems (positive prediction score greater than to 0.5) [30] and PIPs software, to predict proteins in the pathogenicity islands [31]. Gene ontology (GO) functional annotations were generated using the Blast2GO tool [32].

Results

The main objective of this study was to assay the virulence of 1002_*ovis* in a murine model after passage through mice. We thus carried out an in vivo survival assay using BALB/c mice infected with bacteria that did not contact with murine model (Ct) and bacteria recovered (Rc) from mice spleens. In this assay using an infection inoculum of 10^6 CFU, all the animals infected with Rc died within 48 h after infection (Fig. 1a). On the other hand, the control group, infected with Ct, survived the evaluation period (6 days). Similarly, in an assay with a lower infective dose (10^5 CFU), a 100% mortality was observed four weeks post infection with the recovered bacteria (Fig. 1b). Comparison of the Ct and Rc numbers isolated from the spleen within five days of infection (Fig. 1c) showed that the serial passage process affected the potential for spleen colonization during the infection. After four weeks of infection in the assay with 10^5 CFU, bacteria were isolated from the spleen, liver, left and right kidney, only in mice infected with Rc (Fig. 1d). Finally, regarding the clinical signs, in the assay using 10^5 CFU, caseous lesions were detected in different

organs (liver, left kidney and right kidney) of all the animals infected only with Rc (data not shown). Altogether, these results showed that the serial passage process in a murine model increased the virulence potential of strain 1002_*ovis*. In addition, these results confirmed the low virulence of this strain, which was previously suggested based on the composition of its extracellular proteome [8–10].

After passage in BALB/c mice, a dramatic change in the virulence potential of strain 1002_*ovis* was observed. We thus hypothesized that this phenotypic change was visible at the proteome level since *C. pseudotuberculosis* virulence relies on the production of a proteinaceous virulence factor. Thus, considering the importance of extracellular proteins for bacterial virulence, the proteomic analysis was conducted on the extracellular proteomes of 1002_*ovis* recovered from infected mice spleens in comparison to the control condition, using two proteomics approaches: 2-DE and 2D nanoUPLC-HDMS^E. The electrophoretic resolution of the extracellular protein extract of Ct and Rc condition allowed the visualization of spots distributed over pH 3–10 (Fig. 2). A total of 14 spots were found to be differentially expressed between Ct and Rc condition, these spots were excised out of the gel, and identified by MS/MS (Table 1). In the LC/MS analysis, we used the label-free quantitative proteomic to evaluate the relative difference between the proteome of Rc and Ct condition. In this analysis, only proteins which presented $p < 0.05$ and differential expression (log2 ratios) equal or greater than a factor of 1.2 were considered, as described previously [33]. We detected a total of 118 expressed differentially proteins, between Ct and Rc condition (Fig. 3) (Table 2 and Additional file 1). Also, 48 proteins were assigned only to Ct (Additional file 2) and 32 proteins were exclusive to Rc (Table 3) The information about sequence coverage and a number of identified peptides for each protein sequence identified, as well as the information about the native peptide are available at Additional file 3: Table S3.

The proteins identified in both conditions were analyzed by SecretomeP [29] to assess whether these proteins could be exported by non-classical secretion systems. Among the expressed differentially proteins 31% (37 proteins) were predicted as secreted through non-classical secretion systems. In turn, when analyzed the exclusive proteome of each condition 19% (6 proteins) and 27% (13 proteins) were considered to be exported by non-classical secretion systems for recovered and control condition, respectively. The PIPS tool was used to evaluate whether the genes that encode the proteins which were differentially expressed and identified in the exclusive proteome of the Rc condition are included in predicted pathogenicity islands. According these analysis 16 proteins was encoded by genes located on a predicted pathogenicity island; these

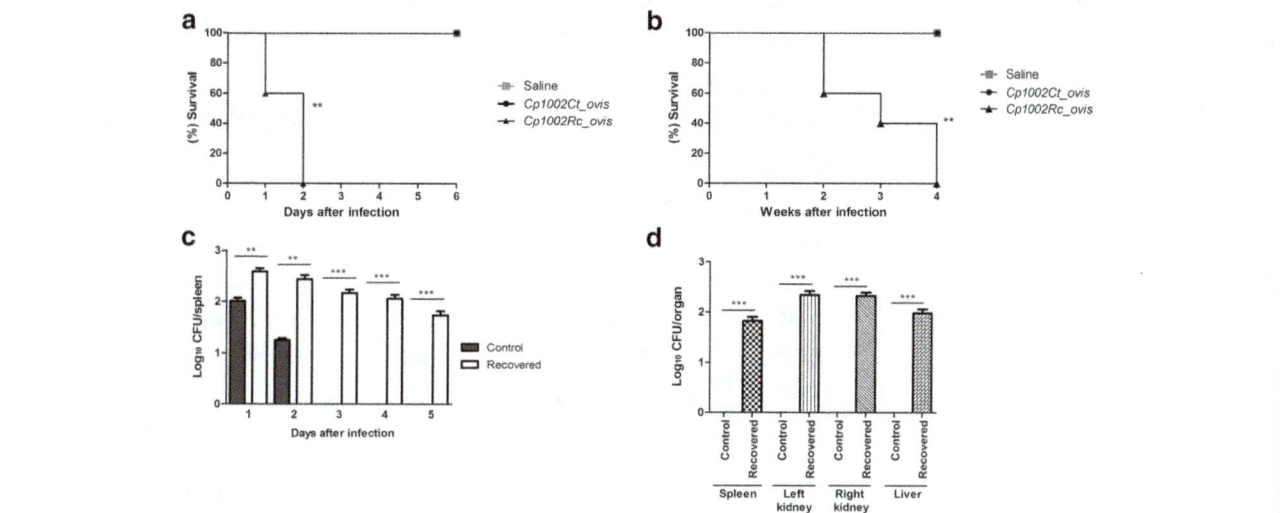

Fig. 1 Survival of Balb/C mice infected with strain 1002_*ovis*. **a** The survival rate was measured to determine the virulence profile of strain 1002_*ovis* control and recovered in mice infected with 10^6 CFU of bacteria Ct = control condition, Rc recovered condition. **b** Survival rates of mice infected with 10^5 CFU of Ct and Rc. **c** CFU in the spleen of BALB/c mice infected with control and recovered condition for the first five days of infection. **d** CFU in the different organs (spleen, left kidney, right kidney and liver) of BALB/c mice infected with control and recovered condition after four weeks of infection. The mortality rates were measured daily. Results represent three independent experiments. *P* values of <0.05 were considered to be statistically significant, and asterisks indicate statistically significant differences

proteins are related to cellular metabolism, pathogenesis, transport pathway, stress response and unknown function (Additional file 4). To classify the proteins identified in functional groups, we used the Blast2Go tool [31]; according to this analysis, the proteins were grouped into 17 biological processes (Fig. 4). Among these proteins, we identified processes that are directly involved in bacterial virulence, such as protein transport, pathogenesis, cell adhesion and stress response (Table 2).

Important factors directly linked to *C. pseudotuberculosis* virulence, like the PLD phospholipase, as well as, the CP40 protease were detected only in the proteome of recovered 1002_*ovis* (Tables 1 and 3). Also, components of several secretion systems were also activated in the bacteria recovered. These include proteins related to

hemin uptake, ATP-binding cassette (ABC) transporters and the Opp transporter, like OppA, OppC, and OppD. Proteins related to detoxification process were also specifically identified in the Rc supernatant: e.g. the glutaredoxin-like protein NrdH, which belongs to the NrdH-redoxins, a family of small protein disulfide oxidoreductases [34], mycothiol glutathione reductase present in Actinobacteria [35] and copper resistance protein CopC (Tables 2 and 3). In addition, we have identified 31 proteins in the recovered condition that also were detected in a strain of *C. pseudotuberculosis* isolated directly from ovine lymph nodes [12] (Tables 2 and 3). Proteins involved in the resistance to antimicrobial agents, such as penicillin-binding proteins, metallo-beta-lactamase, and penicillin-binding protein transpeptidase and proteases like Clp protease involved in the expression of cytotoxins in

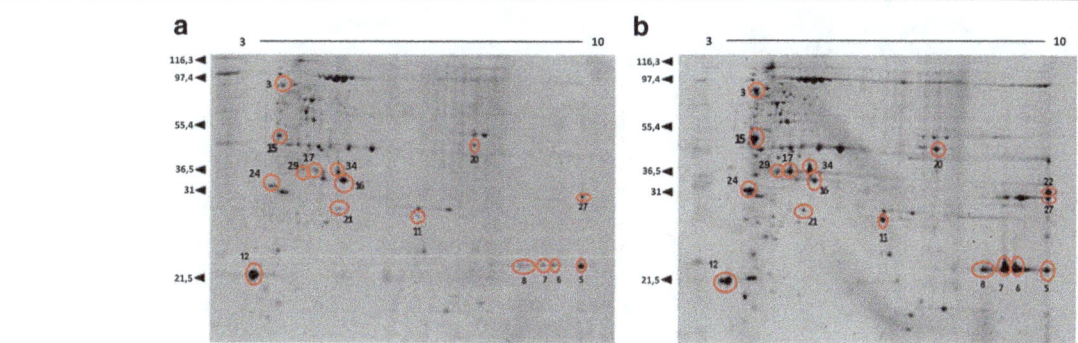

Fig. 2 Two-dimensional electrophoresis of the extracellular proteins 1002_*ovis* after following passage process: **a** Control condition. **b** Recovered condition. Red circle: spot proteins identified by MS/MS

Table 1 List of proteins identified in 1002_ovis control and recovered by 2D-PAGE-MS/MS

Spot	Description	Accession	MW(kDa)/p.l	Peptides Number	Mascot Score	Molecular function
5, 6, 7	Hypothetical protein	ADL20032	24.30/9.24	2	189	Unknown function
11,29	Trypsin-like serine protease	ADL20653	25.72/6.49	2	96	Serine-type endopeptidase activity
15	Hypothetical protein	ADL21714	42.04/5.22	4	159	Catalytic activity
20,34	Corynomycolyl transferase	ADL21610	41.80/7.05	2	58	Transferase activity
16	Cytochrome c oxidase sub II	ADL21302	40.33/6.03	2	96	Cytochrome-c oxidase activity
21	Hypothetical protein	ADL21914	12.30/5.04	2	53	Unknown function
12	Hypothetical protein	ADL19922	19.86/4.30	2	145	Calcium ion binding
8	Hypothetical protein	ADL09626	24.30/9.24	3	228	Unknown
27	Hypothetical protein	ADL20508	31.62/9.52	2	66	Unknown
22	Phospholipase D	ADL19935	34.09/8.91	4	286	Sphingomyelin phosphodiesterase D activity
3	Enolase	ADL20605	45.17/4.68	3	271	Phosphopyruvate hydratase activity
17	Trehalose corynomycolyl transferase B	ADL21814	36.67/6.90	5	245	Transferase activity, transferring acyl groups other than
24	Hypothetical protein	ADL21714	40.90/5.05	3	190	Catalytic activity

Staphylococcus aureus and *Listeria monocytogenes* [36, 37] were found induced in Rc supernatant.

Discussion

To investigate the protein factors that could influence the adaptive processes of *C. pseudotuberculosis* biovar *ovis* during the infection process, we combined a unique bacterial passage experiment in mice with proteomic analyses of 1002_ovis culture supernatants, collected before and after passage. In the first analysis, we observed that strain 1002_ovis (isolated from caprine) exhibited a low virulence potential, which is consistent with

previous reports indicating the low virulence potential of this strain [38, 39]. Although a recent in silico analysis of the 1002_ovis genome predicted various genes involved in virulence [40], studies examining the exoproteome of this strain under laboratory growth conditions failed to detect many of these virulence proteins (e.g., PLD exotoxin or proteins involved in the pathway of cell invasion, detoxification) [8–10].

One explanation for this relies on the fact that after being first isolated, strains 1002_ovis have been maintained, in vitro, under laboratory conditions with extensive passages on the culture medium, which may alter the gene

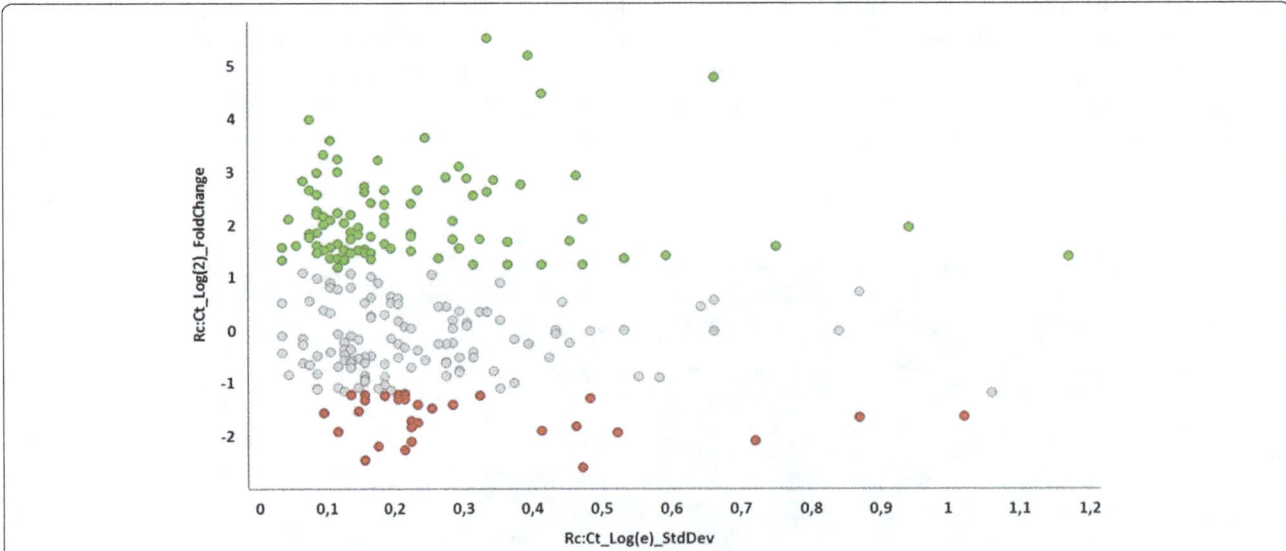

Fig. 3 Volcano Plot show Log(2) Fold Change of the differentially expressed proteins detected by label-free proteomics between the recovered and control condition. Green: Up-regulated proteins; Grey: unchanged proteins; Red: Down-regulated proteins

Table 2 Proteins differentially produced among the recovered and control condition

Accession	Description	Score	Fold Change_Log$^a_{(2)}$	SecretomeP
Transport				
D9Q5H9_CORP1	Periplasmic binding protein LacI	5601,78	3,26	0.612642
D9Q6G4_CORP1	Oligopeptide binding protein oppA[b]	4120,1	3,00	0.892226
D9Q4T5_CORP1	ABC transporter domain containing ATP	1264,05	2,57	0.084974
D9Q7K5_CORP1	Oligopeptide binding protein oppA[b]	33697,17	2,11	0.873687
D9Q5B8_CORP1	Oligopeptide binding protein oppA[b]	852,88	1,88	0.849217
D9Q6C3_CORP1	ABC type metal ion transport system permease	650,43	1,59	0.078043
D9Q796_CORP1	Glutamate binding protein GluB	6254,68	−1,46	0.840325
D9Q7W9_CORP1	Iron(3+)-hydroxamate-binding protein fhuD	2774,62	−1,62	0.824030
Cell division				
D9Q7G1_CORP1	Septum formation initiator protein	2071,46	1,38	0.551153
Cell adhesion				
D9Q5H7_CORP1	Hypothetical protein	115906,3	1,51	0.840443
DNA synthesis and repair				
D9Q7J1_CORP1	GTP binding protein YchF	3487,98	2,68	0.042575
D9Q5F7_CORP1	Chromosome partitioning protein ParB[b]	2467,24	2,44	0.052395
D9Q5G6_CORP1	DNA polymerase III subunit beta	1907,74	1,80	0.071008
D9Q5V6_CORP1	Nucleoid associated protein[c]	68097,59	1,59	0.070074
Transcription				
D9Q6J8_CORP1	DNA directed RNA polymerase subunit	29671,46	1,38	0.094910
D9Q748_CORP1	tRNA rRNA methyltransferase	2467,24	1,27	0.060356
D9Q8L3_CORP1	DNA directed RNA polymerase subunit omega	3784,13	−1,21	0.700214
D9Q6D1_CORP1	DNA directed RNA polymerase subunit beta	2611,89	−1,27	0.067182
D9Q8A5_CORP1	RNA polymerase-binding protein RbpA	10787,51	−1,75	0.103548
Translation				
D9Q584_CORP1	30S ribosomal protein S6	20750,74	4,82	0.047667
D9Q6E4_CORP1	Elongation factor G[b]	16882,71	3,25	0.082321
D9Q5I3_CORP1	Peptidyl prolyl cis trans isomerase[b]	61648,39	2,91	0.142641
D9Q835_CORP1	Phenylalanine tRNA ligase beta subunit	1269,7	2,74	0.064869
D9Q6L0_CORP1	50S ribosomal protein L13	5689,37	2,64	0.101816
D9Q6H2_CORP1	50S ribosomal protein L5[b]	3269,32	2,12	0.076250
D9Q918_CORP1	Proline tRNA ligase[b]	932,79	2,12	0.072151
D9Q6C0_CORP1	50S ribosomal protein L10[b]	27143,51	1,86	0.031374
D9Q6F6_CORP1	50S ribosomal protein L23[b]	6947,79	1,85	0.060878
D9Q6H1_CORP1	50S ribosomal protein L24	27887,33	1,75	0.078408
F9Y2W9_CORP1	Hypothetical protein	3152,39	1,75	0.591013
D9Q6H6_CORP1	30S ribosomal protein S8[c,b]	4941,19	1,56	0.088407
D9Q6F3_CORP1	30S ribosomal protein S10[b]	25117,55	1,54	0.048124
D9Q6G2_CORP1	50S ribosomal protein L29	2467,24	1,44	0.050948
D9Q401_CORP1	50S ribosomal protein L27[b]	2467,24	1,38	0.081399
D9Q7E8_CORP1	50S ribosomal protein L25	1358,05	−1,28	0.037225
D9Q6H8_CORP1	50S ribosomal protein L18	8920,94	−1,31	0.049024
D9Q7S4_CORP1	Homoserine dehydrogenase	698,17	−1,40	0.035138

Table 2 Proteins differentially produced among the recovered and control condition *(Continued)*

D9Q6B7_CORP1	50S ribosomal protein L1	10218.08	−1,63	0.633387
D9Q4T4_CORP1	ATP dependent chaperone protein ClpB	1883,16	−1,80	0.045308
D9Q8N9_CORP1	Aspartate tRNA ligase	1004,33	−2,18	0.092415
D9Q7S2_CORP1	Arginine tRNA ligase	2679,11	−2,44	0.051908
Pathogenesis				
D9Q8M7_CORP1	Metallopeptidase family M24	3213,83	5,55	0.050024
D9Q608_CORP1	Penicillin binding protein transpeptidase[b]	1215,32	3,68	0.859830
D9Q827_CORP1	Metallo beta lactamase superfamily protein[c]	629,38	2,64	0.144158
D9Q721_CORP1	Hypothetical protein[c]	112025	2,24	0.260801
D9Q7K8_CORP1	Trypsin like serine protease	35041,27	1,96	0.648370
D9Q416_CORP1	ATP dependent Clp protease proteolytic[b]	2467,24	1,77	0.087255
D9Q639_CORP1	Secreted hydrolase[b]	22798,13	1,75	0.072385
D9Q588_CORP1	Penicillin binding protein[b]	9951,61	1,26	0.916125
Energy metabolism				
D9Q787_CORP1	Glucose-6-phosphate isomerase	1025,89	4,50	0.058841
D9Q7G0_CORP1	Enolase[b]	53290,95	2,18	0.068928
D9Q651_CORP1	Succinate dehydrogenase flavoprotein	797,48	2,02	0.159059
D9Q4P2_CORP1	Acetate kinase[b]	10828,79	1,96	0.063340
D9Q8G5_CORP1	Aconitate hydratase[b]	4250,81	1,85	0.217637
D9Q4Z7_CORP1	Phosphoenolpyruvate carboxykinase GTP[b]	8764,35	1,66	0.147167
D9Q7X0_CORP1	6 phosphofructokinase	1806,65	1,60	0.052885
D9Q648_CORP1	Dihydrolipoyl dehydrogenase	4110,08	1,57	0.047180
D9Q7T8_CORP1	ATP synthase subunit alpha	2467,24	1,49	0.070875
D9Q752_CORP1	Citrate synthase	6299,21	−1,21	0.116042
D9Q895_CORP1	6-Phosphogluconate dehydrogenase	4246,26	−1,89	0.050906
Lipid metabolism				
D9Q520_CORP1	Glycerophosphoryl diester phosphodieste[c]	2494,25	4,03	0.802154
D9Q718_CORP1	Methylmalonyl CoA carboxyltransferase 1[b]	2467,24	2,16	0.049504
	Amino acid metabolism			
D9Q5X8_CORP1	Aspartokinase[b]	1944,81	2,86	0.043575
D9Q4C2_CORP1	Succinyl CoA Coenzyme A transferase	10894,63	1,63	0.061344
D9Q3L8_CORP1	Glutamine synthetase	320,71	−1,23	0.263700
D9Q8H7_CORP1	Cysteine desulfurase	1689,36	−1,70	0.067087
Stress response				
D9Q929_CORP1	Mycothione glutathione reductase	490,36	2,67	0.085017
D9Q5T5_CORP1	Glyoxalase Bleomycin resistance protein[c]	8420,32	2,21	0.226764
D9Q424_CORP1	DSBA oxidoreductase	12179,8	2,09	0.061566
D9Q566_CORP1	Universal stress protein A[b]	2498,69	1,70	0.034684
D9Q4P4_CORP1	Ferredoxin ferredoxin NADP reductase[b]	1086,71	1,69	0.083585
D9Q824_CORP1	Stress related protein[b]	2467,24	1,54	0.035291
D9Q692_CORP1	Thiol disulfide isomerase thioredoxin	3721,88	−2,25	0.438415
Metabolism of nucleotides and nucleic acids				
D9Q4Y6_CORP1	Deoxycytidine triphosphate deaminase	887,26	2,39	0.216897

Table 2 Proteins differentially produced among the recovered and control condition *(Continued)*

D9Q6J1_CORP1	Adenylate kinase	15629,86	2,21	0.059568
D9Q8L4_CORP1	Guanylate kinase	2467,24	1,34	0.050095
D9Q6T2_CORP1	Ribokinase	890,09	−1,23	0.032324
D9Q4E9_CORP1	Adenylosuccinate lyase	1441,99	−1,54	0.035597
D9Q6P0_CORP1	D methionine binding lipoprotein metQ	11519,67	−1,93	0.817217
Carbohydrate metabolism				
D9Q8V2_CORP1	UDP glucose 4 epimerase[b]	2001,76	3,13	0.094403
D9Q6V6_CORP1	Phosphomannomutase ManB	1730,63	2,05	0.053146
D9Q659_CORP1	Formate acetyltransferase	5456,95	1,54	0.539548
D9Q423_CORP1	Ribose-5-phosphate isomerase B	2467,24	1,38	0.064467
D9Q6V1_CORP1	Mannose-1-phosphate guanylyltransferase	1612,45	−1,21	0.068085
Nitrogen metabolism				
D9Q4Q8_CORP1	Cytochrome c nitrate reductase small	1118,33	2,68	0.901856
Unknow function				
D9Q6T0_CORP1	Hypothetical protein	2277,6	3,62	0.050552
D9Q4R2_CORP1	Hypothetical protein	442,07	3,35	0.866986
D9Q6N1_CORP1	Hypothetical protein	561,84	3,02	0.062141
D9Q8Q4_CORP1	Hypothetical protein[c]	72711,5	2,96	0.974016
D9Q832_CORP1	Hypothetical protein	1774,59	2,90	0.752478
D9Q3S8_CORP1	Hypothetical protein[d]	837,6	2,78	0.231421
D9Q7M9_CORP1	Hypothetical protein	3246,28	2,60	0.147602
D9Q7I6_CORP1	Hypothetical protein	3751,96	2,42	0.707595
D9Q739_CORP1	Hypothetical protein	2845,77	2,28	0.836229
D9Q4C5_CORP1	Hypothetical protein	1339,3	1,83	0.023133
D9Q5C3_CORP1	Hypothetical protein	111234,6	1,49	0.946918
D9Q700_CORP1	Hypothetical protein	2467,24	1,49	0.072810
D9Q657_CORP1	Hypothetical protein	1172,66	1,41	0.830926
D9Q6F2_CORP1	Hypothetical protein	2467,24	1,34	0.061860
D9Q7X5_CORP1	Hypothetical protein	38716,45	−1,21	0.825761
D9Q4T9_CORP1	Hypothetical protein	553,76	−1,28	0.934591
D9Q6R6_CORP1	Hypothetical protein	1457,62	−1,40	0.206908
D9Q890_CORP1	Hypothetical protein	1948,52	−1,51	0.847549
D9Q6M6_CORP1	Hypothetical protein[c]	1935,68	−1,90	0.823541
Others				
D9Q6I3_CORP1	Maltotriose binding protein	5210,9	5,22	0.864851
D9Q4A3_CORP1	DsbG protein	3101,13	2,06	0.814366
D9Q6N9_CORP1	D methionine binding lipoprotein metQ	2665,58	1,79	0.764416
D9Q732_CORP1	Carbonic anhydrase[b]	689,15	1,66	0.130559
D9Q6W6_CORP1	Lipoprotein LpqB	1484,31	1,63	0.670057
D9Q556_CORP1	LSR2 like protein	2714,21	1,49	0.096802
D9Q5Q0_CORP1	UPF0145 protein	2467,24	1,37	0.025009
D9Q7W0_CORP1	Hypothetical protein	2467,24	1,26	0.039678
D9Q701_CORP1	UPF0182 protein	1682,98	1,26	0.869411
D9Q8A3_CORP1	Protein ycel[b]	16885,01	1,21	0.901679

Table 2 Proteins differentially produced among the recovered and control condition *(Continued)*

D9Q5X4_CORP1	Serine aspartate repeat containing protein	528,36	−1,82	0.892317
D9Q826_CORP1	DoxX family protein	697,26	−2,08	0.614317
D9Q7W3_CORP1	Mycothiol acetyltransferase	947,33	−2,11	0.214833
D9Q407_CORP1	Ornithine cyclodeaminase	2566,18	−2,58	0.048247

[a]Fold change - Ratio values to: 1002Rc:11002Ct_Log(2)Ratio ≥ 1.2 proteins with $p < 0.05$
[b]Identified in an isolated of *C. pseudotuberculosis* from ovine lymph nodes [Rees et al. [12]]
[c]Induced in 1002_*ovis* during to stress nitrosative [Pacheco et al. [57], Silva et al. [58]]
[d]Predicted LPXTG cell wall-anchoring motif

Table 3 List of proteins identified in the exclusive proteome of recovered-condition

Accession	Description	Score	Biological process	SecretomeP
D9Q869_CORP1	Esterase[a]	251.44	Others	0.862935
D9Q575_CORP1	Cation transport protein	1961.29	Transport	0.062276
D9Q5N5_CORP1	Uncharacterized iron regulated membrane[a]	46.77	Transport	0.855681
D9Q3T9_CORP1	Pyridoxamine kinase	216.2	Cofactor metabolism	0.083313
D9Q751_CORP1	Phosphoserine aminotransferase	639.64	Amino acid metabolism	0.151778
D9Q537_CORP1	LytR family transcriptional regulator[a]	375.8	Transcription	0.766483
D9Q7F2_CORP1	Multicopper oxidase	74.63	Stress response	0.278840
D9Q525_CORP1	ABC transporter substrate binding lipoprotein	283.38	Transport	0.452814
D9Q6P2_CORP1	Manganese ABC transporter substrate binding[a]	236.6	Transport	0.774461
D9Q4C8_CORP1	Phosphate ABC transporter phosphate binding[a]	125.4	Transport	0.840195
D9Q4L0_CORP1	D alanyl D alanine carboxypeptidase OS	426.74	Others	0.232261
D9Q4T7_CORP1	Hyphotetical protein	157.52	Unknow function	0.349026
D9Q5A9_CORP1	Hyphotetical protein	218.02	Unknow function	0.907333
D9Q476_CORP1	Hyphotetical protein	510.32	Unknow function	0.066368
D9Q5B3_CORP1	Glucosamine-6-phosphate deaminase[b]	524,55	Carbohydrate metabolism	0.079507
D9Q474_CORP1	Glutamate racemase	343,98	Cell wall organization	0.040278
D9Q7N5_CORP1	O-methyltransferase	619,11	DNA process	0.032455
D9Q5N3_CORP1	Gamma type carbonic anhydratase	577,75	Others	0.035357
D9Q4X0_CORP1	Urease accessory protein UreD	333,12	Others	0.055896
D9Q5J0_CORP1	Phospholipase D[b]	40,25	Pathogenesis	0.409585
D9Q8S8_CORP1	Copper resistance protein CopC	4315,26	Stress response	0.964015
D9Q493_CORP1	Glutaredoxin like protein nrdH	725,98	Stress response	0.033036
D9Q6Y6_CORP1	ATP dependent RNA helicase rhlE	1438,25	Transcription	0.060627
D9Q4M0_CORP1	Cell wall channel	4008,59	Transport	0.025882
D9Q4V1_CORP1	CP40	558,79	Pathogenesis	0.926013
D9Q6V9_CORP1	Hyphotetical protein	1278,45	Unknow function	0.953803
D9Q6A8_CORP1	Hyphotetical protein	326,47	Unknow function	0.918886
D9Q485_CORP1	Hyphotetical protein	2795,11	Unknow function	0.890081
D9Q4N2_CORP1	Hypothetical protein[a]	708,75	Unknow function	0.857050
D9Q559_CORP1	Hypothetical protein[a]	475,62	Unknow function	0.472378
D9Q4L8_CORP1	Hyphotetical protein	5324,08	Unknow function	0.038893
D9Q4T0_CORP1	Hyphotetical protein	732,37	Unknow function	0.037132

[a]Induced in 1002_ovis during to stress nitrosative [Pacheco et al. [57], Silva et al. [58]]
[b]Identified in an isolated of *C. pseudotuberculosis* from ovine lymph nodes [Rees et al. [12]]

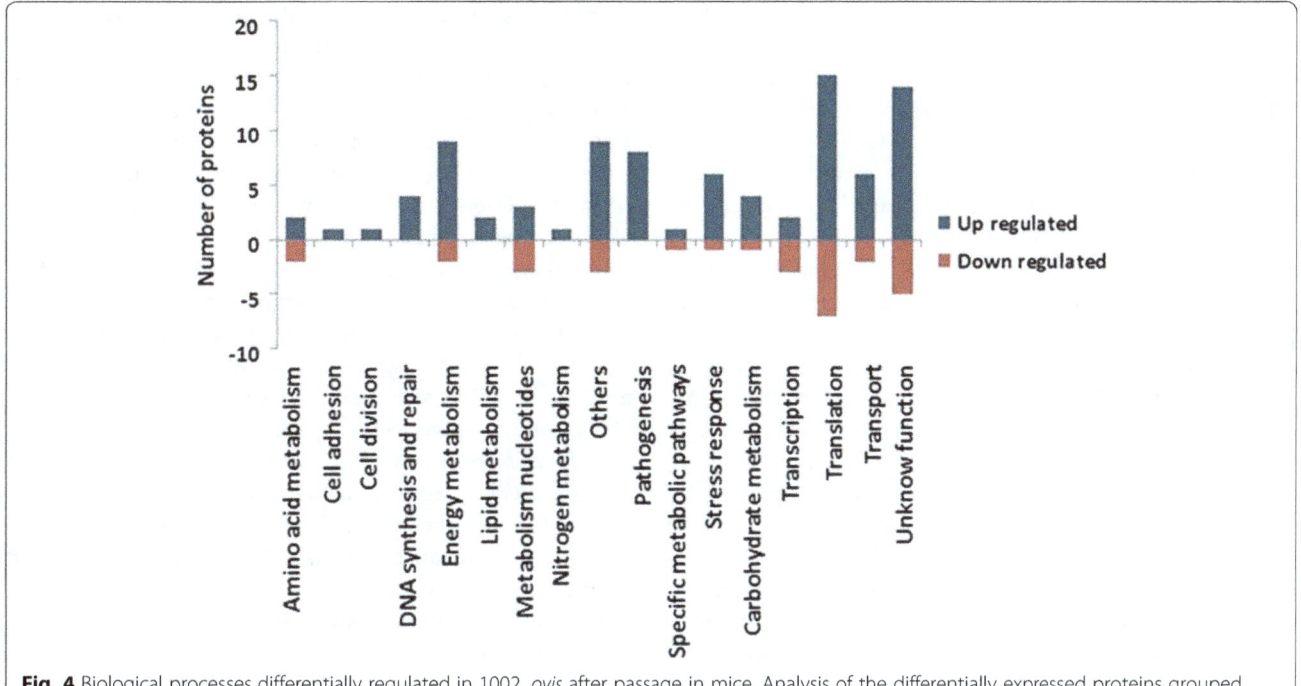

Fig. 4 Biological processes differentially regulated in 1002_*ovis* after passage in mice. Analysis of the differentially expressed proteins grouped into biological processes for strain 1002_*ovis* after passage in mice

expression profile of the strain, especially for effectors related to bacterial virulence. This phenomenon has also been reported in other pathogens such as *Mycobacterium bovis*, *Helicobacter pylori*, *S. aureus*, and *L. monocytogenes*. In vitro passages of these bacteria on culture medium altered both bacterial physiology and virulence profile [41–44]. However, we showed that the bacterial passage process in a murine model changed the virulence potential of strain 1002_*ovis*. Previous reports on experimental serial passages showed that pathogens such as *H. pylori*, *Escherichia coli*, *Xenorhabdus nematophila*, *Arcobacter butzleri*, and *Salmonella enterica* also exhibited altered virulence profiles after in vivo passage in a host, which helped identifying factors that contribute to infectious process [14–19]. Thus, as observed in these pathogens, the recovered condition also showed increased capacity to persist into host, when compared with control condition. The altered physiology and virulence status observed in 1002_*ovis* is supported by our proteomic analyses, where several proteins involved in processes favoring infection and host adaptation were differentially expressed after passage in mice.

Although our study focused on the *C. pseudotuberculosis* extracellular proteins, cytoplasmic proteins were also detected in the proteomic analyses. The presence of cytoplasmic proteins in the extracellular fraction is reported in several other proteomic studies [8–10, 12, 45]. It may be partially due to cell lysis and thus, be considered artifacts. However, cytoplasmic proteins in the culture supernatant may act as *moonlighting* proteins and be

exported via a non-classical secretion pathway [30, 46]. The *moonlighting* proteins are described both Gram-positive and Gram-negative bacteria, and can be detected in different subcellular locations (cytoplasm, membrane, cell surface, and extracellular environment) and exhibit distinct functional behavior depending on the host cell type [46, 47]. Interestingly, some proteins, such as Chromosome partitioning protein ParB, Phosphoenolpyruvate carboxykinase GTP, Methylmalonyl CoA carboxyltransferase 12S subunit, Acetate kinase, and Enolase, induced in the Rc supernatants were identified only in the membrane shaving of *C. pseudotuberculosis* harvested directly from ovine lymph nodes [12].

The passage process in mice was also able to induce other proteins identified in Rc supernatants, and which contribute to the adhesion process. Proteins with an LPTXG domain, which characterizes the cell-wall anchored proteins, were identified and included monomers of membrane pilus. This latter class of proteins is described in pathogenic *Corynebacterium* species and may contribute especially in the process of cellular adhesion [48]. In *Campylobacter jejuni*, serial passages in mice induce the expression of invasiveness and increase the capacity of cell invasion [13]. Components of the Opp system were induced by the passage process, too. The Opp system facilitates the uptake of extracellular peptides, which are further used as carbon and nitrogen sources for bacterial nutrition [49]. Proteins that comprise the Opp system also were induced in a field isolated of *C. pseudotuberculosis* biovar *ovis*, when compared with the strain C231_ovis a laboratory reference strain [12, 50]. In the pathogen *Mycobacterium avium* the *OppA* gene was highly expressed during the infection in a mouse model [51]. We have identified known secreted virulence factors as CP40 serine protease, which previously shown to be necessary for *C. pseudotuberculosis* virulence potential and to induce an immune response [52, 53].

An important factor that precedes the chronic stage of infection by *C. pseudotuberculosis* is the capacity of this pathogen to disseminate within the host, which consequently favors the establishment of the disease [3]. In *C. pseudotuberculosis*, this process is mediated by the action of PLD exotoxin, a major virulence factor of this pathogen [54, 55] that catalyzes the dissociation of sphingomyelin and increases vascular permeability, which contributes to the dissemination process of *C. pseudotuberculosis* in the host. Here, PLD was only detected in the proteome of the Rc condition. This result is noteworthy because, a previous proteomic study performed by our research group, PLD was not identified in the extracellular proteome of 1002_ovis [8–10]. McKean *et al.* [5] showed that *pld* expression is expressed by different environmental factors, thus during the infection and recuperation process 1002_ovis was exposed to different environmental and stimulus, which may have affected the *pld* expression. A study showed that a *pld* mutant strain is indeed unable to disseminate and yields reduced virulence [55]. Here, we observed the presence of caseous lesions in different organs only at the end of experimental infection, only in the group of mice infected with the Rc condition. Altogether, the observations suggest that the expression of PLD can be modified by the passage in the host and can thus change the virulence potential of 1002_ovis.

Another attribute of PLD is its capacity to alter the viability of macrophage cells during the infection [5]. However, before promoting macrophages lysis, *C. pseudotuberculosis* has to be able to resist the hostile environment inside macrophages mainly against reactive oxygen species (ROS) and reactive nitrogen species (RNS). Thus, the induction of proteins involved in detoxification processes in Rc could be contributed for its resistance against ROS and RNS. The inductions of proteins related to oxidative stress also were observed in *Shigella flexneri*, after recuperation process in an in vivo infection model. We detected the mycothione glutathione reductase, a component of the mycothiol system, which is present in *Mycobacterium and Rhodococcus* genera. This system is used as an alternative mechanism of disulphide reduction and contributes to the cytosolic redox homeostasis and the resistance to ROS [35]. Glutaredoxin-like protein, NrdH, which plays an important role in the resistance to ROS, and is present in *C. glutamicum* [34] and *M. tuberculosis* [56] was also detected.

On the other hand, some proteins like dihydroxybiphenyl dioxygenase, Metallo beta lactamase superfamily protein, Formamidopyrimidine DNA glycosylase, MerR family transcriptional regulator, which were induced by 1002_ovis during the exposition to nitric oxide [57, 58] were also found induced in this study in the recovered condition. These proteins are related to different processes of resistance to nitrosative stress, DNA repair, antibiotic resistance, and transcription, these results show a set of proteins involved in the adaptation process of 1002_ovis to nitric oxide, which could contribute to the pathogenic process of this pathogen. Another type of defense of the host immune system against bacterial infection is the utilization of copper [59]. Here, CopC, a protein related to copper resistance, was detected in recovered 1002_ovis. In *M. tuberculosis*, proteins involved in copper resistance are essential to virulence [60, 61]. Thus, the association of this factor related to an antioxidant system with PLD could promote an effective pathway of defense against the action of the innate immune system and consequently contributes to virulence process of *C. pseudotuberculosis*.

Conclusion

In conclusion, the virulence potential and proteomic profiles of strain 1002_ovis undergo dramatic changes after recovery from experimentally infected mice. The proteomic screening outlined, after the serial passage in murine model showed a set of proteins that were induced in the recovered condition. Into this group were detected known secreted virulence factors, as well as some proteins which could contribute in its virulence. Therefore, more study is necessary to show the true role of these proteins in the virulence of *C. pseudotuberculosis*. Altogether, our results demonstrate that in vitro passages alter the expression of *C. pseudotuberculosis* exoproteome leading to a reduced virulence and that a single passage in vivo, in a murine model, can induce significant changes in the *C. pseudotuberculosis* extracellular proteome, contributing to the increase in virulence of this pathogen.

Additional files

Additional file 1: Table S1. Complete list of proteins differentially produced between the recovered and control condition of strain 1002_ovis.

Additional file 2: Table S2. List of proteins identified in the exclusive proteome of control condition.

Additional file 3: Table S3. Total list of peptide and proteins identified by LC-MS^E.

Additional file 4: Table S4. Proteins identified in the recovered condition detected in pathogenicity island.

Acknowledgements
The authors would like to thank to Pará State Genomics and Proteomics Network and Waters Corporation, Brazil.

Funding
The work was supported by the Brazilian Federal Agency for the Support and Evaluation of Graduate Education (CAPES), Pará Research Foundation (FAPESPA), Minas Gerais Research Foundation (FAPEMIG) and the National Council for Scientific and Technological Development (CNPq). Yves Le Loir is the recipient of a PVE grant (71/2013) from Programa Ciências sem Fronteiras.

Authors' contributions
VA, WMS, and FAD designed the experiments. WMS and FAD performed in vivo experiments. WMS, TLPC, and NS performed microbiological analyses and sample preparation for proteomic analysis. GHMFS and WMS conducted the proteomic analysis. WMS and SCS performed bioinformatics analysis of the data. YLL, AM, and HF contributed substantially to data interpretation and revisions. VA, AS, and YLL participated in all steps of the project as coordinators, and critically reviewed the manuscript. All authors read and approved the final manuscript.

Competing interests
The authors declare that they have no competing interests.

Author details
[1]Departamento de Biologia Geral, Instituto de Ciências Biológicas, Universidade Federal de Minas Gerais, Belo Horizonte, Minas Gerais, Brazil. [2]Instituto de Ciências Biológicas, Universidade Federal do Pará, Guamá, Belém, Pará, Brazil. [3]Waters Corporation, Waters Technologies Brazil, MS Applications Laboratory, Alphaville, São Paulo, Brazil. [4]INRA, UMR1253 STLO, 35042 Rennes, France. [5]Agrocampus Ouest, UMR1253 STLO, 35042 Rennes, France. [6]Aquacen, Escola de Veterinária, Universidade Federal de Minas Gerais, Belo Horizonte, Brazil.

References
1. Dorella FA, Pacheco LG, Oliveira SC, Miyoshi A, Azevedo V. *Corynebacterium pseudotuberculosis*: microbiology, biochemical properties, pathogenesis and molecular studies of virulence. Vet Res. 2006;37:201–18.
2. Paton MW, Walker SB, Rose IR, Watt GF. Prevalence of caseous lymphadenitis and usage of caseous lymphadenitis vaccines in sheep flocks. Aust Vet J. 2003;81:91–5.
3. Batey RG. Pathogenesis of caseous lymphadenitis in sheep and goats. Aust Vet J. 1986;63:269–72.
4. Pépin M, Pittet JC, Olivier M, Gohin I. Cellular composition of *Corynebacterium pseudotuberculosis* pyogranulomas in sheep. J Leukoc Biol. 1994;56:666–70.
5. McKean SC, Davies JK, Moore RJ. Expression of phospholipase D, the major virulence factor of *Corynebacterium pseudotuberculosis*, is regulated by multiple environmental factors and plays a role in macrophage death. Microbiology. 2007;153:2203–11.
6. Green ER, Mecsas J. Bacterial Secretion Systems – An overview. Microbiol Spectr. 2016;4:1. Hilbi H, Haas A. Secretive bacterial pathogens and the secretory pathway. Traffic. 2012; 13:1187–1197.
7. Dorella FA, Estevam EM, Pacheco LG, Guimarães CT, Lana UG, Gomes EA, et al. In vivo insertional mutagenesis in *Corynebacterium pseudotuberculosis*: an
8. Pacheco LG, Slade SE, Seyffert N, Santos AR, Castro TL, Silva WM, et al. A combined approach for comparative exoproteome analysis of *Corynebacterium pseudotuberculosis*. BMC Microbiol. 2011;17:12.
9. Silva WM, Seyffert N, Santos AV, Castro TL, Pacheco LG, Santos AR, et al. Identification of 11 new exoproteins in *Corynebacterium pseudotuberculosis* by comparative analysis of the exoproteome. Microb Pathog. 2013;16:37–42.
10. Silva WM, Seyffert N, Ciprandi A, Santos AV, Castro TL, Pacheco LG, et al. Differential Exoproteome analysis of two *Corynebacterium pseudotuberculosis* biovar *ovis* strains isolated from goat (1002) and sheep. Curr Microbiol. 2013;67:460–5.
11. Seyffert N, Silva RF, Jardin J, Silva WM, Castro TL, Tartaglia NR, et al. Serological proteome analysis of *Corynebacterium pseudotuberculosis* isolated from different hosts reveals novel candidates for prophylactics to control caseous lymphadenitis. Vet Microbiol. 2014;174:255–60.
12. Rees MA, Kleifeld O, Crellin PK, Ho B, Stinear TP, Smith AI, Coppel RL. Proteomic Characterization of a Natural Host-Pathogen Interaction: Repertoire of in vivo Expressed Bacterial and Host Surface-Associated Proteins. J Proteome Res. 2015;2:120–32.
13. Fernández H, Vivanco T, Eller G. Expression of invasiveness of *Campylobacter jejuni* ssp. jejuni after serial intraperitoneal passages in mice. J Vet Med B Infect Dis Vet Public Health. 2000;47:635–9.
14. Bleich A, Kohn I, Glage S, Beil W, Wagner S, Mahler M. Multiple in vivo passages enhance the ability of clinical *Helicobacter pylori* isolate to colonize the stomach of *Mongolian gerbils* and to induce gastritis. Lab Anim. 2005;39:221–9.
15. Chapuis É, Pagès S, Emelianoff V, Givauda A, Ferdy JB. Virulence and pathogen multiplication: a serial passage experiment in the hypervirulent bacterial insect-pathogen *Xenorhabdus nematophila*. PLoS One. 2011;31:e15872.
16. Fernandez-Brando RJ, Miliwebsky E, Mejías MP, Baschkier A, Panek CA, Abrey-Recalde MJ, et al. Shiga toxin-producing *Escherichia coli* O157: H7 shows an increased pathogenicity in mice after the passage through the gastrointestinal tract of the same host. J Med Microbiol. 2012;61:852–9.
17. Fernández H, Flores SP, Villanueva M, Medina G, Carrizo M. Enhancing adherence of *Arcobacter butzleri* after serial intraperitoneal passages in mice. Rev Argent Microbiol. 2013;45:75–9.
18. Koskiniemi S, Gibbons HS, Sandegren L, Anwar N, Ouellette G, Broomall S, et al. Pathoadaptive mutations in *Salmonella enterica* isolated after serial passage in mice. PLoS One. 2013;25:e70147.
19. Liu X, Lu L, Liu X, Pan C, Feng E, Wang D, Zhu L, Wang H. Comparative proteomics of *Shigella flexneri* 2a strain using a rabbit ileal loop model reveals key proteins for bacterial adaptation in host niches. Int J Infect Dis. 2015;40:28–33.
20. Moraes PM, Seyffert N, Silva WM, Castro TL, Silva RF, Lima DD, et al. Characterization of the Opp peptide transporter of *Corynebacterium pseudotuberculosis* and its role in virulence and pathogenicity. Biomed Res Int. 2014;2014:489782.
21. Ribeiro D, Rocha FS, Leite KM, Soares SC, Silva A, Portela RW, et al. An iron acquisition-deficient mutant of *Corynebacterium pseudotuberculosis* efficiently protects mice against challenge. Vet Res. 2014;45:28.
22. Moura-Costa LF, Paule BJA, Freire SM, Nascimento I, Schaer R, Regis LF, et al. Chemically defined synthetic medium for *Corynebacterium pseudotuberculosis* culture. Rev Bras Saúde Prod An. 2002;3:1–9.
23. Paule BJ, Meyer R, Moura-Costa LF, Bahia RC, Carminati R, Regis LF, et al. Three-phase partitioning as an efficient method for extraction/concentration of immunoreactive excreted-secreted proteins of *Corynebacterium pseudotuberculosis*. Protein Expr Purif. 2004;34:311–166.
24. Bradford MM. A rapid and sensitive method for the quantitation of microgram quantities of protein utilizing the principle of protein-dye binding. Anal Biochem. 1976;72:248–54.
25. Silva JC, Gorenstein MV, Li GZ, Vissers JP, Geromanos SJ. Absolute quantification of proteins by LCMSE: a virtue of parallel MS acquisition. Mol Cell Proteomics. 2006;5:144–56.
26. Gilar M, Olivova P, Daly AE, Gebler JC. Two-dimensional separation of peptides using RP-RP-HPLC system with different pH in first and second separation dimensions. J Sep Sci. 2005;8:1694–703.
27. Geromanos SJ, Vissers JP, Silva JC, Dorschel CA, Li GZ, Gorenstein MV, et al. The detection, correlation, and comparison of peptide precursor and product ions from data independent LC-MS with data dependant LC-MS/MS. Proteomics. 2009;9:1683–95.
28. Li GZ, Vissers JP, Silva JC, Golick D, Gorenstein MV, Geromanos SJ. Database searching and accounting of multiplexed precursor and product ion spectra

from the data independent analysis of simple and complex peptide mixtures. Proteomics. 2009;9:1696–719.

29. Curty N, Kubitschek-Barreira PH, Neves GW, Gomes D, Pizzatti L, Abdelhay E. Discovering the infectome of human endothelial cells challenged with *Aspergillus fumigatus* applying a mass spectrometry label-free approach. J Proteomics. 2014;31:126–40.

30. Bendtsen JD, Kiemer L, Fausboll A, Brunak S. Non-classical protein secretion in bacteria. BMC Microbiol. 2005;5:58.

31. Soares SC, Abreu VA, Ramos RT, Cerdeira L, Silva A, Baumbach J. PIPS: pathogenicity island prediction software. PLoS One. 2012;7:e30848.

32. Conesa A, Gotz S, García-Gómez JM, Terol J, Talón M, Robles M. Blast2GO: a universal tool for annotation, visualization and analysis in functional genomics research. Bioinformatics. 2005;15:3674–6.

33. Levin Y, Hradetzky E, Bahn S. Quantification of proteins using data-independent analysis (MSE) in simple and complex samples: a systematic evaluation. Proteomics. 2011;11:3273–87.

34. Si MR, Zhang L, Yang ZF, Xu YX, Liu YB, Jiang CY, et al. NrdH Redoxin enhances resistance to multiple oxidative stresses by acting as a peroxidase cofactor in *Corynebacterium glutamicum*. Appl Environ Microbiol. 2014;80:1750–62.

35. Newton GL, Buchmeier N, Fahey RC. Biosynthesis and functions of mycothiol, the unique protective thiol of Actinobacteria. Microbiol Mol Biol Rev. 2008;72:471–94.

36. Frees D, Qazi SN, Hill PJ, Ingmer H. Alternative roles of ClpX and ClpP in *Staphylococcus aureus* stress tolerance and virulence. Mol Microbiol. 2013;48:1565–78.

37. Gaillot O, Pellegrini E, Bregenholt S, Nair S, Berche P. The ClpP serine protease is essential for the intracellular parasitism and virulence of *Listeria monocytogenes*. Mol Microbiol. 2000;35:1286–94.

38. Ribeiro OC, Silva JAH, Oliveira SC, Meyer R, Fernandes GB. Preliminary results on a living vaccine against caseous lymphadenitis. Pesq Agrop Brasileira. 1991;26:461–5.

39. Meyer R, Carminati R, Cerqueira RB, Vale V, Viegas S, Martinez T. Evaluation of the goats humoral immune response induced by the *Corynebacterium pseudotuberculosis* lyophilized live vaccine. Rev Cienc Méd Biol. 2002;1:42–8.

40. Ruiz JC, D'Afonseca V, Silva A, Ali A, Pinto AC, Santos AR. Evidence for reductive genome evolution and lateral acquisition of virulence functions in two *Corynebacterium pseudotuberculosis* strains. PLoS One. 2011;18:e18551.

41. Nascimento IP, Leite LC. The effect of passaging in liquid media and storage on *Mycobacterium bovis*–BCG growth capacity and infectivity. FEMS Microbiol Lett. 2005;1:81–6.

42. Hopkins RJ, Morris Jr JG, Papadimitriou JC, Drachenberg C, Smoot DT, James SP, Panigrahi P. Loss of *Helicobacter pylori* hemagglutination with serial laboratory passage and correlation of hemagglutination with gastric epithelial cell adherence. Pathobiology. 1996;64:247–54.

43. Somerville GA, Beres SB, Fitzgerald JR, DeLeo FR, Cole RL, Hoff JS, Musser JM. In vitro Serial Passage of *Staphylococcus aureus*: Changes in Physiology, Virulence Factor Production, and *agr* Nucleotide Sequence. J Bacteriol. 2002;184:1430–7.

44. Asakura H, Kawamoto K, Okada Y, Kasuga F, Makino S, Yamamoto S, Igimi S. Intra host passage alters SigB-dependent acid resistance and host cell-associated kinetics of *Listeria monocytogenes*. Infect Genet Evol. 2012;12:94–101.

45. Muthukrishnan G, Quinn GA, Lamers RP, Diaz C, Cole AL, Chen S, Cole AM. Exoproteome of *Staphylococcus aureus* reveals putative determinants of nasal carriage. J Proteome Res. 2011;1:2064–78.

46. Henderson B, Martin A. Bacterial virulence in the moonlight: multitasking bacterial moonlighting proteins are virulence determinants in infectious disease. Infect Immun. 2011;79:3476–91.

47. Peng Z, Krey V, Wei H, Tan Q, Vogelmann R, Ehrmann MA, Vogel RF. Impact of actin on adhesion and translocation of *Enterococcus faecalis*. Arch Microbiol. 2014;196:109–17.

48. Rogers EA, Das A, Ton-That H. Adhesion by pathogenic corynebacteria. Adv Exp Med Biol. 2011;715:91–103.

49. Lazzazzera BA, Solomon J, Grossman AD. An exported peptide functions intracellularly to contribute to cell density signaling in B. subtilis. Cell. 1997;13:917–25.

50. Rees MA, Stinear TP, Goode RJ, Coppel RL, Smith AI, Kleifeld O. Changes in protein abundance are observed in bacterial isolates from a natural host. Front Cell Infect Microbiol. 2015;14(5):71.

51. Danelishvili L, Stang B, Bermudez LE. Identification of *Mycobacterium avium* genes expressed during in vivo infection and the role of the oligopeptide transporter OppA in virulence. Microb Pathog. 2014;76:67–76.

52. Wilson MJ, Brandon MR, Walker J. Molecular and biochemical characterization of a protective 40-kilodalton antigen from *Corynebacterium pseudotuberculosis*. Infect Immun. 1995;63:206–11.

53. Silva JW, Droppa-Almeida D, Borsuk S, Azevedo V, Portela RW, Miyoshi A, et al. *Corynebacterium pseudotuberculosis* cp09 mutant and cp40 recombinant protein partially protect mice against caseous lymphadenitis. BMC Vet Res. 2014;20(10):965.

54. Hodgson AL, Tachedjian M, Corner LA, Radford AJ. Protection of sheep against caseous lymphadenitis by use of a single oral dose of live recombinant *Corynebacterium pseudotuberculosis*. Infect Immun. 1994;62:5275–80.

55. McNamara PJ, Bradley GA, Songer JG. Targeted mutagenesis of the phospholipase D gene results in decreased virulence of *Corynebacterium pseudotuberculosis*. Mol Microbiol. 1994;12:921–30.

56. Leiting WU, Jianping XI. Comparative genomics analysis of *Mycobacterium* NrdH redoxins. Microb Pathog. 2010;48:97–102.

57. Pacheco LG, Castro TL, Carvalho RD, Moraes PM, Dorella FA, Carvalho NB, et al. A Role for Sigma Factor σ(E) in *Corynebacterium pseudotuberculosis* Resistance to Nitric Oxide/Peroxide Stress. Front Microbiol. 2012;3:126.

58. Silva WM, Carvalho RD, Soares SC, Bastos IF, Folador EL, Souza GH, et al. Label-free proteomic analysis to confirm the predicted proteome of *Corynebacterium pseudotuberculosis* under nitrosative stress mediated by nitric oxide. BMC Genomics. 2014;15:1065.

59. Samanovic MI, Ding C, Thiele DJ, Darwin KH. Copper in microbial pathogenesis: meddling with the metal. Cell Host Microbe. 2012;16:106–15.

60. Wolschendorf F, Ackart D, Shrestha TB, Hascall-Dove L, Nolan S, Lamichhane S, et al. Copper resistance is essential for virulence of *Mycobacterium tuberculosis*. Proc Natl Acad Sci U S A. 2011;25:1621–6.

61. Rowland JL, Niederweis M. A multicopper oxidase is required for copper resistance in *Mycobacterium tuberculosis*. J Bacteriol. 2013;195:3724–33.

7

Development of a new fluorescent reporter:operator system: location of AraC regulated genes in *Escherichia coli* K-12

Laura E. Sellars[1], Jack A. Bryant[1], María-Antonia Sánchez-Romero[2], Eugenio Sánchez-Morán[3], Stephen J. W. Busby[1] and David J. Lee[1,4]*

Abstract

Background: In bacteria, many transcription activator and repressor proteins regulate multiple transcription units that are often distally distributed on the bacterial genome. To investigate the subcellular location of DNA bound proteins in the folded bacterial nucleoid, fluorescent reporters have been developed which can be targeted to specific DNA operator sites. Such Fluorescent Reporter-Operator System (FROS) probes consist of a fluorescent protein fused to a DNA binding protein, which binds to an array of DNA operator sites located within the genome. Here we have developed a new FROS probe using the *Escherichia coli* MalI transcription factor, fused to mCherry fluorescent protein. We have used this in combination with a LacI repressor::GFP protein based FROS probe to assess the cellular location of commonly regulated transcription units that are distal on the *Escherichia coli* genome.

Results: We developed a new DNA binding fluorescent reporter, consisting of the *Escherichia coli* MalI protein fused to the mCherry fluorescent protein. This was used in combination with a Lac repressor:green fluorescent protein fusion to examine the spatial positioning and possible co-localisation of target genes, regulated by the *Escherichia coli* AraC protein. We report that induction of gene expression with arabinose does not result in co-localisation of AraC-regulated transcription units. However, measurable repositioning was observed when gene expression was induced at the AraC-regulated promoter controlling expression of the *araFGH* genes, located close to the DNA replication terminus on the chromosome. Moreover, in dividing cells, arabinose-induced expression at the *araFGH* locus enhanced chromosome segregation after replication.

Conclusion: Regions of the chromosome regulated by AraC do not colocalise, but transcription events can induce movement of chromosome loci in bacteria and our observations suggest a role for gene expression in chromosome segregation.

Keywords: FROS, GFP, Fluorescent microscopy, Chromosome, Nucleoid, *Escherichia coli*

Background

Bacterial nucleoids are highly compacted structures composed of chromosomal DNA, nucleoid structuring proteins and RNA [1]. The DNA within the *Escherichia coli* K-12 nucleoid is folded into a structure consisting of four independently folded macrodomains, and two non-structured regions [2–4]. Each domain is located at a

* Correspondence: David.lee@bcu.ac.uk
[1]Institute of Microbiology and Infection, School of Biosciences, University of Birmingham, Edgbaston, Birmingham B15 2TT, UK
[4]Department of Life Sciences, Birmingham City University, Edgbaston, Birmingham B15 3TN, UK
Full list of author information is available at the end of the article

distinct position within the cell and the DNA within each domain appears isolated from the rest of the chromosome. Despite this, there is evidence to suggest that, at some level, the nucleoid organisation allows for spatial repositioning of active transcription units and clusters of commonly regulated genes. Qian et al. [5] exploiting a chromatin conformation capture technique, demonstrated that the *E. coli* GalR transcription repressor protein, associated with DNA target sites in different macrodomains, could co-localise. Also, a plasmid-encoded transcription unit can re-locate to particular cellular positions when being actively expressed [6].

To investigate these points, we have exploited the *E. coli* AraC regulon. AraC is a transcription activator that regulates genes involved in the uptake and metabolism of arabinose. AraC binds to its DNA target in the absence of arabinose, and activates transcription of four transcription units, located in three different macrodomains, only in the presence of arabinose [7]. Thus, in this study we have introduced Fluorescent Reporter-Operator System probes (FROS probes) [8–12], adjacent to AraC regulated promoters, to observe their cellular location and any spatial repositioning that occurs upon induction of transcription by arabinose. To facilitate this, we developed a FROS probe based on the *E. coli* MalI DNA binding protein [13, 14], fused to mCherry fluorescent protein, and its cognate DNA target site. In combination with a modified LacI:GFP FROS probe, we have tagged the chromosome of *E. coli* strain MG1655, adjacent to AraC regulated genes, and determined the relative cellular locations by fluorescence microscopy. We show that AraC-regulated genes, within different macrodomains, do not co-localise in the cell. However, we show that the *araFGH* operon, which is near to the replication terminus, is spatially repositioned upon induction of transcription. This was particularly evident in dividing cells, where it was observed that induction of transcription facilitated separation of newly-replicated sister chromatids.

Methods
Bacterial strains, plasmids and growth conditions
All bacterial strains and plasmids used in this study are listed in Additional file 1. For microscopy experiments, strains were grown in M9 minimal media, supplemented with 0.3% fructose and 0.1% casamino acids, at 23 °C for 24 h. Cultures were diluted 1:50 into fresh media and grown for a further 5–6 h until OD_{650} reached approximately 0.1. For cultures supplemented with sugars, a final concentration of 0.3% of the required sugar was added to the culture for 1 min before slides were prepared [15]. For cultures supplemented with erythromycin (20 µg/ml) or rifampicin (50 µg/ml), the antibiotics were added for 15 min prior to the addition of arabinose.

Construction of plasmids for MalI FROS
pLER108, carrying the *malI::mcherry* fusion, is a derivative of pACYC184 and carries resistance to chloramphenicol and contains the *p15A* origin of replication. The *malI* promoter and gene were amplified from the plasmid pACYCMalI using oligo's D63433 and D71192 (Additional file 2) and digested with enzymes HindIII and KpnI and ligated into HindIII and KpnI digested pLER101, creating pLER104. Into this plasmid, the *mCherry* gene, which had been amplified from pmCherry-N1

using oligos D71000 and D71001, was ligated on a KpnI - MfeI digested fragment, resulting in a *malI:mCherry* gene fusion. This fusion was amplified using oligos D71850 and D72002 and the fragment cut with NsiI and HindIII was ligated into pJW15Δ100 to replace the *malI* promoter with the *melR* promoter, creating pLER105. Oligos D77566 and D77567 were used to amplify the promoter and fusion, the fragment was digested with HindIII and MfeI and ligated into pLER101, creating pLER108.

An array of MalI binding sites was created using the iterative PCR based method described by Lau et al., 2003 [16]. Briefly, MalI binding sites were incorporated into pUC19 using oligos with a 5′ end consisting of a MalI binding site and a 3′ end consisting of pUC19 homology (D71689 and D71690). Thus, using pUC19 as a template for PCR, these oligos were used to create a product that could be ligated to form a plasmid containing 2 MalI binding sites, flanked on one side by an XbaI restriction site and on the other side by NheI and HindIII restriction sites. This plasmid was used to generate both vector, by digesting with NheI and HindIII, and insert, by digesting with XbaI and NheI: ligation of these two products generated a new plasmid that contained 4 MalI binding sites separated by a hybrid XbaI/NheI site This was repeated until there were 20 MalI binding sites (MalO), creating pUCMal20.

Construction of gene doctoring donor plasmids
Gene doctoring donor plasmids were derived from pJB32 [17]. These carry the 22 *lac* operator sites (LacO array) or MalO array and a kanamycin cassette, flanked by 500 bp regions of homology from both sides of the insertion site, adjacent to either the *araBAD*, *araJ* or *araFGH* for MalO, or adjacent to either *araBAD*, *araJ* or *dps* for LacO. Oligonucleotides were designed to amplify 500 bp upstream of each insertion site, (Additional file 3) inserting a MfeI site upstream and a XmaI site downstream. This fragment was digested with MfeI and XmaI and ligated into MfeI and XmaI digested pJB32. Oligonucleotides were also designed to amplify 500 bp downstream of each insertion site, and insert a NheI site upstream and SacI site downstream. This product was digested with NheI and SacI and ligated with vector prepared from the previous ligation, digested with the same enzymes. Into the resulting plasmids, the LacO and MalO arrays were inserted: the LacO array was digested from pPM301 on a BglII/NheI fragment and the MalO array was digested from pUCMal20 on an XhoI/NheI. The plasmids that were generated are listed in Additional file 2.

Chromosomal recombination
Gene doctoring was used to make chromosomal modifications using the donor plasmids constructed as described

above [18]. MalO arrays were inserted into the chromosome of MG1655, LacO arrays were inserted into strain DL02. For two colour analysis, the MalO array was inserted into strains already harbouring a LacO array. Candidates were screened for the insert by colony PCR using oligonucleotides designed to bind to the chromosome outside the homology regions. The kanamycin resistance cassette was removed from the chromosome using flippase recombinase (FLP) expressed from plasmid pCP20 [19]. The resulting strains are listed in Additional file 1.

Microscopy

Bacterial cultures were grown for 24 h at 23 °C [20] with aeration in M9 minimal salts media supplemented with 0.3% fructose, 2 mM $MgSO_4$, 0.1 mM $CaCl_2$, 0.1% casamino acids and, if necessary, 17.5 μg/ml chloramphenicol. Cultures were diluted 1:50 and grown under the same conditions until cultures reached OD_{650} 0.1. 1 ml of culture was removed and washed 3 times with PBS then resuspended in 20 μl Hoechst 33,258 solution containing 5 μg/ml Hoechst 33,258 in PBS containing 40% glycerol. 5 μl were loaded onto poly-L-lysine coated slides and a cover slip applied. Slides were imaged using a Nikon Eclipse 90i microscope, Nikon Intensilight C-HGFI lamp, Hamamatsu ORCA ER camera (1344 × 1024 pixels, pixel size 6.45 μm) and Nikon Plan Apo VC 100× Oil immersion lens (Numerical Aperture 1.4), with a final optical magnification of 100×. A DAPI filter set was used for visualising the Hoechst 33,258 stained nucleoid, FITC filter set for GFP and TxRed filter set for mCherry. Cells were also imaged using brightfield. Microscopy was carried out at room temperature, within 30 min of slides being prepared.

Analysis of microscopy

Microscope images were analysed using Image J software. To determine the position of foci within cells, the measuring function was used to measure both the length of the cell and the distance from the focus to the nearest pole. The position of the focus within the cell was then calculated and is presented relative to the length of the cell, which was set at an arbitrary value of 1. For cells containing two foci, the focus nearest to a pole was designated as the '1st of 2 foci', and the distance from the focus to the nearest pole was measured. The distance from the '2nd of 2 foci' to the same pole was then measured. Measurements were taken from at least 300 cells. Where the data are presented on a scatter plot (Fig. 5), the relative position of the first focus is plotted on the x axis, and the relative position of the second focus is plotted on the y axis. To analyse co-localisation, the position of each of the two foci was measured using NIS

elements software (Nikon), which provided a measurement in μm. To determine if the data, pre and post arabinose induction were significantly different, ANOVA or T-tests were done suing Excel software. Cells that had multiple foci of the same colour were not included in this analysis.

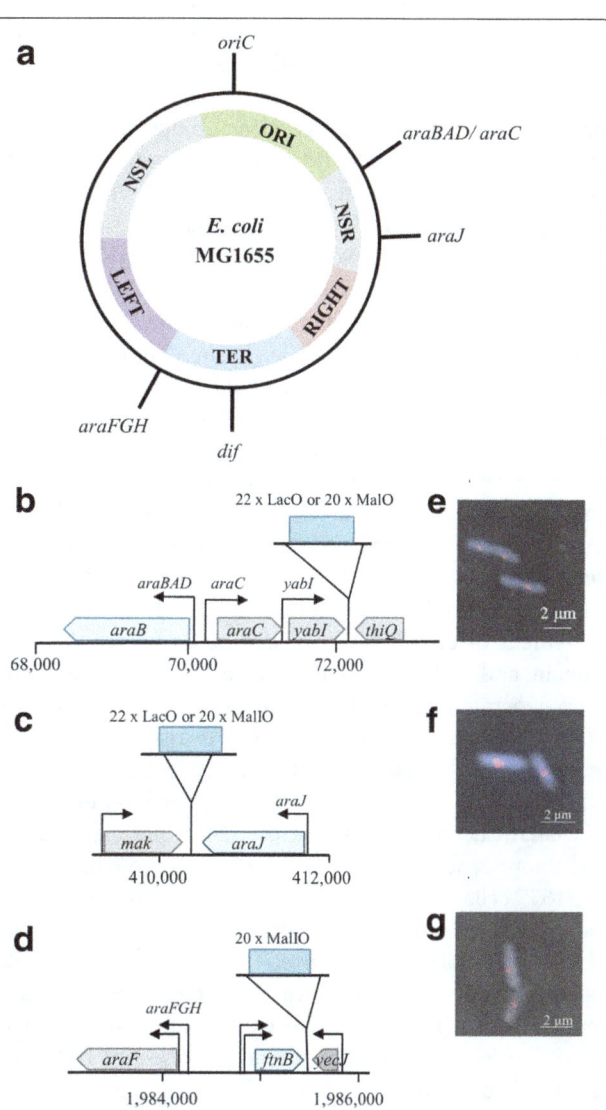

Fig. 1 FROS Tagging AraC regulated promoters. **a** A schematic representation of the circular chromosome of *E. coli* strain MG1655. Macrodomains of chromosome organisation are displayed and the origin of replication (*oriC*) and the region of termination (*dif*) are highlighted [3, 24]. The positions of AraC-regulated promoters and the *dps* gene on the circular chromosome are shown. Multiple *lac* operators (LacO) or MalI DNA binding sites (MalO) were inserted adjacent to *araBAD* (**b**), adjacent to *araJ* (**c**) or adjacent to *araFGH* (**d**). Panels (**e**), (**f**) and (**g**) show examples of fluorescent foci derived from MalI:mCherry binding to a 20 MalO array inserted adjacent to *araBAD*, *araJ* and *araFGH* respectively. The images shown are merged images of MalI;mCherry foci and Hoechst 33,258 stained chromosomes

Results

Mall as a FROS reporter system

Several methodologies have been employed to examine nucleoid structure, one of which is the use of Fluorescent Reporter Operator Systems (FROS) [8–12]. Typically, a DNA binding protein (Reporter), fused to a fluorescent tag, is targeted to an array of DNA target sites (Operators), with resulting fluorescent foci being visualised by microscopy. The *E. coli* K-12 Mall protein is a transcription repressor associated with the Mal operon, and is a member of the GalR/LacI family of DNA binding proteins [13, 14]. The *mall* gene is located on its chromosome, convergent to the *malXY* operon. When expressed, Mall binds to a 16 bp target site at both the *mall* and *malXY* promoters to repress transcription. To generate an array of Mall binding sites, we used an iterative PCR procedure, followed by a cloning approach to build the required number of DNA binding sites in a plasmid. Hence, 20 DNA sites for Mall were incorporated into the Mall operator array (MalO), which was then targeted to specific positions in the chromosome of *E. coli* strain MG1655 using the gene doctoring recombineering method [18]. The array was inserted at three chromosomal targets: adjacent to the *araBAD*, *araJ* and *araFGH* promoter regions (Fig. 1a – d and Additional file 4). The *araJ* and *araBAD* loci are situated on the *E. coli* K-12 chromosome within the non-structured right domain, more than 1 Mbp away from the *araFGH* operon which is within the Ter macrodomain. Hence, the co-localisation and movement of commonly regulated genes within the same domain, and within different domains, could be examined.

To generate a Mall:mCherry fusion protein, the *mall:m-Cherry* gene fusion was cloned downstream of the *melR* promoter in plasmid pACYC184, creating plasmid pLER108. This resulted in constitutive, low level expression of Mall:mCherry. To examine the DNA binding efficiency and fluorescence derived from the fusion protein, wildtype MG1655 cells, and cells carrying the MalO array situated at the *araBAD*, *araJ* and *araFGH* loci were transformed with pLER108. Fluorescence derived from cells in the mid-logarithmic phase of growth was examined using epifluorescence microscopy. In the absence of a MalO array, there are no visible foci and the background fluorescence in the cell was negligible (Additional File 5). The images in Fig. 1e – g show MG1665 cells that contain the MalO array at the *araBAD*, *araJ* and *araFGH* loci respectively. Foci derived from Mall:mCherry bound at the MalO array are clearly observed in each case. Thus Mall:mCherry bound to the MalO array is functional as a reporter:operator system for FROS.

Modification of the LacI:GFP FROS reporter system and comparison with Mall:mCherry

Previous studies have visualised the LacI:GFP fusion protein bound to a large chromosomal target array containing 256 copies of the LacI DNA binding site [11].

Since we demonstrated that Mall foci could be readily visualised bound to an array of 20 Mall DNA binding sites, we sought to reduce the number of LacI binding sites in an array. In our previous work, we observed LacI binding to co-localised plasmids, corresponding to approximately 25 *lacI* target sites (plasmid copy number of 5: each plasmid containing 5 *lacI* DNA binding sites) [6]. Thus, we generated a LacO array, consisting of 22 *lacI* binding sites, which we introduced at the *araBAD* and *araJ* loci in MG1655 cells harbouring a *lacI:gfp* chromosome fusion at the natural *lacI* loci (Fig. 1b and c).

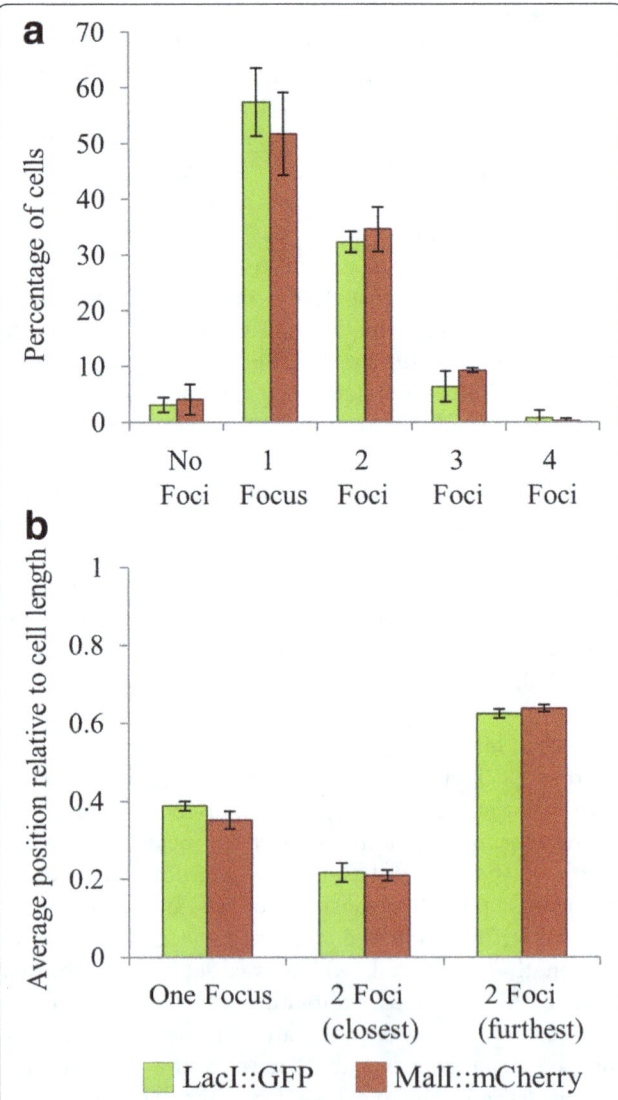

Fig. 2 Comparison of Mall and LacI FROS probes. **a** The number of fluorescent foci were counted in 300 individual cells, grown in M9 minimal medium without arabinose, containing either the MalO or LacO arrays inserted adjacent to *araBAD*. **b** The distance to the nearest cell pole for foci in cells that contained 1 or 2 foci. For 2 foci analysis, the distance of the focus nearest to a pole was measured (closest) and the distance of the second focus to the same pole was then measured (furthest). Error bars represent the standard deviation

Cells harbouring the LacIO or MalO arrays were then examined using epifluorescence microscopy and the number of foci and the position of the foci relative to the length of the cell was determined. The results in Fig. 2a, for the *araBAD* locus, show that the distribution of cells containing foci was comparable when the number of foci derived from the MalI and LacI FROS probes was counted. One focus was observed in the majority of cells, with 2 foci observed in a large proportion of cells which were actively undergoing chromosome segregation. Based on these data, the average number of foci per cell was calculated to be approximately 1.4, which is consistent with our previous measurements of the the average numbers of chromosomes per cell in these growth conditions. When the average position of the foci from cells containing either 1 or 2 foci was then measured, with respect to total cell length, the data derived from the two FROS probes was comparable (Fig. 2b). This indicates that the *araBAD* locus is similarly positioned within the cell when tagged with either the MalI or LacI FROS reporter systems.

Co-localisation of AraC regulated promoters

To assess whether AraC regulated promoters co-localised, strains of *E. coli* were generated that contained a LacI:GFP FROS probe adjacent to the *araBAD* promoter and a MalI:mCherry FROS probe at either the *araJ* or *araFGH* promoters. Individual cells of these strains, grown either in the presence or absence of arabinose, were visualised using fluorescence microscopy (Fig. 3a). Cells containing different numbers of each fluorescent cluster were observed, containing clear and distinct foci derived from GFP and mCherry. To calculate the distance between the MalI:mCherry foci and the LacI:GFP foci, the distance from the GFP focus to the closest pole was measured, and subtracted from the distance of the mCherry focus to the same pole. Hence, the distances between the *araBAD* and *araJ* promoters, and the *araBAD* and *araFGH* promoters were calculated in >500 individual cells, grown in the presence or absence of arabinose (Fig. 3b and c). The principal observation was that the distance between the foci varied substantially

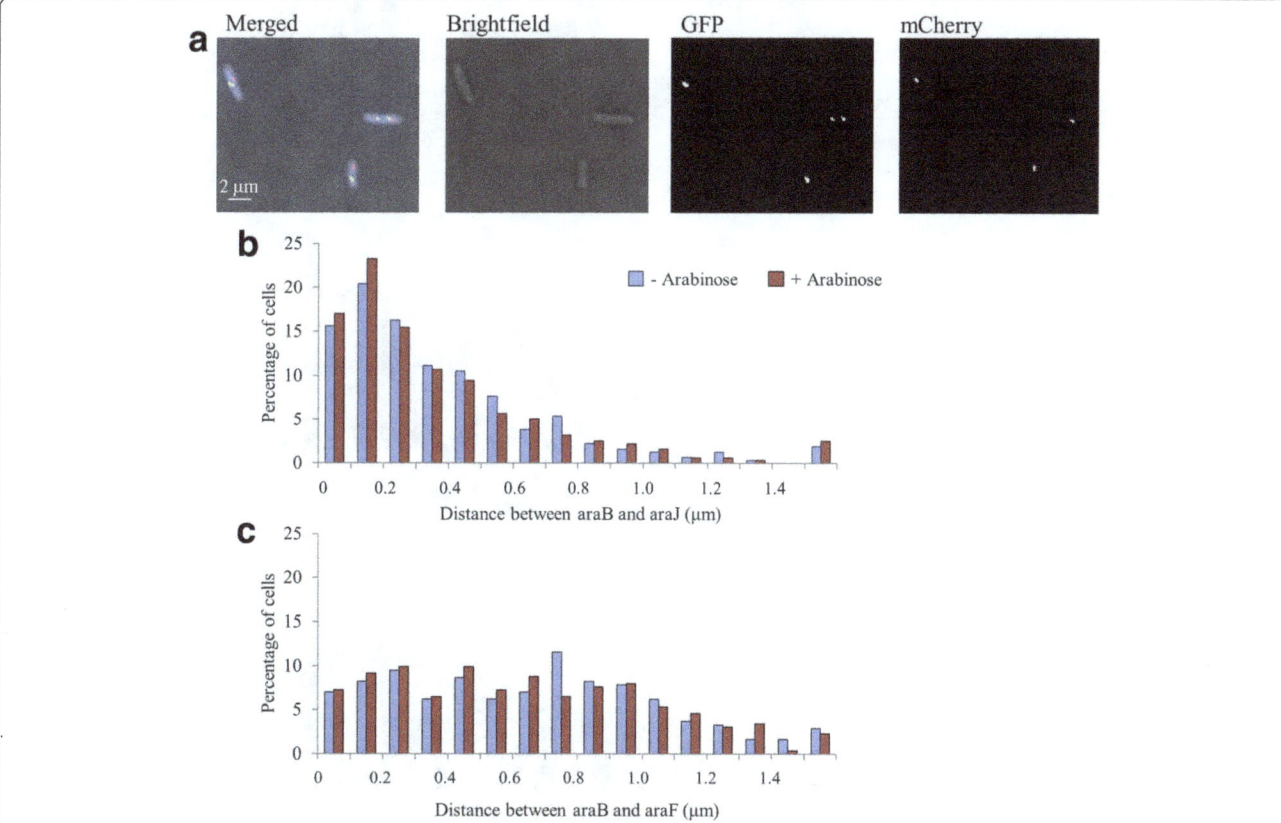

Fig. 3 Colocalisation of genes regulated by AraC. **a** The figure shows a dual fluorescence image of strain LR31, carrying a LacO array at *araBAD* and a MalO array at *araJ* stained with Hoechst 33,258. **b** and **c** The bar charts show the distance between two distal chromosomal locations, each independently tagged with different FROS reporters. **b** Distance measurements between the *araBAD* locus, tagged with a LacO array, and the *araJ* locus, tagged with a MalO array, were calculated in 300 individual cells. Absolute distances between the two chromosomal locations in the presence and absence of the inducer, arabinose, are plotted. **c** Distance measurements between the *araBAD* locus, tagged with a LacO array, and the *araFGH* locus, tagged with a MalO array, were calculated in 300 individual cells. Absolute distances between the two chromosomal locations in the presence and absence of the inducer, arabinose, are plotted

throughout the population, but this did not significantly alter upon addition of arabinose. The range of distances between *araBAD* and *araJ* probes (average 0.37 μm) was less than between the *araBAD* and *araFGH* probes (average 0.64 μm). This was expected since *araBAD* and *araJ* are located within the same macrodomain, whereas the *araBAD* and *araFGH* are in different domains. Thus, unlike previously reported with GalR regulated promoters [5], the AraC regulated promoters do not appear to co-localise in the bacterial nucleoid.

Location and dynamics of AraC regulated promoters

Since AraC-regulated promoters did not appear to co-localise, next we examined whether individual promoter regions were repositioned upon induction. To do this, *E.*

coli strains containing LacI:GFP FROS probes at the *araBAD* and *araJ* loci, and the MalI:mCherry FROS probe at the *araFGH* locus, were grown in the presence or absence of arabinose. Cells were analysed by fluorescence microscopy, and individual, non-dividing cells containing a single fluorescent focus were analysed. The distance from each focus to the nearest cell pole was measured, and this value was divided by the total cell length, thereby providing a position relative to total cell length (Fig. 4). Foci derived from the FROS probes positioned near to the *araBAD* and *araJ* regions did not reposition when the promoters were induced by arabinose (Fig. 4a and b). However, in a small proportion of the cells, the FROS probe adjacent to the *araFGH* operon relocated away from the cell pole towards the centre of the cell

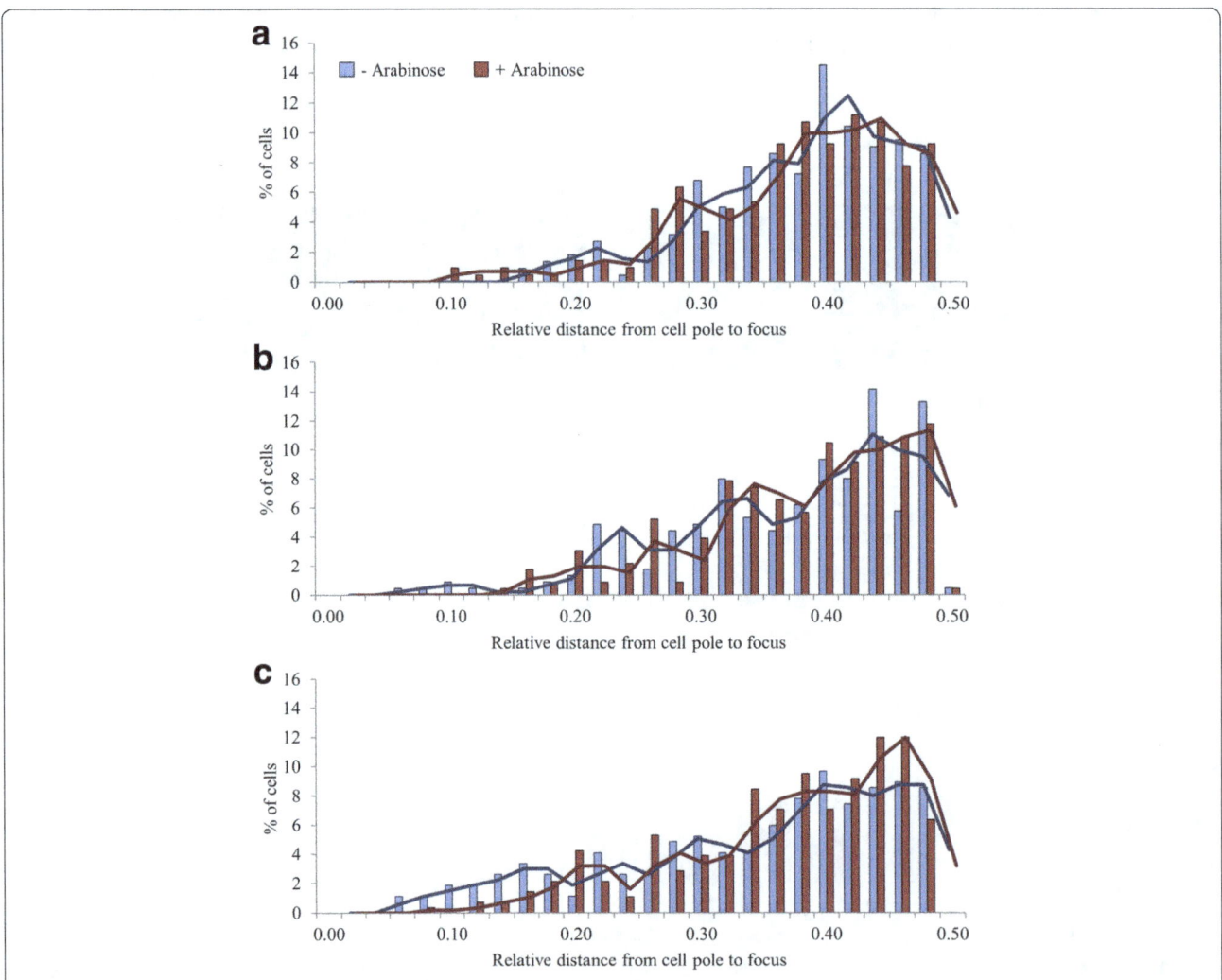

Fig. 4 Relative cellular location of AraC regulated promoters in the presence and absence of inducer. The distances between fluorescent foci and the nearest cell pole was measured in 300 individual cells containing a single fluorescent foci derived from FROS probes adjacent to (**a**) *araBAD*, strain LR06, (**b**) *araJ*, strain LR39 and (**c**) *araFGH*, strain LR38. Distances are plotted, relative to cell length, in the presence and absence of the inducer, arabinose. For these experiments, *araBAD* and *araJ* were tagged with a LacO array and *araFGH* was tagged with a MalO array. The experiment was repeated on 3 separate occasions, with the same outcome observed. Associated *P*-values for uninduced compared to induced cells are: for *araBAD*, 0.556; for *araJ*, 0.252; and for *araF*, 0.005

upon induction (Fig. 4c: redistribution of cells with a focus between 0.05 and 0.18 upon induction with arabinose).

A similar relocation of the *araFGH* locus was observed in cells containing two foci. Fig. 5 shows the relative position of each of the two foci associated with an AraC-regulated promoter. When grown in the presence or absence of arabinose, no discernible repositioning was observed with the probe at the *araBAD* or *araJ* loci. However, for the *araFGH* locus, the focus closest to the cell pole repositioned, with an overall movement away from the cell pole. Thus, the two foci were repositioned relative to each other upon arabinose induction.

To examine further the movement of the *araFGH* locus upon induction, we studied the position of foci in cells at the point of division (Fig. 6a and b). These dividing cells were defined as cells that had two separate nucleoids when stained with Hoechst 33,258 but which did not appear to be two distinct, separate cells when viewed by brightfield microscopy. Such cells accounted for 5–15% of all cells, and in uninduced conditions, approximately 35% of these contained a single *araFGH* focus (Fig. 6a & c). In contrast, when the FROS probe was positioned at the *araBAD* locus, which is more proximal to

the origin of replication, very few cells had a single focus (2%), with 98% of cells containing at least 2 foci. In conditions of growth supplemented with glucose or arabinose, no change in the number of *araBAD* foci in each individual cell was observed. Similarly, no change in the number of fluorescent foci in each individual cell was observed when the FROS probe was located adjacent to the *dps* promoter that was used as a control region of the chromosome, unaffected by arabinose. However, at the *araFGH*, there was a clear reduction in the number of cells containing only one foci locus in the presence of arabinose, but not glucose. The observed shift from 37% of the population containing a single focus to 13% upon induction suggests that expression of the *araFGH* operon assists separation of newly replicated sister chromatids. To test this, cultures were supplemented with arabinose, to induce expression of the *araFGH* operon, and either rifampicin: to inhibit transcription, or erythromycin: to inhibit translation. In both cases, the addition of the inhibitors prevented the separation of foci (Fig. 6d). Treatment of cells with these antibiotics is likely to impact upon the transcription and translation of every gene within the cell. Thus to confirm that the processes of gene expression at the *araFGH* operon are directly responsible for our

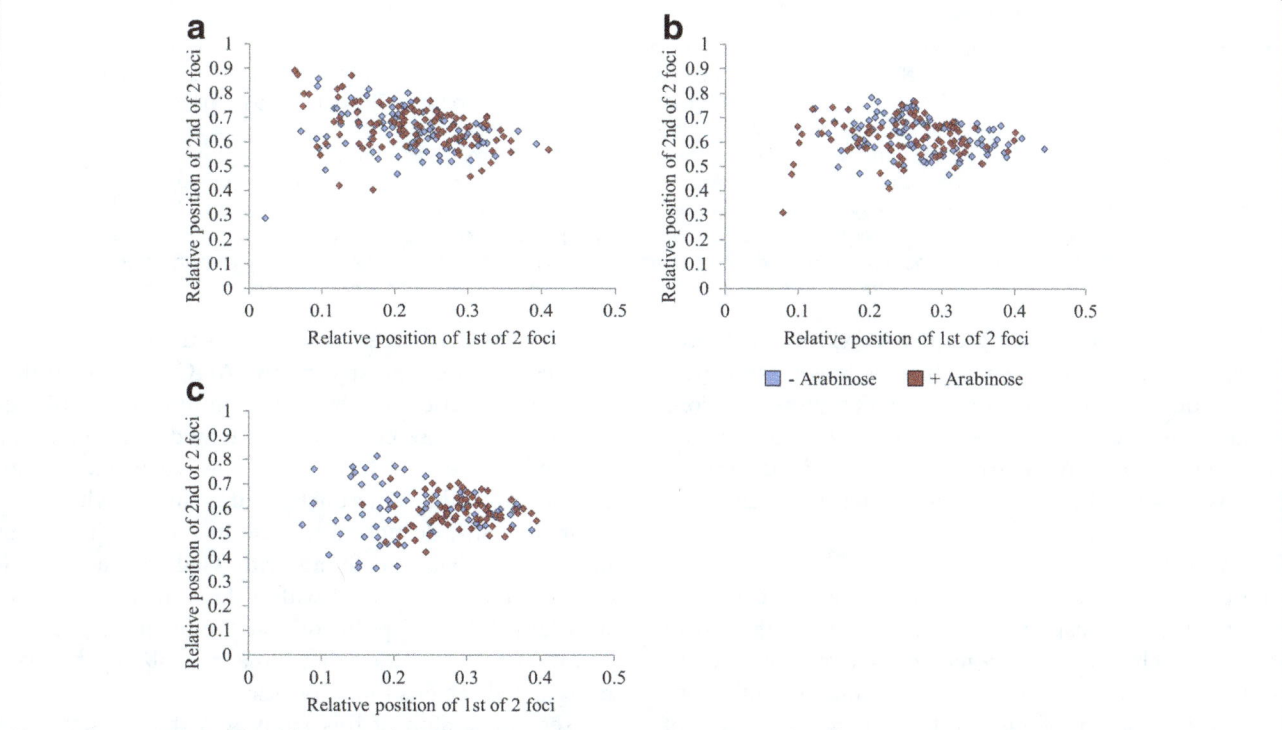

Fig. 5 Relative positions of two fluorescent foci in cells containing FROS probes adjacent to different AraC-regulated promoters in the presence and absence of inducer. The distances between foci were measured in 300 individual cells containing two fluorescent foci derived from FROS probes adjacent to (**a**) *araBAD*, strain LR06, (**b**) *araJ*, strain LR39 and (**c**) *araFGH*, strain LR38. The distance between the focus closest to a cell pole and that cell pole was first measured and calibrated to the relative cell length (1st of 2 foci). The distance from the cell pole to the second focus was then measured relative to cell length. The cellular position of the second focus was then plotted against the position of the first focus from cells grown in the presence and absence of the inducer, arabinose. For these experiments, *araBAD* and *araJ* were tagged with a LacO array and *araFGH* was tagged with a MalO array

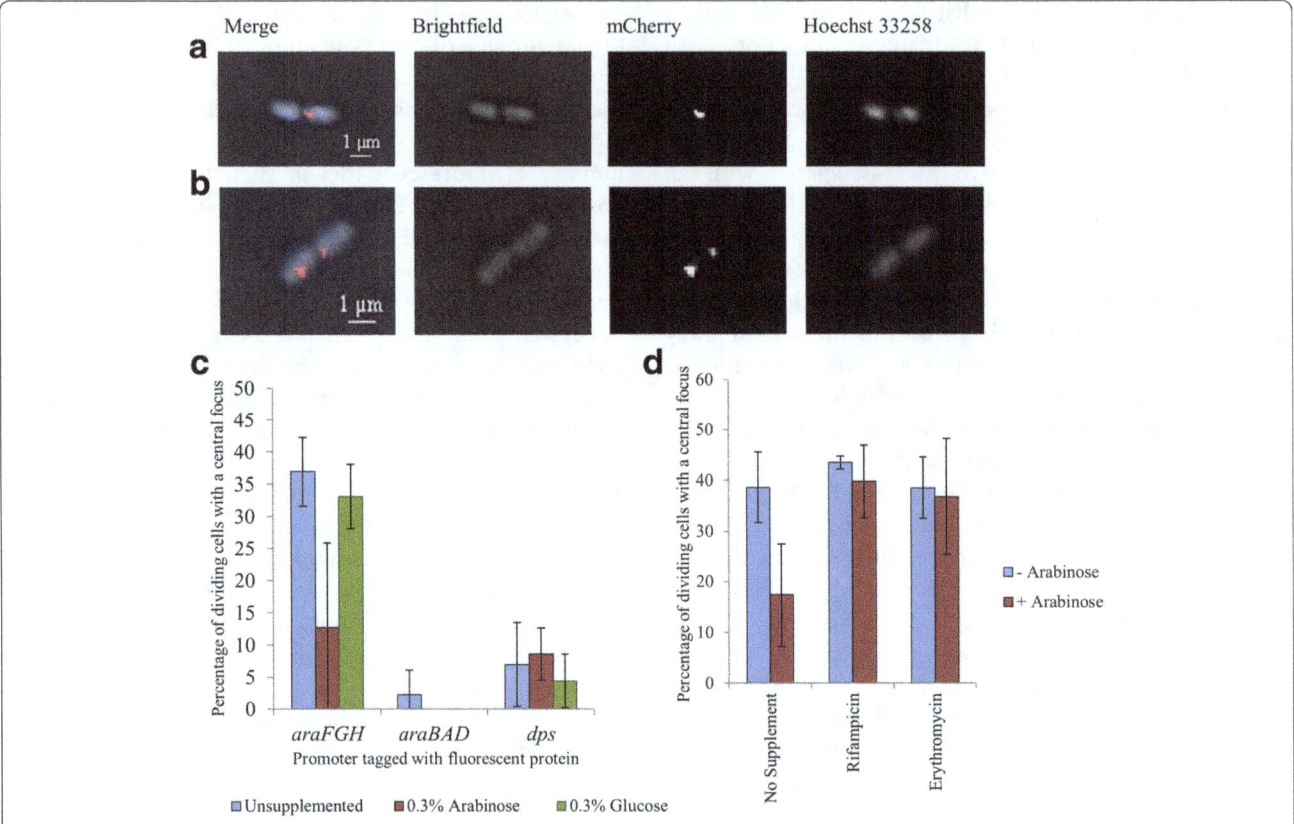

Fig. 6 Gene expression drives chromosome separation. The number of fluorescent foci in cells at the point of division were counted in the presence and absence of the inducer, arabinose. Dividing cells were defined as having two separate nucleoids when stained with Hoechst 33,258 but which were not separate cells when viewed under brightfield microscopy. Cells containing a 20 MalO array adjacent to the *araFGH* locus, predominantly contained either a single centrally located focus (**a**) or two distinct foci (**b**). **c** The number of foci in 300 individual cells were counted, from bacterial cultures grown in minimal medium, supplemented with 0.3% arabinose, or supplemented with 0.3% glucose. For this experiment, *araBAD* and *dps* were tagged with a LacO array and *araFGH* was tagged with a MalO array. For each position, the percentage of dividing cells containing a single central focus is plotted. The values for the *araBAD* tagged strain, induced with arabinose and glucose, were 0%. **d** The impact of inhibiting the processes of transcription of translation on the number of cells containing a single focus derived from a MalO array adjacent to *araFGH*. Growing cultures were supplemented with rifampicin or erythromycin prior to induction with arabinose. The number of cells containing a single focus, from 300 individual cells, is plotted and the error bars represent the standard deviation

observations, direct targeting of the individual DNA promoter elements and ribosome binding sites of the promoters driving expression of fluorescent protein fusions would be necessary. Nevertheless, our data provide compelling evidence that chromosome separation at the *araFGH* locus is enhanced by the processes of gene expression.

Discussion

The aim of this study was to investigate possible transcription factor clustering in a bacterial nucleoid and to investigate changes in response to transcription. Hence, we sought to visualise nucleoid re-organisation and identify co-localisation of distant loci upon expression of commonly regulated genes. To facilitate this, we developed and validated a new fluorescent reporter-operator system, based on the *E. coli* transcription repressor protein, MalI, which was fused to the mCherry fluorescent protein. In combination with a LacI:GFP reporter, we tagged the chromosome of *E. coli* strain MG1655, with

MalI or LacI DNA operator binding site arrays, adjacent to genes that are regulated by the AraC protein, so that the cellular location and transcription induced spatial repositioning of these commonly regulated genes could be monitored. We observed that AraC-regulated genes, located within the same nucleoid domain, or within different domains, do not co-localise in the cell. This is in contrast to what was found with GalR-regulated promoters, which are located within different domains, yet co-localise [5]. We speculate that the ability of GalR to tetramerise may be a driving force in enabling the GalR co-regulated regions to co-localise.

A second finding of this study was that induction of expression of a transcription unit near to the terminus of DNA replication resulted in enhanced separation of newly replicated chromosomes at that locus. We found that this was dependent on both transcription and translation, as inhibition of either prevented separation. We assume that the act of transcription is the driving force behind this observation, since transcription and translation are often coupled in bacteria [21], We suppose that transcription induced supercoiling may drive chromosome separation by enhancing the process of decatenation [22, 23], which is feasible since decatenation is facilitated by topoisomerase enzymes, and is thus impacted by DNA supercoiling.

Conclusion

We have developed resources that facilitate two colour FROS analysis of regions of the chromosome within *Escherichia coli* cells. Our investigations indicate that distal regions on the linear chromosome that are regulated by transcription regulator AraC do not colocalise in the folded nucleoid. Our explorations do however, suggest a role for transcription in facilitating chromosome separation post replication.

Additional files

Additional file 1: Strains used in this study.

Additional file 2: Plasmids used in this study.

Additional file 3: DNA oligonucleotides used in this study.

Additional file 4: Schematic diagram to show the insertion sites of FROS operators adjacent to (a) *araBAD*, (b) *araFGH* and (c) *mntH*.

Additional file 5: MalI:mCherry expressed from plasmid pLER108 in MG1655.

Abbreviations
FLP: Flippase recombinase; FROS: Fluorescent Reporter-Operator System

Acknowledgements
The authors would like to thank Stephen Bevan for assistance with plasmid construction.

Funding
This work was supported by funded by a BBSRC project grant [BB/J006076] and a Leverhulme Trust project grant [RPG-2013-003] to SJWB. The funding bodies were not involved in the design, collection and analysis of data, or the preparation of this manuscript.

Authors' contributions
LES, JAB and MASR performed experiments, and ESM conceived and designed experiments. SJWB and DJL conceived the study, its design and co-ordination, and drafted the manuscript. All authors read and approved the final manuscript.

Competing interests
The authors declare that they have no competing interests.

Author details
[1]Institute of Microbiology and Infection, School of Biosciences, University of Birmingham, Edgbaston, Birmingham B15 2TT, UK. [2]Departamento de Genética, Facultad de Biología, Universidad de Sevilla, 41080 Seville, Spain. [3]School of Biosciences, University of Birmingham, Edgbaston, Birmingham B15 2TT, UK. [4]Department of Life Sciences, Birmingham City University, Edgbaston, Birmingham B15 3TN, UK.

References

1. Dorman CJ. Genome architecture and global gene regulation in bacteria: making progress towards a unified model? Nat Rev Microbiol. 2013;11:349–55.
2. Espeli O, Mercier R, Boccard F. DNA dynamics vary according to macrodomain topography in the E. Coli chromosome. Mol Microbiol. 2008;68:1418–27.
3. Valens M, Penaud S, Rossignol M, Cornet F, Boccard F. Macrodomain organization of the Escherichia Coli chromosome. EMBO J. 2004;23:4330–41.
4. Dame RT, Kalmykowa OJ, Grainger DC. Chromosomal macrodomains and associated proteins: implications for DNA organization and replication in gram negative bacteria. PLoS Genet. 2011;7:e1002123.
5. Qian Z, Dimitriadis EK, Edgar R, Eswaramoorthy P, Adhya S. Galactose repressor mediated intersegmental chromosomal connections in Escherichia Coli. Proc Natl Acad Sci U S A. 2012;109:11336–41.
6. Sanchez-Romero MA, Lee DJ, Sanchez-Moran E, Busby SJ. Location and dynamics of an active promoter in Escherichia Coli K-12. Biochem J. 2012; 441:481–5.
7. Schleif R. AraC protein, regulation of the l-arabinose operon in Escherichia Coli, and the light switch mechanism of AraC action. FEMS Micro Rev. 2010; 34:779–96.
8. Carmi I, Kopczynski JB, Meyer BJ. The nuclear hormone receptor SEX-1 is an X-chromosome signal that determines nematode sex. Nature. 1998;396:168–73.
9. Gasser SM. Visualizing chromatin dynamics in interphase nuclei. Science. 2002;296:1412–6.
10. Kato N, Lam E. Detection of chromosomes tagged with green fluorescent protein in live Arabidopsis Thaliana plants. Genome Biol. 2001;2:11.
11. Robinett CC. Straight a, LiG, Willhelm C, SudlowG, Murray a, Belmont AS. In vivo localization of DNA sequences and visualization of large-scale chromatin organization using lac operator/repressor recognition. J Cell Biol. 1996;135:1685–700.
12. Straight AF, Belmont AS, Robinett CC, Murray AW. GFP tagging of budding yeast chromosomes reveals that protein-protein interactions can mediate sister chromatid cohesion. Curr Biol. 1996;6:1599–608.
13. Lloyd GS, Godfrey RE, Busby SJ. Targets for the MalI repressor at the divergent Escherichia Coli K-12 malX-malI promoters. FEMS Microbiol Lett. 2010;305:28–34.
14. Lloyd GS, Hollands K, Godfrey RE, Busby SJ. Transcription initiation in the Escherichia Coli K-12 malI-malX intergenic region and the role of the cyclic AMP receptor protein. FEMS Microbiol Lett. 2008;288:250–7.
15. Johnson CM, Schleif RF. In vivo induction kinetics of the arabinose promoters in Escherichia Coli. J Bacteriol. 1995;177:3438–42.
16. Lau IF, Filipe SR, Soballe B, Okstad OA, Barre FX, Sherratt DJ. Spatial and temporal organization of replicating Escherichia Coli chromosomes. Mol Microbiol. 2003;49:731–43.
17. Bryant JA, Sellars LE, Busby SJ, Lee DJ. Chromosome position effects on gene expression in Escherichia Coli K-12. Nucleic Acids Res. 2014;42:11383–92.
18. Lee DJ, Bingle LE, Heurlier K, Pallen MJ, Penn CW, Busby SJ, Hobman JL. Gene doctoring: a method for recombineering in laboratory and pathogenic Escherichia Coli strains. BMC Microbiol. 2009;9:252.
19. Cherepanov PP, Wackernagel W. Gene disruption in Escherichia Coli: TcR and KmR cassettes with the option of Flp-catalyzed excision of the antibiotic-resistance determinant. Gene. 1995;158:9–14.
20. Gordon GS, Sitnikov D, Webb CD, Teleman A, Straight A, Losick R, Murray AW, Wright A. Chromosome and low copy plasmid segregation in E. Coli: visual evidence for distinct mechanisms. Cell. 1997;90:1113–21.
21. McGary K. Nudler E RNA polymerase and the ribosome: the close relationship. Curr Opin Microbiol. 2013;16:112–7.
22. Witz G, Stasiak A. DNA supercoiling and its role in DNA decatenation and unknotting. Nucleic Acids Res. 2010;38:2119–33.
23. Liu LF, Wang JC. Supercoiling of the DNA template during transcription. Proc Natl Acad Sci U S A. 1987;84:7024–7.
24. Keseler IM, Mackie A, Peralta-Gil M, Santos-Zavalet A, Gama-Castro S, Bonavides-Martinez C, Fulcher C, Huerta AM, Kothari A, Krummenacker M, Latendresse M, Muniz-Rascado L, Ong Q, Paley S, Schroder I, Shearer AG, Subhraveti P, Travers M, Weerasinghe D, Weiss V, Collado-Vides J, Gunsalus RP, Paulsen I, Karp PD. EcoCyc: fusing model organism databases with systems biology. Nucleic Acids Res. 2013;41:9.

8

Tapping the biotechnological potential of insect microbial symbionts: new insecticidal porphyrins

Ana Flávia Canovas Martinez[1*], Luís Gustavo de Almeida[1], Luiz Alberto Beraldo Moraes[2] and Fernando Luís Cônsoli[1*]

Abstract

Background: The demand for sustainable agricultural practices and the limited progress toward newer and safer chemicals for use in pest control maintain the impetus for research and identification of new natural molecules. Natural molecules are preferable to synthetic organic molecules because they are biodegradable, have low toxicity, are often selective and can be applied at low concentrations. Microbes are one source of natural insecticides, and microbial insect symbionts have attracted attention as a source of new bioactive molecules because these microbes are exposed to various selection pressures in their association with insects. Analytical techniques must be used to isolate and characterize new compounds, and sensitive analytical tools such as mass spectrometry and high-resolution chromatography are required to identify the least-abundant molecules.

Results: We used classical fermentation techniques combined with tandem mass spectrometry to prospect for insecticidal substances produced by the ant symbiont *Streptomyces caniferus*. Crude extracts from this bacterium showed low biological activity (less than 10% mortality) against the larval stage of the fall armyworm *Spodoptera frugiperda*. Because of the complexity of the crude extract, we used fractionation-guided bioassays to investigate if the low toxicity was related to the relative abundance of the active molecule, leading to the isolation of porphyrins as active molecules. Porphyrins are a class of photoactive molecules with a broad range of bioactivity, including insecticidal. The active fraction, containing a mixture of porphyrins, induced up to 100% larval mortality ($LD_{50} = 37.7 \ \mu g.cm^{-2}$). Tandem mass-spectrometry analyses provided structural information for two new porphyrin structures. Data on the availability of porphyrins in 67 other crude extracts of ant ectosymbionts were also obtained with ion-monitoring experiments.

Conclusions: Insect-associated bacterial symbionts are a rich source of bioactive compounds. Exploring microbial diversity through mass-spectrometry analyses is a useful approach for isolating and identifying new compounds. Our results showed high insecticidal activity of porphyrin compounds. Applications of different experiments in mass spectrometry allowed the characterization of two new porphyrins.

Keywords: *Acromyrmex coronatus*, Bacterial symbionts, Sustainable pest control, Symbiosis, Tandem mass spectrometry

* Correspondence: anamartinez@usp.br; fconsoli@usp.br
[1]Laboratório de Interações em Insetos, Departamento de Entomologia e Acarologia, Escola Superior de Agricultura "Luiz de Queiroz", Universidade de São Paulo, Av Pádua Dias 11, 13418–900, Piracicaba, SP, Brazil
Full list of author information is available at the end of the article

Background

Despite the use of new chemistries of synthetic organic pesticides and new technologies such as genetically modified plants that express bacterial entomotoxins [1–3], the continued crop losses to insect pests necessitate the development of new tools to prevent reductions in yield and undesired effects on non-target organisms and the environment [4]. The need for integrative, sustainable management practices has led to the development of incentives and policies to support the use of integrated pest management [5], a strategy based on multiple control techniques, including the use of natural products.

The need for new compounds with insecticidal activity has increased, especially due to the evolution of insect resistance against the majority of existing insecticides and the necessity for target-specific and environmentally friendly molecules [6]. Microbes have proved to be a rich source of new bioactive molecules [7–10]. The diversity of microbes associated with insects and the selective pressures on microbes due to the range of insect habitats, have stimulated research on insect-associated microorganisms as an untapped resource for biotechnological exploitation [11–13]. Microbial symbiosis (sensu de Bary) [14] in leaf-cutting ants is well studied, including the diversity and mode of transmission of bacteria associated with the cuticle of these ants [15–17]. These ectosymbionts play a defensive role in this association by producing bioactive molecules to protect the mutualistic fungi that these ants cultivate as their food resource, from infections with parasitic fungi [18–21]. The mutualistic association of ants with bacteria has been under debate as some report *Pseudonocardia* being the mutualistic bacterium associated with the cuticle of leaf cutting ants [22, 23], while others believe the protective role of cuticle associated bacteria is provided by a diverse community [24–26]. Andersen et al. [27] provided data demonstrating *Pseudonocardia* as the prevalent bacterium growing on the laterocervical plates and pronotum of ants; but their study analysed different colonies belonging only to *Acromyrmex echinator* and a couple of other species from different genera. The prevalence of *Pseudonocardia* in the microbiota associated with the ant cuticle and their efficiency against parasitic fungi of the ant's fungus garden certainly contributes to their role as a mutualist of ants. However, other bacterial species from the cuticle of ants were demonstrated to be more efficient in controlling the growth of the parasitic fungus *Escovopsis* than *Pseudonocardia* associated with *Acromyrmex subterraneus brunneus* [16]. *Streptomyces*-produced candicidin was demonstrated to be a powerful antibiotic against *Escovopsis* while inactive to the ant fungus garden [28]. Both studies indicate other members of the community associated with the cuticle of ants provide the same protective function as *Pseudonocardia,* supporting the proposition that the defensive contribution of the cuticle-associated microbiota is provided by a diverse community. Moreover, the ant-associated microbiota was also reported effective against entomopathogenic fungi [29], showing that these symbionts are a rich source of bioactive molecules [30] to protect the ant mutualist fungi and the ants themselves. Such controversy on the diversity of bacteria growing on the cuticle of ants and on their role in the association with ants could be a result of the varying degree of co-evolution of ants and *Pseudonocardia* [22], although phylogenetic assessment by others did not indicate any topological correspondence between *Pseudonocardia* and their host ants [31].

The identification of metabolites, a key challenge in the search for new bioactive molecules, is based on analytical techniques for structural characterization, such as mass spectrometry (MS) and nuclear magnetic resonance (NMR). Increasing the sensitivity and stability of characterization techniques has made viable studies of complex biological samples by dereplication, allowing the development of databases of metabolites to assist with the identification process [32].

Porphyrins are widely used in photodynamic therapies [33]. After photoactivation, porphyrins transfer energy to oxygen molecules that have changed their energy state (triplet to singlet), increasing oxygen reactivity and leading to cell death [34, 35]. Biotechnological applications of porphyrins are diverse [36–38] and include their successful use as insecticides [38, 39]. Porphyrins are well known for their phototoxic activity [39–42]. Photoactivity is related to the production of very toxic and highly reactive oxygen species (ROS) in the presence of UV or visible radiation. ROS is a term used for molecules and reactive intermediates with highly positive redox potentials [43]. ROS are produced when a photoactive substance, a photosensitizer (PS), is activated by low doses of UV-visible light at an appropriate wavelength. ROS can be produced as free radicals (Type I) or as singlet oxygen $O_2(^1D_g)$ (Type II). The four major ROS studied are superoxide (O_2^-), hydrogen peroxide (H_2O_2), hydroxyl radical (·OH) and singlet oxygen. Singlet oxygen is formed when an electron is removed from π^*2p orbitals of oxygen. The photoactivity observed for porphyrins is Type II [42]. Singlet oxygen is understood to play a major role in this effect, and photoactive molecules are increasingly being used in blood sterilization, cancer therapy, and insect and weed control [44].

The great advantage in the use of porphyrins compared to other photodynamic molecules is their light-absorbing capacity at all wavelengths in the UV-visible spectrum, which allows porphyrin excitation under exposure to natural light [37].

Here, we describe the successful exploration of bacteria-associated insect symbionts as sources of new bioactive molecules, and report the structures of new porphyrins with insecticidal properties against the polyphagous fall armyworm, *Spodoptera frugiperda* (J.E. Smith) (Lepidoptera, Noctuidae), based on tandem MS (neutral-loss experiment).

Results

We used a classical strategy of fermentation and isolation of bioactive compounds to prospect for molecules with insecticidal activity produced by ectosymbionts of the leaf-cutting ant *Acromyrmex coronatus*. Our preliminary screening identified an isolate (IIL-Ac-18dV) that was tentatively identified as *Streptomyces caniferus* because it shared 99.98% identity over 1350 bp of 16S rRNA with the type strain (AB184640). This isolate produced a complex crude extract (Fig. 1) with low insecticidal activity (approximately 10% mortality) against first instars of *S. frugiperda*.

Preliminary experiments by Collision-induced dissociation (CID) of the crude extract of this isolate led to the identification of three chemical classes of compounds in the regions at *m/z* 540–650, *m/z* 730–780 and *m/z* 785–840.

Our experience with isolation of bioactive compounds produced by microorganisms indicates that these compounds are present in low concentrations in most cases. Therefore, we believed that biological activity could not be detected due to the very low concentration of the active compound in initial screenings. To prove this supposition, bioactivity and isolation experiments were conducted with the crude extract, which showed only 10% mortality in the initial screening, and these experiments led to the isolation and identification of an active compound. The higher activity after fractionation and isolation confirmed the complexity of the original crude extract and indicated that the lack of observation of porphyrin signals in the mass spectrum of the original crude extract was likely due to ionization suppression.

Toxicity-driven fractionation of the crude extract from the isolate IIL-Ac-ASP18v allowed the isolation of a fraction that contained a mixture of three porphyrins.

Bioassays of this fraction against third instars of *S. frugiperda* indicated high insecticidal activity, with a LD_{50} of 37.7 μg.cm^{-2} (y = 1.876–1.051×; n = 72; df = 3; χ^2 = 6.0723; P < 0.05). Larvae exposed to the porphyrin fraction turned black, which could be related to the oxidative process in cells and tissues exposed to porphyrins.

The purified porphyrins *coproporphyrin* and *zinc coproporphyrin III* using the estimated LD_{50} concentration caused approximately 30% mortality in third instars of *S. frugiperda*.

Isolation and characterization of porphyrins in the active fraction

Porphyrin signals were observed at *m/z* 655, 669 and 717. The CID of an ion at *m/z* 655 produced a fragmentation profile that allowed the characterization of *coproporphyrin I*, as corroborated by comparison with data in the literature [45, 46]. The CID spectrum of *m/z* 655 produced a neutral loss of 74 Da (*m/z* 581), which corresponded to losses of propionic acid (CH_3CH_2COOH). α-cleavage was observed in losses of 59 Da. Four consecutive losses of 59 Da were observed at *m/z* 596, 537, 478 and 419. Losses of 59 Da revealed radical losses. Consecutive losses of 59 and 73 Da were also observed at *m/z* 596 [M-59 + H]$^+$, *m/z* 523 [M-59-73 + H]$^+$, *m/z* 464 [M-59-73-59 + H]$^+$ and *m/z* 391 [M-59-73-59-73 + H]$^+$. Base peak *m/z* 537 corresponds to a loss of 118 Da, which corresponded to one radical of propionic acid ($\cdot CH_2CH_2COOH$) and three methyl groups ($\cdot CH_3$), while a loss of 132 Da (*m/z* 523) corresponded to losses of four methyl groups and one propionic acid residue. All losses followed radical mechanisms. The CID spectrum at *m/z* 655 and the predicted structure of this molecule are illustrated in Fig. 2a. The theoretical value of log P obtained for *coproporphyrin I* was 5.22.

One loss of 60 Da (*m/z* 657) and three consecutive losses of 59 Da (*m/z* 598, 539 and 480) were observed for an ion at *m/z* 717. The fragmentation profile of the porphyrin at *m/z* 717 differed in showing losses of 72 Da for ions at *m/z*

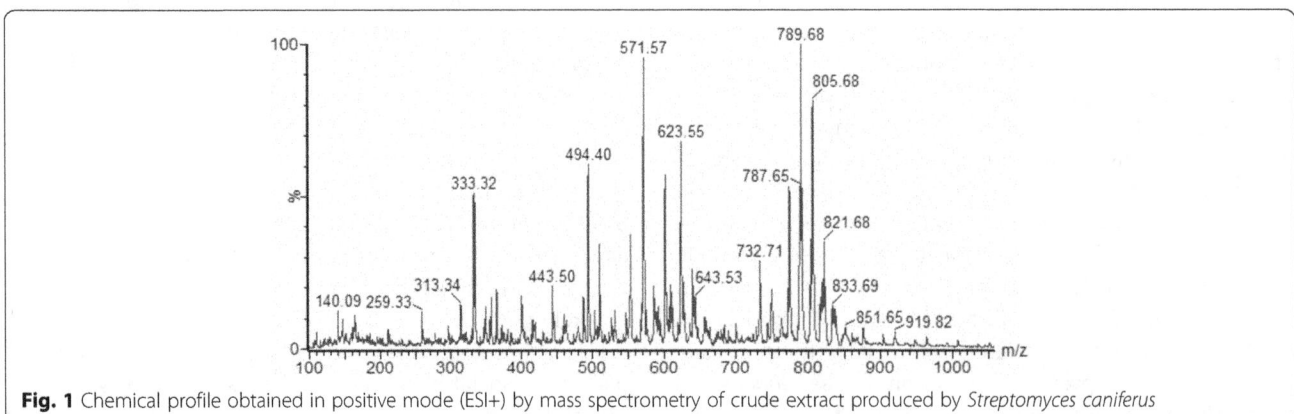

Fig. 1 Chemical profile obtained in positive mode (ESI+) by mass spectrometry of crude extract produced by *Streptomyces caniferus*

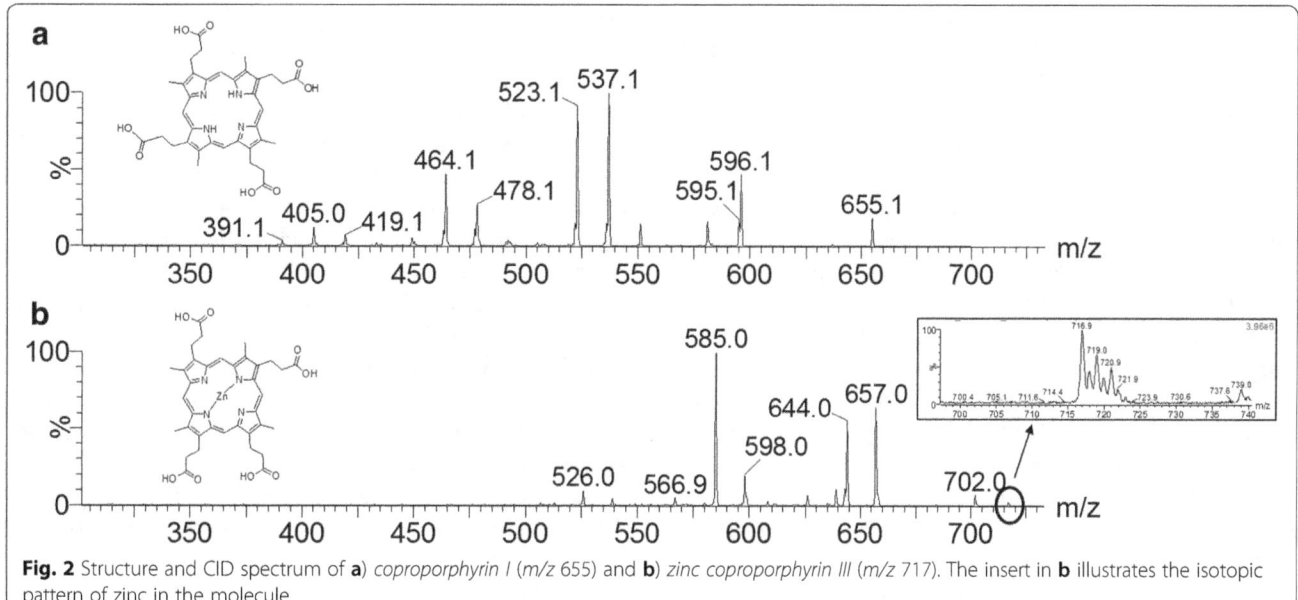

Fig. 2 Structure and CID spectrum of **a**) *coproporphyrin I* (*m/z* 655) and **b**) *zinc coproporphyrin III* (*m/z* 717). The insert in **b** illustrates the isotopic pattern of zinc in the molecule

585 [M-60-72 + H]$^+$ or *m/z* 526 [M-60-59-72 + H]$^+$ or *m/z* 467 [M-60-59-59-72 + H]$^+$. The ion at *m/z* 644 corresponded to a loss of 73 Da ($\cdot$CH$_2$CH$_2$COOH). This structure corresponded to *zinc coproporphyrin III*. The CID spectra, isotopic profile and structure of *zinc coproporphyrin III* are illustrated in Fig. 2b.

Another porphyrin (*porphyrin I*) was available in the active fraction at *m/z* 669. The same fragmentation profile observed for an ion at *m/z* 655 was observed for this porphyrin. The fragmentation profile obtained for *m/z* 669 by CID analysis produced four consecutive α-cleavages, observed in radical losses of 59 Da (*m/z* 610, 551, 492 and 433) and neutral loss of 74 Da (*m/z* 595). Consecutive losses of 59 and 73 Da were also observed:

m/z 610 [M-59 + H]$^+$, *m/z* 537 [M-59-73 + H]$^+$, *m/z* 478 [M-59-73-59 + H]$^+$ and *m/z* 405 [M-59-73-59-73 + H]$^+$. The signal at *m/z* 523 corresponded to loss of 146 Da, i.e. loss of one propionic acid residue, three methyl radicals ($\cdot$CH$_3$) and one ethyl radical ($\cdot$CH$_2$CH$_3$). The theoretical log *P* value for hydrophobicity of *porphyrin I* was 5.64. The CID spectrum at *m/z* 669 and the predicted structure of this molecule are illustrated in Fig. 3a.

Neutral-loss experiments

Neutral-loss experiments for monitoring the availability of porphyrins in 67 crude extracts of insect microbial symbionts led to the identification of

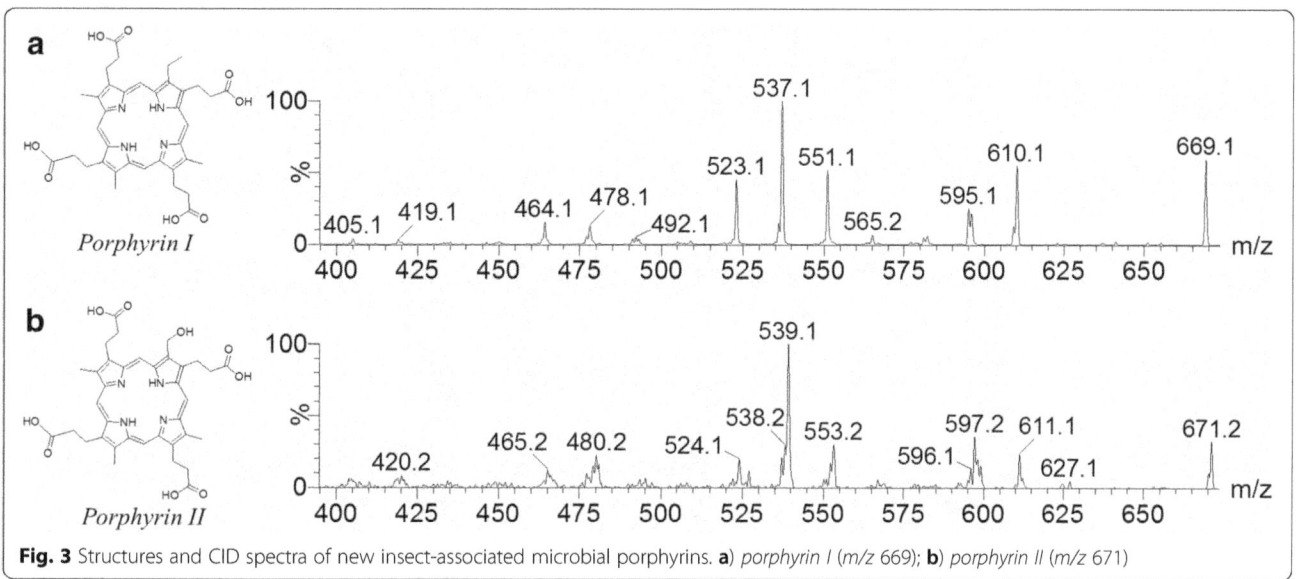

Fig. 3 Structures and CID spectra of new insect-associated microbial porphyrins. **a**) *porphyrin I* (*m/z* 669); **b**) *porphyrin II* (*m/z* 671)

porphyrins in the extracts produced by isolate IIL-Ac-ASP40, putatively identified as *Streptomyces olivochromogenes* (99% identity over a 1350-bp fragment of the 16S rRNA gene) and IIL-Ac-ASP72, putatively identified as *Streptomyces eurocidicus* (98.96% identity over a 1350-bp fragment of the 16S rRNA gene) (unpublished data). Porphyrins were identified by monitoring signals with 118 Da, 132 Da, 74 Da and 60 Da, all products of fragmentation of *coproporphyrin I* (Fig. 2a). Isolates IIL-Ac-ASP40 and IIL-Ac-ASP72 produced two of the same porphyrins produced by isolate IIL-Ac-ASP18v, *coproporphyrin I* at *m/z* 655 and the new *porphyrin I* at *m/z* 669 (Fig. 3a). Neutral-loss experiments on the crude extract of isolate IIL-Ac-ASP18v resulted in the discovery of a new ion at *m/z* 671, allowing the identification of a fourth porphyrin (*porphyrin II*) (Figs. 3b and 4).

The CID experiment resulted in an interesting fragmentation profile for ion *m/z* 671. This structure showed the most distinct fragmentation profile, with neutral losses predominating over radical losses. α-cleavages were observed, with losses of 60 (neutral loss) or 59 (radical loss) Da. The signal at *m/z* 522 corresponds to the loss of 148 Da, i.e. loss of one propionic acid residue, three methyl radicals ($\cdot CH_3$) and one methanol radical ($\cdot CH_2OH$). The theoretical log *P* value for hydrophobicity of *porphyrin II* was 4.1. The CID spectrum at *m/z* 671 and the predicted structure of this molecule are illustrated in Fig. 3b.

Discussion

We demonstrated the potential of insect microbial symbionts to produce new molecules with insecticidal activity against the lepidopteran *S. frugiperda*, describing three new structures of photoactivated porphyrins. We also demonstrated that detailed structure analysis and molecule bioassay-guided isolation of active compounds from complex crude extracts with low biological activity can result in the isolation of new, highly active molecules: the new *porphyrins I* and *II*. The porphyrins that were characterized in this study would not have been identified, nor would their insecticidal potential have been detected if we had followed a traditional approach, as the low proportion of porphyrins in the crude extract did not produce a promising insecticidal activity in the initial screening. Our results indicated that porphyrins can be used for insect control at low concentrations, as we demonstrated using third instars of *Spodoptera frugiperda*.

The insecticidal activity of the purified porphyrins *coproporphyrin* and *zinc coproporphyrin III* was lower than expected from the calculated LD_{50}. This lower than expected mortality could be related to synergism among the different porphyrins in the whole fraction, and/or the additional manipulation for porphyrin purification having allowed some degradation. For example, the antineoplastic activity and in vitro photophysical properties of Photofrin photodynamic therapy, also based on a mixture of porphyrins, are uncertain, owing to the variable pharmacokinetics and photochemistry of the constituents [47].

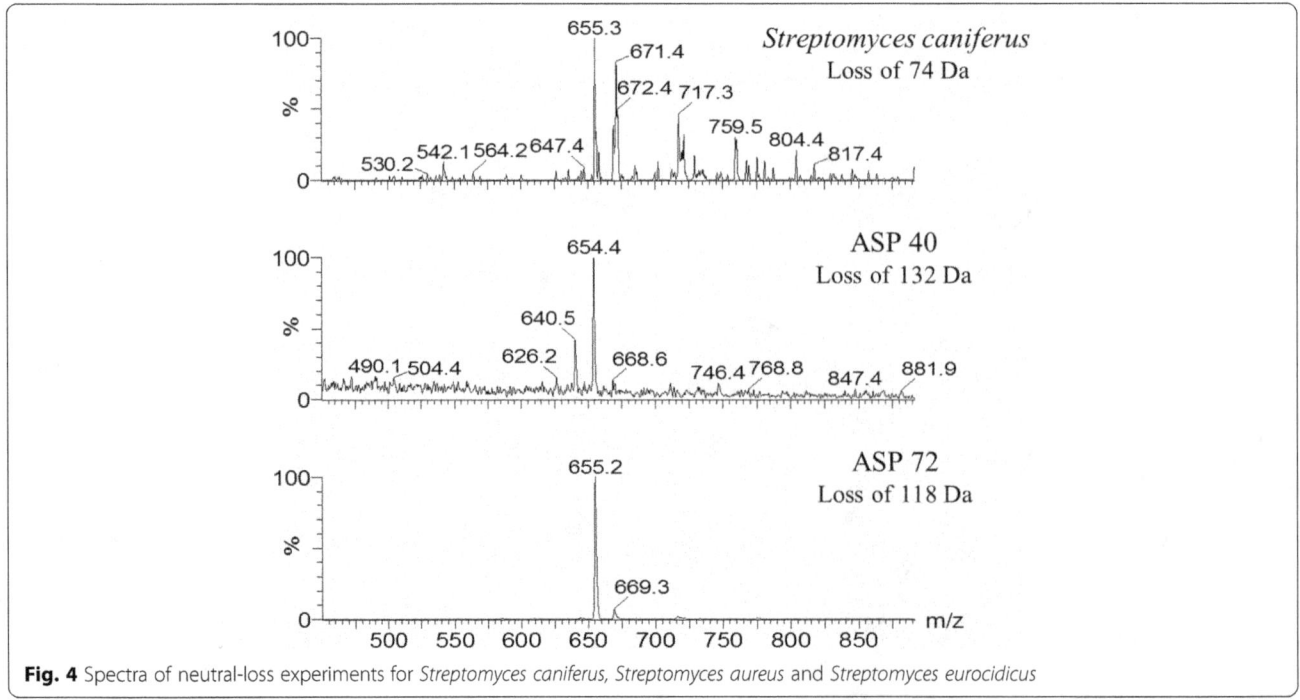

Fig. 4 Spectra of neutral-loss experiments for *Streptomyces caniferus*, *Streptomyces aureus* and *Streptomyces eurocidicus*

We also demonstrated the importance of tandem mass-spectrometry experiments for the discovery of new compounds. Neutral-loss experiments allowed the identification of *porphyrin II* (m/z 671) in a crude extract, which had not been identified in fractions enriched with porphyrins.

Isolate IIL-Ac-ASP18v produces porphyrins with high structural diversity and hydrophobicity. *Porphyrin II* is highly hydrophilic (log P 4.16), while *porphyrin I* (log P 5.64) shows a similar hydrophobicity to *coproporhyrin I* (log P 5.22). The phototoxicity of porphyrins seems to depend on their structure, as their phototoxic activity increases with their hydrophobicity [47]. Therefore, the use of these compounds in photodynamic therapy must rely on the capacity of photosensitizers to move across the cell membrane, which will depend on their hydrophobicity. Molecules with log P values between 8 and 10 are considered moderately or highly hydrophobic, and are likely to be the most toxic as they can diffuse through cell membranes, move into the cell and become integrated into subcellular membranes such as the mitochondrial and lysosomal membranes, Golgi apparatus and rough endoplasmic reticulum. However, even less-hydrophobic compounds can actively or passively diffuse through the cell membrane if electrically charged, due to ionic interactions with groups at the cell surface. Cationic compounds are often located in mitochondria, while anionic compounds are accumulated at the lysosomal level [39].

Ben-Dror et al. [48] employed a series of synthetic porphyrins (log P between 7.4 and 10.5, as estimated with the ChemDraw software) with peripheral modifications that were expected to change the hydrophobicity with no change to the chromophore group. They measured the binding constant to liposomes, and determined that in highly hydrophobic porphyrins, neither the lipophilicity nor the passive binding to liposomes adequately predicted the cell uptake. In these cases, log P is not the only factor that should be used to predict porphyrin activity, and should be evaluated with caution as this may produce misleading conclusions, as in the case of passive uptake into liposomes, or be irrelevant in cases where other uptake mechanisms predominate, as with cells and highly apolar molecules. The singlet-oxygen production showed no significant changes.

Photosensitizers such as porphyrins generally lack cytotoxicity in the absence of light, which could be an advantage in biological systems where rapid breakdown of the photosensitizer after use is necessary. However, for industrial applications this is an undesirable aspect [44].

Similarly to other photosensitizers (e.g. xanthenes), porphyrins have been reported to affect the epithelial membrane of the midgut of insects and to inhibit insect feeding [39]. Amor et al. [41] discussed the efficiency of hematoporphyrin against *Ceratitis capitata* (Mediterranean fruit fly), *Bactrocera* (*Dacus*) *oleae* (olive fruit fly) and *Stomoxys calcitrans* (stable fly). *Ceratitis capitata* showed about 70% mortality after 1 h exposure to 11–12 μmol.mL^{-1} (approximately 6.5 mg.mL^{-1}) of hematoporphyrin; while in our experiment, the active fraction of porphyrins produced by IIL-Ac-18 V showed an LD$_{50}$ of 3.6 mg.mL^{-1} (37.7 μg.cm^{-2}). *Bactrocera* (*D.*) *oleae* showed lower photosensitivity than did *C. capitata*. Additional data on the efficacy of porphyrin mixtures for *C. capitata* control also indicated the importance of the porphyrin structure to their insecticidal activity. The amphiphilic cationic porphyrin DDP (log P 20) produced high mortality, while hematoporphyrin (log P 12) was nearly atoxic [48]. However, high hydrophobicity can also lead to porphyrin aggregation, which would reduce its photoactivity [47]. These results reveal the importance of the porphyrin structure, and log P does not necessarily correlate directly with porphyrin biological activity.

Before field application can be recommended, important aspects of porphyrin-plant interactions must be considered based on experience with the use of tetrapyrrole-dependent photodynamic herbicides (TDPH). TDPH have been considered phytotoxic to certain plants, which are sprayed in the dark a few hours before they are exposed to light. Once exposed to light, the plants can accumulate massive amounts of tetrapyrroles, and damage appears after 20 min, becoming irreversible after 60 min. The toxicity of TDPH to plants is age- and species-dependent: dicotyledonous weeds such as mustard, red-root pigweed, common purslane and lamb's quarter are very susceptible, while monocotyledonous plants such as corn, wheat, barley and oats are not [49, 50]. Therefore, the use of new phorphyrins as described here to control chewing insects such as the fall armyworm in field applications will require additional assays to check for plant toxicity and photodegradation that may occur once the molecules are exposed to light.

Conclusions

Our results demonstrated the importance of bacterial insect symbionts as sources of new and/or bioactive compounds. We used an unusual approach to pursue the isolation of bioactive molecules, by focusing on crude extracts with low insecticidal activity, aiming to identify bioactivity among the least-abundant molecules in the crude extract. The crude extracts from IIL-Ac-ASP18v showed a complex chemical composition after analysis by direct insertion in the mass spectrometer, in positive mode (ESI+). After isolation of the porphyrin fraction, the insecticidal activity induced up to 100% larval mortality (LD$_{50}$ = 37.7 μg.cm^{-2}), proving our strategy successful. The use of sensitive analytical tools such as mass spectrometry allowed the characterization of two new porphyrins.

Methods

Microorganisms: Isolation and identification

Microorganisms were isolated from the leaf-cutting ant *Acromyrmex coronatus* (Hymenoptera, Formicidae) using ISP2 [51], ISP4 [51] and chitin [52] as culture media. The isolates obtained were identified based on 16S rRNA analysis (unpublished data). These isolates were then used to screen for insecticidal activity and porphyrin biosynthesis.

16S rRNA gene sequencing analysis

A selected bacterial isolate (IIL-Ac-18dV) was grown at 28 °C for 24 h under constant agitation (120 rpm) in 1 mL of the same culture medium used in the initial isolation. Cells were precipitated by centrifugation (2000 g × 5 min) and then used for genomic DNA extraction [53]. DNA quality and integrity were assessed after electrophoresis in 0.8% (w/v) agarose gel containing 0.5 μg/mL ethidium bromide at 70 V for 1 h in TAE buffer (40 mM Tris-acetate, 1 mM EDTA, pH 7.2), and spectrophotometry analysis using the A260/280 ratio [54, 55]. DNA samples were subjected to PCR amplification of a fragment of the 16S rRNA gene using 10–20 ng of genomic DNA and the universal primer set 8f (5'-AGA GTT TGA TCC TGG CTC AG-3') and 1491r (5'-GGT TAC CTT GTT ACG ACT T-3') [56] in 1× enzyme buffer, 1.5 mM MgCl$_2$, 0.2 mM dNTPs, 0.32 μM of each primer, and 0.625 U of Taq polymerase (Promega), in a final reaction volume of 25 μL. PCR cycling conditions were 4 min at 95 °C (1×); 95 °C for 1 min, 55 °C for 1 min, and 72 °C for 2 min (35×), followed by a final extension at 72 °C for 10 min (1×). The 16S rRNA fragment obtained was submitted to bidirectional sequencing at the Laboratório de Biologia Molecular de Plantas, Departamento de Ciências Biológicas, ESALQ/USP, in an ABI 3730 DNA Analyzer using the BigDye® Terminator v3.1 Cycle Sequencing kits. The sequence obtained was viewed and edited using Finch TV v1.4.0 (Geospiza Inc.) and forward and reverse reads were assembled in a single ~1350 bp-long read using the Blast2 tool. The 16S rDNA fragment obtained was used in heuristic blast searches against the nucleotide databases of the National Center for Biotechnology Information (NCBI) (http://www.ncbi.nlm.nih.gov/) and EzTaxon [57] and (http://www.ezbiocloud.net/) for the tentative identification of the isolated bacterium. The sequence obtained for the isolate was deposited in GenBank under accession number KX762323.

Crude extract preparation

The selected isolate (IIL-Ac-18dV) and other microorganisms from our bank of isolates were inoculated in ISP2 medium and cultured for 7 d at 28 °C under constant shaking (130 rpm). Crude extracts were obtained by liquid-liquid extraction with one volume of ethyl acetate (Synth), following standard procedures [58].

Isolation of porphyrins for insecticidal bioassay

Isolate IIL-Ac-18dV was inoculated in 4.5 L of ISP2 medium. After liquid-liquid extraction, 180 mg of crude extract was produced. The crude extract was purified using a Sephadex™ LH-20 (particle size range: 27–163 μm; mean diameter: 103 μm; GE Healthcare, Sweden) - packed column (300 × 20 mm), with methanol (J.T. Baker) as the eluent. A total of 29 fractions, including an initial fraction (30 mL) + 27 fractions (12 mL per fraction) + a final fraction (50 mL), were collected. Eluted compounds were monitored for color and UV absorbance (λ = 254 and 356 nm). Pink fractions indicative of porphyrins were separated and analyzed by MS/MS experiments and subjected to biological assays.

Chromatographic conditions for porphyrin purification

Chromatographic analyses were performed in an Ultra High-Performance Liquid Chromatography system (UHPLC Accela 600, Thermo Scientific) equipped with a diode array detector, auto sampler and an ACE 5 column (250 × 4.6 mm; 5 μm). Porphyrin was monitored at λ = 400, 533 and 575 nm. Samples were eluted using a gradient of 0.1% formic acid (phase A) and methanol/ 0.1% formic acid (phase B). The gradient started with 90% phase B and increased linearly to 95% within 3 min, with an additional hold for 7 min in 95% phase B. The flow rate was 1 mL.min^{-1}. Peaks were purified using an ACE 5 column (250 × 7.75 mm, 5 μm) under the same elution gradient at a flow rate of 2.8 mL.min^{-1}.

Bioassay
Screening

The insecticidal activity of isolate IIL-Ac-18dV was evaluated against first instars of the fall armyworm *Spodoptera frugiperda* (Lepidoptera, Noctuidae) by exposing the larvae to a surface-treated artificial diet. The artificial diet used was that of Kasten et al. [59], and insect rearing and handling followed Parra [60]. Bioassays were conducted in sterile 24-well plates filled with 1.25 mL of artificial diet. After the diet solidified, 20 μL of the test solution was applied to the diet surface. In the initial screenings of crude extracts, samples were diluted in methanol to 25 μg.μL^{-1}, resulting in the application of 500 μg of crude extract in each cell (260 μg.cm^{-2}). Methanol was used as the negative control. After the solvent dried completely, each cell was inoculated with ten (10) 24 h-old larvae, totaling 240 larvae/treatment. Plates were maintained under controlled conditions (25 ± 2 °C; 60 ± 10% RH; 14 h photophase), and larval development and mortality were evaluated daily for 3 d. Mortality was assessed by touching the last abdominal segments, and larvae that were unresponsive or showed uncoordinated movements were considered dead.

Porphyrin bioassay
Insecticidal activity of the purified porphyrin fraction produced by isolate IIL-Ac-18dV was evaluated using third instars of *S. frugiperda,* at 1, 2, 8, 16, and 65 µg.cm^{-2}. Bioassays were performed as above, but only three third instars were placed in each well of the plates, totaling 72 larvae/treatment. Plates were maintained under the same conditions, and larval development and mortality assessed as above. The program PoloPlus 1.0 was used to determine the dose-response curve and LD$_{50}$ concentration, by Probit analysis (Leora Software, 1987).

The two major porphyrins (*coproporphyrin* and *porphyrin II*) in the porphyrin fraction were isolated and individually tested for insecticidal activity against third instars of *S. frugiperda* at the estimated LD$_{50}$ concentration for the whole porphyrin fraction, as described above.

MS and MS/MS experiments
The crude extract and porphyrin-enriched fractions were analyzed by direct insertion in a Xevo™ TQ-S (Waters Corporation) mass spectrometer coupled with Acquity™ Ultra High-Performance Liquid Chromatography (UPLC™, Waters). The mass spectrometer operated with electrospray ionization in the positive mode (ESI+). The flow rate was 0.15 mL.min^{-1}. The capillary voltage and the spray voltage were set to 3.3 kV and 60 V, respectively; the desolvation temperature was 350 °C; and argon was used as the collision gas.

Porphyrin characterization
Fractions obtained from the crude extract were analyzed by direct insertion in a mass spectrometer operating in positive (ESI+) and negative (ESI−) modes, in a full scan experiment, and fractions with similar chemical compositions were pooled. The porphyrins present in the active fraction were characterized with collision-induced dissociation (CID) experiments employing *coproporphyrin* I, due to the availability of the fragmentation profile in the literature [45, 46]. Several collision energies were tested (20–70 V) before 50 V was selected as the collision energy. The concentration of the porphyrin fractions was 5 µg.mL^{-1} in methanol. The hydrophobicity of the porphyrins identified was also estimated, by determining the partition coefficient between an organic solvent and water, or log *P*, a measurement normally used to predict the ability of a molecule to diffuse into biomembranes. Theoretical values of log *P* for the porphyrins produced by isolate IIL-Ac-ASP18v were calculated using the commercial software Chem DrawPro 8.0 (CambridgeSoft Corporation).

Neutral-loss experiment
Neutral-loss experiments were performed in order to search for porphyrins in crude extracts from 67 antectosymbiont isolates (crude extract preparations were obtained as described above). Selected neutral losses were 60 Da, 74 Da, 118 Da and 132 Da. These neutral losses were selected based on the main fragments found in CID experiments for *coproporphyrin I*, the major porphyrin present in the active fraction of isolate IIL-Ac-18dV and previously described [45, 46], which allowed the dereplication studies for this compound. The collision energy employed was 50 V. The concentration of the crude extracts was 5 µg.mL^{-1} in methanol. Ions that showed two losses were analyzed by daughter scan experiments to obtain structural information.

LC-MS conditions
The analyses of isomers of porphyrins and of the crude extracts obtained under acidic conditions were conducted on a LC-MS in a Xevo TQ-S (Waters Corporation) Mass Spectrometer coupled with Acquity Ultra High-Performance Liquid Chromatography (UPLC, Waters). Samples were eluted using a gradient of 0.1% formic acid (phase A) and methanol/0.1% formic acid (phase B). The gradient started with 40% phase B and increased linearly to 95% within 5 min, with an additional hold for 2 min in 95% phase B. The flow rate was 0.4 mL.min^{-1}. The mass spectrometer operated with electrospray ionization in the positive mode (ESI+). The capillary voltage and spray voltage were set at 3.0 kV and 40 V, respectively; with desolvation temperature 300 °C, source temperature 120 °C, argon used as the collision gas, and collision gas flow 0.2 mL.min^{-1}.

Abbreviations
CID: Collision-induced dissociation; Da: Daltons; DPP: diketopyrrolopyrrole; LD: lethal dose; MS: mass spectrometry; PS: photosensitizer; ROS: reactive oxygen species; rRNA: ribosomal ribonucleic acid; UV: ultraviolet

Acknowledgments
We thank the Fundação de Amparo à Pesquisa do Estado de São Paulo (FAPESP) for providing a fellowship to AFCM (Process 2014/21584-3) and a research grant to FLC (Process 2011/50877-0).

Funding
We thank the Fundação de Amparo à Pesquisa do Estado de São Paulo (FAPESP) for providing a fellowship to AFCM (Process 2014/21584-3) and a research grant to FLC (Process 2011/50877-0).

Authors' contributions

FLC and AFCM designed the experiments. AFCM analyzed and interpreted all mass spectra. LGA performed the bioassays. AFCM and LABM elucidated the molecular structures. AFCM wrote and FLC edited the initial draft of the manuscript. All authors read and approved the final version of the manuscript.

Competing interests

The authors declare that they have no competing interests.

Author details

[1]Laboratório de Interações em Insetos, Departamento de Entomologia e Acarologia, Escola Superior de Agricultura "Luiz de Queiroz", Universidade de São Paulo, Av Pádua Dias 11, 13418–900, Piracicaba, SP, Brazil. [2]Laboratório de Espectrometria de Massas Aplicada a Produtos Naturais, Departamento de Química, Faculdade de Filosofia, Ciências e Letras de Ribeirão Preto, Universidade de São Paulo, Av Bandeirantes 3900, 14040–901, Ribeirão Preto, SP, Brazil.

References

1. Catarino R, Ceddia G, Areal FJ, Park J. The impact of secondary pests on Bacillus thuringiensis (Bt) crops. Plant Biotechnol J. 2015;13(5):601–12.
2. Popp J, Petö K, Nagy N. Pesticide productivity and food security. Agron Sustain Dev. 2013;33(1):243–55.
3. Barrows G, Sexton S, Zilberman D. Agricultural biotechnology: the promise and prospects of genetically modified crops. J Econ Perspect. 2014;28(1):99–120.
4. Gerwick BC, Sparks TC. Natural products for pest control: an analysis of their role, value and future. Pest Manag Sci. 2014;70(8):1169–85.
5. Lefebvre M, Langrell SRH, Gomez-y-Paloma S. Incentives and policies for integrated pest management in Europe: a review. Agron Sustain Dev. 2015;35(1):27–45.
6. Sparks TC. Insecticide discovery: an evaluation and analysis. Pest Biochem Physiol. 2013;107(1):8–17.
7. Poulsen M, Oh DC, Clardy J, Currie CR. Chemical analyses of wasp-associated Streptomyces bacteria reveal a prolific potential for natural products discovery. PLoS One. 2011;6(2):e16763.
8. Oh DC, Poulsen M, Currie CR, Clardy J. Dentigerumycin: a bacterial mediator of an ant-fungus symbiosis. Nat Chem Biol. 2009;5:391–3.
9. Oh DC, Poulsen M, Currie CR, Clardy J. Sceliphrolactam, a polyene macrocyclic lactam from a wasp-associated Streptomyces sp. Org Lett. 2011;13(4):752–5.
10. Carr G, Poulsen M, Klassen JL, Hou Y, Wyche TP, Bugni TS, et al. Microtermolides a and B from termite-associated Streptomyces sp. and structural revision of vinylamycin. Org Lett. 2012;14(11):2822–5.
11. Crawford JM, Clardy J. Bacterial symbionts and natural products. Chem Commun. 2011;47(27):7559–66.
12. Brachmann AO, Bode HB. Identification and bioanalysis of natural products from insect symbionts and pathogens. Adv Biochem Eng Biotechnol. 2013;135:123–55.
13. Douglas AE. Symbiotic microorganisms: untapped resources for insect pest control. Trends Biotechnol. 2013;25(8):338–42.
14. De Bary A. The phenomenon of symbiosis. Karl J. Trubner: Strasbourg; 1879. 364 p.
15. Van Born S, Billen J, Boomsma JJ. The diversity of microorganisms associated with Acromyrmex leafcutter ants. BMC Evol Biol. 2002;2:9–20.
16. Zucchi TD, Guidolin AS, Cônsoli FL. Isolation and characterization of actinobacteria ectosymbionts from Acromyrmex subterraneus brunneus (hymenoptera, Formicidae). Microbiol Res. 2011;166(1):68–76.
17. Poulsen M, Bot ANM, Currie CR, Nielsen MG, Boomsma JJ. Within-colony transmission and the cost of a mutualistic bacterium in the leaf cutting ant Acromyrmex octospinosus. Funct Ecol. 2003;17(2):260–9.
18. Haeder S, Wirth R, Herz H, Spiteller D. Candicidin-producing Streptomyces support leaf-cutting ants to protect their fungus garden against the pathogenic fungus Escovopsis. Proc Natl Acad Sci U S A. 2009;106(12):4742–6.
19. Barke J, Seipke RF, Grüschow S, Heavens D, Drou N, Bibb M, et al. A mixed community of actinomycetes produce multiple antibiotics for the fungus farming ant Acromyrmex octospinosus. BMC Biol. 2010;8:109–19.
20. Seipke RF, Barke J, Brealey C, Hill L, Yu DW, Goss RJM, et al. A single Streptomyces symbiont makes multiple antifungals to support the fungus farming ant Acromyrmex octospinosus. PLoS One. 2011;6(8):e22028.
21. Schoenian I, Spiteller M, Ghaste M, Wirth R, Herz H, Spiteller D. Chemical basis of the synergism and antagonism in microbial communities in the nests of leaf-cutting ants. Proc Natl Acad Sci U S A. 2010;108(5):1955–60.
22. Cafaro MJ, Currie CR. Phylogenetic analysis of mutualistic filamentous bacteria associated with fungus-growing ants. Can J Microbiol. 2005;51:441–6.
23. Poulsen M, Cafaro M, Boosma JJ, Currie CR. Specificity of the mutualistic association between actinomycete bacteria and two sympatric species of Acromyrmex leaf-cutting ants. Mol Ecol. 2005;14(11):3597–604.
24. Kost C, Lakatos T, Böttcher I, Arendholz WR. Redenbach, Wirth R. Non-specific association between filamentous bacteria and fungus-growing ants. Naturwissenschaften. 2007;94(10):821–8.
25. Mueller UG. Symbiont recruitment versus ant-symbiont co-evolution in the attine ant-microbe symbiosis. Curr Opinion Microbiol. 2012;15(3):269–77.
26. Sen R, Ishak HD, Estrada D, Dowd SE, Hong E, Mueller UG. Generalized antifungal activity and 454-screening of Pseudnocardia and Amycolatopsis bacteria in nests of fungus-growing ants. Proc Natl Acad Sci U S A. 2009; 106(42):17805–10.
27. Andersen SB, Hansen LH, Sapountzis P, Sorensen SJ, Boosma JJ. Specificity and stability of the Acromyrmex-Pseudonocardia symbiosis. Mol Ecol. 2013;22:4307–21.
28. Haeder S, Wirth R, Herz H, Spiteller D. Candicidin-producing Streptomyces support leaf-cutting ants to protect their fungus garden against the pathogenic fungus Escovopsis. Proc Natl Acad Sci U S A. 2009;106:4742–6.
29. Mattose TC, Moreira DDO, Samuels RI. Symbiotic bacteria on the cuticle of the leaf-cutting ant Acromyrmex subterraneus protect workers from attack by entomopathogenic fungi. Biol Lett. 2012;8(3):461–4.
30. Klassen JL. Microbial secondary metabolites and their impacts on insect symbiosis. Curr Opin Insect Sci. 2014;4:15–22.
31. Mueller UG, Dash D, Rabeling C, Rodrigues A. Coevolution between Attine ants and actinomycete bacteria: a reevaluation. Evolution. 2008;62(11):2894–912.
32. Hooft JJJ, de Vos RCH, Ridder L, Vervoort J, Bino RJ. Structural elucidation of low abundant metabolites in complex sample matrices. Metabolomics. 2013;9(5):1009–18.
33. Ferreira DP, Conceição DS, Calhelha RC, Sousa T, Socoteanu R, Ferreira ICFR, et al. Porphyrin dye into biopolymeric chitosan films for localized photodynamic therapy of cancer. Carbohydr Polym. 2016;151(20):160–71.
34. Boscencu R, Oliveira AS, Ferreira DP, Ferreira LFV. Synthesis and spectral evaluation of some unsymmetrical mesoporphyrinic complexes. Int J Mol Sci. 2012;13:8112–25.
35. O'Connor AE, Gallagher WM, Byrne AT. Porphyrin and nonporphyrin photosensitizers in oncology: preclinical and clinical advances in photodynamic therapy. J Photochem Photobiol. 2009;85(5):1053–74.
36. Malikt Z, Hanania J, Nitzan Y. Bactericidal effects of photoactivated porphyrins – an alternative approach to antimicrobial drugs. J Photochem Photobiol B. 1990;5(3–4):281–93.
37. Vatansever F, de Melo WCMA, Avci P, Vecchio D, Sadasivam M, Gupta A, et al. Antimicrobial strategies centered around reactive oxygen species – bactericidal antibiotics, photodynamic therapy, and beyond. FEMS Microbiol Rev. 2013;37(6):955–89.
38. Alves E, Faustino MAF, Neves MGPMS, Cunha A, Nadais H, Almeida A. Potential applications of porphyrins in photodynamic inactivation beyond the medical scope. J Photochem Photobiol C Photochem Rev. 2015;22:34–57.
39. Amor TB, Jori G. Sunlight-activated insecticides: historical background and mechanisms of phototoxic activity. Insect Biochem Molec Biol. 2000;30:915–25.
40. Buda V, Luksiene Z, Radziute S, Kurilcil N, Jursenas S. Search of photoinsecticides: effect of hematoporphyrin dimethyl ether on leaf mining pest Liriomyza bryoniae (Diptera: Agromyzidae). Agron Res. 2006;4:141–6.
41. Amor TB, Tronchin M, Bortolotto L, Verdiglione R, Joril G. Porphyrins and related compounds as photoactivatable insecticides 1. Phototoxic activity of

hematoporphyrin toward *Ceratitis capitata* and *Bactrocera oleae*. Photochem Photobiol. 1998;67(2):206–11.

42. Amor TB, Bortolotto L, Jori G. Porphyrins and related compounds as photoactivatable insecticides 2. Phototoxic activity meso-substituted porphyrins. Photochem Photobiol. 1998;68(3):314–8.

43. Silva EFF, Serpa C, Dabrowski JM, Monteiro CJP, Formosinho SJ, Stochel G, et al. Mechanisms of singlet-oxygen and superoxide-ion generation by porphyrins and bacteriochlorins and their implications in photodynamic therapy. Chem Eur J. 2010;16(30):9273–86.

44. De-Rosa MC, Crutchley JR. Photosensitized singlet oxygen and its applications. Coord Chem Rev. 2002;233–234(1):351–71.

45. Danton M, Lim CK. Porphyrin profiles in blood, urine and faeces by HPLC/ electrospray ionization tandem mass spectrometry. Biomed Chromatogr. 2006;20(6–7):612–21.

46. Bu W, Myers N, McCarty JD, O'Neill T, Hollar S, Stetson PL, et al. Simultaneous determination of six urinary porphyrins using liquid chromatography–tandem mass spectrometry. J Chromatogr B. 2003;783(2):411–23.

47. Jones LR, Grossweiner LI. Singlet oxygen generation by Photofrin® in homogeneous and light-scattering media. J Photochem Photobiol B. 1994;26(3):249–56.

48. Ben-Dror S, Bronshtein I, Wiehe A, Roder B, Senge MO, Ehrenberg B. On the correlation between hydrophobicity, liposome binding and cellular uptake of porphyrin sensitizers. Photochem Photobiol. 2006;82:695–701.

49. Rebeiz CA, Montazer-Zouhoor A, Hopen HJ, Wu SM. Photodynamic herbicides: 1. Concept and phenomenology Enzyme Microb Technol. 1984;6(9):390–6.

50. Rebeiz CA, Reddy KN, Nandihalli UB, Velu J. Tetrapyrrole dependent photodynamic herbicides. Photochem Photobiol. 1990;52(6):1099–117.

51. Shirling EB, Gottlieb D. Methods for characterizations of *Streptomyces* species. Int J Syst Evol Microbiol. 1966;16:313–40.

52. Hsu SC, Lockwood JL. Powdered chitin agar as a selective medium for enumeration of actinomycetes in water and soil. J Appl Microbiol. 1975; 29(3):422–6.

53. Sunnucks P, Hales DF. Numerous transposed sequences of mitochondrial cytochrome oxidase I-II in aphids of the genus *Sitobion* (Hemiptera: Aphididae). Mol Biol Evol. 1996;13(3):510–24.

54. Gallagher SR, Desjardins PR. Quantitation of DNA and RNA with absorption and fluorescence spectroscopy. Curr Protoc Protein Sci. 2007;A-3D doi:10. 1002/0471140864.

55. Sambrook J, Russel DW. Molecular cloning: a laboratory manual 3rd Ed. New York: Cold Spring Harbor Laboratory Press; 2001.

56. Weisburg WG, Barns SM, Pelletier DA, Lane DJ. 16S ribosomal DNA amplification for phylogenetic study. J Bacteriol. 1991;173(2):697–703.

57. Kim OS, Cho YJ, Lee K, Yoon SH, Kim M, Na H, et al. Introducing EzTaxon-e: a prokaryotic 16S rRNA Gene sequence database with phylotypes that represent uncultured species. Int J Syst Evol Microbiol. 2012;62:716–21.

58. Melo IS, Sanhueza RMV. Métodos de seleção de microrganismos antagônicos a fitopatógenos: manual técnico. Embrapa-CNPMA: Jaguariúna; 1995.

59. Kasten P, Precetti AA, Precetti CM, Parra JRP. Dados biológicos comparativos de *Spodoptera frugiperda* (J.E. Smith, 1797) em duas dietas artificias e substrato natural. Rev Agric. 1978;53:68–78.

60. Parra JRP. Criação de insetos para estudo com patógenos. In: Alevs SB, editor. Controle microbiano de insetos. Piracicaba: FEALQ; 1986. p. 348–73.

9

RpoN2- and FliA-regulated *fliTX* is indispensible for flagellar motility and virulence in *Xanthomonas oryzae* pv. *oryzae*

Chao Yu, Huamin Chen, Fang Tian, Fenghuan Yang and Chenyang He[*] (ID)

Abstract

Background: Bacterial blight of rice caused by *Xanthomonas oryzae* pv. *oryzae* (Xoo) is one of the most important crop diseases in the world. More insights into the mechanistic regulation of bacterial pathogenesis will help us identify novel molecular targets for developing effective disease control strategies. A large flagellar gene cluster is regulated under a three-tiered hierarchy by σ^{54} factor RpoN2 and its activator FleQ, and σ^{28} factor FliA. A hypothetical protein gene *fliTX* is located upstream of *rpoN2*, however, how it is regulated and how it is related to bacterial behaviors remain to be elucidated.

Results: Sequence alignment analysis indicated that FliTX in Xoo is less well conserved compared with FliT proteins in *Escherichia coli*, *Salmonella typhimurium*, and *Pseudomonas fluorescens*. Co-transcription of *fliTX* with a cytosolic chaperone gene *fliS* and an atypical PilZ-domain gene *flgZ* in an operon was up-regulated by RpoN2/FleQ and FliA. Significantly shorter filament length and impaired swimming motility were observed in Δ*fliTX* compared with those in the wildtype strain. Δ*fliTX* also demonstrated reduced disease lesion length and *in planta* growth in rice, attenuated ability of induction of hypersensitive response (HR) in nonhost tobacco, and down-regulation of type III secretion system (T3SS)-related genes. *In trans* expression of *fliTX* gene in Δ*fliTX* restored these phenotypes to near wild-type levels.

Conclusions: This study demonstrates that RpoN2- and FliA-regulated *fliTX* is indispensible for flagellar motility and virulence and provides more insights into mechanistic regulation of T3SS expression in Xoo.

Keywords: *Xanthomonas oryzae* pv. *oryzae*, Flagellar motility, Pathogenicity, Induction of hypersensitive response, T3SS

Background

Bacterial leaf blight caused by *Xanthomonas oryzae* pv. *oryzae* (Xoo) is a major bacterial disease of rice in Asian countries, which can lead to 20%–50% yield loss in rice production [1]. Xoo has been used as a model pathogen to study the molecular mechanism of bacterial pathogenesis in monocotyledonous plants [2, 3]. Now, we have learned that Xoo produces multiple virulence factors, such as exopolysaccharide (EPS), extracellular enzymes, adhesins, and the type III secretion system (T3SS) and its effectors [1, 4, 5]. HrpG and HrpX are the two master regulators to control the expression of *hrp* genes and type III effector genes [6]. Moreover, some other regulators controlling the expression of these

virulence factors have also been identified [6–8]. One of the important regulators is alternative sigma factor σ^{54} encoded by *rpoN2* [9, 10]. Deletion of *rpoN2* significantly reduces virulence and flagellar motility, yet how exactly RpoN2 regulates these virulence phenotypes in Xoo remains unknown [10].

The flagellum is the main motor organ in bacteria, which helps bacteria move toward favorable conditions and become infectious [11–13]. The flagellum consists of three parts, the basal body, the hook, and the filament. The regulatory network of flagellar gene transcription is quite complicated and fascinating. In *Escherichia coli* and *Salmonella typhimurium*, over 650 genes involved in flagellum assembly are organized into a hierarchy of three classes [14–16]. The FlhDC encoded by the class I gene *flhDC* is the master regulator and controls the transcription of class II genes [17]. The class II gene products include most of flagellum structural components and alternative sigma factor FliA

* Correspondence: hechenyang@caas.cn
State Key Laboratory for Biology of Plant Diseases and Insect Pests, Institute of Plant Protection, Chinese Academy of Agricultural Sciences, Beijing 100193, China

(σ^{28}). FliA regulates the transcription of class III genes, which encode the hook-associated proteins FlgK and FlgL, the anti-σ^{28} factor FlgM, the flagellar cap FliD, the flagellin FliC and other proteins involving in chemosensory signal transduction [18, 19]. The flagellar gene cluster of *Pseudomonas aeruginosa* has a four-tiered hierarchy of transcriptional regulation. Class I genes encode the σ^{54} factor RpoN and σ^{54}-dependent transcriptional activator FleQ. Class II genes include the two-component system *fleSR* and the σ^{28} factor *fliA*. The transcription of *fleSR* and *fliA* are regulated by RpoN and FleQ. Class III genes are regulated by FleR and are necessary for completion of the basal-body hook structure. Class IV genes are transcribed by FliA and encode the flagellin and some chemotaxis proteins [20, 21].

FliT is a key chaperone in the flagellar assembly and operation, which interacts with several flagellar proteins, including the filament-cap FliD, the export apparatus components FliI (ATPase), FliJ and FlhA, and the master regulator FlhDC [22–27]. FliT binds to the cognate substrates to not only prevent them from degradation and aggregation in the cytoplasm, but also efficiently transfer them to the export apparatus [28]. The structural analysis has showed that FliT adopts an auto-inhibited conformation, in which both the substrate- and FlhA-binding sites are occluded. Formation of FliT-substrate complex activates its binding to FlhA and thus targeting of the complex to the export gate [29]. In addition, FliT acts as a negative regulator of flagellar regulon and inhibits the binding of FlhDC to the promoter DNA [27, 30]. Interestingly, deletion of *fliT* does not affect the swimming ability in *S. typhimurium*, but significantly reduces motility properties in *P. fluorescens* [23, 31]. Moreover, disruption of *fliT* induces the expression of *Salmonella* pathogenicity island 1 (SPI1) genes, implying the potential role of FliT in bacterial virulence [32].

Our previous study has showed that over 60 contiguous flagellar genes forming a large gene cluster in Xoo PXO99^{A} encode proteins with various functions, including structural components, protein export apparatus, regulatory factors, post-translational modification enzymes, and chemotaxis proteins [10]. These genes were tightly regulated under a three-tiered hierarchy by σ^{54} factor RpoN2, and transcriptional activator FleQ, and σ^{28} factor FliA. Interestingly, a hypothetical protein gene *PXO_06168*, named as *fliTX*, has been revealed upstream of *rpoN2* and downstream of *fliS*, which is very similar location of the *fliT* genes in the genome of *S. typhimurium* and *P. fluorescens*. However, how *fliTX* is regulated and how it is related to bacterial behaviors, such as flagellar motility and virulence, remain to be elucidated.

In this study, we characterized the regulation and biological functions of *fliTX*. Promoter activities and quantitative real-time polymerase chain reaction (qRT-PCR) assays demonstrated that the transcription of *fliTX* was up-regulated by RpoN2, FleQ and FliA. In frame deletion of *fliTX* led to significant changes in flagellar motility, pathogenicity on rice, hypersensitivity on tobacco, and T3SS-related gene transcription, suggesting that FliTX plays key roles in controlling flagellar motility and virulence in Xoo.

Results

Identification, deletion and complementation of *fliTX*

Our previous study showed that there is a flagellar regulon containing over 60 contiguous genes in the genome of Xoo strain PXO99^{A}, which are regulated by RpoN2 and FleQ [10]. Upstream of *rpoN2*, there were five genes encoding a filament cap protein FliD (PXO_06166), a cytosolic chaperone FliS (PXO_06167), a hypothetical protein FliTX (PXO_06168), a non-canonical PilZ-domain protein FlgZ (PXO_06169), and a DNA-binding response regulator (PXO_06170) (Fig. 1a). The intergenic distances of neighboring genes are 150 bp, 9 bp, −1 bp, and 71 bp, respectively. Reverse transcription polymerase chain reaction (RT-PCR) analysis was performed to determine whether these five genes form an operon. The fragments containing junctions of *fliS-fliTX* and *fliTX-flgZ* were obtained using the Xoo cDNA as the template (Fig. 1a), indicating that *fliS*, *fliTX* and *flgZ* are co-transcribed in an operon. Sequence alignment analysis indicated that FliTX was less well conserved compared with FliT proteins in *Escherichia coli*, *Salmonella typhimurium*, and *Pseudomonas fluorescens* (Fig. 1b). To further identify the role of FliTX in Xoo, an in-frame deletion mutant Δ*fliTX* and its complementary strain Δ*fliTX*-C were generated as described in the Materials and Methods. DNA sequencing analysis showed that the corresponding region of *fliTX* was completely deleted in Δ*fliTX*.

fliTX is transcriptionally up-regulated by RpoN2, FleQ and FliA

To identify whether and how *fliTX* is regulated in Xoo, the promoter activities of the *fliS-fliTX-FlgZ* operon were examined by measuring β-galactosidase activity of the *fliSp-lacZ* fusion in Δ*rpoN2*, Δ*fleQ*, Δ*fliA*, and the relevant complementary strains. β-galactosidase activity of *fliSp-lacZ* was significantly reduced in Δ*rpoN2*, Δ*fleQ*, and Δ*fliA* compared with that in the wild type, and restored in the relevant complementary strains (Fig. 2a). qRT-PCR analysis showed that transcripts of *fliS*, *fliTX*, and *FlgZ* were dramatically decreased in Δ*rpoN2*, Δ*fleQ*, and Δ*fliA* compared with that in wild type (Fig. 2b), indicating that the transcription of the *fliS-fliTX-FlgZ* operon was regulated by RpoN2, FleQ, and FliA. Consistent with our previous report [10], the transcription of *fliA* was also significantly decreased in Δ*rpoN2* and Δ*fleQ* (Fig. 2b). These results strongly suggest that RpoN2/FleQ regulate the transcription of the *fliS-fliTX-FlgZ* operon via FliA in Xoo.

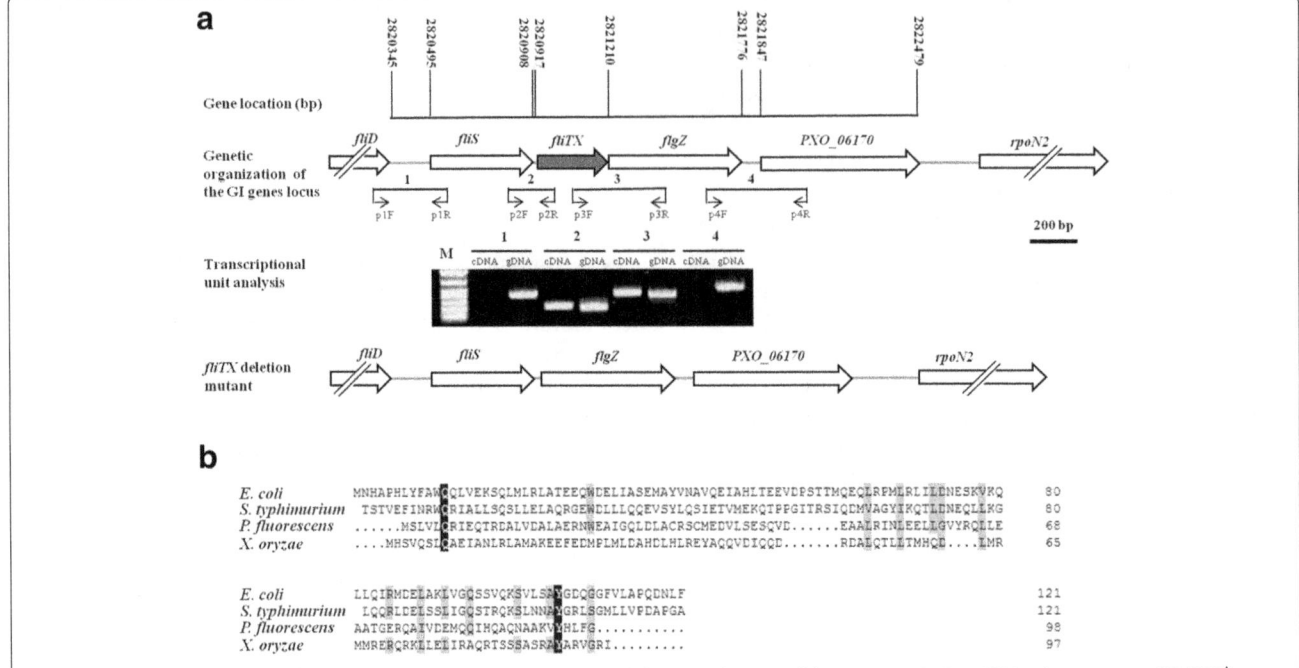

Fig. 1 Bioinformatics analysis of *fliTX* in *Xanthomonas oryzae* pv. *oryzae*. **a** Schematic diagram of the region including *fliTX* in the genome of PXO99^A. Open arrows indicate length, location and orientation of the ORFs. The short lines below the arrows indicate the location and length of RT-PCR products. The lower element shows the RT-PCR analysis of RNA isolated from PXO99^A. RT-dependent amplification of DNA fragments suggested that the *fliS*, *fliTX* and *flgZ* were transcribed in one operon. The lowest element shows the *fliTX* was in-frame deleted in Δ*fliTX*. **b** Sequence alignment of FliTX was performed using DNAMAN software. The amino acid sequences of FliT were obtained from the National Center for Biotechnology Information (NCBI) website. *E. coli*: *Escherichia coli* strain MG1655; *S. typhimurium Salmonella typhimurium* LT2 strain; *P. fluorescens*: *Pseudomonas fluorescens* strain F113; *X. oryzae*: *Xanthomonas oryzae* pv. *oryzae* strain PXO99^A. The amino acid residues highlighted with black means the homology level is 100%

FliTX is required for flagellar motility and filament production

Since *fliTX* is located within the flagellar regulon, the function of FliTX in flagellar filament assembly and flagellum-dependent motility was investigated. The swimming ability of Δ*fliTX* was detected on the 0.25% agar semisolid plates. Compared with the wild type strain, Δ*fliTX* showed a much smaller swimming zone, and the defect was restored to near wild-type level in the complementary strain containing a plasmid to express the full length of *fliTX in trans* (Fig. 3a). To further determine whether deletion of *fliTX* affected the flagellar biogenesis in Xoo, the single-polar flagellum of various Xoo strains were observed by transmission electron microscope (TEM). The average length of the flagellum on Δ*fliTX* was significantly shorter than that on the wild type, and it was recovered to wild-type level in the complementary strain (Fig. 3b). These results indicate that FliTX is necessary for flagellar biogenesis and motility in Xoo.

Δ*fliTX* shows reduced pathogenicity and bacterial growth in rice

To demonstrate the function of *fliTX* in virulence, the pathogenicity of various Xoo strains on susceptible rice

cultivar IR24 was tested by the leaf clipping method, and lesion lengths were measured 14 days post inoculation. Compared with the wild type, Δ*fliTX* caused much shorter disease lesion, which were restored in the complementary strain (Fig. 4a and b). Measuring bacterial population in the diseased leaves of rice showed that deletion of *fliTX* significantly led to reduced bacterial population in rice leaves, but complementation with *fliTX* in *trans* restored the bacterial growth *in planta* to near wild-type levels (Fig. 4c). These findings reveal that FliTX is required for virulence of Xoo in rice.

Δ*fliTX* is impaired in the ability to elicit hypersensitive response (HR) in tobacco

To unveil the role of FliTX in Xoo when interacting with nonhost plants, HR-inducing ability of Xoo strains on tobacco leaves was tested. Wild type strain induced the typical programmed cell death due to the hypersensitive responses (HR) in the non-host tobacco plants. In contrast, Δ*fliTX* completely lost such an ability to elicit HR, while the complementary strain caused a similar phenotype as the wild type strain (Fig. 5a). We then hypothesized that FliTX protein might be able to elicit HR. To test it, the recombinant protein FliTX-His$_6$ was first expressed in *E. coli* strain BL21 and extracted from the

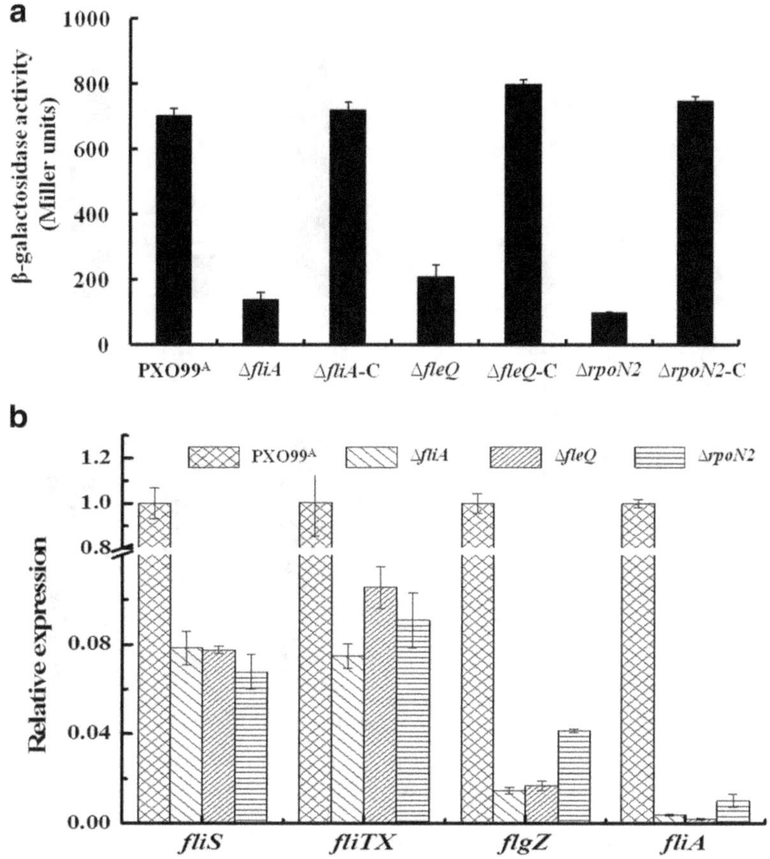

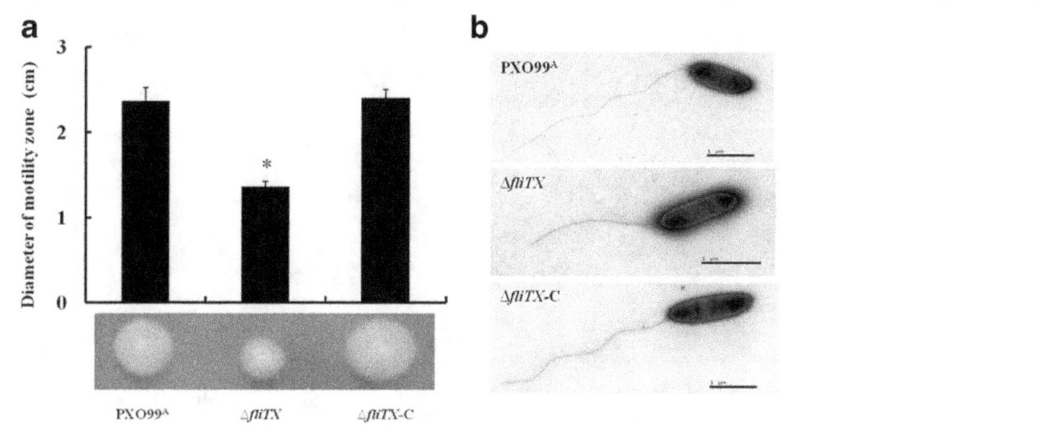

Fig. 2 Regulation of *fliTX* transcription in *Xanthomonas oryzae* pv. *oryzae*. **a** β-galactosidase activity assay. Activities of the *fliS* promoter in Xoo strains were detected. The experiments were repeated three times, independently. **b** qRT-PCR analysis of genes in *fliS* operon and *fliA* in Xoo strains. The data represents the relative expression level of genes in PXO99^A, Δ*fliA*, Δ*fleQ* and Δ*rpoN2*. The error bar represents standard deviations from three biological repeats

Fig. 3 Flagellar motility and filament production of *Xanthomonas oryzae* pv. *oryzae* strains. **a** Assay of swimming motility for PXO99^A, Δ*fliTX* and Δ*fliTX-C* strains. The swimming zones are recorded after bacterial growth for 4 days on the semisolid plates at 28 °C. Error bars indicate stand deviation. Statistical significance is presented by asterisk ($P < 0.05$, Student's *t* test). **b** Observation of filament for PXO99^A, Δ*fliTX* and Δ*fliTX-C* strains using transmission electron microscopy

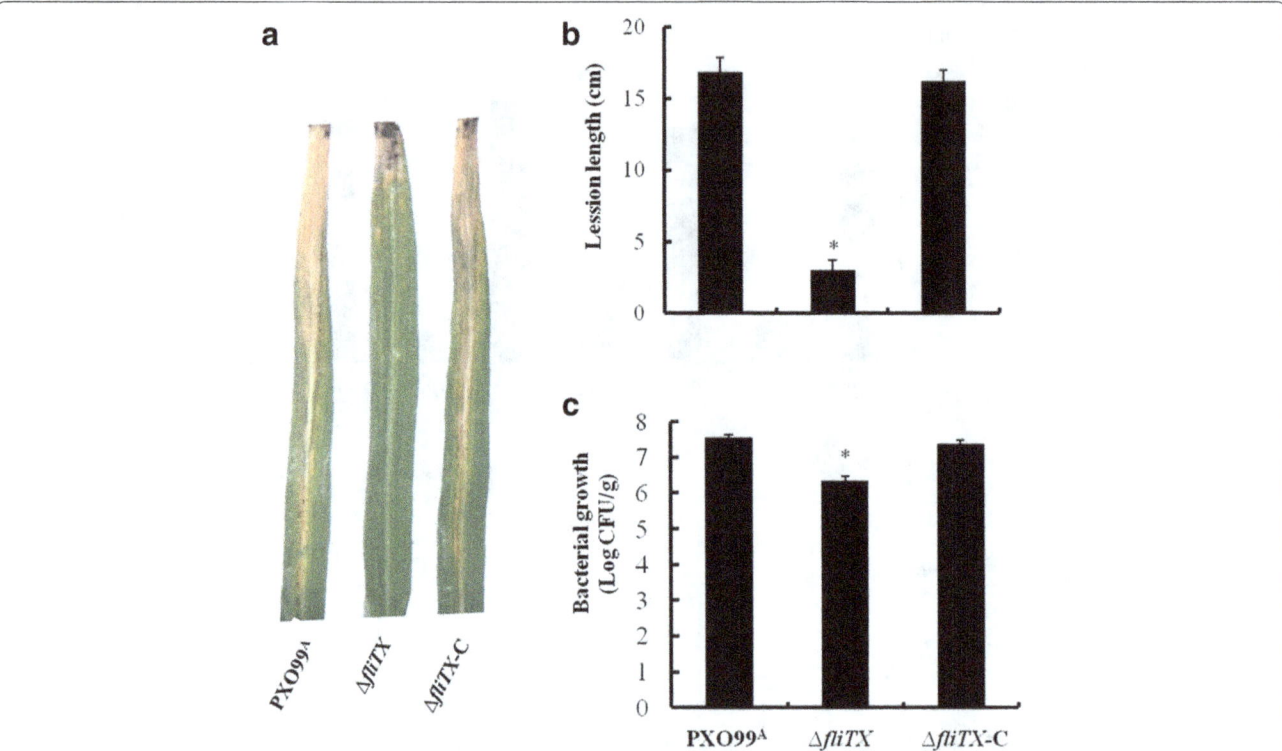

Fig. 4 Virulence of *Xanthomonas oryzae* pv. *oryzae* strains in rice. **a** PXO99^A, Δ*fliTX* and Δ*fliTX-C* strains were inoculated into 6-week-old rice leaves by using the leaf-clipping method. The disease symptoms were observed at 14 days post-inoculation. **b** The lesion lengths were recorded from 10 inoculated leaves for every strain. **c** Bacterial numbers in the top 20 cm of each lesion leaf were scored. Data represent the mean and standard deviations of three independent experiments, and the asterisk above the bars denote statistically significant differences ($P < 0.05$, Student's *t* test)

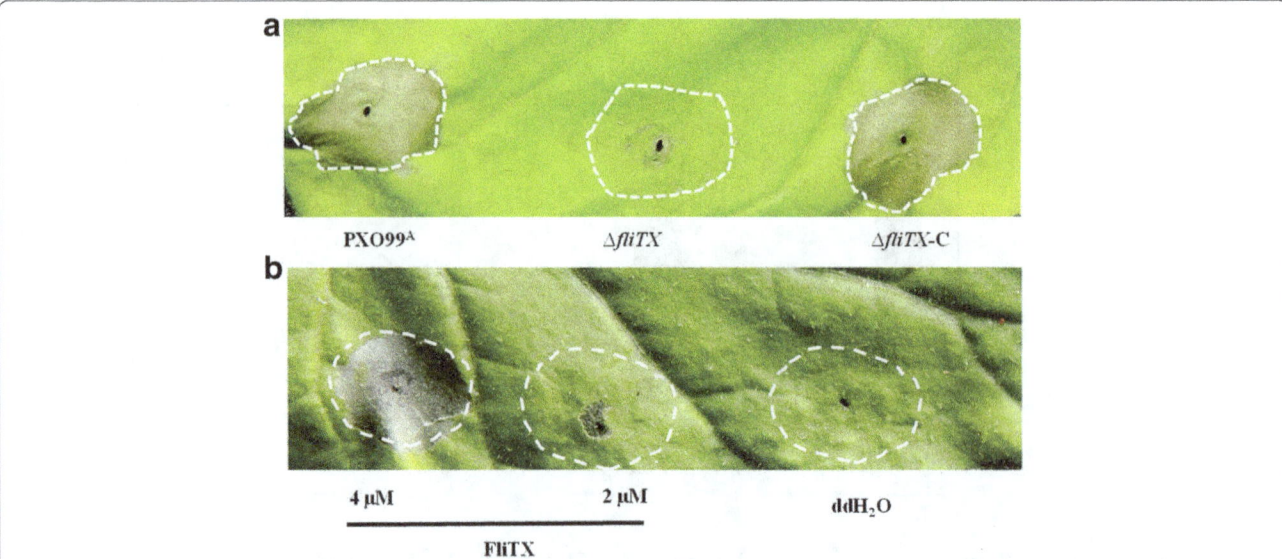

Fig. 5 Hypersensitive cell death in tobacco induced by *Xanthomonas oryzae* pv. *oryzae* strains and FliTX protein. Cell suspensions of Xoo strains at OD_{600} of 0.1 (**a**) or recombinant FliTX protein (**b**) were infiltrated onto 6-week-old tobacco leaves. The ddH$_2$O was used as control. The HR symptoms were detected and photographed at 24 h post-inoculation. At least four independent experiments were performed with similar results

soluble fraction by using pre-equilibrated Ni2_resin. SDS-PAGE analysis demonstrated that FliTX-His$_6$ was about 14 KDa in size (Additional file 1: Figure S1). Then, the purified protein was infiltrated into tobacco leaves at two different concentrations. HR in tobacco was strongly induced when FliTX-His$_6$ was applied at the concentration of 4 μM, while no HR was observed when the concentration was reduced to 2 μM (Fig. 5b). The negative control of sterilized double-distilled water (ddH$_2$O) did not cause HR either. These observations demonstrate that FliTX protein plays an important role in Xoo to elicit HR in nonhost tobacco.

ΔfliTX was attenuated in T3SS-related gene expression

The HR-inducing ability on nonhost and pathogenicity on host (Hrp) is closely related to T3SS in pathogenic bacteria [33, 34]. To understand the function of FliTX in T3SS in Xoo, transcripts of T3SS-related *hrp* genes were measured through qRT-PCR analysis. Compared with that in the wild type, transcription levels of *hrpG*, *hrpX*, *hrpE* and *hpa1* were significantly decreased in Δ*fliTX*, and restored near to wild-type level in the complementary strain (Fig. 6a). Moreover, promoter activities of *hrpG*, *hrpX* and *hpa1* revealed through flow cytometry analysis were dramatically reduced in Δ*fliTX* compared with that of the wild type. All promoter activities were restored to wild-type levels in the complementary strain (Fig. 6b). Since HrpG controls the transcript of other *hrp* genes via regulating *hrpX* expression in *Xanthomonas* [35], these results suggest that FliTX positively regulates the expression of T3SS in Xoo through the master regulator HrpG.

Discussion

In the current study, we identified a novel flagellar gene *fliTX*, determined its expression patterns, and assessed its functions in motility and virulence on rice through bioinformatics and genetic analysis. We demonstrated that the transcription of *fliTX* was positively regulated by RpoN2/FleQ and FliA. We also revealed that *fliTX* was indispensible for bacterial phenotypes, including flagellar motility, pathogenicity in rice, induction of HR in tobacco, and T3SS-related gene expression. Therefore, our identification of FliTX provides more insights into mechanistic regulation of motility and virulence in Xoo.

An over 60 contiguous gene containing cluster has been shown to putatively encode flagellar proteins with various functions, including structural components, protein export apparatus, regulatory factors, post-translational modification enzymes/proteins, and chemotaxis proteins in Xoo [9, 10]. The flagellar assembly and operation are tightly regulated under a three-tiered hierarchy by RpoN2/FleQ and FliA [10]. Based on the gene location and transcription feature, we found that *fliTX* was

transcribed in the *fliS-fliTX-flgZ* operon regulated by RpoN2/FleQ and FliA (Figs.1 and 2). This is quite different from the *fliD-fliS-fliT* operon in other pathogenic bacteria including *E. coli* and *S. typhimurium* [36, 37]. The significantly reduced transcripts of *fliA* in Δ*rpoN2* and Δ*fleQ* (Fig. 2b) suggest that regulation of transcription of the *fliS-fliTX-flgZ* operon by RpoN2/FleQ might be through FliA under a three-tiered hierarchy.

The varied functions of FliT in flagellar motility have been shown in several pathogenic bacteria. For example, FliT has been described as the filament-capping protein FliD substrate-specific chaperone in *S. typhimurium* [24]. Deletion of *fliD* inhibited the assembly of flagellin molecules onto the hooks, resulting in failure to filament biogenesis [37, 38]. However, no difference in swimming ability was observed between wildtype and the mutant [23]. In contrast, Δ*fliT* showed normal flagellar filaments but attenuated swimming motility in *P. fluorescens* F113 [31], suggesting that FliT might not act as a FliD chaperone. Our current observation that FliT was less conserved in the strains of *E. coli*, *S. typhimurium*, *P. fluorescens* and Xoo (Fig. 1) implicates that FliT may function differentially in flagellar motility in various bacteria. In this study, in frame deletion of *fliTX* led to significantly abnormal filaments and reduced swimming motility (Fig. 3), demonstrating that FliTX plays important roles in filament assembly and motility in Xoo. Further experiments are required to determine whether FliTX functions as a FliD chaperone to affect the flagellar motility in Xoo.

The role of FliT in bacterial virulence has only been reported in *S. typhimurium*, in which the transcription of T3SS-containning SPI1 was induced upon disruption of *fliT*, and the repressive effect of *fliT* on SPI1 genes was completely abolished in Δ*flhDC* [32], indicating that FliT negatively regulates the virulence and related gene expression in the FlhDC-dependent manner in *Salmonella*. In this study, we demonstrated that in frame deletion of *fliTX* resulted in dramatically reduced lesion length and bacterial growth in rice (Fig. 4), and impaired HR-inducing ability in tobacco (Fig. 5a). It has been known that the T3SS plays critical roles in conferring pathogenicity on the host and triggering the HR on nonhost plants by delivering effector proteins into plant cells [33, 34]. Meanwhile, in the current study, we showed that the expression of two T3SS regulator genes, *hrpG and hrpX*, were attenuated in the Δ*fliTX* mutant (Fig. 6). Thus, this study provides for the first time the experimental evidence that FliTX functions to promote the bacterial virulence via regulating T3SS gene expression in Xoo. Based on our previous demonstration that RpoN2 positively regulates the virulence on rice through an unknown manner [10], and the current observation that *fliTX* is up-regulated by RpoN2 and required for

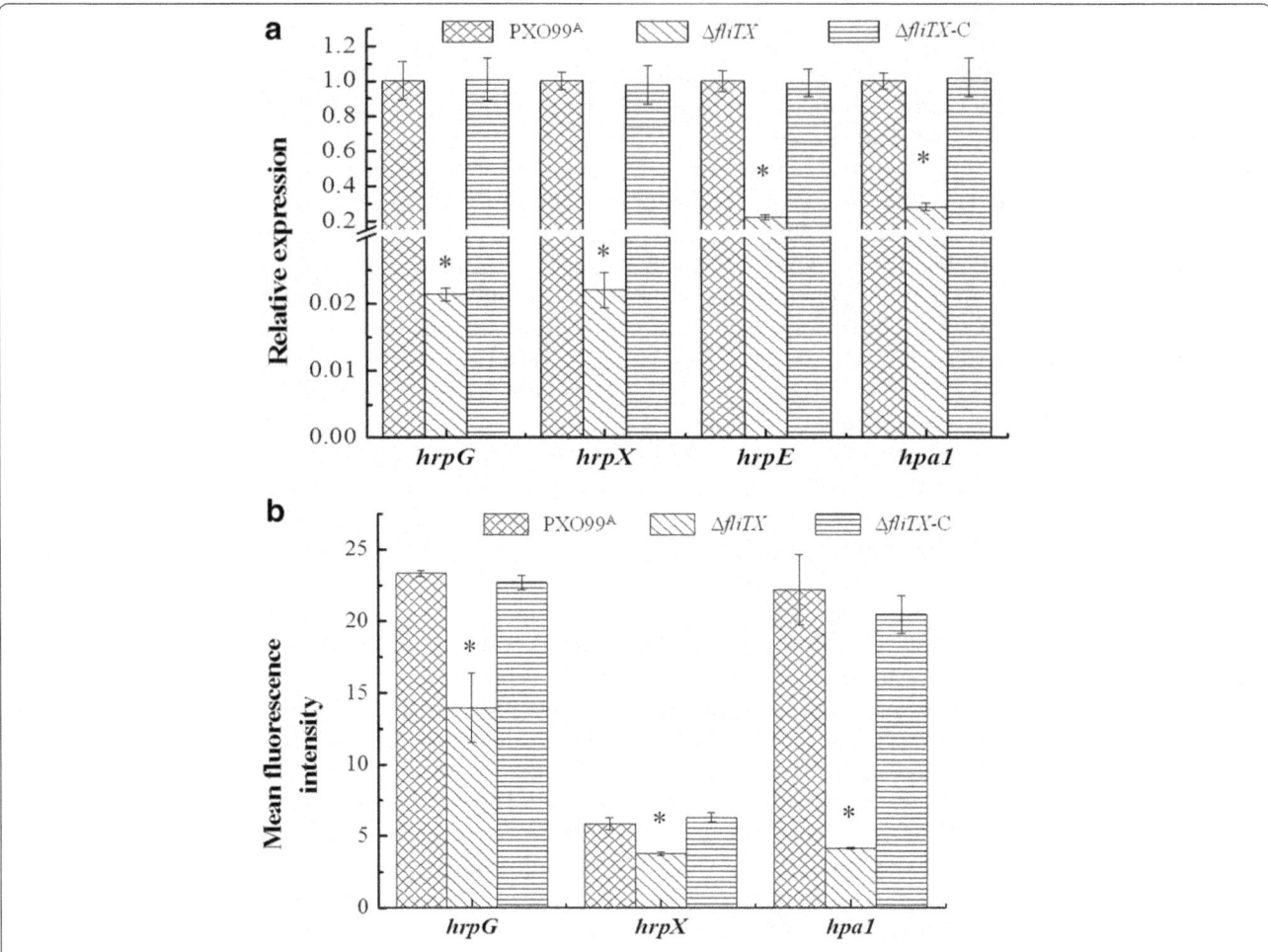

Fig. 6 Transcription of T3SS-related genes in *Xanthomonas oryzae* pv. *oryzae* strains. **a** The relative expression of T3SS-related genes was detected by qRT-PCR in PXO99^A, Δ*fliTX* and Δ*fliTX-C* strains. Fold changes of each gene was calculated using the 2$^{-\Delta\Delta Ct}$ method. **b** Promoter activity of *hrpG*, *hrpX* and *hpa1* in PXO99^A, Δ*fliTX* and Δ*fliTX-C* strains. The promoter of *hrpG*, *hrpX* and *hpa1* were ligated to pPROBE-AT, a broad-host-range vector carrying a promoter-less *gfp* gene, resulting in plasmids pPhrpG, pPhrpX and pPhpa1, respectively. These plasmids, pPhrpG, pPhrpX and pPhpa1, were transferred to *fliTX* deletion mutant, complementary strain and wildtype strains by electroporation. Green fluorescent protein mean fluorescence intensity was determined for gated populations of bacterial cells by flow cytometry. Error bars represent standard deviations from three biological repeats, and asterisk indicates *P* < 0.05 by Student's *t* test

the virulence in Xoo, it is reasonable to speculate that FliTX works in the RpoN2-dependent pathway to promote the bacterial pathogenesis in rice. More investigations are needed to confirm this hypothesis and further understand the regulatory mechanism of virulence by FliTX in Xoo.

For the assembly of bacterial flagellum for motility, the flagellar type III export apparatus utilizes both ATP and proton motive force to cross the cytoplasmic membrane and export flagellar proteins from the cytoplasm to the cell membrane [28]. FliT acts as the specific chaperone of the filament-capping protein FliD that is protected from degradation and aggregation in the cytoplasm and efficiently transferred to the distal end of the flagellar structure [26]. However, it remains mysterious whether and how the FliT protein is secreted. In

addition, we showed that the recombinant FliTX protein induced HR in monhost tobacco leaves (Fig. 5b), implicating a potential role of FliTX in inducing plant defense responses. Therefore, it is required to further demonstrate whether and how FliTX is secreted into the plant cells during the induction of HR.

Conclusions

The *fliTX* gene is transcriptionally up-regulated by RpoN2/FleQ and FliA, and necessary for flagellar assembly and motility in Xoo. Deletion of *fliTX* led to significantly reduced virulence in rice, attenuated ability of induction of HR in tobacco and decreased *hrp* gene expression. RpoN2/FleQ- and FliA-regulated FliTX controls the bacterial pathogenesis via T3SS regulation with the unknown manner(s).

Methods

Bacterial strains and culture conditions

Xanthomonas oryzae pv. *oryzae* wildtype strain PXO99[A] and derived mutants were grown in peptone sucrose agar (PSA) medium [39] or M210 liquid medium [40] at 28 °C, *Escherichia coli* DH5α and BL21 strains were grown in Luria-Bertani (LB) medium at 37 °C. The antibiotics used were ampicillin (Ap), gentamicin (Gm), kanamycin (Km) and spectinomycin (Sp) at concentrations of 100, 50, 50, and 50 μg/mL, respectively. The bacterial strains and plasmids used in this study are listed in Table 1.

Bioinformatics analysis of *fliTX*

The domain organization of FliTX was analyzed using online software available at the SMART Website (http://smart.embl-heidelberg.de/). The amino acid sequences of FliTX were obtained from the National Center for Biotechnology Information (NCBI) website. BLASTP was using for searching the homology in *Xanthomonas* species. Relevant sequence alignment was performed using the DNAMAN software (Lynnon Biosoft, San Ramon, USA).

Generation of *lacZ* fusion and assay for β-galactosidase activity

The promoter region (−309 to −1) of *fliS* was amplified from PXO99[A] genomic DNA using specific primers fliSpF/R (Additional file 2: Table S1), and ligated into the *Hind*III and *Bam*HI sites of the pHT304BZ vector containing a promoterless *lacZ* reporter gene [41]. Recombinant pHTpS was verified by DNA sequencing (Beijing Genomics Institute, Beijing) and treated with HindIII and KpnI, and the fragment containing *fliS* promoter region and the promoterless *lacZ* reporter gene was obtained and then cloned into plasmid pHM1 [42]. Finally,

Table 1 Bacterial strains and plasmids used in this study

Strain or plasmid	Relevant characteristics[a]	Source or Reference
Escherichia coli		
DH5α	supE44 ΔlacU169(Φ80lacZΔM15) hsdR17 recA1 endA1 gyrA96 thi-1 relA1	[49]
BL21	For protein expression	Novagen
Xanthomonas oryzae pv. *oryzae*		
PXO99[A]	Wildtype strain, Philippine race 6	Lab collection
Δ*fliTX*	*fliTX* gene deletion mutant derived from PXO99[A]	This study
Δ*fliTX-C*	Complementary strain of Δ*fliTX*, Ap[r]	This study
Δ*rpoN2*	*rpoN2* gene deletion mutant derived from PXO99[A], Gm[r]	Our lab
Δ*rpoN2-C*	Complementary strain of Δ*rpoN2*, Ap[r]	Our lab
Δ*fleQ*	*fleQ* gene deletion mutant derived from PXO99[A], Gm[r]	Our lab
Δ*fleQ-C*	Complementary strain of Δ*fleQ*, Ap[r]	Our lab
Δ*fliA*	*fliA* gene deletion mutant derived from PXO99[A], Gm[r]	Our lab
Δ*fliA-C*	Complementary strain of Δ*fliA*, Ap[r]	Our lab
Plasmid		
pMD18-T	Cloning vector, Ap[r]	TaKaRa, Tokyo
pKMS1	Suicidal vector carrying *sacB* gene for non-marker mutagenesis, Km[r]	[45]
pBBR1MCS-4	Broad-host range expression vector, Ap[r]	[50]
pHT304BZ	Promoterless *lacZ* vector, Ap[r]	[41]
pHTpS	pHT304BZ derivative carrying the promoter region of *fliS*, Ap[r]	This study
pHM1	Broad-host range expression vector, Sp[r]	[42]
pH-*fliSp-lacZ*	pHM1 derivative carrying the promoter region of *fliS* and promoterless *lacZ*, Sp[r]	This study
pET-28a	Expression vector to generate a N-terminal His₆ tag, Km[r]	Haigene
pET-*fliTX*	pET-28a derivative carrying *fliTX*, Km[r]	This study
pPROBE-AT	broad-host-range vector carrying a promoter-less *gfp* gene, Ap[r]	[47]
pPhrpG	pPROBE-AT derivative carrying the promoter region of *hrpG* and promoterless *gfp*, Ap[r]	This study
pPhrpX	pPROBE-AT derivative carrying the promoter region of *hrpX* and promoterless *gfp*, Ap[r]	This study
pPhpa1	pPROBE-AT derivative carrying the promoter region of *hpa1* and promoterless *gfp*, Ap[r]	Our lab

[a]Ap[r], Km[r], Sp[r], and Gm[r] indicate resistant to ampicillin, kanamycin, spectinomycin and gentamicin, respectively

the recombinant plasmid pH-*fliSp-lacZ* was generated and introduced into PXO99^A and derived mutants. The resultant strains contained pH-*fliSp-lacZ* were selected by resistance to spectinomycin and verified by polymerase chain reaction (PCR). For β-galactosidase assay, these Xoo strains were cultured in M210 liquid medium at 28 °C and 200 rpm, and till an optical density (OD_{600}) of 1.0, cells were collected by centrifugation at 12,000 g. The β-galactosidase activity was determined using the β-Galactosidase Enzyme Assay System (Promega, Wisconsin, USA). The experiments were repeated three times, independently.

RNA isolation and qRT-PCR analysis

RNA isolation and qRT-PCR analysis were performed as described previously with some modifications [8]. Briefly, Xoo strains were grown in M210 liquid medium at 28 °C till OD_{600} of 0.8, and harvested by centrifugation at 12,000 g for analysis of gene expression. For T3SS-related gene assays, the harvested bacterial cells were sub-cultured in XOM2 medium [43] overnight at 28 °C and collected again. Total RNA was extracted with RNAprep pure Cell/Bacteria Kit (Tiangen, Beijing, China) and treated with DNase and cDNA was synthesized from total RNA using the FastQuant RT Super Mix (Tiangen, Beijing, China). RT-qPCR was performed using Quant qRT-PCR kit (Tiangen, Beijing, China) in Applied Biosystem's 7500 (Applied Biosystems, Foster City, CA, USA) with gene specific primers, and *gyrB* was used as a reference gene (Additional file 2: Table S1). The relative expression ratio was calculated using $2^{-\Delta\Delta Ct}$ method [44]. These experiments were performed in three biological replicates and triplicate PCR.

Protein expression and purification

The FliTX expression and purification were performed as described previously [8]. Briefly, the coding region for *fliTX* was amplified by PCR with primers TXF/R (Additional file 2: Table S1) and ligated to the middle vector pMD18-T for verification by DNA sequencing. Then the *fliTX* fragment was digested from verified pMD18-T with corresponding restriction enzymes and ligated to pET28a, resulting in pET-*fliTX*. The recombinant plasmid was transformed into *E. coli* BL21 strain for protein expression. For protein purification, the BL21 strain carrying pET-*fliTX* was cultured in LB liquid medium at 37 °C to OD_{600} of 0.6, and isopropyl-thiogalactopyranoside at a final concentration of 0.1 mM was added to induce *fliTX* expression. After 6 h cultured, the BL21 cells were collected by centrifugation and re-suspended in 0.1× PBS. The crude cell extracts were obtained by sonication and centrifuged at 12,000 g for 10 min in 4 °C. The supernatant containing the soluble proteins was mixed with pre-equilibrated Ni2_resin

(GE Healthcare, Piscataway, NJ, USA) for 1 h at 4 °C. Finally, the FliTX protein combined to Ni was eluted with elution buffer (20 mM Tris-HCl, 350 mM NaCl, 0.5 mM EDTA, 10% glycerol, 5 mM $MgCl_2$ and 100 mM imidazole, pH 8.0) and dialyzed with 0.1× PBS. The purified FliTX was detected by sodium dodecyl sulfate polyacrylamide gel electrophoresis (SDS-PAGE) and adjusted to 10 μM with 0.1 × PBS for the next experiments.

Gene deletion and complementation

An in-frame gene deletion mutant Δ*fliTX* derived from PXO99^A was constructed through homologous recombination by using the suicide vector pKMS1 [45]. About 900 bp upstream and 800 bp downstream fragments of the *fliTX* gene were amplified by PCR from Xoo genomic DNA using primers fliTXLF/R and fliTXRF/R, respectively. The PCR products were first cloned into the middle vector pMD18-T (Takara, Dalian, China) and verified by sequencing. Then the upstream and downstream fragments of *fliTX* were digested with corresponding restriction enzymes from the middle vectors and ligated into pKMS1. The final vector pKMS1 containing upstream and downstream fragments of *fliTX* was introduced into Xoo by electroporation. The transformants were first selected on NAN medium (1% tryptone, 0.1% yeast extract, 0.3% peptone and 1.5% agar) with Km, and after continuous transfer cultured in NBN medium (1% tryptone, 0.1% yeast extract and 0.3% peptone) at least five times. Finally, the Δ*fliTX* mutant was selected on NAS medium (1% tryptone, 0.1% yeast extract, 0.3% peptone, 10% sucrose and 1.5% agar) and further confirmed by PCR analysis. For complementation strain construction, the full length of *fliTX* was amplified by PCR with primers fliTXF/R and inserted into vector pMD18-T. After verifying by sequencing, *fliTX* was digested from pMD18-T and ligated into pBBR1MCS-4. The final vector pBBR1MCS-4 containing *fliTX* was electroporated into Δ*fliTX* and confirmed by PCR analysis, resulting in the Δ*fliTX* complementary strain (Δ*fliTX*-C). The primers are listed in Additional file 2: Table S1.

Motility assay and electron microscopy visualization of filament

For motility assay, bacterial strains were cultured in M210 liquid medium till reached OD_{600} of 1.0 and harvested by centrifugation at 12,000 g for 5 min. Cells were re-suspended in equal volume of ddH_2O. Two microliters of bacterial suspension were spotted onto semisolid plates (0.03% peptone, 0.03% yeast extract and 0.25% agar) and incubated at 28 °C. The diameters of the swimming zone were recorded after 4 days. The experiments were repeated three times with five replicates for each time. The TEM assay was performed as described previously [46]. Briefly, bacterial strains were grown on PSA plates at 28 °C for 48 h, and cells were collected and re-suspended with ddH$_2$O, then one drop of suspension was deposited onto grids coated with Formvar (Standard Technology, Ormond Beach, FL, USA). The grids with bacteria were stained with 2% uranyl acetate for 30 s, and air drying for 10 min. The bacterial flagella were observed by TEM using Hitachi H-7500 electron microscope.

Pathogenicity test

As described above, bacterial strains were cultured in M210 liquid medium at 28 °C and 200 rpm till reached OD$_{600}$ of 1.0, and collected by centrifugation at 12,000 g for 5 min, and re-suspended with equal volume of ddH$_2$O. For the disease lesion length assay, bacterial cells were inoculated into leaves of 8-week-old rice (*Oryza sativa ssp. indica*) cultivar IR24 using the leaf-clipping method [8], and the lesion length was measured at 14 days post-inoculation. For the bacterial population assay, the top 20 cm of inoculated rice leaves were collected and weighted, then ground into 1 mL of ddH$_2$O. The ground mixtures were optional diluted and spread onto the PSA plates. The bacterial colonies were counted after cultured in incubator with 28 °C for 72 h. At least 10 leaves were inoculated for each strain, and the experiments were repeated three times, independently.

Assay for induction of HR in tobacco

Xoo strains were grown in M210 liquid medium at 28 °C to OD$_{600}$ of 1.0, and collected by centrifugation at 7000 g for 10 min. The cells were re-suspended with ddH$_2$O, and adjusted to OD$_{600}$ of 0.1. Then these bacterial cells or purified FliTX protein were inoculated into leaves of 6-week-old tobacco (*Nicotiana benthamiana*) using a needleless syringe. The HR symptoms were detected and photographed at 24 h post-inoculation. The experiments were repeated three times, independently.

Flow cytometry detection

The plasmid pPhpa1 containing the *hpa1* promoter region and a promoterless *gfp* gene was constructed in our previous studies [40]. Here two near 200-bp fragments containing the promoter region of *hrpG* or *hrpX* were PCR amplified using the primers hrpGpF/R or hrpXpF/R (Additional file 2: Table S1), and ligated to pPROBE-AT, a broad-host-range vector carrying a promoter-less *gfp* gene [47], resulting in plasmids pPhrpG and pPhrpX, respectively. These plasmids, pPhpa1, pPhrpG and pPhrpX, were transferred to *fliTX* deletion mutant, complementary strain and wildtype strains by electroporation. The transformed strains were cultured in M210 liquid medium to OD$_{600}$ of 1.0 and transferred to XOM2 medium for 12 h at 28 °C. The cells were collected by centrifugation at 12,000 g for 5 min and re-suspended with 0.1× PBS. The promoter activities of

hpa1, *hrpG* and *hrpX* were detected using a FACS-Caliber flow cytometer (BD Bioscience, CA, USA) as previously described [48]. Xoo wildtype carrying a promoterless pBROBE-AT was used as a negative control. The experiments were repeated three times, independently.

Data analysis

The values of β-galactosidase activity, gene expression level, motility zone, disease lesion length and bacterial population were presented as means ± SD (standard deviations). Student's *t* test was performed with statistical significance set at the 0.05 confidence level.

Additional files

Additional file 1: Figure S1. Coomassie blue staining of the FliTX protein expressed and extracted from *E. coli* strain BL21. M: Molecular marker; 1: FliTX in the soluble fraction; 2: purified FliTX; 3: FliTX in the insoluble fraction.

Additional file 2: Table S1. The primers used in this study.

Abbreviations

Ap: Ampicillin; ddH$_2$O: Sterilized double-distilled water; EPS: Exopolysaccharide; Gm: Gentamicin; HR: Hypersensitive response; Km: Kanamycin; LB: Luria-Bertani; NCBI: National Center for Biotechnology Information; OD$_{600}$: Optical density; PCR: Polymerase chain reaction; PSA: Peptone sucrose agar; qRT-PCR: Quantitative real-time polymerase chain reaction; RT-PCR: Reverse transcription polymerase chain reaction; SD: Standard deviations; SDS-PAGE: Sodium dodecyl sulfate polyacrylamide gel electrophoresis; Sp: Spectinomycin; SPI1: *Salmonella* pathogenicity island 1; T3SS: Type III secretion system; TEM: Transmission electron microscope; Xoo: *Xanthomonas oryzae* pv. *oryzae*

Acknowledgements

We would like to thank John J. Srok at University of Wisconsin-Milwaukee for editing the language.

Funding

This work was supported by the grants from the National Basic Research Program of China (2011CB100700) and the Nature Science Foundation of China (31600105).

Authors' contributions

CY and CYH designed the experiments; CY performed the experiments; CY, HMC, FT, FHY and CYH analyzed the data; CY, FT and CYH wrote the manuscript; All authors read and approved the final manuscript.

Competing interests

The authors declare that they have no competing interests.

References

1. Nino-Liu DO, Ronald PC, Bogdanove AJ. *Xanthomonas oryzae* pathovars: model pathogens of a model crop. Mol Plant Pathol. 2006;7(5):303–24.
2. Martin GB, Bogdanove AJ, Sessa G. Understanding the functions of plant disease resistance proteins. Annu Rev Plant Biol. 2003;54:23–61.
3. Ronald PC. The molecular basis of disease resistance in rice. Plant Mol Biol. 1997;35(1–2):179–86.
4. Das A, Rangaraj N, Sonti RV. Multiple adhesin-like functions of *Xanthomonas oryzae* pv. *oryzae* are involved in promoting leaf attachment, entry, and virulence on rice. Mol Plant-Microbe Interact. 2009;22(1):73–85.

5.   White FF, Yang B. Host and pathogen factors controlling the rice-*Xanthomonas oryzae* interaction. Plant Physiol. 2009;150(4):1677–86.

6.   Tsuge S, Terashima S, Furutani A, Ochiai H, Oku T, Tsuno K, Kaku H, Kubo Y. Effects on promoter activity of base substitutions in the cis-acting regulatory element of HrpXo regulons in *Xanthomonas oryzae* pv. *oryzae*. J. Bacteriol. 2005;187(7):2308–14.

7.   He YW, Wu J, Cha JS, Zhang LH. Rice bacterial blight pathogen *Xanthomonas oryzae* pv. *oryzae* produces multiple DSF-family signals in regulation of virulence factor production. BMC Microbiol. 2010;10:187.

8.   Yang FH, Tian F, Sun L, Chen HM, Wu MS, Yang CH, He CY. A novel two-component system PdeK/PdeR regulates c-di-GMP turnover and virulence of *Xanthomonas oryzae* pv. *oryzae*. Mol Plant-Microbe Interact. 2012;25(10):1361–9.

9.   Salzberg SL, Sommer DD, Schatz MC, Phillippy AM, Rabinowicz PD, Tsuge S, Furutani A, Ochiai H, Delcher AL, Kelley D, et al. Genome sequence and rapid evolution of the rice pathogen *Xanthomonas oryzae* pv. *oryzae* PXO99^A. BMC Genomics. 2008;9:204.

10. Tian F, Yu C, Li HY, Wu XL, Li B, Chen HM, Wu MS, He CY. Alternative sigma factor RpoN2 is required for flagellar motility and full virulence of *Xanthomonas oryzae* pv. *oryzae*. Microbiol Res. 2015;170:177–83.

11. Haefele DM, Lindow SE. Flagellar motility confers epiphytic fitness advantages upon *Pseudomonas syringae*. Appl Environ Microbiol. 1987; 53(10):2528–33.

12. Giron JA, Torres AG, Freer E, Kaper JB. The flagella of enteropathogenic *Escherichia coli* mediate adherence to epithelial cells. Mol Microbiol. 2002; 44(2):361–79.

13. Tans-Kersten J, Brown D, Allen C. Swimming motility, a virulence trait of *Ralstonia solanacearum*, is regulated by FlhDC and the plant host environment. Mol Plant-Microbe Interact. 2004;17(6):686–95.

14. Aldridge P, Hughes KT. Regulation of flagellar assembly. Curr Opin Microbiol. 2002;5(2):160–5.

15. Kutsukake K, Ohya Y, Iino T. Transcriptional analysis of the flagellar regulon of *Salmonella typhimurium*. J Bacteriol. 1990;172(2):741–7.

16. Macnab RM. Type III flagellar protein export and flagellar assembly. Biochim Biophys Acta. 2004;1694(1–3):207–17.

17. Liu X, Matsumura P. The FlhD/FlhC complex, a transcriptional activator of the *Escherichia coli* flagellar class II operons. J Bacteriol. 1994;176(23):7345–51.

18. Karlinsey JE, Tanaka S, Bettenworth V, Yamaguchi S, Boos W, Aizawa SI, Hughes KT. Completion of the hook-basal body complex of the *Salmonella typhimurium* flagellum is coupled to FlgM secretion and fliC transcription. Mol Microbiol. 2000;37(5):1220–31.

19. Sorenson MK, Ray SS, Darst SA. Crystal structure of the flagellar sigma/anti-sigma complex sigma(28)/FlgM reveals an intact sigma factor in an inactive conformation. Mol Cell. 2004;14(1):127–38.

20. Dasgupta N, Wolfgang MC, Goodman AL, Arora SK, Jyot J, Lory S, Ramphal R. A four-tiered transcriptional regulatory circuit controls flagellar biogenesis in *Pseudomonas aeruginosa*. Mol Microbiol. 2003;50(3):809–24.

21. McCarter LL. Regulation of flagella. Curr Opin Microbiol. 2006;9(2):180–6.

22. Bange G, Kummerer N, Engel C, Bozkurt G, Wild K, Sinning I. FlhA provides the adaptor for coordinated delivery of late flagella building blocks to the type III secretion system. Proc Natl Acad Sci U S A. 2010;107(25):11295–300.

23. Bennett JC, Thomas J, Fraser GM, Hughes C. Substrate complexes and domain organization of the *Salmonella* flagellar export chaperones FlgN and FliT. Mol Microbiol. 2001;39(3):781–91.

24. Fraser GM, Bennett JC, Hughes C. Substrate-specific binding of hook-associated proteins by FlgN and FliT, putative chaperones for flagellum assembly. Mol Microbiol. 1999;32(3):569–80.

25. Imada K, Minamino T, Kinoshita M, Furukawa Y, Namba K. Structural insight into the regulatory mechanisms of interactions of the flagellar type III chaperone FliT with its binding partners. Proc Natl Acad Sci U S A. 2010; 107(19):8812–7.

26. Minamino T, Kinoshita M, Imada K, Namba K. Interaction between FliI ATPase and a flagellar chaperone FliT during bacterial flagellar protein export. Mol Microbiol. 2012;83(1):168–78.

27. Yamamoto S, Kutsukake K. FliT acts as an anti-FlhD2C2 factor in the transcriptional control of the flagellar regulon in *Salmonella enterica* serovar *typhimurium*. J Bacteriol. 2006;188(18):6703–8.

28. Minamino T. Protein export through the bacterial flagellar type III export pathway. Biochim Biophys Acta. 2014;1843(8):1642–8.

29. Khanra N, Rossi P, Economou A, Kalodimos CG. Recognition and targeting mechanisms by chaperones in flagellum assembly and operation. Proc Natl Acad Sci U S A. 2016;113(35):9798–803.

30. Sato Y, Takaya A, Mouslim C, Hughes KT, Yamamoto T. FliT selectively enhances proteolysis of FlhC subunit in FlhD4C2 complex by an ATP-dependent protease, ClpXP. J Biol Chem. 2014;289(47):33001–11.

31. Capdevila S, Martinez-Granero FM, Sanchez-Contreras M, Rivilla R, Martin M. Analysis of *Pseudomonas fluorescens* F113 genes implicated in flagellar filament synthesis and their role in competitive root colonization. Microbiology. 2004;150(Pt 11):3889–97.

32. Hung CC, Haines L, Altier C. The flagellar regulator fliT represses *Salmonella* pathogenicity island 1 through flhDC and fliZ. PLoS One. 2012;7(3):e34220.

33. Alfano JR, Collmer A. The type III (Hrp) secretion pathway of plant pathogenic bacteria: trafficking harpins, Avr proteins, and death. J Bacteriol. 1997;179(18):5655–62.

34. Buttner D, Bonas U. Common infection strategies of plant and animal pathogenic bacteria. Curr Opin Plant Biol. 2003;6(4):312–9.

35. Wengelnik K, Van den Ackerveken G, Bonas U. HrpG, a key hrp regulatory protein of *Xanthomonas campestris* pv. *vesicatoria* is homologous to two-component response regulators. Mol Plant-Microbe Interact. 1996;8(8):704–12.

36. Kutsukake K, Ide N. Transcriptional analysis of the flgK and fliD operons of *Salmonella typhimurium* which encode flagellar hook-associated proteins. Mol Gen Genet. 1995;247(3):275–81.

37. Yokoseki T, Kutsukake K, Ohnishi K, Iino T. Functional analysis of the flagellar genes in the fliD operon of *Salmonella typhimurium*. Microbiology. 1995; 141(Pt 7):1715–22.

38. Homma M, Kutsukake K, Iino T, Yamaguchi S. Hook-associated proteins essential for flagellar filament formation in *Salmonella typhimurium*. J Bacteriol. 1984;157(1):100–8.

39. Tsuchiya K, Mew TW, Wakimoto S. Bacteriological and pathological characteristics of wild type and induced mutants of *Xanthomonas campestris* pv. *oryzae*. Phytopathology. 1982;72:43–6.

40. Fan SS, Tian F, Li JY, Hutchins W, Chen HM, Yang FH, Yuan X, Cui ZN, Yang CH, He CY. Identification of phenolic compounds that suppress the virulence of *Xanthomonas oryzae* on rice via the type III secretion system. Mol Plant Pathol. 2016; doi:10.1111/mpp.12415.

41. Lereclus D, Agaisse H, Gominet M, Salamitou S, Sanchis V. Identification of a *Bacillus thuringiensis* gene that positively regulates transcription of the phosphatidylinositol-specific phospholipase C gene at the onset of the stationary phase. J Bacteriol. 1996;178(10):2749–56.

42. Hopkins CM, White FF, Choi SH, Guo A, Leach JE. Identification of a family of avirulence genes from *Xanthomonas oryzae* pv. *oryzae*. Mol Plant-Microbe Interact. 1992;5(6):451–9.

43. Tsuge S, Furutani A, Fukunaka R, Oku T, Tsuno K, Ochiai H, Inoue Y, Kaku H, Kubo Y. Expression of *Xanthomonas oryzae* pv. *oryzae* hrp genes in XOM2, a novel synthetic medium. J Gen Plant Pathol. 2002;68(4):363–71.

44. Livak KJ, Schmittgen TD. Analysis of relative gene expression data using real-time quantitative PCR and the 2(T)(−Delta Delta C) method. Methods. 2001;25(4):402–8.

45. Li YR, Zou HS, Che YZ, Cui YP, Guo W, Zou LF, Chatterjee S, Biddle EM, Yang CH, Chen GY. A novel regulatory role of HrpD6 in regulating hrp-hrc-hpa genes in *Xanthomonas oryzae* pv. *oryzicola*. Mol Plant-Microbe Interact. 2011;24(9):1086–101.

46. Li HY, Yu C, Chen HM, Tian F, He CY. PXO_00987, a putative acetyltransferase, is required for flagellin glycosylation, and regulates flagellar motility, exopolysaccharide production, and biofilm formation in *Xanthomonas oryzae* pv. *oryzae*. Microb Pathog. 2015;85:50–7.

47. Miller WG, Leveau JH, Lindow SE. Improved *gfp* and *inaZ* broad-host-range promoter-probe vectors. Mol Plant-Microbe Interact. 2000;13(11):1243–50.

48. Yamazaki A, Li J, Zeng Q, Khokhani D, Hutchins WC, Yost AC, Biddle E, Toone EJ, Chen X, Yang CH. Derivatives of plant phenolic compound affect the type III secretion system of *Pseudomonas aeruginosa* via a GacS-GacA two-component signal transduction system. Antimicrob Agents Chemother. 2012;56(1):36–43.

49. Hanahan D. Studies on transformation of *Escherichia coli* with plasmids. J Mol Biol. 1983;166:557–80.

50. Kovach ME, Elzer PH, Hill DS, Robertson GT, Farris MA, Roop RM 2nd, Peterson KM. Four new derivatives of the broad-host-range cloning vector pBBR1MCS, carrying different antibiotic-resistance cassettes. Gene. 1995;166:175–6.

10

Ribosomal subunit protein typing using matrix-assisted laser desorption ionization time-of-flight mass spectrometry (MALDI-TOF MS) for the identification and discrimination of *Aspergillus* species

Sayaka Nakamura[1], Hiroaki Sato[1][*] 🆔, Reiko Tanaka[2], Yoko Kusuya[2], Hiroki Takahashi[2] and Takashi Yaguchi[2]

Abstract

Background: Accurate identification of *Aspergillus* species is a very important subject. Mass spectral fingerprinting using matrix-assisted laser desorption ionization time-of-flight mass spectrometry (MALDI-TOF MS) is generally employed for the rapid identification of fungal isolates. However, the results are based on simple mass spectral pattern-matching, with no peak assignment and no taxonomic input. We propose here a ribosomal subunit protein (RSP) typing technique using MALDI-TOF MS for the identification and discrimination of *Aspergillus* species. The results are concluded to be phylogenetic in that they reflect the molecular evolution of housekeeping RSPs.

Results: The amino acid sequences of RSPs of genome-sequenced strains of *Aspergillus* species were first verified and compared to compile a reliable biomarker list for the identification of *Aspergillus* species. In this process, we revealed that many amino acid sequences of RSPs (about 10–60%, depending on strain) registered in the public protein databases needed to be corrected or newly added. The verified RSPs were allocated to RSP types based on their mass. Peak assignments of RSPs of each sample strain as observed by MALDI-TOF MS were then performed to set RSP type profiles, which were then further processed by means of cluster analysis. The resulting dendrogram based on RSP types showed a relatively good concordance with the tree based on β-tubulin gene sequences. RSP typing was able to further discriminate the strains belonging to *Aspergillus* section *Fumigati*.

Conclusions: The RSP typing method could be applied to identify *Aspergillus* species, even for species within section *Fumigati*. The discrimination power of RSP typing appears to be comparable to conventional β-tubulin gene analysis. This method would therefore be suitable for species identification and discrimination at the strain to species level. Because RSP typing can characterize the strains within section *Fumigati*, this method has potential as a powerful and reliable tool in the field of clinical microbiology.

Keywords: *Aspergillus*, Ribosomal subunit proteins, Matrix-assisted laser desorption ionization time-of-flight mass spectrometry

* Correspondence: sato-hiroaki@aist.go.jp
[1]Research Institute for Sustainable Chemistry, National Institute of Advanced Industrial Science and Technology (AIST), Higashi 1-1-1, Tsukuba, Ibaraki 305-8565, Japan
Full list of author information is available at the end of the article

Background

Aspergillus is a saprophytic genus found in diverse environments [1]. Some species, typically *A. fumigatus*, are causative agents of aspergillosis [2]. Because the degree of virulence and susceptibility to antifungal agents are known to vary among species [3], accurate identification of *Aspergillus* species is a very important subject, especially in the field of clinical mycology.

The identification of fungal species has up to now been based on the morphological characteristics of colonies and filaments as observed by microscopy [4]. However, the morphology-based method suffers several drawbacks. It requires specialized skills and knowledge and is tedious and time-consuming work. Nevertheless, some strains lack obvious characteristic features under laboratory conditions. To achieve objective identification, molecular biological methods based on the DNA sequences of particular genes are increasingly being adopted. The internal transcribed spacer (ITS) regions between 18S rRNA, 5.8S rRNA, and 28S rRNA are regarded as "barcode regions" and are frequently used as biomarkers for species identification [5]. DNA sequences that code housekeeping proteins such as β-tubulin [6] and calmodulin [7] are also often used for detailed molecular studies. To improve the resolution of species discrimination, combinations of multi-genes have been attempted. For example, a combination of two genes (β-tubulin and calmodulin) [8] or four genes (β-tubulin, calmodulin, ITS and large-subunit rDNA, and RNA polymerase II) [9] has been used to characterize *A. fumigatus* strains. Multi-locus sequence typing (MLST), focusing on seven types of gene fragments, has also been applied to characterize *A. fumigatus* strains [10]. These DNA-based methods provide a more objective evaluation than the traditional morphological method.

In the field of medical microbiology, much attention has been paid recently to the mass spectrometric technique of matrix-assisted laser desorption ionization time-of-flight mass spectrometry (MALDI–TOF MS) as a tool for the rapid identification of fungal isolates [11]. MALDI–TOF MS has major clinical advantages, since it requires much smaller samples and the total process from sample preparation to data analysis is very rapid. This method is a type of mass spectral fingerprinting, for which mass spectral databases are commercially available from several mass spectrometer companies. The rapid identification and discrimination of clinical *A. fumigatus* isolates has been reported using this method [12, 13]. Several research groups have further attempted to discriminate *Aspergillus* isolates at the species and strain level [14–16]. However, Welker [17] has pointed out the following problems with this method in his review.

(i) the general finding that the proteome is very dynamic in living cells and hence protein pattern expectedly could be subject to changes in response to growth conditions,

(ii) doubts whether differences and similarities in mass spectral patterns are completely consistent with the established taxonomy,

(iii) a lack of comprehensive databases covering all clinically relevant species.

Furthermore, in the author's opinion, the reported mass spectra of fungal samples sometimes show too few peaks when sample preparation is performed using the recommended protocol proposed by the manufacturers.

To overcome these problems relating to mass spectral fingerprinting, we have proposed a method using ribosomal proteins as biomarkers for microorganism analysis by MALDI–TOF MS [18–25]. Ribosomal proteins are typical housekeeping proteins and are abundantly present in microorganisms' cells. Prokaryotic (bacterial) ribosomes consist of 57 kinds of ribosomal subunit proteins (RSPs), whereas eukaryotic ribosomes typically consist of 78 RSPs. The combination of subunit proteins and their structures are not influenced by culture conditions. Because most RSPs are basic proteins with higher proton affinity (i.e., easily producing $[M + H]^+$ ions) and their masses are distributed in the range of ca. 4 - 30 kDa, RSPs can be easily observed in MALDI mass spectra [26]. We have reported that the identification of bacterial species and classification at the strain level can be accomplished based on the expressed mass types of RSPs [18–25]. The masses of RSPs used as biomarkers can be estimated from translational amino acid sequences of genome-sequenced strains, which can be obtained from public databases such as UniProt Knowledgebase (UniProtKB) [27]. Our proposed method is a form of molecular typing like MLST, based on bioinformatics. The biomarker RSPs are a complex of typical housekeeping proteins. Since the sequence variation of RSPs observed as the peak shift on the MALDI mass spectra results from molecular evolution, the results of identification and discrimination of microorganisms are assumed to phylogenetic ally. This is the crucial difference between our proposed RSP typing as "phylogenetic" method and the conventional mass spectral fingerprinting as "chemotaxonomic" method.

The aim of our project is to extend the RSP-based method to the identification of eukaryotic fungi. As the first step, we have investigated the actual state of information of RSPs of fungi registered in public protein databases through the characterization of ribosomal protein fractions extracted from genome-sequenced *A. fumigatus* strains as a model [28]. In our previous paper [28], we revealed that more than half of the amino acid sequences of RSPs registered in the public databases were incorrect, due chiefly to mis-annotation of exon/intron structures. We were able to successfully correct the sequence errors using a combination of *in silico* inspection by sequence homology analysis and MALDI–TOF MS measurements. Post-translational modifications such as acetylation and methylation could also be verified. In this way, the expressed masses of RSPs observed under 16,000 Da could finally be confirmed.

As the next step, this paper describes the results of comparable characterization of RSPs of eleven *Aspergillus* species to establish biomarker references for the reliable identification of *Aspergillus* species. First, verification and correction of the amino acid sequences of RSPs and confirmation of post-translational modifications common to all sample strains were performed to accurately determine the expressed mass, as described in our previous paper [28]. RSPs with appropriate intensity commonly observed in each strain were then selected as reliable biomarkers for the identification of *Aspergillus* species. The selected RSPs of each strain were categorized into "RSP types" based on their mass and used to construct a dendrogram. The resulting dendrogram was compared with that arrived at using the DNA-based method; the reliability of the species identification and discrimination of this method was then assessed.

Results and discussion
Characterization of RSPs of genome-sequenced *Aspergillus* strains

In our previous paper [28], the amino acid sequences of RSPs in *A. fumigatus* strains were verified to compile the reference mass list of expressed RSPs. We noted that more than half of the amino acid sequences in the public databases, such as UniProtKB [27] and the NCBI protein database [29] were incorrect. These errors could be corrected by a combination of *in silico* inspection using sequence homology analysis and verification of actual expressed masses of RSPs by MALDI–TOF MS measurements. In this study, by applying this strategy, the amino acid sequences of RSPs of ten genome-sequenced strains of *Aspergillus* species were further verified and compared to build a reliable biomarker list for the identification of *Aspergillus* species. The sample strains used in this study are summarized in Table 1. For RSPs with amino acid sequences not registered in the public databases, corresponding gene sequences were manually identified in the genome sequence by referring to gene sequences of *A. fumigatus* strains. The genome sequence of *A. viridinutans* is not yet published, but the authors have annotated the RSP sequences manually. Finally, 26 kinds of RSPs whose molecular weights were under 16,000 Da were selected that are common to those already verified for *A. fumigatus* strains [28]. Supporting Information Additional file 3:

Table 1 List of sample strains used in this study

species	strain names [a]
A. fumigatus	* IFM 53842 (= A1163), *IFM 54229 (= Af293), IFM 57323[NT]
N. fischeri	* IFM 57324[T] (= NRRL 181[T])
A. lentulus	* IFM 54703[T], IFM 47547, IFM 58399, IFM 60648, IFM 61392, IFM 62073, IFM 62096
A. viridinutans	* IFM 47045[T]
A. felis (former *A. viridinutans*)	IFM 59564, IFM 60053, IFM 62093
A. pseudoviridinutans (former *A. viridinutans*)	IFM 55266, IFM 62075
A. wyomingensis (former *A. viridinutans*)	IFM 62083
A. udagawae	*IFM 46973[T], IFM 46972, IFM 5058, IFM 51744, IFM 53868, IFM 61606, IFM 62070, IFM 62100
A. clavatus	* IFM 60676[NT] (= NRRL 1[NT])
A. niger	* CBS 513.88
A. kawachii	* IFO 4308
A. flavus	* IFM 60677 (= NRRL 3357)
A. oryzae	* IFM 59475 (=RIB 40)
A. nidulans	* IFM 60678 (= FGSC A4)

a) * genome sequenced strain.

Figures SI-1 to SI-10 show the mass spectra of each genome-sequenced strain (except for *A. fumigatus*, for which the mass spectra were reported in ref. [28]). Additional file 1: Table SI-1 summarizes accession numbers in public protein databases of ribosomal protein biomarkers of genome sequenced strains used in this study. Here, the names of RSPs are adopted from the yeast nomenclature system [30] to prevent confusion in the RSP's nomenclature. Supporting Information Additional file 2: Table SI-2 summarizes the data of the RSPs of ten genome-sequenced strains such as the accession number, post-translational modifications, corrected amino acid sequences, and corrected exon/intron structures (the corrected sequences of *A. fumigatus* A1163 and Af293 have been reported in ref [28]).

Table 2 shows the number and ratio of incorrect or not registered sequences among 26 RSPs for nine newly analyzed genome-sequenced strains (the result of *A. viridinutans*, annotated by the authors, was not added to this list). Depending on the species, about 10 - 60% RSPs needed to be corrected or newly added. The source of sequence errors in prokaryotic bacteria was chiefly due to misidentification of the start codon [20]. The main reason in *Aspergillus* fungi seemed to be due to misidentification of the exon/intron structure, resulting in incorrect CDS as well as an incorrect stop codon caused by frame shift. Because this type of error is unique to prokaryotes, a similar

Table 2 The number of corrected RSPs and their ratio to analyzed RSPs

Species	Strains	The number of the corrected RSPs	The ratio of the corrected RSPs to the analyzed RSPs (%)
N. fischeri	IFM 57324^{T} (= NRRL 181^{T})	11	42
A. lentulus	IFM 54703^{T}	14	54
A. udagawae	IFM 46973^{T}	15	58
A. clavatus	IFM 60676^{NT} (= NRRL 1^{NT})	9	35
A. niger	CBS 513.88	4	15
A. kawachii	IFO 4308	11	42
A. flavus	IFM 60677 (= NRRL 3357)	7	27
A. oryzae	IFM 59475 (=RIB 40)	3	12
A. nidulans	IFM 60678 (= FGSC A4)	4	15

problem in the annotations of RSP genes might have occurred in other fungi.

The post-translational modifications were then confirmed by referring to already-reported modifications in eukaryotic RSPs. Details of the assignments of each modification are described in the Supporting Information (Additional file 4: Figures SI-11 to SI-14). Acetylation (S16, S21, S24, S28, L31, and L35), methylation (L42), and two hydroxylations (S23) have been reported in several papers (see citations in ref. [28]) and have also observed in A. fumigatus [28]. A mass shift of these RSPs from the calculated sequence mass after taking into account N-terminal methionine loss was commonly observed in all sample strains (+42 Da for acetylation, +14 Da for methylation, and +32 Da for two hydroxylations). This result suggests these modifications to be evolutionarily-conserved modifications, at least in Aspergillus species. In addition, S27 showed a common +28 Da shift, suggesting two methylations. Although this modification has, to our knowledge, not been reported before, two methylations of S27 were concluded to be common modifications in Aspergillus species.

Species identification using the RSP types

The amino acid sequences and the theoretical mass of RSPs thus determined mostly varied among species. This finding strongly suggests that species identification can be performed using RSPs as biomarkers. To make a reference table for the RSP typing, each RSP was classified into different types based on mass. For example, S29 has five types of different expressed mass, of which the peaks are distributed approximately from m/z 6570 to m/z 6650, as shown in Fig. 1 (the whole-range mass spectra are shown in Supporting Information Additional file 3: Figures SI-1

to SI-10). For S29, Type I was first allocated to A. fumigatus A1163 observed at m/z 6647, which was common with that of N. fischeri NRRL 181^{T}, A. lentulus IFM 54703^{T}, A. viridinutans IFM 47045^{T}, and A. udagawae IFM 46973^{T}. Interestingly, these species belong to Aspergillus section Fumigati. The mass of S29 of A. clavatus NRRL 1^{NT} was different from Type I, so it was allocated to Type II. In the same manner, S29 of A. niger CBS 513.88 and A. kawachii IFO 4308 were allocated to Type III, that of A. flavus NRRL 3357 and A. oryzae RIB 40 to Type IV, and that of A. nidulans FGSC A4 to Type V. The type classification was conducted in the same way in other RSPs. Table 3 summarizes the mass and types of RSPs of each genome-sequenced strain, in which the post-translational modifications were taken into consideration as affecting the mass in this list.

The distribution of the RSP types shown in Table 3 was then processed using the unweighted pair group method with arithmetic mean (UPGMA) cluster analysis using a categorical coefficient. Fig. 2 compares the dendrogram based on the RSP types (Fig. 2a) with that based on the β-tubulin gene sequence (Fig. 2b). Among the Aspergillus species used in this study, A. fumigatus, N. fischeri, A. lentulus, A. viridinutans, and A. udagawae are known to be genetically related species, belonging to section Fumigati [31] (note that Neosartorya is a teleomorph of Aspergillus). Interestingly, these five species form a cluster in the dendrograms based on both the RSP type (Fig. 2a) and the β-tubulin gene sequence (Fig. 2b). In these sample strains, eight RSPs (L30, L34, L39, L43, S26, S27, S28, and S29) of the analyzed 26 RSPs (31%) matched completely.

Aspergillus clavatus has been suggested to be closest to section Fumigati as shown in the dendrogram based on β-tubulin (Fig. 2b). RSP typing (Fig. 2a) also shows the closest position of A. clavatus next to section Fumigati. Five RSPs (S26, S27, S28, L34, and L39) of common eight RSPs in section Fumigati are also shared with A. clavatus.

The positions of A. niger, A. kawachii, A. flavus, A. oryzae, and A. nidulans are distinct from the cluster of section Fumigati. The RSP types of these species show little in common with section Fumigati species: only L39 of A. oryzae and A. flavus was common. Of these species, pairs of 'A. niger and A. kawachii' and 'A. flavus and A. oryzae' are known to be very closely interrelated: A. flavus and A. oryzae are assumed to be ecotypes of the same species because they have only 350 unique genes, even when comparing the total genome sequences [32]. In the types of 26 RSPs used in this study, 21 out of 26 RSPs agreed between A. niger and A. kawachii and also 100% of RSPs agreed between A. flavus and A. oryzae. The ratio of conformity between species thus appears to be reflected in the relatedness between species.

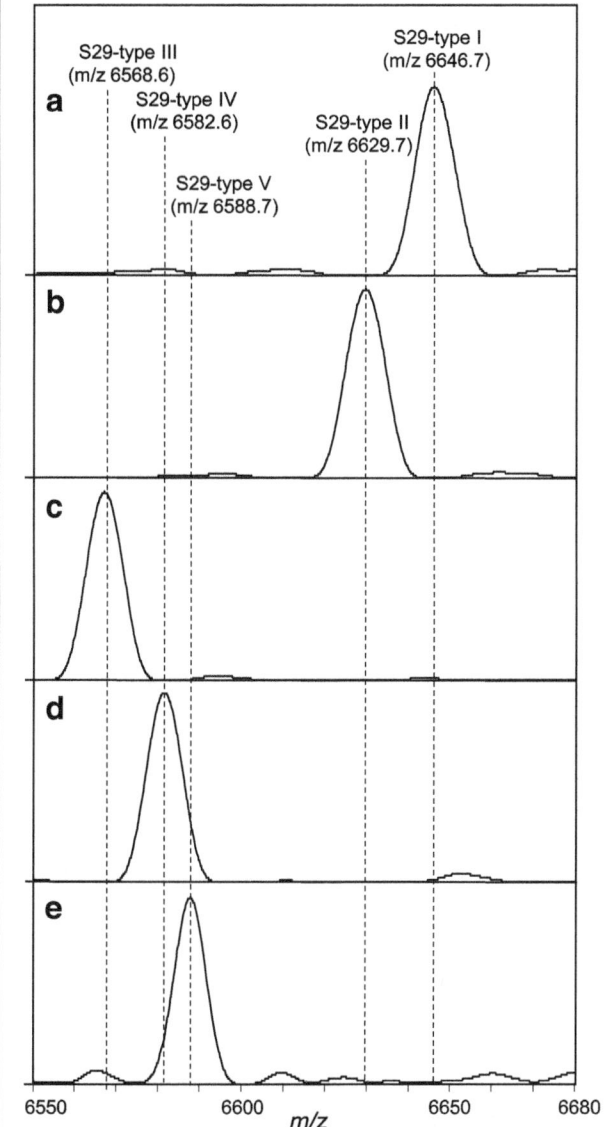

Fig. 1 Expanded view of S29 (*m/z* 6550-6680) regions. **a** *A. fumigatus* A1163, **b** *A. clavatus* NRRL 1[NT], (C) *A. niger* CBS 513.88, (D) *A. flavus* NRRL 3357, and (E) *A. nidulans* FGSC A4

As discussed above, the dendrogram based on the RSP types shows relatively good concordance with the tree based on β-tubulin gene sequences (Fig. 2b) and with the tree arrived at by genomic analysis of *Aspergillus* fungi [1]. This result suggests that RSP typing can perform accurate species identification that reflects molecular evolution. Therefore, the dendrogram constructed by the RSP typing can be considered as a kind of phylogenetic tree.

The effectiveness of RSP typing for discriminating section *Fumigati* strains

Because sensitivity to antifungal agents differs between *A. fumigatus* and other species within section *Fumigati*, accurate discrimination of these strains is very important

[31]. However, traditional morphological analysis does not always accurately locate species within section *Fumigati*, so misidentifications often occur. To overcome this problem, Yaguchi *et al.* have characterized the species within section *Fumigati* by molecular phylogenetic analysis using multiple genes [33]. On the other hand, as described in the previous section, RSP typing shows the potential of species discrimination within section *Fumigati*. To reveal the effectiveness of this method, the strains of *A. fumigatus*, *A. lentulus*, *A. viridinutans*, *A. felis*, *A. pseudoviridinutans*, *A. wyomingensis*, *A. udagawae*, and *N. fischeri* belonging to section *Fumigati* were characterized.

Table 4 shows the RSP types of the sample strains assigned using the RSP reference list (Table 3). In this section, 18 RSPs with clearly separated peaks observed for all strains were adopted from 26 RSPs. Some RSP peaks were not matched to the reference due to peak shift or not detected. These RSPs are designated as N (Not matched) in Table 4.

The distribution of the RSP types within species tended to be consistent, and the variation was assumed to be small. This allowed the typing of sample strains to be conducted using the mass list of the genome-sequenced strain. Fig. 3 shows the result of UPGMA cluster analysis based on the RSP typing profile. In the dendrogram based on RSP typing, every species formed one general cluster.

The RSP types of *A. lentulus* strains completely match those of the type strain. Because *A. lentulus* was originally regarded as a sibling species of *A. fumigatus*, it has proved difficult to discriminate them morphologically [34]. Several mass spectrometric strategies based on mass spectral fingerprinting have been able to discriminate these species [35, 36]; however, these reports do not record the criteria used for discrimination. Our proposed RSP typing, on the other hand, can discriminate these species based on variations in RSPs. The types of 9 RSPs (L40, S30, L38, L42, L32, L26, L27, S23 and S16) differed from that of *A. fumigatus*, which can be used for discrimination between *A. fumigatus* and *A. lentulus*. Interestingly, *N. fischeri* is very close to the *A. lentulus* cluster, in which only L42 was different from that of *A. lentulus*. This species is close to the *A. fumigatus* cluster in the case of the β-tubulin gene as shown in Fig. 2b. Such differences are likely to be caused by the difference in genes used: the tree compiled using RSP typing (Fig. 3a) is constructed based on a combination of the types of 18 RSPs that reflect 18 different house-keeping genes.

RSP typing has demonstrated that *A. udagawae* strains form a relatively clear-cut cluster. Although some RSP peaks were not detected clearly, moderate concordance with the reference mass of the type strain was confirmed. Five RSPs (L39, S29, S28, L30, and L34) are further

Table 3 The expressed masses and the types of ribosomal protein biomarkers of genome-sequenced strains used in this study

RP name	A. fumigatus IFM 53842 = A1163	A. fumigatus IFM 54229 = Af293	N. fischeri IFM 57324[T] = NRRL 181[T]	A. lentulus IFM 54703[T]	A. viridinutans IFM 47045[T]	A. udagawae IFM 46973[T]
L40	I (6002.3)	I	II (6016.3)	II	II	II
L39	I (6151.2)	I	I	I	I	I
S29	I (6646.7)	I	I	I	I	I
S30	I (6789.1)	I	II (6807.1)	II	III (6793.1)	II
L29	I (7456.6)	I	I	I	I	II (7470.6)
S28	I (7710.0)	I	I	I	I	I
S27	I (8765.3)	I	I	I	I	I
S31	I (9134.9)	I	I	I	I	II (9120.9)
L38	I (9153.8)	I	II (9169.8)	II	II	II
S21	I (10038.1)	II (10052.2)	I	III (10035.1)	IV (10008.1)	V (10024.1)
L43	I (10025.8)	I	I	I	I	I
L37	I (10386.9)	I	I	II (10387.9)	III (10358.9)	II
L30	I (11171.1)	I	I	I	I	I
L36	I (11869.8)	I	II (11861.8)	III (11847.8)	II	IV (11875.8)
L42	I (12028.3)	I	II (12009.2)	III (11979.2)	IV (11995.2)	V (11965.2)
L33	I (12215.1)	I	I	I	II (12233.1)	III (12206.1)
L34	I (13164.5)	I	I	I	I	I
S26	I (13338.7)	I	I	I	I	I
L31	I (13919.1)	I	II (13937.1)	II	II	III (13937.1)
L35	I (14532.0)	I	II (14518.0)	II	a)	III (14548.0)
L32	I (14836.6)	I	II (14822.6)	II	II	III (14880.6)
L26	I (14979.4)	I	II (14982.5)	II	II	III (14954.4)
S24	I (15226.6)	I	I	I	II (15212.6)	I
L27	I (15682.6)	I	II (15654.5)	II	II	III (15667.5)
S23	I (15802.5)	I	II (15786.5)	II	II	II
S16	I(15883.4)	I	II (15853.4)	II	II	II

RP name	A. clavatus IFM 60676[NT] = NRRL 1[NT]	A. niger CBS 513.88	A. kawachii IFO 4308	A. flavus NRRL 3357	A. oryzae RIB 40	A. nidulans FGSC A4
L40	II	II	II	III (6043.3)	III	IV (6057.3)
L39	I	II(6123.2)	II	I	I	III (6123.2)
S29	II (6629.7)	III	III (6568.6)	IV (6582.6)	IV	V (6588.7)
S30	IV (6790.0)	V (6745.0)	VI (6729.0)	VII (6691.9)	VII	VIII (6761.1)
L29	III (7433.6)	IV (7368.4)	IV	V (7414.5)	V	VI (7387.5)
S28	I	II (7681.9)	II	III (7708.0)	III	IV (7754.0)
S27	I	II (8735.3)	II	III (8792.4)	III	IV (8807.3)
S31	II	III (9101.8)	III	IV (9090.8)	IV	V (9111.9)
L38	III (9160.8)	IV (9073.7)	IV	V (9173.9)	V	VI (9128.8)
S21	VI (9922.9)	VII (10027.1)	VIII (10041.2)	IX (10050.1)	IX	X (10064.1)
L43	II (10011.8)	IV (9997.7)	IV	IV	IV	V (10011.8)
L37	IV (10279.7)	V (10451.9)	VI (10435.9)	VII (10436.9)	VII	VIII (10371.8)
L30	II (11169.1)	III (11210.1)	III	IV (11223.1)	IV	V (11259.2)
L36	V (11850.8)	VI (11831.8)	VI	III	III	VII (11757.7)

Table 3 The expressed masses and the types of ribosomal protein biomarkers of genome-sequenced strains used in this study
(Continued)

L42	VI (11921.1)	VII (12025.2)	VII	VIII (12029.2)	VIII	IX (11994.2)
L33	II	IV (12250.1)	IV	V (12299.2)	V	VI (12216.1)
L34	I	II (13147.4)	II	III (13178.5)	III	IV (13133.4)
S26	I	II (13381.7)	II	II	II	IV (13349.7)
L31	IV (13809.9)	V (13878.0)	V	VI (13981.2)	VI	VII (13847.9)
L35	IV (14517.0)	V (14384.8)	VI (14411.9)	VII (14536.9)	VII	VIII (14458.0)
L32	IV (14852.6)	V (14891.6)	V	VI (14794.5)	VI	VII (14838.5)
L26	IV (14915.3)	V (14877.2)	VI (14891.2)	VII (14999.3)	VII	VIII (15084.6)
S24	III (15130.5)	IV (15163.5)	IV	V (15157.5)	V	VI (15232.7)
L27	IV (15636.5)	V (15610.4)	V	VI (15613.4)	VI	VII (15642.5)
S23	II	III (15739.4)	III	IV (15752.4)	IV	V (15757.5)
S16	III (15849.4)	IV (15881.4)	IV	V (15835.3)	V	VI (15852.3)

[a] The amino acid sequence of L35 in *A. viridinutans* IFM 47045[T] was not obtained from the draft genome sequences

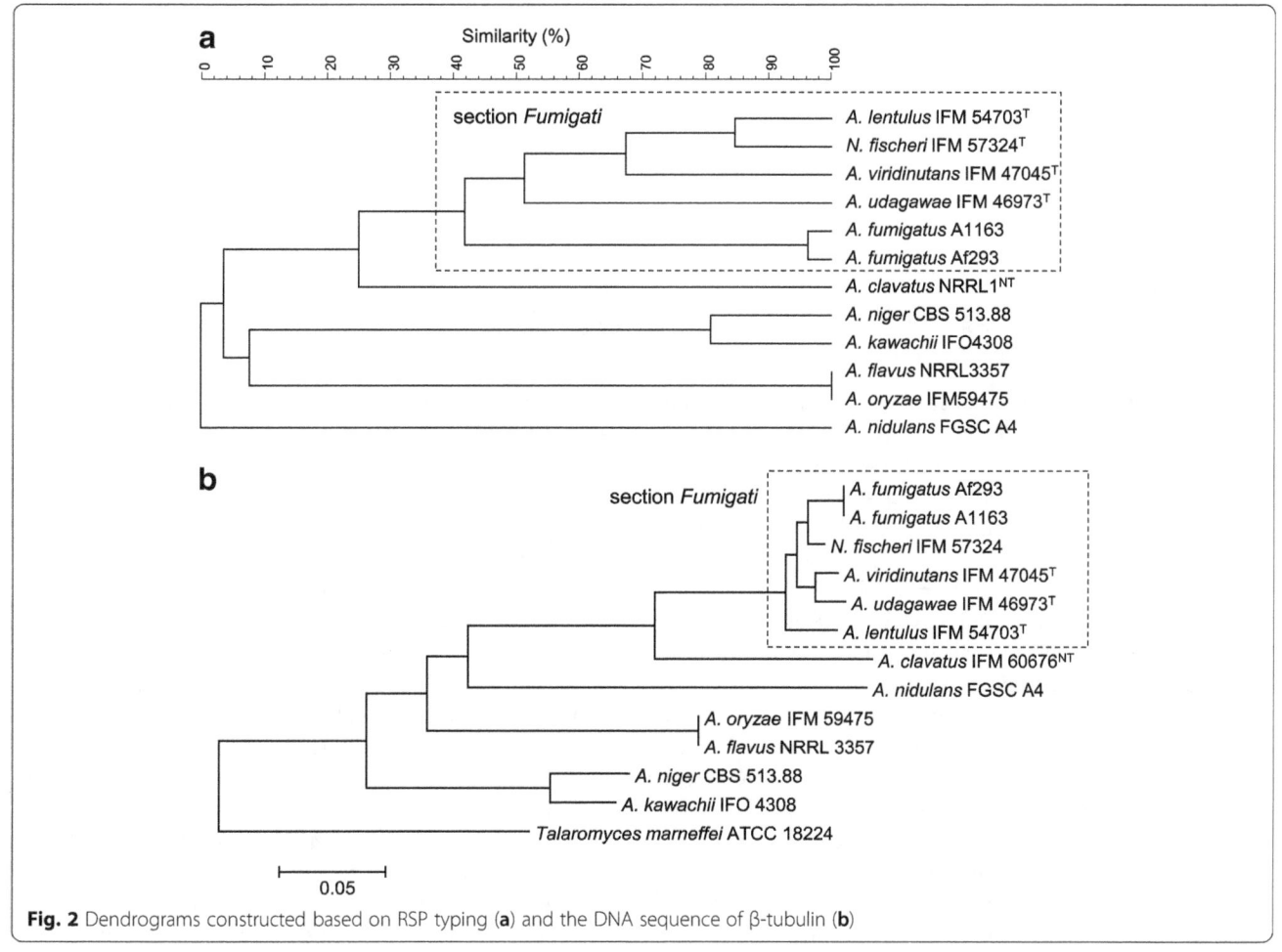

Fig. 2 Dendrograms constructed based on RSP typing (**a**) and the DNA sequence of β-tubulin (**b**)

Table 4 The RSP typing profile of the sample strains belonging to *Aspergillus* section *Fumigati*

Sample strains [a]	The types of biomarker RSPs																	
	L40	L39	S29	S30	S28	S31	L38	L30	L42	L33	L34	S26	L32	L26	S24	L27	S23	S16
IFM 53842 (Afu)	I	I	I	I	I	I	I	I	I	I	I	I	I	I	I	I	I	I
IFM 54229 (Afu)	I	I	I	I	I	I	I	I	I	I	I	I	I	I	I	I	I	I
IFM 57323[NT] (Afu)	I	I	I	I	I	I	I	I	I	I	I	I	I	I	I	I	I	I
IFM 54703[T] (Al)	II	I	I	II	I	I	II	I	III	I	I	I	II	II	I	II	II	II
IFM 47457 (Al)	II	I	I	II	I	I	II	I	III	I	I	I	II	II	I	II	II	II
IFM 58399 (Al)	II	I	I	II	I	I	II	I	III	I	I	I	II	II	I	II	II	II
IFM 60648 (Al)	II	I	I	II	I	I	II	I	III	I	I	I	II	II	I	II	II	II
IFM 61392 (Al)	II	I	I	II	I	I	II	I	III	I	I	I	II	II	I	II	II	II
IFM 62073 (Al)	II	I	I	II	I	I	II	I	III	I	I	I	II	II	I	II	II	II
IFM 62096 (Al)	II	I	I	II	I	I	II	I	III	I	I	I	II	II	I	II	II	II
IFM 47045[T] (Av)	II	I	I	III	I	I	II	I	IV	II	I	I	II	II	II	II	II	II
IFM 55266 (Ap)[b]	II	I	I	II	I	II	N	I	II	III	I	I	II	II	N	III	II	II
IFM 62075 (Ap)	II	I	I	II	I	II	II	I	II	III	I	I	II	II	I	III	II	II
IFM 59564 (Afe) [b]	II	I	I	II	I	II	I	N	III	II	I	I	II	N	N	III	II	II
IFM 62093 (Afe)	II	I	I	II	I	II	I	N	III	II	I	I	II	N	N	III	II	II
IFM 60053 (Afe)	II	I	I	II	I	II	I	N	III	II	I	I	II	N	N	III	II	N
IFM 62083 (Aw) [b]	II	I	N	II	I	II	N	I	N	II	I	I	II	N	I	I	II	II
IFM 46972[T] (Au)	II	I	I	II	I	II	II	I	V	III	I	N	III	III	I	III	II	II
IFM 46973 (Au)	II	I	I	II	I	II	II	I	V	III	I	I	III	III	I	III	II	II
IFM 5058 (Au)	II	I	I	II	I	II	II	I	V	III	I	I	III	III	I	III	II	II
IFM 51744 (Au)	II	I	I	II	I	II	II	I	V	III	I	N	III	III	N	III	II	N
IFM 53868 (Au)	II	I	I	II	I	N	II	I	V	III	I	I	III	III	I	III	II	II
IFM 61606 (Au)	II	I	I	II	I	II	II	I	V	III	I	N	III	III	I	III	II	II
IFM 62070 (Au)	II	I	I	II	I	N	II	I	V	III	I	I	III	III	I	III	II	II
IFM 62100 (Au)	II	I	I	II	I	II	II	I	V	III	I	I	III	III	I	III	II	II
IFM 57324[T] (Nf)	II	I	I	II	I	I	II	I	II	I	I	I	II	II	I	II	II	II

[a] Abbreviations: Afu; *A. fumigatus*, Afe; *A. felis*, Al; *A. lentulus*, Ap; *A. pseudoviridinutans*, Av; *A. viridinutans*, Au; *A. udagawae*, Aw; *A. wyomingensis* and Nf; *N. fischeri*

[b] *A. felis*, *A. pseudoviridinutans* and *A. wyomingensis* are former *A. viridinutans*

matched completely to those of *A. fumigatus*, whereas three RSPs (L42, L32, and L26) are totally different from the other species. These RSPs can therefore be used to discriminate *A. udagawae* from other species. The three strains (IFM 5058, IFM 51744, and IFM 53868) were formerly reported as variant isolates of *A. fumigatus* (or *Aspergillus* sp.) but were re-identified, using multiple genes (β-tubulin, hydrophobin, and calmodulin), as *A. udagawae* [33]. In RSP typing, these strains are included in the *A. udagawae* cluster, supporting this re-identification.

In contrast to the high conformity of RSPs in *A. lentulus* and *A. udagawae*, former *A. viridinutans* strains (*A. felis*, *A. pseudoviridinutans*, and *A. wyomingensis*) do not form an obvious cluster. Interestingly, the type strain of *A. viridinutans* (IFM 47045[T]) is located separately from other former *A. viridinutans* strains. The diversity of *A. viridinutans* has already been reported, and this species was divided into some species [37–39]. The strains IFM 55266

and 62075, IFM 59564, 60053 and 62093, and IFM 62083 were re-identified as *A. pseudoviridinutans*, *A. felis* and *A. wyomingensis*, respectively. In each species, the high conformity of RSPs was indicated.

Conclusions

In the context of the ongoing conversion, on the basis of fungal taxonomy, from morphology to molecular phylogeny, molecular biological methods have been adopted for the identification and discrimination of fungal strains. To avoid misidentification of closely-related species, especially within section *Fumigati*, the reliability of identification results increases on increasing the numbers of evaluation points (*i.e.*, the numbers of genes or proteins). This gives MALDI-TOF MS great potential, since many proteomic peaks that assist with the identification of fungal species are processed. Our proposed RSP typing represents the next generation of mass spectral identification/discrimination of

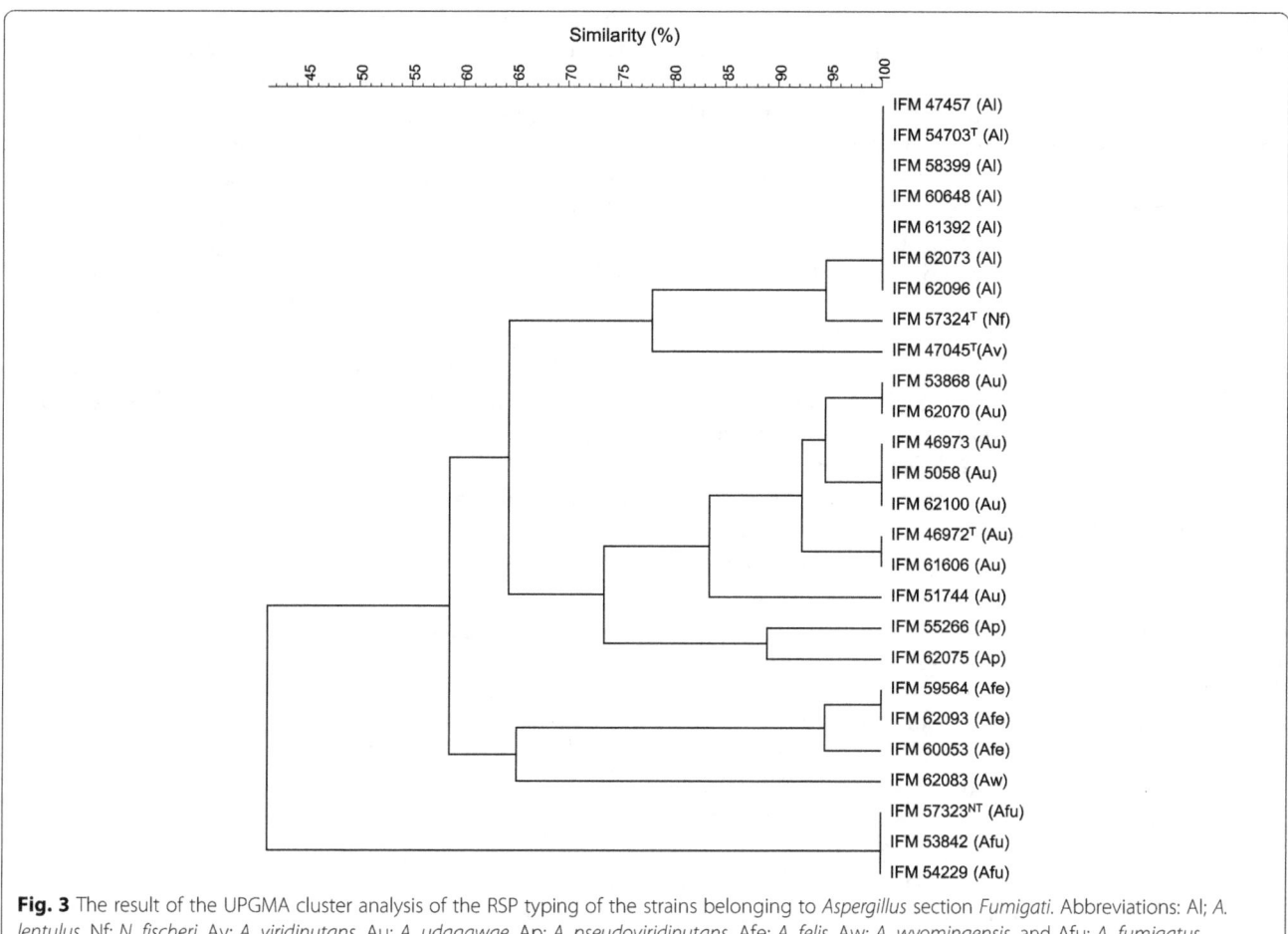

Fig. 3 The result of the UPGMA cluster analysis of the RSP typing of the strains belonging to *Aspergillus* section *Fumigati*. Abbreviations: Al; *A. lentulus*, Nf; *N. fischeri*, Av; *A. viridinutans*, Au; *A. udagawae*, Ap; *A. pseudoviridinutans*, Afe; *A. felis*, Aw; *A. wyomingensis*, and Afu; *A. fumigatus*

fungal strains, in that it supersedes current mass spectral fingerprinting, which is simple pattern-matching without peak assignment.

The merits of RSP typing are (1) it requires no commercial database and (2) it can be used to perform phylogenetic analysis. As for the first point, RSP typing requires reference to the RSP biomarker lists, via the internet, constructed from the public protein databases. At this time, of course, commercial mass spectral databases are more substantial than available RSP information. However, as whole genome-sequencing of fungal species progresses, information on RSPs is expected to expand exponentially in the near future. Although we initially encountered a confused situation as concerns the protein information registered on the public protein databases, we have successfully corrected the errors in the amino acid sequences and the names of representative *Aspergillus* RSPs. The sequences and expressed mass of RSPs of other fungal species can now be easily verified and corrected by homology analysis using the sequence list summarized in Supporting Information Additonal file 2: Table SI-2.

The second benefit of this method is valuable, because the identification results have a phylogenetic rationale:

they relate to a combination of more than a dozen house-keeping proteins. This method can eliminate the influence of growth and experimental conditions, if only the RSP peaks are observed. RSPs are one of the most expressed proteins, and RSP fractions are easily collected by cell-grinding and ultracentrifugation. The discriminatory power of the RSP typing appears to be comparable with the conventional β-tubulin gene analysis. This method would therefore be suitable for species identification and discrimination at strain to species level. Because RSP typing can characterize the strains within section *Fumigati*, this method is potentially a powerful and reliable tool in the field of clinical microbiology.

Methods
Cell culture and preparation of ribosomal protein samples
The strains of the 14 species used in this study are summarized in Table 1. All sample strains were provided by Chiba University's Medical Mycology Research Center (Chiba, Japan) and were grown in potato dextrose broth (PDB) medium at 25 °C for three days. After incubation, harvesting and the preparation of the ribosomal fractions were similar to the methods described in our previous

paper [28]. The cultured mycelia were harvested by centrifugation and ground between zirconia silica beads. After removing the beads and cell debris by centrifugation, the fungus lysates were subjected to ultracentrifugation. The resulting ribosome fraction was solubilized in 20 - 50 μL of 50% acetonitrile containing 1% trifluoroacetic acid (TFA), and then subjected to MALDI–TOF MS measurement. Detailed sample preparation procedures are shown in Supporting Information Additional file 5: Figure SI-15.

MALDI–TOF MS measurements

Sample preparation, apparatus, and MALDI–TOF MS data acquisition methods were similar to those described in our previous papers [18–25]. The ribosomal protein sample solution (ca. 1 μL) was spotted onto the MALDI target. About 1 μL of a sinapinic acid matrix solution at a concentration of 20 mg/mL in 50% acetonitrile with 1% TFA was then overlaid and dried in air. The MALDI–TOF MS measurements were performed using an AXIMA CFR-plus time-of-flight mass spectrometer (Shimadzu/Kratos, Kyoto, Japan) in positive linear mode. At least nine mass spectra for each sample were collected by each of three repeated measurements for each of three sample spots (total 3 spots × 3 measurements). External mass calibration was carried out using three peaks of ACTH (human, 1-24) ([M + H]$^+$, m/z 2932.6) and myoglobin ([M + H]$^+$, m/z 16952.6 and [M + 2H]$^{2+}$, m/z 8476.8) as references.

Calculation of the theoretical mass of RSPs

The amino acid sequence of each RSP was obtained from UniProtKB [27]. Because the genome sequence of *A. viridinutans* IFM 54703^T, sequenced by Chiba University's Medical Mycology Research Center, was not registered on the public databases at this time, in-house draft genome sequence data were used. The annotated RSP sequences of *A. viridinutans* have been deposited at DDBJ under the accession numbers LC213039-LC213063. The amino acid sequences of L39 and S21 of *A. niger* were not determined from shotgun sequences in the database. Therefore, these sequences were determined and have been deposited at DDBJ under the accession numbers LC255002 and LC215003. The sequence mass of each RSP was predicted using a Compute pI/Mw tool on the ExPASy proteomics server [40]. After taking into account the post-translational modifications as effected in our previous paper [28] (such as N-terminal methionine loss, acetylation, methylation and hydroxylation), the theoretical mass of each expressed RSP was calculated as [M + H]$^+$ ion. Detailed construction procedure of the ribosomal protein biomarker list together with peak assignments are shown in Supporting Information Additional file 5: Figure SI-16.

Phylogenetic analysis of RSPs

The observed masses of each RSP were compared with the reference masses constructed in this study. The matching of the average observed masses to the reference masses was judged from errors within 150 ppm. The results of mass matching were designated as RSP types. The RSP typing profiles for each sample strain were processed using UPGMA to build a dendrogram cluster for analysis employing a categorical coefficient, using BioNumerics software (version 3.5; Applied Maths, Kortrijk, Belgium).

The partial DNA sequence of β-tubulin was obtained from the UniProtKB and the alignment of the sequences was performed using ClustalW [41] software. The dendrogram was constructed using MEGA6 [42] software.

Additional files

Additional file 1: Table SI-1. The accession number (TrEMBL) of ribosomal protein biomarkers of genome-sequenced strains used in this study.

Additional file 2: Table SI-2. Corrected amino acid sequences and relating information of RSPs.

Additional file 3: Mass spectra of genome sequenced sample strains used in this study. **Figure SI-1.** Mass spectra of RSPs of *N. fischeri* NRRL 181^T. **Figure SI-2.** Mass spectra of RSPs of *A. lentulus* IFM 54703^T. **Figure SI-3.** Mass spectra of RSPs of *A. viridinutans* IFM 47045^T. **Figure SI-4.** Mass spectra of RSPs of *A. udagawae* IFM 46973^T. **Figure SI-5.** Mass spectra of RSPs of *A. clavatus* NRRL 1^{NT}. **Figure SI-6.** Mass spectra of RSPs of *A. niger* CBS 513.88. **Figure SI-7.** Mass spectra of RSPs of *A. kawachii* IFO 4308. **Figure SI-8.** Mass spectra of RSPs of *A. flavus* NRRL 3357. **Figure SI-9.** Mass spectra of RSPs of *A. oryzae* RIB 40. **Figure SI-10.** Mass spectra of RSPs of *A. nidulans* FGSC A4.

Additional file 4: Post-translational modifications. **Figure SI-11.** Peak shift (+42 Da) of S24 with acetylation. (a) *A. fumigatus* A1163, (b) *A. viridinutans* IFM 47045^T, (c) *A. clavatus* NRRL 1^{NT}, (d) *A. niger* CBS 513.88, (e) *A. flavus* NRRL 3357, and (f) *A. nidulans* FGSC A4. **Figure SI-12.** Methylation of L42. (1) Peak shift of +14 Da from sequence mass in L42 of the *Aspergillus* species; (a) *A. fumigatus* A1163, (b) *N. fischeri* NRRL 181^T, (c) *A. lentulus* IFM 54703^T, (d) *A. viridinutans* IFM 47045^T, (e) *A. udagawae* IFM 46973^T, (f) *A. clavatus* NRRL 1^{NT}, (g) *A. niger* CBS 513.88, (h) *A. flavus* NRRL 3357, and (i) *A. nidulans* FGSC A4. (2) Amino acid sequences around Lys-55. **Figure SI-13.** Dihydroxylation of S23. (1) Peak shift of +32 Da from sequence mass in L23 of *Aspergillus* species; (a) *A. fumigatus* A1163, (b) *N. fischeri* NRRL 181^T, (c) *A. niger* CBS 513.88, (d) *A. flavus* NRRL 3357, and (e) *A. nidulans* FGSC A4. (2) Amino acid sequences around Pro-64. **Figure SI-14.** Peak shift (+28 Da) of S27 with dimethylation. (a) *A. fumigatus* A1163, (b) *A. niger* CBS 513.88, (c) *A. flavus* NRRL 3357, and (d) *A. nidulans* FGSC A4.

Additional file 5: Figure SI-15. Detailed experimental protocols. Detailed sample preparation procedures. **Figure SI-16.** Detailed construction procedure of the ribosomal protein biomarker list together with peak assignments.

Abbreviations

CDS: Coding DNA sequences; MALDI-TOF MS: Matrix-assisted laser desorption ionization time-of-flight mass spectrometry; MLST: Multi-locus sequence typing; RSP: Ribosomal subunit protein; UPGMA: Unweighted pair group method with arithmetic mean

Acknowledgements

This work was supported in part by a research grant from the Institute for Fermentation, Osaka (IFO), JSPS Kakenhi Grant Number 25430198, and the National Bioresource Project (Pathogenic Microbes) in Japan (http://www.nbrp.jp/).

Authors' contribution

SN, HS, RT, and TY conceived the study. RT, and TY cultured fungal strains and prepared the RSP samples. SN and HS operated MALDI-TOFMS and did the mass spectral data analysis. RT, YK, HT, and TY did gene sequencing and all authors participated gene data analysis. SN and HS drafted the manuscript. RT, HT, and TY helped to draft the manuscript. All authors read and approved the final manuscript.

Competing interests

The authors declare that they have no competing interests.

Author details

[1]Research Institute for Sustainable Chemistry, National Institute of Advanced Industrial Science and Technology (AIST), Higashi 1-1-1, Tsukuba, Ibaraki 305-8565, Japan. [2]Medical Mycology Research Center, Chiba University, 1-8-1 Inohana, Chuo-ku, Chiba 260-8673, Japan.

References

1. Gibbons JG, Rokas A. The function and evolution of the *Aspergillus* genome. Trends Microbiol. 2013;21(1):14–22.
2. Latge JP. *Aspergillus fumigatus* and aspergillosis. Clin Microbiol Rev. 1999; 12(2):310–54.
3. Buzina W. *Aspergillus* - Classification and Antifungal Susceptibilities. Curr Pharm Des. 2013;19(20):3615–28.
4. McClenny N. Laboratory detection and identification of *Aspergillus* species by microscopic observation and culture: the traditional approach. Med Mycol. 2005;43:S125–8.
5. Hinrikson HP, Hurst SF, Lott TJ, Warnock DW, Morrison CJ. Assessment of ribosomal large-subunit D1-D2, internal transcribed spacer 1, and internal transcribed spacer 2 regions as targets for molecular identification of medically important *Aspergillus* species. J Clin Microbiol. 2005;43(5):2092–103.
6. Glass NL, Donaldson GC. Development of primer sets designed for use with the PCR to amplify conserved genes from filamentous ascomycetes. Appl Environ Microbiol. 1995;61(4):1323–30.
7. Hong SB, Go SJ, Shin HD, Frisvad JC, Samson RA. Polyphasic taxonomy of *Aspergillus fumigatus* and related species. Mycologia. 2005;97(6):1316–29.
8. Serrano R, Gusmao L, Amorim A, Araujo R. Rapid identification of *Aspergillus fumigatus* within the section *Fumigati*. BMC Microbiol. 2011;11:7.
9. Peterson SW. Phylogenetic analysis of *Aspergillus* species using DNA sequences from four loci. Mycologia. 2008;100(2):205–26.
10. Bain JM, Tavanti A, Davidson AD, Jacobsen MD, Shaw D, Gow NAR, Odds FC. Multilocus sequence typing of the pathogenic fungus *Aspergillus fumigatus*. J Clin Microbiol. 2007;45(5):1469–77.
11. Chalupova J, Raus M, Sedlarova M, Sebela M. Identification of fungal microorganisms by MALDI-TOF mass spectrometry. Biotechnol Adv. 2014; 32(1):230–41.
12. De Carolis E, Posteraro B, Lass-Florl C, Vella A, Florio AR, Torelli R, Girmenia C, Colozza C, Tortorano AM, Sanguinetti M, et al. Species identification of *Aspergillus*, *Fusarium* and Mucorales with direct surface analysis by matrix-assisted laser desorption ionization time-of-flight mass spectrometry. Clin Microbiol Infect. 2012;18(5):475–84.
13. Iriart X, Lavergne RA, Fillaux J, Valentin A, Magnaval JF, Berry A, Cassaing S. Routine Identification of Medical Fungi by the New Vitek MS Matrix-Assisted Laser Desorption Ionization-Time of Flight System with a New Time-Effective Strategy. J Clin Microbiol. 2012;50(6):2107–10.
14. Del Chierico F, Masotti A, Onori M, Fiscarelli E, Mancinelli L, Ricciotti G, Alghisi F, Dimiziani L, Manetti C, Urbani A, et al. MALDI-TOF MS proteomic phenotyping of filamentous and other fungi from clinical origin. J Proteome. 2012;75(11):3314–30.
15. Ranque S, Normand AC, Cassagne C, Murat JB, Bourgeois N, Dalle F, Gari-Toussaint M, Fourquet P, Hendrickx M, Piarroux R. MALDI-TOF mass spectrometry identification of filamentous fungi in the clinical laboratory. Mycoses. 2014;57(3):135–40.
16. Hettick JM, Green BJ, Buskirk AD, Kashon ML, Slaven JE, Janotka E, Blachere FM, Schmechel D, Beezhold DH. Discrimination of *Aspergillus* isolates at the species and strain level by matrix-assisted laser desorption/ionization time-of-flight mass spectrometry fingerprinting. Anal Biochem. 2008;380(2):276–81.
17. Welker M. Proteomics for routine identification of microorganisms. Proteomics. 2011;11(15):3143–53.
18. Sun L, Teramoto K, Sato H, Torimura M, Tao H, Shintani T. Characterization of ribosomal proteins as biomarkers for matrix-assisted laser desorption/ionization mass spectral identification of *Lactobacillus plantarum*. Rapid Commun Mass Spectrom. 2006;20(24):3789–98.
19. Teramoto K, Sato H, Sun L, Torimura M, Tao H, Yoshikawa H, Hotta Y, Hosoda A, Tamura H. Phylogenetic classification of Pseudomonas putida by MALDI-MS using ribosomal proteins as biomarkers. Anal Chem. 2007;79(22):8712–9.
20. Teramoto K, Sato H, Sun L, Torimura M, Tao H. A simple intact protein analysis by MALDI-MS for characterization of ribosomal proteins of two genome-sequenced lactic acid bacteria and verification of their amino acid sequences. J Proteome Res. 2007;6(10):3899–907.
21. Hotta Y, Teramoto K, Sato H, Yoshikawa H, Hosoda A, Tamura H. Classification of genus *Pseudomonas* by MALDI-TOF MS based on ribosomal protein coding in *S10-spc-alpha* operon at strain level. J Proteome Res. 2010;9(12):6722–8.
22. Hotta Y, Sato J, Sato H, Hosoda A, Tamura H. Classification of the genus *Bacillus* based on MALDI-TOF MS analysis of ribosomal proteins coded in *S10* and *spc* operons. J Agric Food Chem. 2011;59(10):5222–30.
23. Sato H, Teramoto K, Ishii Y, Watanabe K, Benno Y. Ribosomal protein profiling by matrix-assisted laser desorption/ionization time-of-flight mass spectrometry for phylogenety-based subspecies resolution of *Bifidobacterium longum*. Syst Appl Microbiol. 2011;34(1):76–80.
24. Hotta Y, Sato H, Hosoda A, Tamura H. MALDI-TOF MS analysis of ribosomal proteins coded in *S10* and *spc* operons rapidly classified the *Sphingomonadaceae* as alkylphenol polyethoxylate-degrading bacteria from the environment. FEMS Microbiol Lett. 2012;330:23–9.
25. Sato H, Torimura M, Kitahara M, Ohkuma M, Hotta Y, Tamura H. Characterization of the *Lactobacillus casei* group based on the profiling of ribosomal proteins coded in S10-spc-alpha operons as observed by MALDI-TOF MS. Syst Appl Microbiol. 2012;35(7):447–54.
26. Ryzhov V, Fenselau C. Characterization of the protein subset desorbed by MALDI from whole bacterial cells. Anal Chem. 2001;73(4):746–50.
27. UniProt Knowledgebase (UniProtKB) [http://www.uniprot.org/]
28. Nakamura S, Sato H, Tanaka R, Yaguchi T. Verification of Ribosomal Proteins of *Aspergillus fumigatus* for use as Biomarkers in MALDI-TOF MS identification. Mass Spectrometry (Tokyo). 2016;5:A0049.
29. National Center for Biotechnology Information (NCBI) The protein database [https://www.ncbi.nlm.nih.gov/protein/]
30. Mager WH, Planta RJ, Ballesta JPG, Lee JC, Mizuta K, Suzuki K, Warner JR, Woolford J. A new nomenclature for the cytoplasmic ribosomal proteins *Saccharomyces cerevisiae*. Nucleic Acids Res. 1997;25(24):4872–5.
31. Lamoth F. *Aspergillus fumigatus*-Related Species in Clinical Practice. Front Microbiol. 2016;7:8.
32. Payne GA, Nierman WC, Wortman JR, Pritchard BL, Brown D, Dean RA, Bhatnagar D, Cleveland TE, Machida M, Yu J. Whole genome comparison of Aspergillus flavus and A-oryzae. Med Mycol. 2006;44:S9–S11.
33. Yaguchi T, Horie Y, Tanaka R, Matsuzawa T, Ito J, Nishimura K. Molecular Phylogenetics of Multiple Genes on *Aspergillus* Section *Fumigati* Isolated from Clinical Specimens in Japan. Medical Mycology Journal. 2007;48(1):37–46.
34. Balajee SA, Gribskov JL, Hanley E, Nickle D, Marr KA. *Aspergillus lentulus* sp nov., a new sibling species of *A-fumigatus*. Eukaryot Cell. 2005;4(3):625–32.

35. Verwer PEB, van Leeuwen WB, Girard V, Monnin V, van Belkum A, Staab JF, Verbrugh HA, Bakker-Woudenberg I, van de Sande WWJ. Discrimination of *Aspergillus lentulus* from *Aspergillus fumigatus* by Raman spectroscopy and MALDI-TOF MS. Eur J Clin Microbiol Infect Dis. 2014;33(2):245–51.

36. Pinel C, Arlotto M, Issartel JP, Berger F, Pelloux H, Grillot R, Symoens F. Comparative proteomic profiles of *Aspergillus fumigatus* and *Aspergillus lentulus* strains by surface-enhanced laser desorption ionization time-of-flight mass spectrometry (SELDI-TOF-MS). BMC Microbiol. 2011;11:10.

37. Novakova A, Hubka V, Dudova Z, Matsuzawa T, Kubatova A, Yaguchi T, Kolarik M. New species in *Aspergillus* section *Fumigati* from reclamation sites in Wyoming (USA) and revision of *A. viridinutans* complex. Fungal Divers. 2014;64(1):253–74.

38. Barrs VR, van Doorn TM, Houbraken J, Kidd SE, Martin P, Pinheiro MD, Richardson M, Varga J, Samson RA. *Aspergillus felis* sp nov., an Emerging Agent of Invasive Aspergillosis in Humans, Cats, and Dogs. Plos One. 2013;8(6):11.

39. Sugui JA, Peterson SW, Figat A, Hansen B, Samson RA, Mellado E, Cuenca-Estrella M, Kwon-Chung KJ. Genetic Relatedness versus Biological Compatibility between *Aspergillus fumigatus* and Related Species. J Clin Microbiol. 2014;52(10):3707–21.

40. Compute pI/Mw tool [http://www.expasy.org/tools/pi_tool.html]

41. Larkin MA, Blackshields G, Brown NP, Chenna R, McGettigan PA, McWilliam H, Valentin F, Wallace IM, Wilm A, Lopez R, et al. Clustal W and clustal X version 2.0. Bioinformatics. 2007;23(21):2947–8.

42. Tamura K, Stecher G, Peterson D, Filipski A, Kumar S. MEGA6: Molecular Evolutionary Genetics Analysis Version 6.0. Mol Biol Evol. 2013;30(12): 2725–9.

Staphylococcal enterotoxin B administration in pregnant rats alters the splenic lymphocyte response in adult offspring rats

Ping Zhou[1†], Xin-sheng Zhang[2†], Zhi-ben Xu[1†], Shu-xian Gao[1], Qing-wei Zheng[1], Ming-zhu Xu[3], Lin Shen[4], Feng Yu[5] and Jun-chang Guan[1*]

Abstract

Background: Our previous study suggested that SEB exposure in pregnant rats could lead to the change of T cells subpopulation in both peripheral blood and thymus of the offspring rats. However, rarely is known about the influence of SEB exposure in pregnant rats on T cell subpopulation in the spleens of offspring rats.

Results: SEB was intravenously administered to the pregnant rats at gestational day 16 in this study. The percentages, in vivo and in vitro responses of CD4 and CD8 T cells were investigated with flow cytometry. The prenatal SEB exposure obviously increased splenic CD4 T cell percentages of both neonates and adult offspring rats, and obviously reduced splenic CD8 T cell percentages of both the fifth day neonates and adult offspring rats. After spleens in the adult offspring rats were re-stimulated with SEB in vivo or in vitro, in vivo SEB stimulation could lead to the marked decrease of splenic CD4 T cell percentage and the marked increase of splenic CD8 T cell percentage. While in vitro SEB stimulation to the cultured splenocytes markedly decreased the proliferation of the splenic lymphocytes and the CD4 T cell percentage, and had no influence on CD8 T cell percentage.

Conclusion: The prenatal SEB exposure could alter the percentages of CD4/CD8 T cell subpopulation and the response of CD4 and CD8 T cells to the in vivo and in vitro secondary SEB stimulation in the splenocytes of adult offspring rats.

Keywords: Staphylococcal enterotoxin B, Splenic lymphocyte, Pregnancy, Offspring, Rat

Background

In five types (A to E) of staphylococcal enterotoxin [1], staphylococcal enterotoxin B (SEB) as an important superantigen (SAg) has been widely studied. SEB can cross-link major histocompatibility complex class II molecules on the antigen-presenting cell with the β chain of the T cell receptor and activate vigorous fractions of the T cell population at high frequency [2, 3], which has no need of classical processing and presentation of antigen [4]. The immune response of T cells to SEB displays a biphasic change [5, 6] which consists of an early activation presented as T cell proliferation and a second anergy due to apoptosis of the appropriate T cells. Ultimately, the hyper-response and immunosuppression of T cells following SEB exposure may lead to illness and disease in mammals [7, 8]. A variety of literatures [9–12] have demonstrated the influence of SEB exposure on T cells during adulthood or neonatal period in animal experiments. Our previous study suggested that SEB exposure in pregnant rats could influence the T cells subpopulation in both peripheral blood and thymus of the offspring rats [13–15], but rarely is known about how SEB exposure in pregnant rats to influence the splenic T cell subpopulation of offspring rats. Therefore, SEB was intravenously injected to the

* Correspondence: guanjc2013@126.com
†Equal contributors
[1]Department of Microbiology and Anhui Key Laboratory of Infection and Immunity, Bengbu Medical College, 2600 Dong Hai Avenue, Bengbu, Anhui 233030, People's Republic of China
Full list of author information is available at the end of the article

maternal rats at gestational day (GD) 16 in this study. The percentages, in vivo and in vitro responses of CD4/CD8 T cells to SEB were investigated with flow cytometry in the spleens of offspring rats born to maternal rats exposed SEB during pregnancy.

Methods
Animals
Three-month-old Sprague–Dawley rats used in this study were fed with rodent chow and filtered tap water ad libitum and housed under controlled conditions at 23 °C ±2 °C and a constant 12 h light/12 h dark cycle. Each female rat was placed in contact with an adult male rat for mating. After 15 h, a plug was evaluated in female rats. The day of the plug observed initially in the vagina was considered day 1 of gestation (GD). Then, the female rats were kept in separate cages and randomly separated into the phosphate buffer saline (PBS) group and the SEB group. Twelve pregnant rats (four of them used in our previous papers [13, 15]) were used for present study in each group. In the SEB group, the pregnant rats were injected i.v. once with 0.3 ml 50 µg/ml SEB (Sigma-Aldrich, St Louis, MO) in 0.2 M PBS. The pregnant rats in PBS group were injected with the same volume of PBS. Then, the pregnant rats were reared as above and allowed to give birth naturally. Some neonatal offspring rats between days 0 and 5 after delivery were used to analyze T cell subpopulation in the spleens, the others were fed to adult offspring rats (about 3 to 5 months) for the analysis of T cell subpopulation and lymphocyte proliferation in the spleens. All surgery was performed under sodium pentobarbital anesthesia, and euthanasia was accomplished with CO_2.

In vivo response of adult offspring rats to SEB re-stimulation
When neonates were fed to adult offspring rats in the PBS and SEB groups, the adult offspring rats were administrated with either SEB or PBS in the same way as pregnant rats and continued to rear for 5 days. Then, the rats were anaesthetized and spleens were harvested to prepare for the splenocyte suspensions.

Preparation of splenocyte suspensions
The spleens of the neonatal and adult offspring rats were minced and pressed through a 100-µm fine wire mesh screen. Cells mixture with PBS were collected in 5 ml centrifugal tube and centrifuged at 400 g for 10 min at 4 °C. After removing the supernatant, cell pellet was acquired and added with lysis buffer (Beyotime Biotechnology, China) to break red blood cells, according to the manufacturer's instructions. Then, the cells were washed three times with balanced PBS solution

containing 2% fetal bovine serum (FBS) (HyClone, UT) by centrifugation at 400 g for 10 min at 4 °C. Finally, splenocytes were mixed in staining buffer (PBS containing 2% FBS and 0.02% NaN3) and counted 1×10^6/ml cells for detecting T cell subpopulation.

Splenic lymphocyte culture
The spleens of the adult offspring rats in two groups were aseptically removed and splenic lymphocytes were isolated by standard Ficoll–Paque density gradient from each animal. The cells (1×10^5/ml) were co-cultured with either 100 ng/ml SEB or 5 µg/ml concanavalin (Con) A (Sigma-Aldrich, St Louis, MO) for 3 days in 96-well and 24-well culture plates in 200 µl RPMI medium (Gibco BRL, USA) containing 10% FBS, L-glutamine, penicillin and streptomycin in a humidified incubator in 5% CO2 at 37 °C. For analysis of the T cells subpopulation, the lymphocytes in 24-well culture plates were harvested at the end of culture days 1, 2, & 3 and stained for the analysis of T cell subpopulation with flow cytometry as shown below. For the experiment of the lymphocyte proliferation, 1 µCi ^{3}H-thymidine was added to the lymphocytes in 96-well culture plates 5 h before the end of culture days 1, 2, & 3. Thymidine incorporation by cells was determined using a cell harvester and 1450 MICROBETA liquid scintillation counter (PerkinElmer®, Waltham, MA, USA). Proliferation was measured as radioactivity incorporation [presented as counts per minute (CPM)].

Flow cytometry cell analysis
Splenocyte suspensions were incubated and stained with fluorescently labeled antibodies of anti–CD3-FITC, anti–CD4-APC, anti–CD8a-PE (eBioscience, USA) in the dark at room temperature for 30 min. For determination of T cell subpopulation, flow cytometry was performed on a FACS calibur (Becton Dickinson, Heidelberg, Germany). Images of stained cells were analyzed with CellQuest analysis software (BD Biosciences, Franklin Lakes, NJ, USA).

Statistical analysis
Statistical analysis was performed with SPSS software. To evaluate the difference of splenic T cells in the adult offspring rats, independent T-test was used. Turkey's-b in one-way ANOVA was employed to evaluate the different significances of T cells in the neonatal offspring rats, as well as the proliferation and response of splenocytes re-stimulated with SEB in the adult offspring rats. Data are expressed as the mean ± SEM. Statistical significance was defined as $p < 0.05$.

Results

Influence of SEB exposed prenatally on splenic CD4/CD8 T cells of offspring rats

Compared with the PBS group, CD4 T cell percentage was obviously increased in the spleens of neonatal rats between days 0 and 5 after delivery (Fig. 1a), while CD8 T cell percentage was significantly decreased in the fifth-day neonates in the SEB group, but not different between the PBS and SEB groups between days 0 and 4 after delivery (Fig. 1b). In the adult offspring rats, it was revealed that the prenatal SEB exposure caused the markedly increased percentage of splenic CD4 T cells (Fig. 1c) and the markedly reduced percentage of splenic CD8 T cells (Fig. 1d).

In vivo response of CD4/CD8 T cells to SEB re-stimulation

Five days after in vivo PBS or SEB administration to the adult offspring rats, the percentage of T cell subpopulation in the spleens were detected with flow cytometry. In the PBS group, it indicated that SEB administration led to the markedly higher percentage of CD4 T cells (Fig. 2a) and the markedly lower percentage of CD8 T cells (Fig. 2b) than those of PBS administration in the adult offspring rats. While the response trends of CD4

(Fig. 2a) and CD8 (Fig. 2b) T cells to secondary SEB administration in the SEB group were completely contrary to those in the PBS group.

In vitro response of the splenic lymphocytes of adult offspring rats to SEB

After the splenic lymphocytes of adult offspring rats were in vitro co-cultured with either SEB or ConA for 3 days, the percentages of CD4 and CD8 T cells were examined. In the PBS group, the CD4 T cell percentage with SEB stimulation was markedly higher than those with ConA stimulation in each cultured day (Fig. 3a, b). While in the SEB group, in vitro SEB stimulation markedly reduced the CD4 T cell percentage compared with that with ConA stimulation. However, neither SEB nor ConA stimulation in each cultured day altered the percentage of CD8 T cells in the PBS and SEB groups (Fig. 3b, d).

Proliferation of the splenic lymphocytes

After in vitro co-cultured with either SEB or ConA for 3 days, proliferation of the splenic lymphocytes of adult offspring rats was measured by ^{3}H-thymidine incorporation. With the increase of stimulation time, the

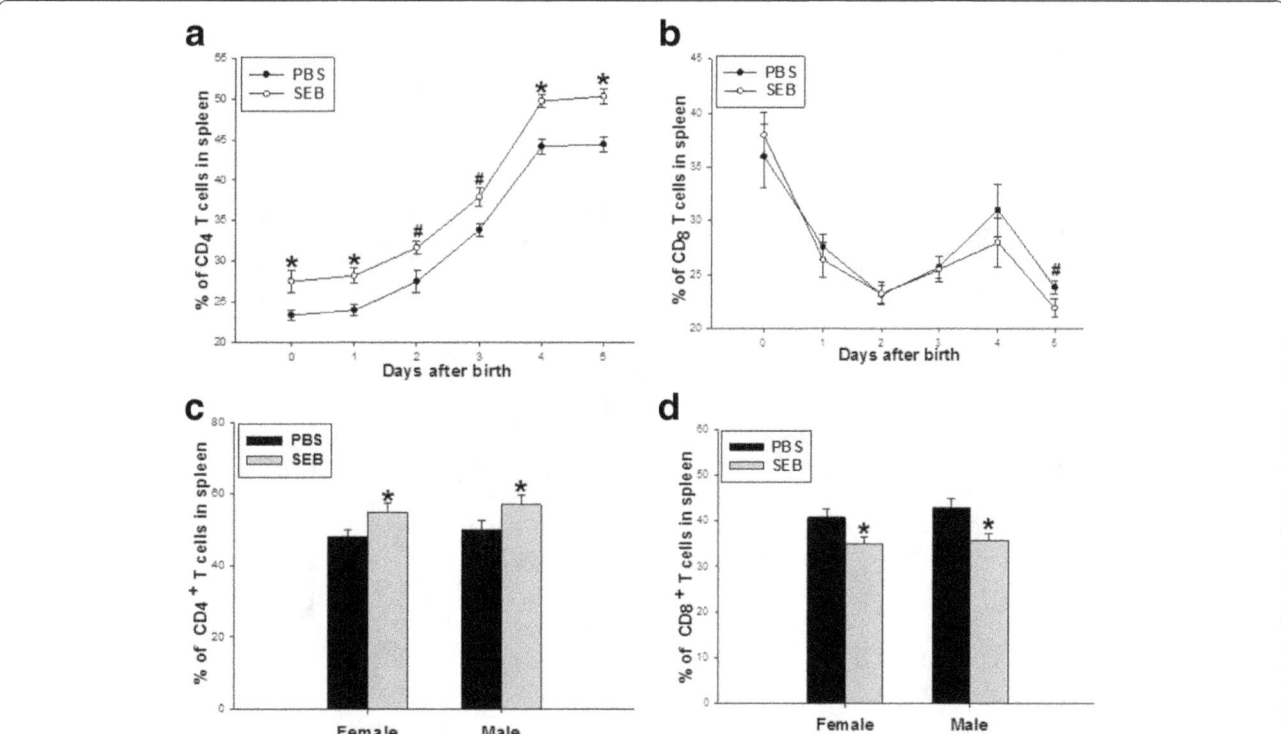

Fig. 1 Effect of prenatal SEB exposure on CD4/CD8 T cells in the spleens of offspring rats. The spleens of the neonatal rats between days 0 and 5 after delivery and the adult offspring rats were harvested in the PBS and SEB groups. The percentages of both CD4 and CD8 T cells in the spleens of the neonatal rats (**a, b**) and the adult offspring rats (**c, d**) were analyzed by flow cytometry. Values were calculated with data from 12 independent experiments [In each group, 72 of neonatal offspring rats and 24 of adult offspring rats (half male and half female) were used.]. Each experiment of the neonatal rats included 1–2 neonatal rats of the same litter from same mother. Data represent mean ± SE. Compared with the PBS group at each time point: # $P < 0.05$; * $P < 0.01$

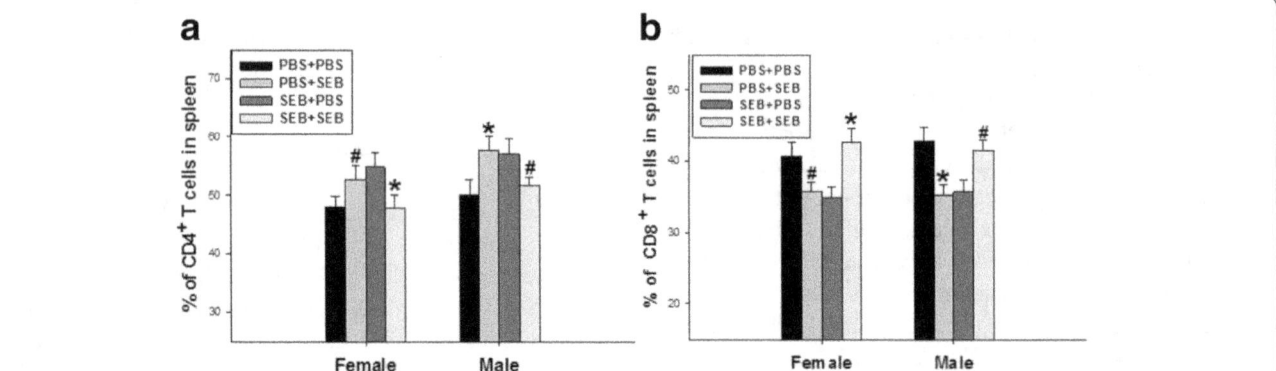

Fig. 2 Effect of the secondary SEB administration on CD4/CD8 T cells in the spleens of adult offspring rats exposed prenatally to SEB. The adult offspring rats in the PBS and SEB groups were injected i.v. with either SEB (named as PBS + SEB, SEB + SEB) or PBS (named as PBS + PBS, SEB + PBS), separately. Five days after administration, the splenocytes of adult female and male offspring rats were harvested. The percentages of both CD4 (**a**) and CD8 (**b**) T cells were analyzed by flow cytometry. Values were calculated with data from 10 independent experiments [Twenty adult offspring rats (half male and half female) was used in each group.]. Data represent mean ± SE. Compared with PBS + PBS: # $P < 0.05$; Compared with SEB + PBS: * $P < 0.05$

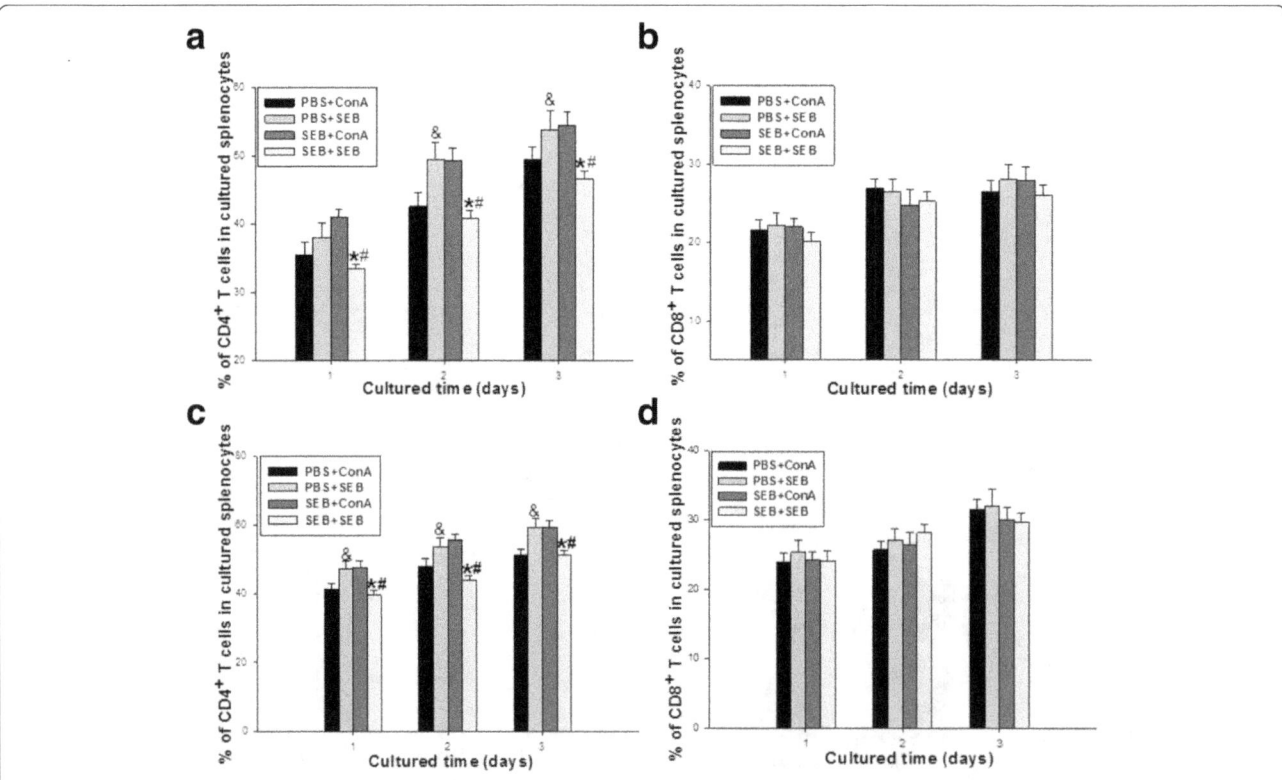

Fig. 3 In vitro response of the splenic lymphocytes of adult offspring rats to SEB. After the splenic lymphocytes of adult offspring rats were acquired in the PBS and SEB groups and in vitro co-cultured in RPMI medium with either SEB (named as PBS + SEB, SEB + SEB) or ConA (named as PBS + ConA, SEB + ConA) for 3 days, the percentages of CD4 and CD8 T cells in the splenocytes of adult female (**a**, **b**) and male (**c**, **d**) offspring rats were analyzed by flow cytometry. Values were calculated with data from 10 independent experiments [Twenty adult offspring rats (half male and half female) was used in each group.]. Data represent mean ± SE. Compared with PBS + ConA: & $P < 0.05$; Compared with SEB + ConA: * $P < 0.05$; Compared with PBS + SEB: # $P < 0.05$

proliferation of lymphocytes stimulated by SEB and ConA was significantly increased in the lymphocytes of adult female (Fig. 4a) and male (Fig. 4b) offspring rats in the PBS and SEB groups. While in the PBS group, SEB stimulation significantly increased the lymphocyte proliferation compared with that stimulated by ConA in each cultured day, but in the SEB group, the proliferation of lymphocytes stimulated by SEB was significantly lower than that by ConA in each cultured day.

Discussion

In this study, the data indicated that SEB exposure in pregnant rats markedly altered the percentages of both CD4 and CD8 T cells, as well as the response characteristics of CD4/CD8 T cells to secondary SEB administration, not only in vivo but also in vitro, in the splenocytes of adult offspring rats. As far as we know, the present study firstly investigates the effect of SEB exposure in pregnant rats on the changes and responses in splenic T cell subpopulation of offspring rats.

SEB exposure to naive mice could selectively lead to specific anergy of Vβ8 T cells [16–18] of not only the peripheral but also central immune compartments [5, 19]. Although our previous study [15] suggested that SEB exposure in pregnant rats could affect thymic T cell subpopulation in the central compartment of the offspring rats, rarely is known about the influence of SEB exposure on T cell subpopulation in the spleens. The present study showed that SEB exposure in pregnant rats could cause the increased percentage of splenic CD4 T cells between days 0 and 5 after delivery and the decreased percentage of splenic CD8 T cells in the fifth day in the neonatal offspring rats. While in the spleens of adult offspring rats, SEB exposure in pregnant rats was able to lead to the increased CD4 T cells and the decreased CD8 T cells, in accordance with other data from direct SEB stimulation in adult mice [5, 20]. These results suggest that SEB exposure in pregnant rats was able to cause the decreased percentage of CD8 T cells accompanied by a relative increase in the percentage of CD4 T cells and could imprint the changes of the CD4 and CD8 T cell percentage from neonatal to adult offspring rats. The trend of these changes was similar to those of both CD4 and CD8 T cells in the thymus [15] and peripheral blood [13] of offspring rats in our previous study, suggesting the consistency of changes between central and peripheral compartments about the effect of prenatal exposure of SEB on the CD4 and CD8 T cells. Many studies [21, 22] have suggested that the imprinting effects are caused by the physicochemical and biological factors exposed during pregnancy, and are the developmental origins of health and disease. A lot of studies indicate that SEB exposure is involved in the pathogenicity of some immunological diseases [9, 23]. Whether the imprinted alteration of T cells caused by prenatal SEB exposure is associated with these diseases remains further study in the future. In addition, whether male or female adult offspring rats, there had no difference of splenic T cell percentage between PBS and SEB groups. These data suggest the effect of SEB exposure in pregnant rats on T cell subpopulation did not display gender difference.

Since prenatal SEB exposure could lead to the changes of splenic T cells in the offspring rats, another question raised immediately was whether prenatal exposure of SEB could influence T cell function in the spleens of the

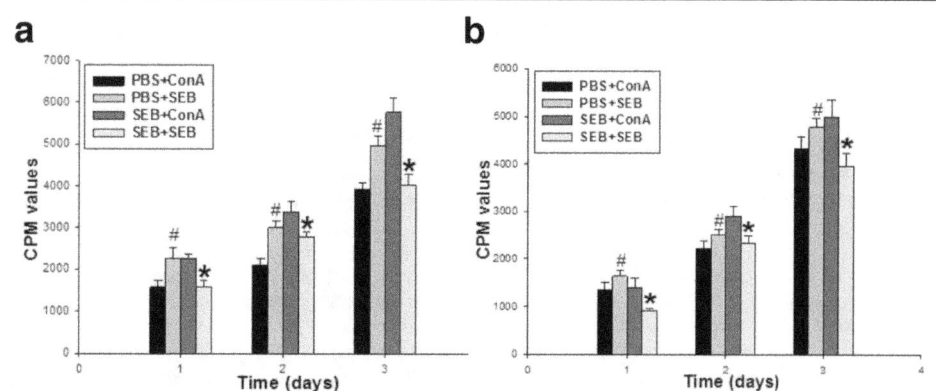

Fig. 4 Effect of SEB on the lymphocyte proliferation of adult offspring rats. Three days after the splenocytes of adult offspring rats in the PBS and SEB groups were in vitro co-cultured in RPMI medium with either SEB (named as PBS + SEB, SEB + SEB) or ConA (named as PBS + ConA, SEB + ConA), the proliferation was measured by ^{3}H-thymidine incorporation in the splenic lymphocytes of adult female (**a**) and male (**b**) offspring rats. Data are expressed as mean ± SE and the measure unit of [^{3}H] thymidine incorporation is counts per min (CPM). The results are representative of 10 independent cultures with each condition in triplicate [Twenty adult offspring rats (half male and half female) was used in each group]. Compared with PBS + ConA: # $P < 0.05$; Compared with SEB + ConA: * $P < 0.05$

offspring rats. To address this question, the responses of splenocytes to the in vivo and in vitro secondary SEB stimulation were further investigated in the adult offspring rats. Five days after the in vivo secondary SEB administration to the adult offspring rats exposed SEB during pregnancy, interesting results were found that the SEB re-stimulation significantly reduced CD4 T cells with an increased CD8 T cell percentage in the spleen of the adult offspring rats, which was contrary to the results from the primed SEB administration during pregnancy in the adult offspring rats. It may be due to the possibility that the primed SEB administration during pregnancy abrogated the response of CD4 T cells (anergy) despite these cells existed in significant percentage [17, 24]. Furthermore, the SEB administration to the in vitro cultured splenocytes in the adult offspring rats was also able to induce the significantly decreased CD4 T cell percentage compared with that of the ConA administration, but had no influence on the CD8 T cell percentage. These data suggest that SEB exposure in pregnant rats could lead to the hypo-responsiveness or anergy of CD4 T cells in the splenocytes of adult offspring rats, which is consistent with other results from direct SEB administration in the cultured splenocytes of the adult mice [17, 25, 26]. Taken together, these data suggest that SEB exposure in pregnant rats could alter the secondary response of CD4 /CD8 T cells to both in vivo and in vitro SEB re-stimulation in the splenocytes of adult offspring rats.

Conclusions

SEB exposure in pregnant rats could induce the decrease in the percentage of CD8 T cells accompanied by a relative increase in the percentage of CD4 T cells and could imprint the changes of the CD4 and CD8 T cell percentages from the neonatal to adult offspring rats. Furthermore, SEB exposure in pregnant rats could alter the secondary response of CD4 and CD8 T cells to both in vivo and in vitro SEB re-stimulation in the splenocytes of adult offspring rats.

Abbreviations
Con A: Concanavalin A; CPM: Counts per minute; FBS: Fetal bovine serum; GD: Gestational day; PBS: Phosphate buffer saline; SAg: Superantigen; SEB: Staphylococcal enterotoxin B

Acknowledgements
We want to thank Dr. Jiu Jiang (Drexel University College of Arts and Sciences, Philadelphia, PA, USA) for his assistance in FACS analysis and with the English.

Funding
This work was supported by grants from the National Nature Science Foundation of China (81571454, 81200482), the Key Scientific Research Foundation of the Higher Education Institutions of Anhui Province (KJ2015A211), Foundation for Innovative Research Groups of Anhui province (2016–40), National (201510367005) and Anhui Province (201510367013, AH201410367007) Undergraduate Training Programs for Innovation and Entrepreneurship, and the Scientific Research Foundation of Bengbu Medical College (BYKL1405ZD).

Authors' contributions
JG planned and designed the experiments. PZ, XZ and ZX performed the experiments. SG analyzed the data. QZ drafted the manuscript. MX and LS helped to perform the experiment. JG and FY finalized the manuscript. All authors read and approved the manuscript.

Competing interests
The authors declare that they have no competing interests.

Author details
[1]Department of Microbiology and Anhui Key Laboratory of Infection and Immunity, Bengbu Medical College, 2600 Dong Hai Avenue, Bengbu, Anhui 233030, People's Republic of China. [2]Editorial Board of Journal of Bengbu Medical College, Bengbu Medical College, Bengbu, Anhui 233030, People's Republic of China. [3]Department of Life Sciences, Bengbu Medical College, Bengbu, Anhui 233030, People's Republic of China. [4]Scientific Research Center, Bengbu Medical College, Bengbu, Anhui 233030, People's Republic of China. [5]Huzhou University Schools of Medicine and Nursing Science, Huzhou, Zhejiang 313000, People's Republic of China.

References
1. Bergdoll MS: Enterotoxins. Staphylococci and staphylococcal infections; 1983
2. Heriazon A, Zhou P, Borojevic R, Foerster K, Streutker CJ, Ng T, Croitoru K. Regulatory T cells modulate staphylococcal enterotoxin B-induced effector T-cell activation and acceleration of colitis. Infect Immun. 2009;77(2):707–13.
3. Buza JJ, Burgess SC. Different signaling pathways expressed by chicken naive CD4(+) T cells, CD4(+) lymphocytes activated with staphylococcal enterotoxin B, and those malignantly transformed by Marek's disease virus. J Proteome Res. 2008;7(6):2380–7.
4. Bell SJ, Buxser SE. Staphylococcal enterotoxin B modulates V beta 8+ TcR-associated T-cell memory against conventional antigen. Cell Immunol. 1995;160(1):58–64.
5. MacDonald HR, Baschieri S, Lees RK. Clonal expansion precedes anergy and death of V beta 8+ peripheral T cells responding to staphylococcal enterotoxin B in vivo. Eur J Immunol. 1991;21(8):1963–6.
6. Seth A, Stern LJ, Ottenhoff TH, Engel I, Owen MJ, Lamb JR, Klausner RD, Wiley DC. Binary and ternary complexes between T-cell receptor, class II MHC and superantigen in vitro. Nature. 1994;369(6478):324–7.
7. Bernal A, Proft T, Fraser JD, Posnett DN. Superantigens in human disease. J Clin Immunol. 1999;19(3):149–57.
8. Paiva CN, Pyrrho AS, Lannes-Vieira J, Vacchio M, Soares MB, Gattass CR. Trypanosoma cruzi sensitizes mice to fulminant SEB-induced shock: overrelease of inflammatory cytokines and independence of Chagas' disease or TCR Vbeta-usage. Shock. 2003;19(2):163–8.
9. Kohno H, Sakai T, Tsuneoka H, Imanishi K, Saito S. Staphylococcal enterotoxin B is involved in aggravation and recurrence of murine experimental autoimmune uveoretinitis via Vbeta8 + CD4+ T cells. Exp Eye Res. 2009;89(4):486–93.
10. Kurella S, Yaciuk JC, Dozmorov I, Frank MB, Centola M, Farris AD. Transcriptional modulation of TCR, Notch and Wnt signaling pathways in SEB-anergized CD4+ T cells. Genes Immun. 2005;6(7):596–608.
11. Watson AR, Janik DK, Lee WT. Superantigen-induced CD4 memory T cell anergy. I. Staphylococcal enterotoxin B induces Fyn-mediated negative signaling. Cell Immunol. 2012;276(1–2):16–25.
12. Watson AR, Mittler JN, Lee WT. Staphylococcal enterotoxin B induces anergy to conventional peptide in memory T cells. Cell Immunol. 2003;222(2):144–55.
13. Zhang T, Yu FL, Yang WX, Ruan MM, Yue ZY, Liu Y, Liu TT, Zhou P, Xia H, Guan JC. Staphylococcal enterotoxin B administration during pregnancy imprints the increased CD4:CD8 T-cell ratio in the peripheral blood from neonatal to adult offspring rats. J Med Microbiol. 2015;64:1–6.
14. Guan J, Liu Y, Kong X, Zhu X, Yu FL, Lin N, Liu CS, Zhang T. Effect of maternal staphylococcal enterotoxin B administration during pregnancy on CD3^{+} TCR Vβ8^{+}T cells of adult offspring rats. Nan Fang Yi Ke Da Xue Xue Bao. 2012;32(9):1230–3.

15. Yang WX, Zhu X, Yu FL, Zhang T, Lin N, Liu TT, Zhao FF, Liu Y, Xia H, Guan JC. Decreased Vβ8.2 T-cells in neonatal rats exposed prenatally to Staphylococcal enterotoxin B are further deleted by restimulation in an in vitro cultured thymus. Mol Med Rep. 2014;10:989–94.

16. Goettelfinger P, Roussin R, Lecerf F, Berrih-Aknin S, Fattal-German M. T cell deletion and unresponsiveness induced by intrathymic injection of staphylococcal enterotoxin B. Transpl Immunol. 2000;8(1):39–48.

17. Kawabe Y, Ochi A. Programmed cell death and extrathymic reduction of Vbeta 8+ CD4+ T cells in mice tolerant to Staphylococcus aureus enterotoxin B. Nature. 1991;349(6306):245–8.

18. Li J, Zhou YB, Zhang Y, Zhang J, Chen WF. Cloning deletion of mouse medullary CD4 SP thymocyte subgroups induced by superantigen staphylococcal enterotoxin B. Beijing Da Xue Xue Bao. 2008;40(6):557–61.

19. Webb S, Morris C, Sprent J. Extrathymic tolerance of mature T cells: clonal elimination as a consequence of immunity. Cell. 1990;63:1249–56.

20. Baschieri S, Lees RK, Lussow AR, MacDonald HR. Clonal anergy to staphylococcal enterotoxin B in vivo: selective effects on T cell subsets and lymphokines. Eur J Immunol. 1993;23(10):2661–6.

21. Ruchat SM, Hivert MF, Bouchard L. Epigenetic programming of obesity and diabetes by in utero exposure to gestational diabetes mellitus. Nutr Rev. 2013;71:S88–94.

22. Kaplan JL, Shi HN, Walker WA. The role of microbes in developmental immunologic programming. Pediatr Res. 2011;69(6):465–72.

23. Fukushima H, Hirano T, Shibayama N, Miwa K, Ito T, Saito M, Sumida H, Oyake S, Tsuboi R, Oka K. The role of immune response to Staphylococcus aureus superantigens and disease severity in relation to the sensitivity to tacrolimus in atopic dermatitis. Int Arch Allergy Immunol. 2006;141(3):281–9.

24. Kawabe Y, Ochi A. Selective Anergy of Vb8+, CD4+ T Cells in Staphylococcus Enterotoxin B-primed Mice. J Exp Med. 1990;172:1065–70.

25. Aroeira LS, Mouton CG, Toran JL, Ward ES, Martínez C. Anti-Vbeta8 antibodies induce and maintain staphylococcal enterotoxin B-triggered Vbeta8+ T cell anergy. Eur J Immunol. 1999;29(2):437–45.

26. Migita K, Ochi A. Induction of clonal anergy by oral administration of staphylococcal enterotoxin B. Eur J Immunol. 1994;24(9):2081–6.

12

Growth phase-dependent expression profiles of three vital H-NS family proteins encoded on the chromosome of *Pseudomonas putida* KT2440 and on the pCAR1 plasmid

Zongping Sun[1,2], Delyana Vasileva[2], Chiho Suzuki-Minakuchi[2], Kazunori Okada[2], Feng Luo[1], Yasuo Igarashi[1] and Hideaki Nojiri[1,2]* (iD)

Abstract

Background: H-NS family proteins are nucleoid-associated proteins that form oligomers on DNA and function as global regulators. They are found in both bacterial chromosomes and plasmids, and were suggested to be candidate effectors of the interaction between them. TurA and TurB are the predominantly expressed H-NS family proteins encoded on the chromosome of *Pseudomonas putida* KT2440, while Pmr is encoded on the carbazole-degradative incompatibility group P-7 plasmid pCAR1. Previous transcriptome analyses suggested that they function cooperatively, but play different roles in the global transcriptional network. In addition to differences in protein interaction and DNA-binding functions, cell expression levels are important in clarifying the detailed underlying mechanisms. Here, we determined the precise protein amounts of TurA, TurB, and Pmr in KT2440 in the presence and absence of pCAR1.

Results: The intracellular amounts of TurA and TurB in KT2440 and KT2440(pCAR1) were determined by quantitative western blot analysis using specific antibodies. The amount of TurA decreased from the log phase (~80,000 monomers per cell) to the stationary phase (~20,000 monomers per cell), while TurB was only detectable upon entry into the stationary phase (maximum 6000 monomers per cell). Protein amounts were not affected by pCAR1 carriage. KT2440(pCAR1pmrHis), where histidine-tagged Pmr is expressed under its original promotor, was used to determine the intracellular amount of Pmr, which was constant (~30,000 monomers per cell) during cell growth. Quantitative reverse transcription PCR demonstrated that the transcriptional levels of *turA* and *turB* were consistent with protein expression, though the transcriptional and translational profiles of Pmr differed.

Conclusion: The amount of TurB increases as TurA decreases, and the amount of Pmr does not affect the amounts of TurA and TurB. This is consistent with our previous observation that TurA and TurB play complementary roles, whereas Pmr works relatively independently. This study provides insight into the molecular mechanisms underlying reconstitution of the transcriptional network in KT2440 by pCAR1 carriage.

Keywords: H-NS family proteins, Nucleoid-associated proteins, Plasmid, *Pseudomonas*

* Correspondence: anojiri@mail.ecc.u-tokyo.ac.jp
[1]Research Center of Bioenergy & Bioremediation, College of Resources and Environment, Southwest University, No.2 Tiansheng Road, BeiBei District, Chongqing 400715, China
[2]Biotechnology Research Center, The University of Tokyo, 1-1-1 Yayoi, Bunkyo-ku, Tokyo 113-8657, Japan

Background

Nucleoid-associated proteins (NAPs) play important roles in chromatin compaction and global regulation in bacterial cells [1–3]. Among the best characterized NAPs are the histone-like protein H1 (H-NS) family proteins. They have two structurally independent domains connected by a flexible linker, an N-terminal domain involved in dimerization/oligomerization, and a C-terminal domain in charge of DNA-binding [4]. MvaT homologs of *Pseudomonas* were experimentally shown to complement *Escherichia coli* H-NS-deficient phenotypes, and are now recognized as members of the H-NS family despite low sequence similarity [5]. *Pseudomonas putida* KT2440 harbors five genes encoding H-NS family proteins, namely PP_1366 (TurA), PP_3765 (TurB), PP_0017 (TurC), PP_3693 (TurD), and PP_2947 (TurE) [6]. Our previous report revealed that several bacteria carry the same types of H-NS family proteins on both their plasmids and chromosome [7]. An MvaT homolog, Pmr, is encoded on the carbazole-degradative incompatibility (Inc) P-7 group plasmid pCAR1, whose hosts are mainly *Pseudomonas* [8–12].

turA and *turB* are predominantly transcribed in log and stationary phases, respectively, in KT2440 and KT2440(pCAR1) cells, whereas *pmr* is actively transcribed in KT2440(pCAR1) cells [13]. TurA, TurB, and Pmr can form homo- and hetero-oligomers in vitro based on the N-terminal domain, while the coupling ratios among them differ [14, 15]. In addition, though the TurA-, TurB-, and Pmr-binding regions detected in vivo were almost identical, the regulons of the three proteins vary significantly [16]. While pCAR1 carriage altered the global transcriptional network in KT2440 cells [17–19], our previous results suggested that the three H-NS family proteins function cooperatively, but their respective roles are not equivalent [16]. To understand the molecular mechanisms underlying reconstitution of the transcriptional network in KT2440 by pCAR1 carriage, it is important to determine the detailed roles of these proteins.

To this end, in addition to the DNA-binding and dimerization/oligomerization functions, it is important to determine the intracellular amounts of the three H-NS family proteins and how they cooperatively compact the genome and regulate gene expression. The transcriptional levels of *turA*, *turB*, and *pmr* were previously determined [13], but the transcriptional and translational levels sometimes differ among H-NS family proteins [20]. In addition, NAPs show various expression patterns throughout the growth phases [21]. Thus, in the present study, we quantified the intracellular amounts of TurA, TurB, and Pmr in KT2440 and/or KT2440(pCAR1) cells during the growth phases by western blot analysis using specific antibodies. We provide basic knowledge about the cooperative regulatory network of the H-NS family proteins in KT2440 and KT2440(pCAR1) cells.

Methods

Bacterial strains, plasmids, and media

Bacterial strains and plasmids used in this study are listed in Table 1. *E. coli* BL21(DE3), used for the overexpression of histidine (His)-tagged TurA, TurB, and Pmr, was cultured in lysogeny broth (LB) [22] at 25 or 30 °C. *E. coli* DH5α, used for plasmid construction, was cultured in LB at 37 °C. *Pseudomonas* strains were cultured at 30 °C in LB filtered by Stericups, 0.22-μm pore size filters (Merck Millipore, Darmstadt, Germany). The medium was supplemented with 50 μg/mL kanamycin (Km) or 30 μg/mL chloramphenicol (Cm) where necessary. Solid medium was prepared by the addition of 1.6% (w/v) agar powder (Nacalai Tesque, Kyoto, Japan).

Determination of total cell number

To cultivate *P. putida* strains, a single colony from an overnight-incubated LB agar plate was inoculated into 5 mL of fresh LB for pre-cultivation. When pCAR1-harboring strains were used, the carbazole-degrading ability of the cells was confirmed with 0.1% carbazole-containing minimal agar [23] before pre-cultivation. After overnight pre-cultivation, the second pre-cultivation was performed in 100 mL of LB. Optical density at 600 nm (OD_{600}) was adjusted to 0.03 and cells were incubated for 4 h. Then the portion of the resultant culture was transferred into 400 mL of fresh LB, where the OD_{600} was re-adjusted to 0.03, and cultivation was started. The OD_{600} was monitored and cells were harvested every 2 h after inoculation. At the same time, aliquots (0.5 mL) of the culture were harvested to determine the total number of cells. The aliquots were mixed with an equal volume of 0.3% (w/v) crystal violet and a volume of PBS buffer (2.56 g/L Na_2H-$PO_4 \cdot 7H_2O$, 8 g/L NaCl, 0.2 g/L KCl, 0.2 g/L KH_2PO_4, pH 7.4) was added for appropriate dilution. Ten microliters of the mixture were applied to a C-Chip DHC-N01 hemocytometer (NanoEnTek, Seoul, Korea) to count the number of cells using an optical microscope (BX53; Olympus, Tokyo, Japan).

Expression and purification of TurA, TurB, and Pmr

TurA, TurB, and Pmr were overexpressed as C-terminal His-tagged forms using pET-C-His-turA, pET-C-His-turB, and pET-C-His-pmr, respectively, and purified to homogeneity as described previously [14]. These proteins were used as standards in western blot analysis.

Preparation of antibodies

Polyclonal antibodies against TurA and TurB were prepared by Sigma-Aldrich (St. Louis, MO, USA). The antibodies were produced in rabbits by injecting synthesized peptides corresponding to residues 61–77 of TurA and 12–29 of TurB. The resultant serum was used in western blot analysis. Anti-His antibody (Medical and Biological

Table 1 Bacterial strains and plasmids used in this study

Strain or plasmid	Relevant characteristics	Source or reference
Bacterial strains		
Escherichia coli		
BL21(DE3)	F⁻ *ompT hsdS*(r$_B^-$ m$_B^-$) *gal dcm* λ (DE3)	Novagen
DH5α	F⁻ φ80d*lacZ*ΔM15 Δ(*lacZYA-argF*)U169 *endA1 recA1 hsdR17*(r$_K^-$m$_K^+$) *deoR thi-1 supE44 gyrA96 relA1* λ⁻ *phoA*	Toyobo
Pseudomonas putida		
KT2440	Naturally Cm^r	[34]
KT2440(pCAR1)	KT2440 harboring pCAR1	[17]
KT2440(pCAR1pmrHis)	KT2440(pCAR1) carrying gene encoding His-tagged Pmr under original promotor	[13]
Plasmids		
pET-C-His-turA	pET-26b(+) with NdeI-XhoI fragment containing *turA*	[14]
pET-C-His-turB	pET-26b(+) with NdeI-SalI fragment containing *turB* ligated into the NdeI-XhoI site	[14]
pET-C-His-pmr	pET-26b(+) with NdeI-XhoI fragment containing *pmr*	[13]
pT7Blue T-vector	Ap^r, *lacZα*, T7 promoter, f1 origin, pUC/M13 priming sites	Novagen
pTuniv16S	pT7Blue T-vector with PCR fragment amplified from total DNA of *Pseudomonas resinovorans* CA10 with the primer set, univ16S-F and univ16S-R.	[19]
pTturA	pT7Blue T-vector with PCR fragment amplified from total DNA of KT2440 with the primer set, PP_1366-F and PP_1366-R.	This study
pTturB	pT7Blue T-vector with PCR fragment amplified from total DNA of KT2440 with the primer set, PP_3765-F and PP_3765-R.	This study
pTpmr2	pT7Blue T-vector with PCR fragment amplified from pET-C-His-pmr with the primer set, pmr-F-2 and pmr-R-2.	This study

Laboratories, Nagoya, Japan) was used for detection of C-terminal-His-tagged Pmr in KT2440(pCAR1pmrHis).

Preparation of whole cell protein for quantification of TurA, TurB, and His-tagged Pmr

Preparation of whole cell protein was performed according to the methods of Azam et al. and Ohniwa et al. [21, 24] with modifications. KT2440, KT2440(pCAR1), or KT2440(pCAR1pmrHis) cells (10^9–10^{10}) were harvested by centrifugation at various time intervals and washed with PBS buffer. Cell pellets were suspended in 250 μL of Buffer A containing 40 mM Tris-HCl (pH 8.1) supplemented with 25% sucrose, followed by the addition of 50 μL of Buffer B (40 mM Tris-HCl [pH 8.1], 10 mM EDTA [pH 8.1], and 3 mg/mL lysozyme) and incubated for 20 min on ice. Lysis was then performed by addition of 100 μL of Buffer C (10 mM Tris-HCl [pH 8.2], 10 mM EDTA [pH 8.0], 1% Brij 58, and 0.4% sodium deoxycholate) in the presence of 0.75% detergent NP-40 and 0.5 mM phenylmethylsulfonyl fluoride for 20 min on ice. The mixtures were supplemented with 50 μL of 2 M KCl, 2.5 μL of 2 M MgCl$_2$, 10 μg of RNase, and 12.5 μg of DNaseI and incubated for 10 min at 37 °C. Finally, sonication was performed for 10 min (10 s on/30 s off; output power, high-level) using a Bioruptor II (Cosmo Bio, Tokyo, Japan). Protein concentration was determined using Protein Assay Dye Reagent Concentrate (Bio-Rad Laboratories, Hercules, CA, USA).

Western blot analysis

To confirm the specificity of anti-TurA and anti-TurB antibodies, purified His-tagged TurA, TurB, and Pmr were electrophoresed on a 15% tricine-SDS-PAGE gel and then transferred onto a Sequi-Blot polyvinylidene difluoride (PVDF) membrane (Bio-Rad Laboratories) at 4 °C, overnight. The blot was blocked at room temperature with 5% enhanced chemiluminescence (ECL) blocking agent (GE Healthcare, Little Chalfont, UK) in TBST buffer (10 mM Tris-HCl [pH 8.0], 0.146 M NaCl, 0.05% Tween-20) for 1 h, washed with TBST (three times for 5 min each), and then incubated with 5000-fold diluted anti-TurA or 1000-fold diluted anti-TurB antibodies in TBST supplemented with 5% ECL blocking agent for 1 h at room temperature. After an additional washing step with TBST (three times for 15 min each), the blot was incubated with a horseradish peroxidase-conjugated goat anti-rabbit secondary antibody (Sigma-Aldrich; 10,000-fold diluted in TBST containing 5% ECL blocking agent) for 1 h at room temperature. After an additional washing step with TBST (three times for 15 min each), the blot was probed with 6 mL of Immobilon Western Chemiluminescent HRP Substrate (Merck Millipore) for 5 min and then detected using LAS1000 (Fujifilm, Tokyo, Japan) image analyzer.

To quantify TurA and TurB, whole cell lysates containing 80 μg of protein from KT2440 and KT2440(pCAR1) were separated by 13% glycine-SDS-PAGE. To quantify His-tagged Pmr, whole cell lysate containing 40 μg of protein from KT2440(pCAR1pmr-His) was separated by 13% glycine-SDS-PAGE. Different amounts of purified His-tagged TurA, TurB, or Pmr, for which a linear relationship between protein concentration and signal intensity had been confirmed, were used as standards in each assay. PVDF membrane transfer and development of the blots were performed essentially as described above, except that the dilution ratio of anti-TurA, anti-TurB, and anti-His antibodies were 3000-fold, 2000-fold, and 5000-fold, respectively. The images were obtained using LAS1000 or LAS500 (GE Healthcare) image analyzers. The intensity of the protein bands was determined using ImageJ software (http://rsbweb.nih.gov/ij/). RNA polymerase α subunit was employed as a control for sample loading. Anti-RNA polymerase α subunit antibody (NeoClone, Madison, WI, USA; 2000-fold dilution in TBST with 5% ECL blocking agent) and ECL peroxidase-labeled anti-mouse antibody (GE Healthcare; 10,000-fold dilution in TBST with 5% ECL blocking agent) were used as primary and secondary antibodies, respectively, with 5 μg of whole cell protein samples.

RNA extraction and cDNA synthesis

RNA extraction was performed according to a previous report [19]. First-strand cDNA synthesis was conducted with 1.25 μg of purified total RNA, 250 ng of random primers (Invitrogen, Carlsbad, CA, USA), 1 × First Strand Buffer (Invitrogen), 40 U of RNase OUT (Invitrogen), 5 mM dithiothreitol (Invitrogen), and 200 U of SuperScript III Reverse Transcriptase (Invitrogen). After the RNA and random primers had been denatured at 65 °C for 5 min and annealed at 4 °C for 2 min, the remaining reagents were added, and the mixture was incubated at 25 °C for 5 min, 50 °C for 60 min, and 70 °C for 15 min. To degrade the RNA template, 6.67 μL of 1 N NaOH were added and the reaction mixture was incubated at 65 °C for 30 min, then the mixture was neutralized with 6.67 μL of 1 N HCl.

Quantitative reverse transcription (qRT)-PCR

qRT-PCR was performed using Power SYBR green PCR master mix (Applied Biosystems, Foster City, CA, USA) and the ABI 7300 real-Time PCR System (Applied Biosystems) according to a previous method [19] with some modifications. The primers used for qRT-PCR are shown in Table 2. The reaction conditions were as follows: 95 °C for 10 min for enzyme activation and 40 cycles of 95 °C for 5 s, 60 °C for 10 s, and 72 °C for 35 s. A melting-curve analysis was performed to verify the amplification specificity. For a standard curve, a series of 10-fold dilutions of the target PCR product ligated into the pT7Blue T-vector (Novagen, Madison, WI, USA) were used. 16S rRNA was used as an internal standard. All of the reactions were carried out at least in triplicate, and the data were normalized using the average of the internal standard.

Results and discussion

Design of anti-TurA and anti-TurB antibodies and confirmation of their specificity

Considering that the three MvaT homologs have high amino acid sequence identity (50–60%), peptides that show relatively low identity should be used as antigens to avoid cross-reaction. The highlighted sequences of TurA and TurB shown in Fig. 1 (indicated by the green boxes) were used for the preparation of anti-TurA and anti-TurB antibodies. Specificity of these antibodies was confirmed with purified His-tagged TurA, TurB, and Pmr using western blotting (Fig. 2).

We also produced anti-Pmr antibodies using the sequences indicated by the pink boxes in Fig. 1, but failed to detect Pmr in KT2440(pCAR1) cell lysates (data not shown). Thus, we used the KT2440(pCAR1pmrHis) [13] strain, where His-tagged Pmr is expressed under its original promotor, with specific anti-His antibody to quantify the intracellular amount of Pmr. Note that there was little possibility of the C-terminal His-tags affecting the DNA-binding ability, sequence preference, or three-dimensional structure of Pmr. The structure of MvaT, an H-NS family protein of *Pseudomonas aeruginosa* PAO1, determined by NMR clearly suggested that the C-terminus is not involved in DNA-binding [25]. In addition, binding sites of His-tagged Pmr were successfully determined by modified chromatin precipitation

Table 2 Primers used for qRT-PCR

Name	Sequences (5'→3')	References
univ16S-F	ACACGGTCCAGACTCCTACG	17
univ16S-R	TACTGCCCTTCCTCCCAACT	17
PP_1366-F	AACTGGAGTTCGAAGGCAAA	13
PP_1366-R	GAGGTGCCTTGCTCAGTTTC	13
PP_3765-F	ATATCATCGCCATCCTCGAC	13
PP_3765-R	TGCGGGTTCTGATAGACCTT	13
pmr-F-2	TCGCGATTCTTGATCCGGAC	This study
pmr-R-2	CCTTGGTCTCAACGAGCTCA	This study

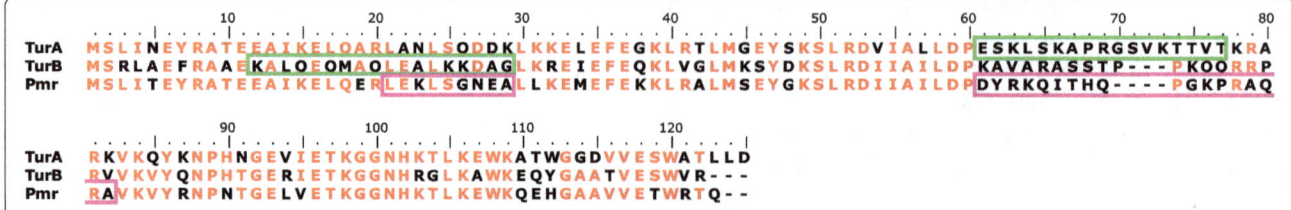

Fig. 1 Amino acid sequence alignment of TurA, TurB, and Pmr. The alignment was performed using the ClustalW program (version 2.1, http://clustalw.ddbj.nig.ac.jp/). Identical amino acids between at least two proteins are shown in red. Residues 61–77 of TurA and residues 12–29 of TurB (highlighted in green boxes) were used as antigens to produce the anti-TurA and anti-TurB antibodies used in this study. Residues 21–29 and 61–78 of Pmr (highlighted in the pink boxes) were also used as antigens to produce anti-Pmr antibody, but failed to detect Pmr in KT2440 (pCAR1) cell lysates (data not shown)

assay coupled with microarray technology using the same strain [13, 16]. Therefore, the His-tag introduced at the C-terminus is expected to have a negligible effect on the function of Pmr.

Intracellular amounts of TurA, TurB, and his-tagged Pmr
To quantify the intracellular amount of the three H-NS family proteins using western blotting, whole cell protein extraction was performed with KT2440, KT2440(pCAR1), and KT2440(pCAR1pmrHis) cells cultured in LB. Three independent biological replicates were prepared to quantify total amounts of DNA-bound and free-state TurA, TurB, and His-tagged Pmr. The representative western blot raw data are provided as Additional file 1. A loading control using anti-RNA polymerase α subunit antibody was included to ensure that equal amounts of protein were loaded in each lane. As the growth of the cultures was monitored by OD_{600} measurement and cell counting (Figs. 3a and 4a), the intracellular amounts of the proteins of interests could be calculated per cell. RNA extraction was conducted with cells from the same culture at the same sampling time to investigate the relationship between transcription and translation.

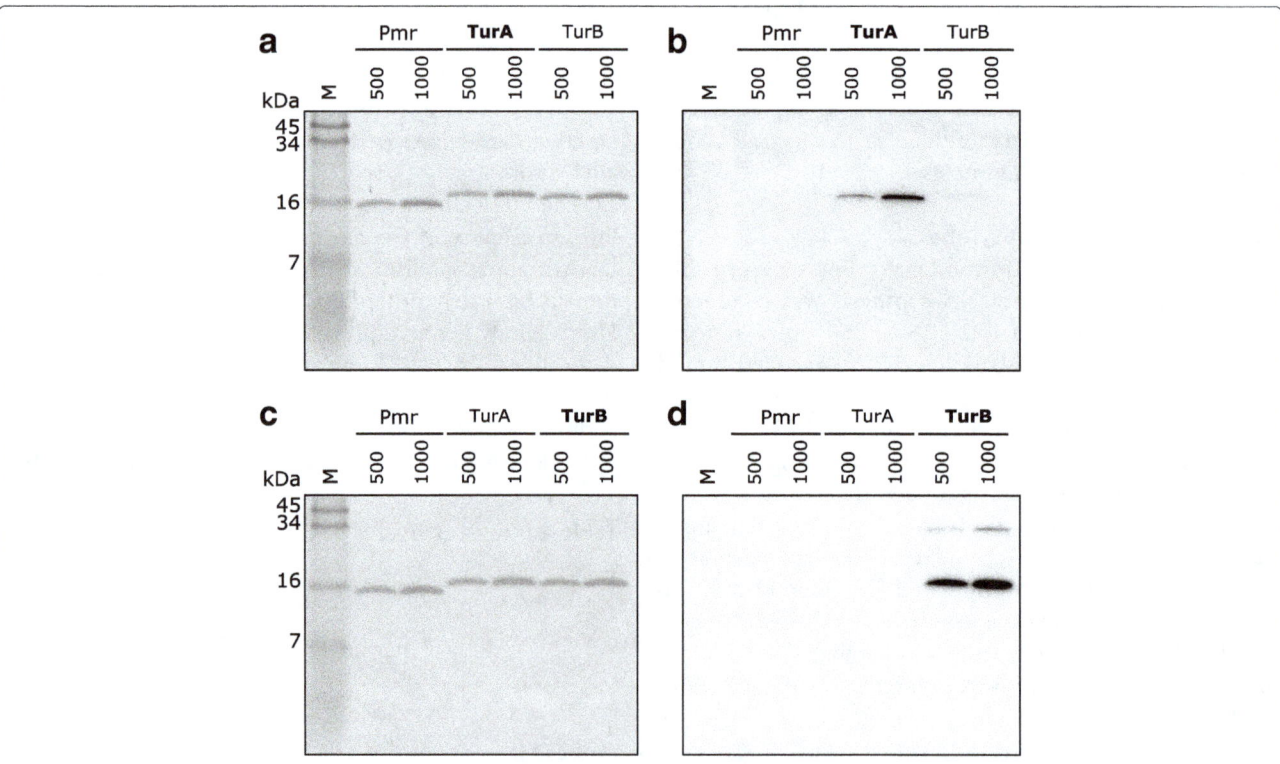

Fig. 2 Confirmation of anti-TurA and anti-TurB antibody specificity with purified His-tagged TurA, TurB, and Pmr. Panels **a** and **c** show the results of tricine-SDS-PAGE analyses of 500 or 1000 ng of purified TurA, TurB, and Pmr, where M represents the protein mass marker. Panels **b** and **d** show the results of western blotting using anti-TurA (**b**) and anti-TurB antibodies (**d**). Note that the gels shown in panels **a** and **c** were used for western blotting in panels **b** and **d**, respectively

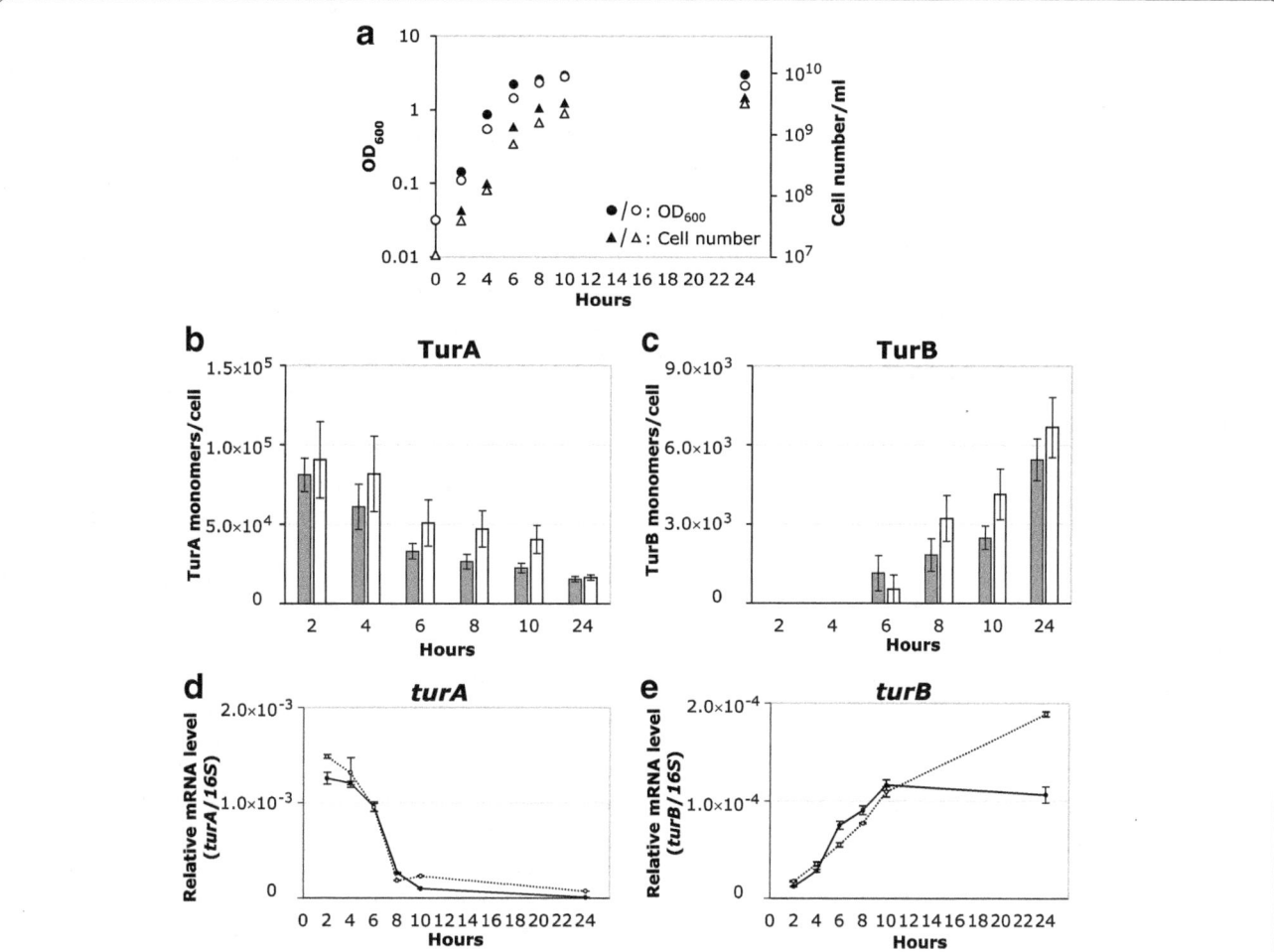

Fig. 3 Quantification of TurA and TurB monomers per cell in KT2440 and KT2440(pCAR1) cells. Panel **a** shows the growth curves of KT2440 and KT2440(pCAR1) in LB. The OD_{600} of the culture was measured every 2 h (circles) and cell number per milliliter was counted at the same time (triangles). Filled symbols represent the results of KT2440 and open symbols represent those of KT2440(pCAR1). In panels **b** and **c**, the numbers of TurA (**b**) and TurB (**c**) monomers per cell are shown. Values and error bars correspond to averages and standard deviations of results from at least three independent biological replicates. Gray bars represent the results of KT2440 and white bars represent those of KT2440(pCAR1). Note that the amount of TurB was below the detection limit at 2 h and 4 h in both KT2440 and KT2440(pCAR1). In panels **d** and **e**, corresponding mRNA levels of *turA* (**d**) and *turB* (**e**) genes are shown. Values and error bars correspond to averages and standard deviations of results from at least three independent technical replicates. Filled symbols and lines represent the results of KT2440 and open symbols and dot lines represent those of KT2440(pCAR1)

TurA amount declined from ~80,000 monomers per cell at the early log phase to ~20,000 monomers per cell at the late stationary phase in both KT2440 and KT2440(pCAR1) (Fig. 3b). A relatively rapid decrease was observed in the early stationary phase: The number of TurA monomers was reduced almost by half during the first 6 h of cultivation. This tendency was similar to the transcriptional profile of *turA* (Fig. 3d). In contrast, TurB was below the detection limit during the log phase in both KT2440 and KT2440(pCAR1), but was detected upon entry into the stationary phase (Fig. 3c). The intracellular number of TurB monomers increased from ~1000 (6 h) to ~6000 (24 h) per cell. Transcriptional levels of the *turB* gene similarly increased

toward the stationary growth phase (Fig. 3e). Note that the transcriptional profiles of *turA* and *turB* in KT2440 and KT2440(pCAR1) were consistent with previous reports [13, 26]. TurA is present mainly in cells during log phase, whereas both TurA and TurB are present in cells during the stationary phase. pCAR1 carriage did not lead to obvious differences in the transcriptional and translational levels of TurA and TurB in KT2440.

Similarly, we determined cell growth and intracellular amounts of His-tagged Pmr in KT2440(pCAR1pmrHis) (Fig. 4). As was shown in Fig. 4b, the His-tagged Pmr amount was relatively constant during growth at around 30,000 monomers per cell, with some fluctuations. These results differ from those in our previous study, in which

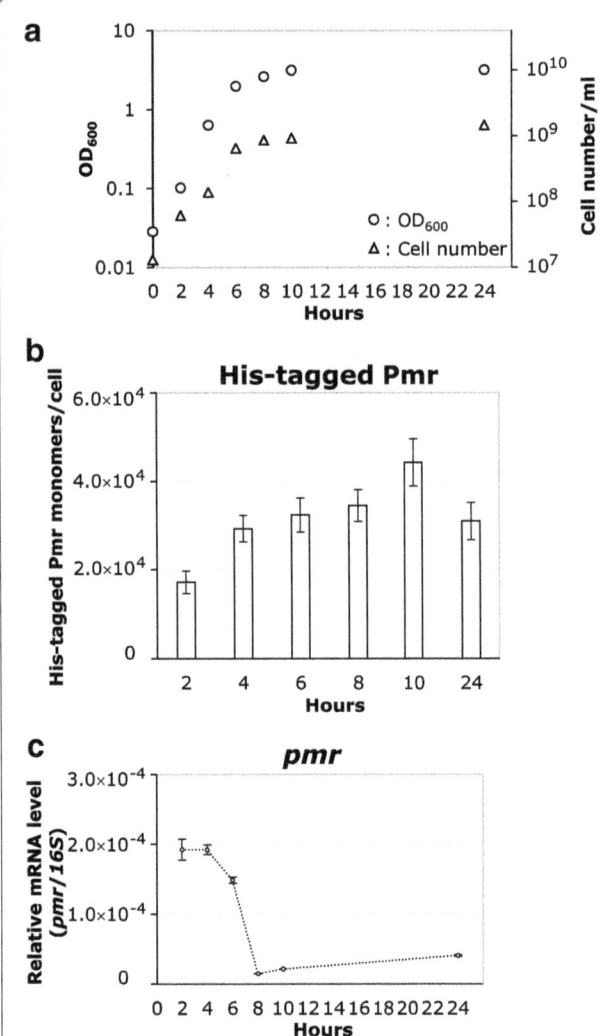

Fig. 4 Quantification of His-tagged Pmr monomers per cell in KT2440(pCAR1pmrHis) cells. Panel **a** shows the growth curve of KT2440(pCAR1pmrHis) in LB. The OD$_{600}$ (circles) and cell number per milliliter (triangles) of the culture were measured every 2 h. Panel **b** shows the number of His-tagged Pmr monomers per cell. Values and error bars correspond to averages and standard deviations of results from at least three independent biological replicates. Panel **c** shows the corresponding mRNA levels of the *pmr* gene. Values and error bars correspond to averages and standard deviations of results from at least three independent technical replicates

KT2440(pCAR1pmrHis) cells until the late stationary phase. This tendency differs from the chromosomally-encoded TurA protein (Fig. 3). In the case of Sfh, an H-NS family protein encoded on the IncHI1 self-transmissible plasmid pSf-R27 originally found in *Shigella flexneri* serotype 2a strain 2457T, the *sfh* transcription level decreased to nearly 60% in the early stationary phase while the Sfh protein increased almost 2.5-fold as the culture entered the stationary phase [20]. The results in this study suggest the possibility that the protein levels of Pmr and Sfh are subject not only to transcriptional but also post-transcriptional control, such as degradation. To clarify the detailed mechanism maintaining the appropriate amount of Pmr, we need to clarify the mechanisms underlying this regulation.

Our results clearly indicate that TurB increases in the H-NS family protein pool in cells as TurA decreases, suggesting a functional relationship between these two proteins. Actually, our previous transcriptome analyses using gene disruptants implied that TurA and TurB play rather complementary roles as global transcriptional regulators in response to plasmid carriage, because the similarity of the TurA and TurB regulons was comparatively higher [16]. This relationship is similar to that of MvaT and MvaU in *P. aeruginosa*, where MvaU binds almost the same regions as MvaT on the genome and can partially complement the repressing function of MvaT in an *mvaT*-deficient strain [27, 28]. Considering that MvaT and MvaU affect each others amounts [27], the amount of TurA and TurB can be changed in *turB* and *turA* disruptants, respectively. On the other hand, in the pCAR1-harboring KT2440 cells, the relatively high expression of Pmr had no effect on the amounts of TurA and TurB, implying the possibility that Pmr works relatively independently. This is consistent with our previous observation that the Pmr regulon differs significantly from those of TurA and TurB [16]. The function of Pmr appears to differ from that of Sfh, which can replace the function of chromosomally encoded H-NS and acts as a molecular backup [29, 30]. It is possible that the expression level of Pmr is not affected by the absence of TurA or TurB in contrast to Sfh whose protein amounts increased when H-NS or its chromosomally encoded homolog StpA was absent [31].

In addition to Pmr, pCAR1 contains genes encoding two other highly transcribed NAPs: Phu, which is an HU subunit homolog, and Pnd, which is an NdpA homolog [32]. When *pmr* and *phu* or *pmr* and *pnd* were simultaneously disrupted, stability and transfer frequency of pCAR1 were significantly decreased, suggesting that Pmr acts synergistically with Phu and Pnd [32]. Quantification of expression levels of Phu and Pnd, as well as their functional characterizations, is necessary to clarify the whole landscape of cooperative functions of NAPs encoded on pCAR1.

the band intensity of His-tagged Pmr increased from 2 to 10 h after inoculation [13]. Because the expression levels of His-tagged Pmr were monitored in western blots of soluble proteins in our previous study, we consider the current results to be more accurate. The transcriptional profile of *pmr* detected in this study revealed a rapid decrease in *pmr* transcription at around the transition state to stationary phase (Fig. 4c), a similar pattern to that described by Yun et al. [13]. Notably, while the transcription of *pmr* decreased rapidly 8 h after inoculation, Pmr protein levels were stable in

Conclusions

Our results revealed that the expression levels of TurA and His-tagged Pmr were one order of magnitude higher than that of TurB when pCAR1-free or -harboring KT2440 cells were cultured under the same conditions. Furthermore, TurA and His-tagged Pmr were highly expressed throughout the whole growth cycle with different translational profiles, while TurB was only detectable in the early stationary phase.

Information on protein expression levels provides a basis to explore the detailed manners of the functions of H-NS family proteins, i.e., how they form homo- and hetero-oligomers on DNA. Recently, preference of target DNA sequences of MvaT in *P. aeruginosa* [25] and the structure of dimerization/oligomerization domain of TurB [33] were reported. We are now performing kinetic studies on the DNA-protein and protein-protein binding properties of TurA, TurB, and Pmr. Together with the current results, these results will clarify the molecular basis of how these proteins form oligomers on the *Pseudomonas* genome.

Additional file

Additional file 1: Representative western blotting results of the three H-NS family proteins. The amounts of TurA, TurB, and His-tagged Pmr were monitored at various points in the growth curve of KT2440 (A), KT2440(pCAR1) (B), and KT2440(pCAR1pmrHis) (C). Results of western blotting analyses using anti-TurA (left panels in A and B), anti-TurB (right panels in A and B), and anti-His antibodies (C) are shown in the black boxes. The loading amounts of whole cell protein were 80 μg for KT2440 and KT2440(pCAR1) and 40 μg for KT2440(pCAR1pmrHis) per lane. Specific amounts of purified His-tagged proteins (shown in red) were used to quantify the number of protein molecules. For each strain, 5 μg of whole cell protein were loaded and detected by anti-RNA polymerase α subunit (RNAP) antibody (green boxes) as a loading control.

Abbreviations

Cm: Chloramphenicol; ECL: Enhanced chemiluminescence; His: Histidine; H-NS: Histone-like protein H1; Inc: Incompatibility; Km: Kanamycin; LB: Lysogeny broth; NAPs: Nucleoid-associated proteins; OD_{600}: Optical density at 600 nm; PVDF: Polyvinylidene difluoride; qRT: Quantitative reverse transcription

Acknowledgements
ZS thanks the China Scholarship Council (Grant Number: 201506990055).

Funding
This study was supported by Kato Memorial Bioscience Foundation (to CSM) and the Exploratory Research for Advanced Technology (ERATO) of Japan Science and Technology Agency (JST).

Authors' contributions

ZS, DV, CSM, and HN designed the experiments and ZS performed them. KO, FL, YI, and HN supervised the study. ZS, DV, CSM, and HN wrote the manuscript. All authors read and approved the final version of the manuscript.

Competing interests
The authors declare that they have no competing interest.

References

1. Dame RT. The role of nucleoid-associated proteins in the organization and compaction of bacterial chromatin. Mol Microbiol. 2005;56:858–70.
2. Dorman CJ. Nucleoid-associated proteins and bacterial physiology. Adv Appl Microbiol. 2009;67:47–64.
3. Dillon SC, Dorman CJ. Bacterial nucleoid-associated proteins, nucleoid structure and gene expression. Nat Rev Microbiol. 2010;8:185–95.
4. Ali SS, Xia B, Liu J, Navarre WW. Silencing of foreign DNA in bacteria. Curr Opin Microbiol. 2012;15:175–81.
5. Tendeng C, Soutourina OA, Danchin A, Bertin PN. MvaT proteins in *Pseudomonas* spp.: a novel class of H-NS-like proteins. Microbiology. 2003;149:3047–50.
6. Renzi F, Rescalli E, Galli E, Bertoni G. Identification of genes regulated by the MvaT-like paralogues TurA and TurB of *Pseudomonas putida* KT2440. Environ Microbiol. 2010;12:254–63.
7. Shintani M, Suzuki-Minakuchi C, Nojiri H. Nucleoid-associated proteins encoded on plasmids: occurrence and mode of function. Plasmid. 2015;80:32–44.
8. Nojiri H, Sekiguchi H, Maeda K, Urata M, Nakai SI, Yoshida T, et al. Genetic characterization and evolutionary implications of a *car* gene cluster in the carbazole degrader *Pseudomonas* sp. strain CA10. J Bacteriol. 2001;183:3663–79.
9. Maeda K, Nojiri H, Shintani M, Yoshida T, Habe H, Omori T. Complete nucleotide sequence of carbazole/dioxin-degrading plasmid pCAR1 in *Pseudomonas resinovorans* strain CA10 indicates its mosaicity and the presence of large catabolic transposon Tn*4676*. J Mol Biol. 2003;326:21–33.
10. Shintani M, Yano H, Habe H, Omori T, Yamane H, Tsuda M, et al. Characterization of the replication, maintenance, and transfer features of the IncP-7 plasmid pCAR1, which carries genes involved in carbazole and dioxin degradation. Appl Environ Microbiol. 2006;72:3206–16.
11. Takahashi Y, Shintani M, Yamane H, Nojiri H. The complete nucleotide sequence of pCAR2: pCAR2 and pCAR1 were structurally identical IncP-7 carbazole degradative plasmids. Biosci Biotechnol Biochem. 2009;73:744–6.
12. Nojiri H. Structural and molecular genetic analyses of the bacterial carbazole degradation system. Biosci Biotechnol Biochem. 2012;76:1–18.
13. Yun CS, Suzuki C, Naito K, Takahashi Y, Sai F, Terabayashi T, et al. Pmr, a Histone-like protein H1 (H-NS) family protein encoded by the IncP-7 plasmid pCAR1, is a key global regulator that alters host function. J Bacteriol. 2010;192:4720–31.
14. Suzuki C, Yun CS, Umeda T, Terabayashi T, Watanabe K, Yamane H, et al. Oligomerization and DNA-binding capacity of Pmr, a Histone-like protein H1 (H-NS) family protein encoded on IncP-7 carbazole-degradative plasmid pCAR1. Biosci Biotechnol Biochem. 2011;75:711–7.
15. Suzuki C, Kawazuma K, Horita S, Terada T, Tanokura M, Okada K, et al. Oligomerization mechanisms of an H-NS family protein, Pmr, encoded on the plasmid pCAR1 provide a molecular basis for functions of H-NS family members. PLoS One. 2014;9:e105656.
16. Yun CS, Takahashi Y, Shintani M, Takeda T, Suzuki-Minakuchi C, Okada K, et al. MvaT family proteins encoded on IncP-7 plasmid pCAR1 and the host chromosome regulate the host transcriptome cooperatively by differently. Appl Environ Microbiol. 2016;82:832–42.
17. Miyakoshi M, Shintani M, Terabayashi T, Kai S, Yamane H, Nojiri H. Transcriptome analysis of *Pseudomonas putida* KT2440 harboring the completely sequenced IncP-7 plasmid pCAR1. J Bacteriol. 2007;189:6849–60.
18. Shintani M, Takahashi Y, Tokumaru H, Kadota K, Hara H, Miyakoshi M, et al. Response of the *Pseudomonas* host chromosomal transcriptome to carriage of the IncP-7 plasmid pCAR1. Environ Microbiol. 2010;12:1413–26.
19. Takahashi Y, Shintani M, Takase N, Kazo Y, Kawamura F, Hara H, et al. Modulation of primary cell function of host *Pseudomonas* bacteria by the conjugative plasmid pCAR1. Environ Microbiol. 2015;17:134–55.
20. Deighan P, Beloin C, Dorman CJ. Three-way interactions among the Sfh, StpA and H-NS nucleoid-structuring proteins of *Shigella flexneri* 2a strain 2457T. Mol Microbiol. 2003;48:1401–16.

21. Azam TA, Iwata A, Nishimura A, Ueda S, Ishihama A. Growth phase-dependent variation in protein composition of the *Escherichia coli* nucleoid. J Bacteriol. 1999;181:6361–70.

22. Sambrook J, Russell DW. Molecular cloning: a laboratory manual. 3rd ed. Cold Spring Harbor: Cold Spring Harbor Laboratory Press; 2001.

23. Shintani M, Habe H, Tsuda M, Omori T, Yamane H, Nojiri H. Recipient range of IncP-7 conjugative plasmid pCAR2 from *Pseudomonas putida* HS01 is broader than from other *Pseudomonas* strains. Biotechnol Lett. 2005;27:1847–53.

24. Ohniwa RL, Ushijima Y, Saito S, Morikawa K. Proteomic analyses of nucleoid-associated proteins in *Escherichia coli*, *Pseudomonas aeruginosa*, *Bacillus subtilis*, and *Staphylococcus aureus*. PLoS One. 2011;6:e19172.

25. Ding P, McFarland KA, Jin S, Tong G, Duan B, Yang A, et al. A novel AT-rich DNA recognition mechanism for bacterial xenogeneic silencer MvaT. PLoS Pathog. 2015;11:e1004967.

26. Yuste L, Hervás AB, Canosa I, Tobes R, Jimémez JI, Nogales J, et al. Growth phase-dependent expression of the *Pseudomonas putida* KT2440 transcriptional machinery analysed with a genome-wide DNA microarray. Environ Microbiol. 2006;8:165–77.

27. Vallet-Gely I, Donovan KE, Fang R, Joung JK, Dove SL. Repression of phase-variable *cup* gene expression by H-NS-like proteins in *Pseudomonas aeruginosa*. Proc Natl Acad Sci U S A. 2005;102:11082–7.

28. Castang S, McManus H, Turner KH, Dove SL. H-NS family members function coordinately in an opportunistic pathogen. Proc Natl Acad Sci U S A. 2008;105:18947–52.

29. Doyle M, Fookes M, Ivens A, Mangan MW, Wain J, Dorman CJ. An H-NS-like stealth protein aids horizontal DNA transmission in bacteria. Science. 2007;315:251–2.

30. Dillon SC, Cameron AD, Hokamp K, Lucchini S, Hinton JC, Dorman CJ. Genome-wide analysis of the H-NS and Sfh regulatory networks in *Salmonella* Typhimurium identifies a plasmid-encoded transcription silencing mechanism. Mol Microbiol. 2010;76:1250–65.

31. Doyle M, Dorman CJ. Reciprocal transcriptional and posttranscriptional growth-phase-dependent expression of *sfh*, a gene that encodes a paralogue of the nucleoid-associated protein H-NS. J Bacteriol. 2006;188:7581–91.

32. Suzuki-Minakuchi C, Hirotani R, Shintani M, Takeda T, Takahashi Y, Matsui K, et al. Effects of three different nucleoid-associated proteins encoded on the IncP-7 plasmid pCAR1 on the host *Pseudomonas putida* KT2440. Appl Environ Microbiol. 2015;81:2869–80.

33. Suzuki-Minakuchi C, Kawazuma K, Matsuzawa J, Vasileva D, Fujimoto Z, Terada T, et al. Structural similarities and differences in H-NS family proteins revealed by the N-terminal structure of TurB in *Pseudomonas putida* KT2440. FEBS Lett. 2016;590:3583–94.

34. Bagdasarian M, Lurz R, Rückert B, Franklin FC, Bagdasarian MM, Frey J, et al. Specific-purpose plasmid cloning vectors. II. Broad host range, high copy number, RSF1010-derived vectors, and a host-vector system for gene cloning in *Pseudomonas*. Gene. 1981;16:237–47.

13

Insights into the *Geobacillus stearothermophilus* species based on phylogenomic principles

S. A. Burgess[1,4*], S. H. Flint[1], D. Lindsay[2], M. P. Cox[3] and P. J. Biggs[3,4*]

Abstract

Background: The genus *Geobacillus* comprises bacteria that are Gram positive, thermophilic spore-formers, which are found in a variety of environments from hot-springs, cool soils, to food manufacturing plants, including dairy manufacturing plants. Despite considerable interest in the use of *Geobacillus* spp. for biotechnological applications, the taxonomy of this genus is unclear, in part because of differences in DNA-DNA hybridization (DDH) similarity values between studies. In addition, it is also difficult to use phenotypic characteristics to define a bacterial species. For example, *G. stearothermophilus* was traditionally defined as a species that does not utilise lactose, but the ability of dairy strains of *G. stearothermophilus* to use lactose has now been well established.

Results: This study compared the genome sequences of 63 *Geobacillus* isolates and showed that based on two different genomic approaches (core genome comparisons and average nucleotide identity) the *Geobacillus* genus could be divided into sixteen taxa for those *Geobacillus* strains that have genome sequences available thus far. In addition, using *Geobacillus stearothermophilus* as an example, we show that inclusion of the accessory genome, as well as phenotypic characteristics, is not suitable for defining this species. For example, this is the first study to provide evidence of dairy adaptation in *G. stearothermophilus* - a phenotypic feature not typically considered standard in this species - by identifying the presence of a putative *lac* operon in four dairy strains.

Conclusions: The traditional polyphasic approach of combining both genotypic and phenotypic characteristics to define a bacterial species could not be used for *G. stearothermophilus* where many phenotypic characteristics vary within this taxon. Further evidence of this discordant use of phenotypic traits was provided by analysis of the accessory genome, where the dairy strains contained a putative *lac* operon. Based on the findings from this study, we recommend that novel bacterial species should be defined using a core genome approach.

Keywords: *Geobacillus*, Thermophile, Dairy, Comparative genomics

Background

The *Geobacillus* genus contains Gram-positive, rod-shaped, spore-forming bacteria that have an optimum growth temperature of 55–65 °C [1]. Members of the *Geobacillus* genus were originally classified in Group 5 of the *Bacillus* genus [2]. In 2001, based on a combination of 16S ribosomal RNA (rRNA) sequence analysis, fatty acid composition and DNA-DNA hybridization (DDH), some members of Group 5 were reclassified into the new genus *Geobacillus*, with the word *Geobacillus* meaning "soil or earth small rod" [1]. Recently it was proposed that the *Geobacillus* genus be separated into two genera based on a comparative genomics analysis, which we explore further here [3]. There is extensive interest in the *Geobacillus* genus for biotechnological purposes such as for bioremediation, the production of thermostable enzymes, and biofuels [4–7]. In addition, *Geobacillus* spp. are common spoilage organisms in food manufacturing plants and products [8–14]. *Geobacillus* spp. have been isolated from temperate as well as hot environments including hot springs, oilfields, deep sea sediments, sugar refineries, canned foods, dehydrated vegetables and dairy

* Correspondence: s.burgess1@massey.ac.nz; p.biggs@massey.ac.nz
[1]School of Food and Nutrition, Massey University, Palmerston North, New Zealand
[3]Statistics and Bioinformatics Group, Institute of Fundamental Sciences, Massey University, Palmerston North, New Zealand
Full list of author information is available at the end of the article

factories. The species *G. stearothermophilus* was first described in 1920 and was isolated from canned cream-style corn. *G. stearothermophilus* is a common contaminant of dairy products, particularly milk powder and has also been isolated from dried soups and vegetables. Until the 1980s *G. stearothermophilus* was regarded as the only known obligate thermophile of the *Bacillus* genus [15, 16].

According to the LPSN bacterio.net [17], as of April 2017, there were sixteen *Geobacillus* species (*G. caldoxy-losilyticus, G. galactosidasius, G. icigianus, G. jurassicus, G. kaustophilus, G. lituanicus, G. stearothermophilus, G. subterraneus, G. thermantarcticus, G. thermocatenulatus, G. thermodenitrificans, G. thermoglucosidasius, G. thermoleovorans, G. toebii, G. uzenensis* and *G. vulcani*) described with validly published names [1, 18–27]. However, the classification of many of these species remains uncertain. To date over 60 *Geobacillus* genomes have been sequenced, mainly to identify genes that could be used in different biotechnological applications [3]. Of these, there are eleven species with genome sequences of the type strain (*G. caldoxylosilyticus* NBRC 10776, *G. icigianus* DSM 28325, *G. jurassicus* DSM 15726, *G. kaustophilus* NBRC 102445, *G. stearothermophilus* ATCC 12980, *G. subterraneus* DSM 13552, *G. thermoantarcticus* M1, *G. thermodenitrificans* DSM 465 *G. thermoglucosidasisus* NBRC 107763, *G. thermoleovorans* DSM 5366, and *G. toebii* DSM 14590) [3, 28, 29]. Recent studies have shown that it is possible for a comparative genomics approach to resolve the taxonomy of this important genus [3, 30]. However, the question still remains as to the most appropriate genomics tool for the classification of new species.

Despite the advances of the post-genomics age, there is still no consensus as to what characterizes a bacterial species [31, 32]. However, in describing a new bacterial species, the two methods on which the most emphasis has been placed are 16S rRNA gene sequence analysis and DDH, alongside various phenotypic methods [33]. However, in some cases, including the *Geobacillus* genus, the sequence similarity of the 16S rRNA is >97% between species despite being distinct when the overall genome DNA similarity is analyzed using DDH [34–38]. Therefore the identification of new *Geobacillus* species generally relies on other approaches, such as DDH.

In general, DDH is also fraught with challenges as a method for the differentiation of bacterial species because it is laborious and there is a lack of reproducibility, reciprocation, and calibration of the method with a reference strain of a known DDH value [39–41]. In the case of new *Geobacillus* species, DDH values between studies show large variations [21, 27], which has led to the reclassification of some species of *Geobacillus*. Dinsdale et al. [21] showed that some of the previously published species were in fact synonymous with current

species and should no longer be considered valid. For example, the described species *G. kaustophilus* [1, 38, 42], *G. lituanicus* [23] and *G. vulcani* [43] were shown to be synonymous with *G. thermoleovorans*. In addition, the described species *G. gargensis* [27] was synonymous with *G. thermocatenulatus*. Most of the disagreement in assigning new species to the *Geobacillus* genus comes from the DDH values used to distinguish strains being very different between studies. More recently it was proposed that the strains of *G. kaustophilus* and *G. thermoleovorans* should both be designated to the *G. thermoleovorans* species [3].

Other housekeeping genes, such as *recN, recA, rpoB, gyrB, parE* and *spo0A*, have been evaluated as alternatives to the 16S rRNA gene for identifying *Geobacillus* species, all with limited success [37, 44–46]. Of the genes analyzed, *recN* appears to be the most reliable, with a higher taxonomic resolution compared with 16S rDNA [46]. However, the taxonomic resolution between some species of *Geobacillus* is still poor (for example, between *G. subterraneus* and *G. uzenensis*). This is not surprising given that house-keeping genes are well conserved between closely related species, and relying on one or a few genes does not depict the real diversity of the entire genome.

In the era of next generation sequencing it is likely that DDH will become outdated. This is already apparent with the proposal to use comparative genomics approaches to demarcate new species with genomic DNA as the type material archived alongside live cultures [47, 48]. There are a number of different ways in which whole genome sequence data can be used in taxonomy; for example, average nucleotide identity (ANI), tetranucleotide frequency, core genome analysis, pan genome analysis, and multilocus sequence typing (MLST) [49]. There appear to be two schools of thought on how a genomics based method should be incorporated into prokaryotic taxonomy. Firstly, there is a traditional polyphasic approach that incorporates both genomic as well as phenotypic characteristics [50]. In this case, the most likely substitute for DDH is ANI [33, 47]. It has been shown that an ANI value of <95–96% generally corresponds well with the thresholds of <70% for DDH and <97–98% for 16S rRNA gene identity for defining new species [40, 51]. Secondly, there is a reliance on a genomic approach only, simply using a core genome analysis or a combination of core genome and ANI [52, 53].

Until recently, none of the broader taxonomic studies on the *Geobacillus* genus have included *G. stearothermophilus* strains of dairy origin as part of their comparison. Traditionally both a genotypic and phenotypic analysis is carried out to identify a new species. However, the relationship between phenotype and genotype is not always straightforward. This is particularly well exemplified with dairy strains of *G. stearothermophilus*, which show unique physiological characteristics such as their metabolism (e.g. the ability to utilize lactose), and the fatty acid profile from the type strain *G. stearothermophilus* ATCC 12980 [54]. Differences in phenotypic traits may therefore result from niche adaptation, possibly mediated by differential gene expression, without major changes to the genome as a whole.

The aims of this study were two-fold: firstly, to establish whether the species boundaries of the *G. stearothermophilus* taxon could exclusively be determined by whole-genome sequence analyses, and secondly to determine whether the genomes of the dairy strains of *G. stearothermophilus* provide evidence of niche adaptation in ways that deviate from the standard phenotypic spectrum of the species. To pursue these goals, we compared the genome sequences of 63 *Geobacillus* strains, including twelve *G. stearothermophilus* strains, of which four were isolated from a dairy manufacturing environment.

Results

To gain an understanding of how the *G. stearothermophilus* strains isolated from dairy manufacture are related to other *Geobacillus* species, two different phylogenomic approaches were taken: ANI and a comparison of the core genomes. In addition, these methods were evaluated for their ability to replace the traditional methods of DDH, 16S rRNA sequence analysis and phenotypic characteristics to define a bacterial taxon, using *G. stearothermophilus* as an exemplar. The genomes of 63 *Geobacillus* strains (including ten type strains) were compared, of which eight strains were originally isolated from a dairy manufacturing environment or food product. Four of the eight strains were *G. stearothermophilus*, three of which were isolated from a New Zealand milk powder manufacturing and the fourth was isolated in the Netherlands from buttermilk powder [55, 56]. Within the *Geobacillus* genus the *G. stearothermophilus* type strain ATCC1290 had the smallest genome (2.63 Mb) compared with the dairy strain *G. caldoxylosilyticus* B4119 which has the largest genome size (3.95 Mb) within the *Geobacillus* genus (Additional file 1: Table S1). The genome sizes of the *G. stearothermophilus* dairy strains ranged from 2.77 to 3.02 Mb.

Phylogenetic relationships within the *Geobacillus* genus based on core genome comparisons

A core genome analysis was used to determine phylogenetic relationships within the *Geobacillus* genus and to establish the species boundary of the *G. stearothermophilus* taxon. The core genome was defined using the program OrthoMCL, in which each orthologous group contained only one gene from each genome. In addition, to be included in the core genome, the length range

(between the smallest and the largest) of the amino acid sequences within each cluster was not allowed to vary by more than 20%. A phylogenetic network was then generated using the concatenated sequence of those orthologous genes (Fig. 1). Core genome comparisons separated the *Geobacillus* genus (Subset A, Table 1) into sixteen main groups and several sub-groups. Genomes of strains isolated from a dairy environment, indicated by asterisks, included strains of *G. thermoglucosidasius*, *G. caldoxylosilyticus*, *G. kaustophilus* and the focus of this study *G. stearothermophilus*.

To analyze the relationship of the *G. stearothermophilus* taxon more closely, comparison of the core genome was carried out on two smaller groups of *Geobacillus* taxa (Subset B and C, Table 1, Fig. 1b and c). There is a clear delineation between the *G. stearothermophilus* cluster and other closely related *Geobacillus* taxa (Groups 1–5, Fig. 1b). Within the *G. stearothermophilus* taxon three of the dairy strains (all from the same manufacturing plant) clustered together, showing no sequence diversity between strains A1 and P3 (Fig. 1c).

Defining taxa in *Geobacillus* on the basis of ANI calculations

As stated above, the most feasible substitute for DDH is ANI [33, 47]. To examine the use of ANI for demarcating species of the *Geobacillus* genus, ANIm frequencies were calculated for all of the sequenced genomes of the *Geobacillus* genus (Additional file 1: Table S2) and visualized using a heat-map (Fig. 2). Two ANI values were calculated for each pair of genomes with one being the subject and the other the query, and vice versa. The heat-map was non-symmetrical as a result of greater differences between the ANIm value and its reciprocal value for some pairs of genomes. When the difference between two ANIm values is greater than 0.5% around the 95% threshold it could potentially place ambiguity around the taxonomic position of a strain. However, this was not seen in this study, where the difference in two ANIm values between two members of the same taxon was always less than 0.5% (data not shown), so that there were clear demarcations between taxa (as designated by a red box in Fig. 2). The *G. stearothermophilus* strains had ANIm values >95% grouping them within the same taxon.

Phenotypic characteristics as taxonomic determinants

To date, descriptions of novel bacterial species have included unique phenotypic characteristics. However, many descriptions are based on only a small population of strains, and in some cases, only one strain. When a larger population is examined, phenotypic characteristics can often vary between strains of the same taxon [32]. This was seen within the *G. stearothermophilus* taxon

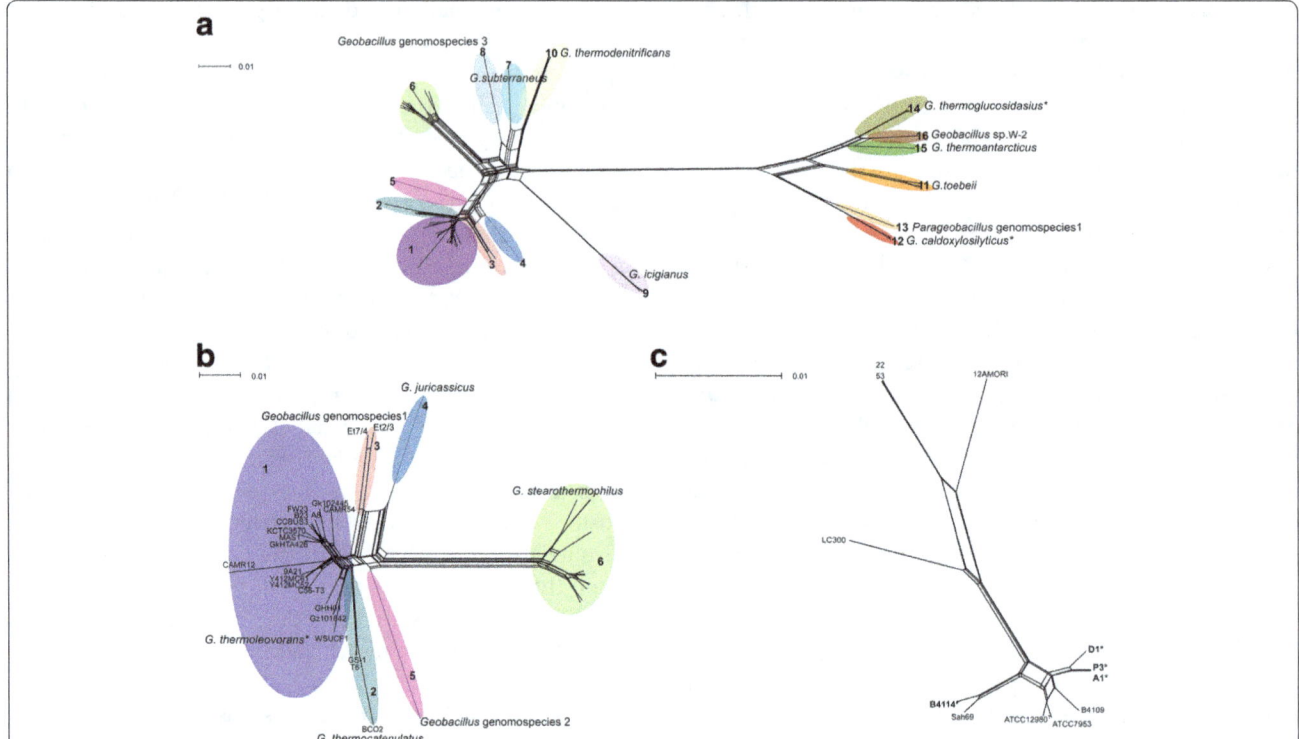

Fig. 1 Core genome sequence comparisons. The phylogenetic networks were generated using the Neighbor-Net algorithm in SplitsTree (v. 4.13.1). The orthologous groups were defined using the program OrthoMCL (v. 2.0.9) and the analyses were based on those genes that have orthologous gene members with a length range less than or equal to 20%. Those groups marked with an asterisk contain a strain(s) that originate from a dairy environment. **a**. Includes all of the *Geobacillus* genomes where *G. thermoleovorans* group 1 refers to strains B23, CCB_US3_UF5, CAMR5420, FW23, MAS1, NBRC 102445, A8, Y412MC52, Y412MC61, 9A21, C56-T3, GHH01, WSUCF1 and HTA426, KCTC 3570, and *G. zalihae* NBRC 101842; Group 2 refers to "*G. thermocatenocatulatus*" strains GS-1, T6, and BC02; Group 3 refers to *Geobacillus* genomospecies 1 strains Et7/4 and Et2/3; Group 4 refers to *G. juracassicus*; Group 5 refers to *Geobacillus* genomospecies 2 strain PSS1; *G. stearothermophilus* Group 6 includes strains ATCC 12980, ATCC 7953, A1, P3, D1, B4114, 22, 53, Sah69, 12AMORI, LC300 and B4109; *G. subterraneus* Group 7 includes strains KCTC 3922 and K; Group 8 refers to *Geobacillus* genomospecies 3 strain JF8; *G. icigianus* Group 9 refers to strains PSS2 and G1w1T; *G. toebii* Group 10 includes strains NBRC 107807, WCH70 and B4110; *G thermodenitrificans* Group 11 includes strains DSM 465, NG80–2, PA3 and G11MC16; *G. caldoxylosilyticus* Group 12 includes strains CIC9, B4119 and NBRC 10776; Group 13 refers to *Parageobacillus* genomospecies 1 strain NUB3621; *G. thermoglucosidasius* Group 14 includes strains NBRC 10776, C56YS93, TNO-09 and Y4.1MC1; *G. thermoantarcticus* Group 15 includes strain M1; Group 16 includes strain W-2. **b**. Includes Groups 1–6. **c**. Includes *G. stearothermophilus* Group 6 only

(Table 2), where the use of phenotypic characteristics was not a reliable taxonomic determinant. Several phenotypic characteristics were different between the dairy strains and that described for the *G. stearothermophilus* species, as well as differences identified between the dairy strains themselves (Table 2).

Unique accessory genes required for adaptation to a dairy environment

Recently the genomes of four dairy strains of *G. stearothermophilus* have been sequenced [54, 55]. The accessory genomes of these four strains were analyzed to determine whether the presence or absence of genes or

Table 1 Number of genes and amino acids used in the OrthoMCL clustering

Subset[a]	Members of group	Same length and same sequence		Same length and different sequence		Cluster length[b]		Total number of core genes
		Genes	Amino acids	Genes	Amino acids	Genes	Amino acids	
A	All *Geobacillus*	1	116	58	7678	390[c]	81,812	391
B	Groups 1 – 6[d]	7	865	131	19,180	478[c]	99,197	485
C	*G. stearothermophilus* group	138	19,490	992	277,724	1524[c]	446,984	1662

[a]Refer also to Fig. 1
[b]To be in an orthologous cluster, genes had to have a length range of 20% across all cluster members, and only one member per strain
[c]This number of genes was used in the Neighbor-Net analysis
[d]Refer to Fig. 1 for strains included in each group
[e]Includes strains ATCC 12980, ATCC 7953, LC300, 12AMORI, 22, 53, Sah69, A1, P3, D1, B4109, B4114

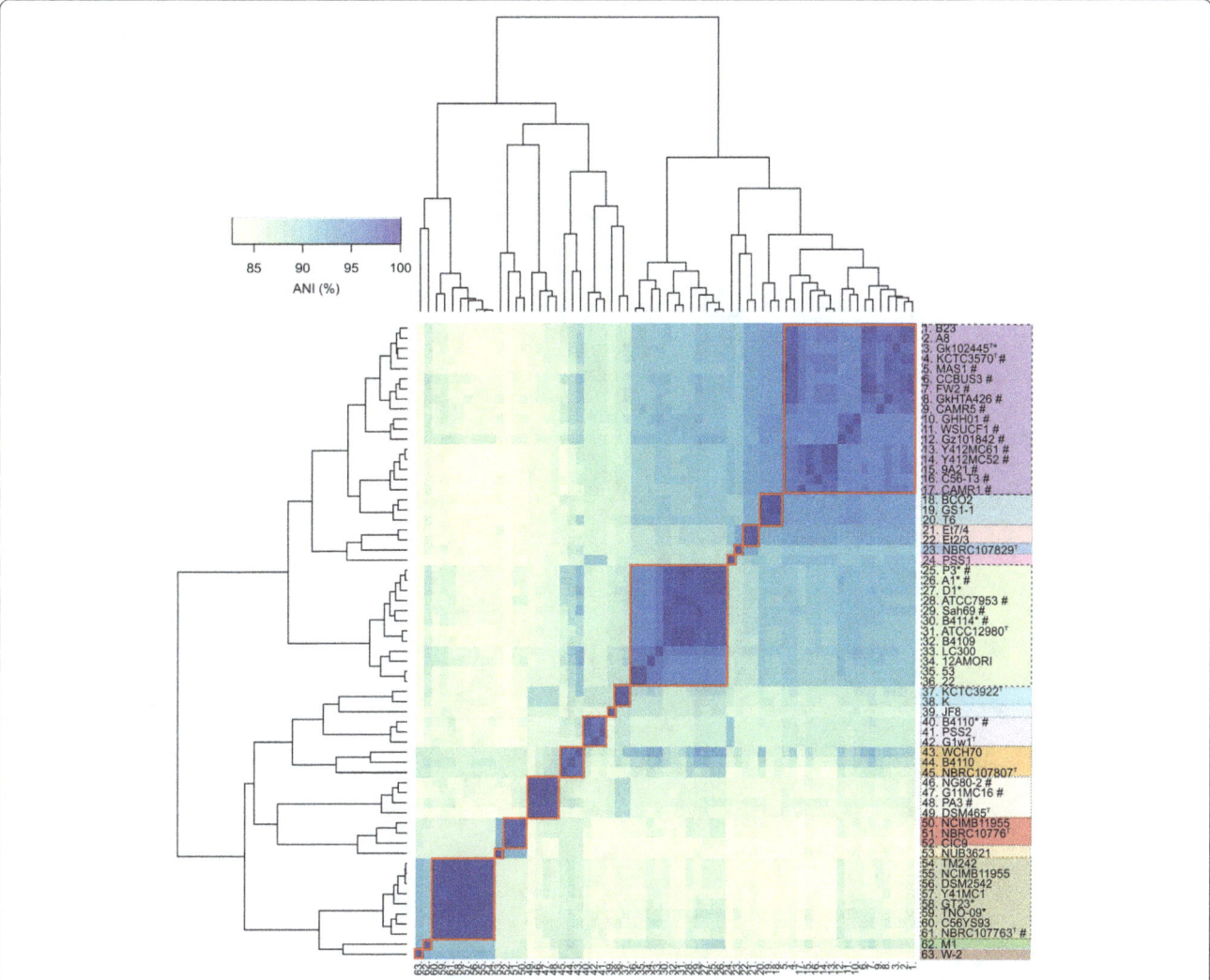

Fig. 2 Heat map comparison of the ANIm values. Those strains marked with an asterisk were isolated from a dairy environment and those strains marked with a hash were placed in a different order for the reciprocal pairwise comparison by the dendrogram option using the heatmap.2 function in R. Those ANI values greater than 95%, grouping the strains within the same species, were enclosed by a *red box*

gene clusters could account for any of the phenotypic differences observed between the dairy strains and the type strain (ATCC 1294). A putative *lac* operon was identified in the dairy strains of *G. stearothermophilus* that was not found in any of the other *Geobacillus* genomes analysed, with the exception of *G. stearothermophilus* strain Sah69 that originates from soil. For all four dairy strains and strain Sah69, the putative *lacA*, *lacB* and *lacC* genes showed highest homology (95–99% amino acid identity) with *Bacillus smithii* and the *lacE*, *lacF* and *bglC* genes showed highest homology (70–79% amino acid identity) with *Bacillus cereus*. The gene organisation of these *lac* operons were compared with the *lac* operon of *Staphylococcus aureus*, and as seen in Fig. 3, they are missing the *lacG* gene, which encodes a galactosidase, required for splitting lactose into galactose

and glucose. Instead of a galactosidase, they contained a gene encoding a glucosidase, annotated as *bglC*. However, the two enzymes LacG and BglC are closely related, and in *Lactococcus lactis*, it has been shown that a glucosidase enzyme can act as a galactosidase under certain conditions [57, 58]. The dairy strain B4114 contained an additional gene within this putative *lac* operon, which is homologous (85% amino acid identity) to the *B. smithii gatA* gene, which is predicted to encode subunit IIA of a sugar phosphotransferase system [59]. The putative *lac* operon was also unique to the *G. stearothermophilus* taxon. The other dairy strains examined (*G. kaustophilus* NBRC 102445, *G. thermoglucosidasisus* strains TNO and GT23, and *G. caldoxylosilyticus* B4119) did not contain this putative *lac* operon (data not shown).

Table 2 Phenotypic characteristics[a] of *G. stearothermophilus* strains

Characteristic	A1[b,g]	D1[b,g]	P3[b,g]	Dairy[c]	ATCC 7953[d]	ATCC 12980[e]	Species description[f]
Motility	–	–	+	v	n/d	n/d	+
Acid from:							
Adonitol	–	–	–	–	–	-*n/d	–
Amidon[g]	–	+.	–	v	+*	v*	n/d
Arabinose	w	–	w	w/–	–	–	v
Cellubiose[g]	+	+	+	+	v	-*	–
Fructose	+	+	+	+	+	+	+
Galactose[g]	w	+	+	w/+	–	v	–
Gentiobiose[h]	–	–	–	v	-*	v	–
Glucose	n/d	n/d	n/d	+	+	+	+
Glycerol	n/d	n/d	n/d	-	+*	v	+
Glycogen[g,h]	–	–	–	v	+*	+	+
Inositol	–	–	–	–	–	–	–
Inulin[g]	–	–	–	–	–	v	+
Lactose[g]	+	+	+	+	–	-*	–
Maltose	n/d	n/d	n/d	+	+	+	+
Mannitol	–	–	–	–	Var	–	Var
Mannose[h]	+	+	+	v	+*	+	+
Melezitose[h]	–	+	–	v	+*	+	+
Melibiose	+	+	+	v	+*	+	+
MethylD-glucoside[g, h]	–	–	–	v	n/d	+	+
Raffinose[, h]	n/d	n/d	n/d	v	v	+	+
Rhamnose	–	–	–	–	–	–	–
Ribose	–	–	–	–	-*	–	–
Salicin[g, h]	+	+	+	v	-*	v	–
Sorbitol	–	–	–	–	–	–	–
Sucrose	+	+	+	+	+	+	+
Trehalose[h]	+	–	+	v	+*	+	+
D-Turanose[h]	–	–	–	v	+*	+	v
Xylose	–	–	–	–	–	-*	v
Utilization of:							
Citrate	–	–	–	–	-	n/d	–
Formate				n/d			–
Lactate				n/d			–
Hydrolysis of							
Casein	–	–	–	-	v	n/d	v
Esculin	+	+	+	+	v	n/d	v
Gelatin[g]	–	–	–	–	+	n/d	+
Starch[g]	–	–	–		+	n/d	+

Table 2 Phenotypic characteristics[a] of *G. stearothermophilus* strains *(Continued)*

Nitrate reduction	+	+	+	+	+	n/d	v
Phenylalanine deamination	–	–	–	–	n/d	n/d	n/d
L-Pyroglutamic acid[h]	–	+	–	v	n/d	n/d	n/d
p-Nitrophenyl-β-D-glucoside[g]	+	–	+	v	n/d	n/d	n/d

[a]Abbreviations are as follows: v, variable; w, weak reaction; n/d, not described
[b]Data from this present study and Burgess et al. [54]
[c]Data from Flint et al. [10] and Burgess et al. [54]
[d]Data from this present study (marked as *), Baldock [95], Humbert et al. [96] and Jung et al. [97]
[e]Data from this present study (marked as *), Walker and Wolf [98], Logan and Berkeley [99] and Flint et al. [10]
[f]As described by Logan et al. [70]
[g]Phenotypic characteristic that is variable between the dairy strains
[h]Phenotypic characteristic that is different between the dairy strains and that described for the *G. stearothermophilus* species

Discussion

Traditionally the taxonomic classification of bacterial species has relied on 16S rDNA sequence analysis, DDH similarity values and phenotypic characteristics. It is challenging to classify strains to a species within the *Geobacillus* genus based solely on the 16S rRNA gene due to its high sequence similarity across the genus. It is likely that this has resulted in the mis-identification of many *Geobacillus* strains as demonstrated here and elsewhere [3, 20, 21]. Several strains analyzed in this study were previously mis-identified as *G. stearothermophilus*, for example, strain BGSC 9A21. This strain was isolated prior to the 1980s when it was believed that *G. stearothermophilus* was the only obligate thermophile of the *Bacillus* genus [60]. Although, the 16S rRNA gene sequence of this strain is approximately 98% to the type strain of *G. stearothermophilus* ATCC 12980, it is also 98–99% similar to other type strains of the *Geobacillus* genus and based on other genomic evidence is actually more closely related to *G. thermoleovorans* as demonstrated in this study. Generally isolates with <97% identity for the 16S rRNA gene are regarded as separate species [30, 33]. More recently, it was proposed that this threshold for demarcating species should be increased to 98.65% [48]. In reality, setting a threshold based on 16S rRNA gene similarity, let alone such a specific number, does not work.

The taxonomic classification of *Geobacillus* species is also uncertain, due to differences in DNA-DNA hybridization (DDH) similarity values between studies. Novel bacterial taxon descriptions also rely on phenotypic descriptions, but phenotypic characteristics may vary within a taxon. To circumvent these issues, a comparative genomics approach was taken to determine whether genome sequence data could replace the traditional methods of 16S rRNA sequence analysis, DDH, and phenotypic characteristics for defining bacterial taxa, using *G. stearothermophilus* as an exemplar.

The *Geobacillus* genus could be divided into sixteen taxa, based on both a core genome comparison and ANI, for those *Geobacillus* strains that had genome sequences available at the time of analysis. Of these, twelve appear to have validly published names (*G. caldoxylosilyticus, G. icigianus, G. juricassicus, G. stearothermophilus, G. subterraneus, G. thermoantarcticus, G. thermocatenulatus, G. thermodenitrificans, G. thermoglucosidasius, G. thermoleovorans, G. toebeii* and *G. vulcani*). Previous studies disagree on whether *G. thermocatenulatus* can be regarded as a separate species [21, 61, 62] and

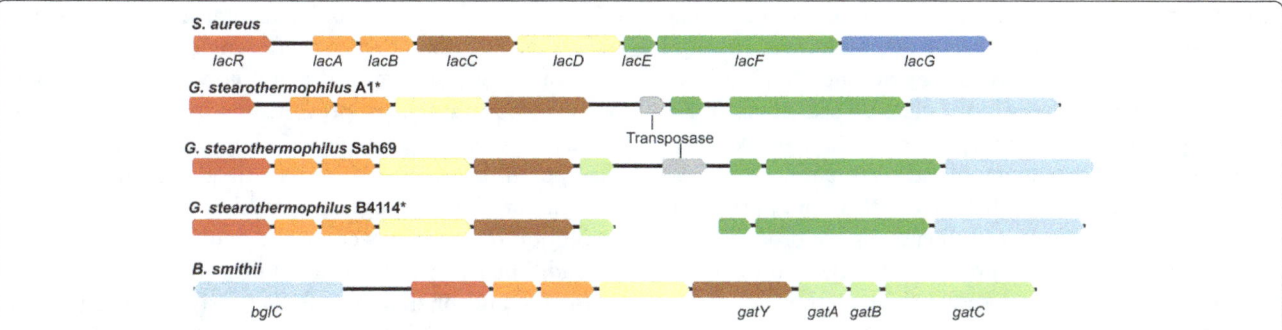

Fig. 3 Comparison of the organisation of the *lac* genes. Annotations are based on the assigned KEGG KO for each gene. Colours represent those genes belonging that to the same KO group and/or KEGG enzyme entry. The *lac* operon in *S. aureus* and the putative *lac* operons in strains A1, Sah69, B4114 as well as *B. smithii* (which showed the highest similarity to the putative *lacA, lacB* and *lacC* genes from strain A1). Those strains marked with an asterisk were isolated from a dairy environment. The gene organisation of the putative *lac* operon in strains P3 and D1 was syntenic with that of A1. The *gatABC* operon encodes a galactitol transport system and *gatY* a component of the of the GatYZ tagatose aldolase as described byVan der Heiden et al. [82]. GatY and LacD both belong to the same enzyme group (EC 4.1.2.40)

analysis of further *G. thermocatenulatus* strains as well as the type strain will be required to determine its taxonomic position. The taxonomic position of *G. zalihae* is also unclear. It bordered on the ANIm demarcation threshold from genomes in the *G. thermoleovorans* group (95.8–96.1%), although it formed a sub-group within Group 1, which may indicate that it is a subspecies of *G. thermoleovorans* rather than a separate species. In contrast, other studies describe *G. zalihae* as a genomospecies [3]. This highlights a need for clearer guidelines on how whole genome sequence analyses are interpreted to identify novel species.

In this present study, a phylogenetic network was generated for making core genome comparisons. An advantage of using a phylogenetic network, as opposed to a branching phylogenetic tree, is that it can show any ambiguous signal as to the taxonomic relationship between strains [63]. Ambiguous signal can arise from events such as gene duplication, gene transfer, different rates of mutation and recombination [64]. A comparative genomics approach used to re-examine the taxonomy of the *Geobacillus* genus demonstrated that the *Geobacillus* genus could be divided into two clades, and proposed that clade II be considered as the new genus *Parageobacillus*. This is also consistent with our results where a phylogenetic network generated using 332 core genes, showed a clear delineation between Groups 1–10 and Groups 11–16. However, distinct clades within a bacterial genus are not unusual [52, 65, 66]; separation of the *Geobacillus* genus into two genera should also be made on additional criteria, such as a discrete set of phenotypic characteristics separating the two clades. There were differences between our study and the recent analysis of Aliyu et al. [3], which compared a larger number of core genes (*n* = 1048). This is unexpected, given they examined a larger number of genomes, so the number of core genes might be expected to be lower compared with this present study. The most likely explanation is that the criteria used for defining the core genome in this present study were more stringent than that used in Aliyu et al. [3]. A core genome comparison of *Geobacillus* spp. was carried out by Studholme [30]; however, that analysis only included genome sequences in the *G. thermoleovorans*, *G. kaustophilus* and *G. thermocatenulatus* group. The groupings found were similar to those identified here using the OrthoMCL clustering, providing evidence that core genome comparisons are broadly comparable between research groups (although we note that Studholme [30] did not describe their method for determining the core genome).

The main focus of our study was on *G. stearothermophilus*. Compared with Groups 1–5 (Fig. 1.), *G. stearothermophilus* formed a discrete group, resulting in a clear delineation between *G. stearothermophilus* and the other *Geobacillus* taxa based on both core genome

sequence analysis and ANI. Core genome sequence comparisons provided genomic evidence that the dairy strains of *G. stearothermophilus* fell within the same clade as other members of the *G. stearothermophilus* taxon. Within *G. stearothermophilus*, distinct groups were defined by both the core genome and ANI analyses, perhaps indicative of subspecies.

There is no one school of thought on how genomics based methodologies should be incorporated into prokaryotic taxonomy. One approach is to find a substitution for DDH, such as ANI. The use of ANI for defining new species is not without its problems [64]. Two key issues are that the genome sequences of many type strains are not available, and there are many strains that have been incorrectly identified to a given species. In the analysis of Richter and Rossello-Mora [64], it was found that for those genomes with validly published names, only 45% actually belonged to the same species as the type strain (as defined by other means such as DDH). As of 31 July 2013, there were 10,546 validly published bacterial species names, but only 14.9% of these had genome sequences available for the type strain [47]. This issue has arisen within the *Geobacillus* genus when in defining the new species *G. icigianus*, Bryanskaya et al. [19] carried out an ANI analysis, which included the genome sequences of only two type strains. In this present study, it was also shown that some genomes with validly published names did not belong to the same species as the type strain. For example, based on a *recN* sequence analysis *G. vulcani* PSS1 did not belong to the same clade as the type strain *G. vulcani* DSM 3174. Although *G. vulcani* is a validly published name, it has previously been shown to be synonymous with *G. thermoleovorans* [21] and evidence is provided here that *G. vulcani* PSS1 is a novel species, as also supported by Aliyu et al. [3].

Another issue faced when using ANI is that it takes into account the entire genome, including accessory genes. Accessory genes are generally carried by mobile elements and acquired via horizontal gene transfer as a means of adapting to a specific environment [67]. For this reason, we believe ANI is not good measure of phylogeny. Importantly, as previously expressed by others, the use of ANI in replacing DDH appears to be a case of manipulating a new method to fit an old method [49, 68], rather than taking advantage of the much greater resolution of other aspects of the new dataset.

Traditionally, a polyphasic approach, combining both genotypic and phenotypic characteristics, is used for defining new species. In incorporating a genomics approach into prokaryotic taxonomy, it has been suggested that a polyphasic approach should still be used [50]. This could not be used for *G. stearothermophilus* because of the range of phenotypic variation observed between strains. Other phenotypic characteristics such as the fatty acid content have also been shown to differ between *G. stearothermophilus* strains [54]. Importantly, discernible phenotypic characteristics are dependent on certain genes being expressed; for example, changes in the growth conditions can change the manifestation of certain phenotypic traits. Unless strict standards are in place, it can be difficult to reproduce certain phenotypic characteristics between laboratories, such as bacterial cell components (for example, fatty acids) [50].

A description of *G. stearothermophilus* has not been republished since 1986 by Claus and Berkeley [69]; therefore, Logan et al. [70] advise that this description is likely to have encompassed a variety of thermophilic bacilli strains that would now be regarded as separate taxa. In addition, it did not take into account phenotypic differences that could occur between strains as a result of adaptation of to specific environmental niches (e.g. lactose utilization).

Further evidence of this discordant use of phenotypic traits was provided by analysis of the accessory genome, where the dairy strains contained a putative *lac* operon not found in the other genomes of *G. stearothermophilus*. The presence and absence of other gene clusters required for the utilization of different carbohydrates is not unusual in the *Geobacillus* genus. Zeigler [71] analysed ten *Geobacillus* genomes and found there was variation in the number of gene clusters predicted to be involved in plant polysaccharide degradation both within and between different taxa. This supports the notions derived in this current study that inclusion of the accessory genome is not a good measure of phylogeny because of their environmental specificity and therefore should not be used for describing new species.

It has been suggested that where there are important phenotypic differences between strains of the same species (as defined by the core genome), they should be described as "biovars" of a species, instead of using phenotypic differences as a measure of taxonomy [53]. In the same study it was found that within a population of *Rhizobium leguminosarum*, the accessory genome and the ability to utilize different carbon sources differed. The authors also use the *Bacillus cereus* group, as an example, suggesting that *Bacillus anthracis* and *Bacillus thuringenisis* be named as *Bacillus cereus* biovar *anthracis* and biovar *thuringenisis* respectively. This group of bacteria show a high degree of similarity based on their chromosomal DNA, raising the question as to whether they are separate species, as they can only be differentiated by their virulence characteristics [72]. Using the biovar concept, the dairy strains of *G. stearothermophilus* could be named *G. stearothermophilus* biovar *lactis*.

Conclusions
Two comparative genomics approaches were evaluated for their ability to define a bacterial species, in this case

G. stearothermophilus. Both genomic approaches (core genome comparisons and ANI) grouped the twelve strains of *G. stearothermophilus* together, with the core genome comparison demonstrating variation between eleven of the strains, particularly between the dairy and non-dairy strains. Comparison of the genomes was able to resolve differences between species of the *Geobacillus* genus that cannot be determined using the traditional approach of 16S rRNA gene sequence analysis. However, although ANI was able to be used for demarcating taxa, it should not be used for determining phylogenetic relationships as it takes into account the accessory genome. When strains belonging to the same species are isolated from different environments, they may contain a different set of accessory genes as a way of adapting to a specific environment. This was seen in this present study where the dairy strains contained a unique set of genes that are probably required for lactose metabolism. A polyphasic approach for defining a bacterial species by combining genomic data with a broad range of phenotypic data would therefore not work for the *G. stearothermophilus* taxon due to the range of phenotypic variation observed between strains. Based on the findings from this study, we recommend that novel bacterial species should be defined using a core genome approach. However, for any genomic approach to become routine, all of the type strains would need to be sequenced first.

Methods
Genome sequences
The genome sequences of four dairy strains of *G. stearothermophilus*: three strains (A1, P3 and D1) isolated from a New Zealand milk powder manufacturing plant and one strain (B4114) isolated from buttermilk powder, [55, 56] were compared with the genome sequences of 59 other strains of *Geobacillus* (Additional file 1: Table S1) [4–7, 28, 29, 73–88]. All of the genomes were parsed and re-annotated using Prokka v. 1.10 with default parameters [89].

Average nucleotide identity (ANI)
The ANI between two genomes has been proposed as an *in-silico* method to replace DDH [64]. This study used the default parameters in the JSpecies software package v. 1.2.1 to calculate the ANI using the program MUMmer (ANIm) between each pair of *Geobacillus* genomes. The ANIm values were used to compare the relationships between the *Geobacillus* genomes by generating a heat-map. The heat-map was generated using the heatmap.2 function included in the gplots library of the statistics software package R v. 3.2.0, visualized in Rstudio v. 0.98.1103.

Core genome comparisons

The program OrthoMCL v. 2.0.9 [90] was used to determine the core genome. Comparison of the core genome was based on predicted amino acid sequences from 'perfect sets' of orthologous gene clusters (i.e., for a given gene, there were no paralogues identified within a genome), as previously described [91]. The length range of the amino acid sequences within a cluster, used in this analysis, did not vary by more than 20% of the length of the longest gene. This value allows some variation, without being too flexible, in the length of the protein amongst all cluster members. Variation in predicted protein length may occur, for example, from the actual gene starting at a different start codon from that of the predicted annotation. The core genes were aligned individually using MUSCLE v. 3.8.31 [92] and concatenated. The Neighbor-Net algorithm [93] in SplitsTree v. 4.13.1 was used to generate a Neighbor-Net with the aligned sequences.

Phenotypic characteristics

Biochemical assays were carried out as described in Burgess et al. [54]. Motility was determined using the hanging drop method, as described by Harrigan [94], using cultures of *G. stearothermophilus* strains (A1, P3 and D1) grown in tryptic soya broth for 8 h at 55 °C.

Acknowledgements

We thank Roberto Kolter (Harvard Medical School, Boston, MA, USA) and Hera Vlamakis (Broad Institute, Cambridge, MA, USA) for their valued discussion on aspects of this project, and Haoran Wang (Massey University, Palmerston North, New Zealand) for her help with some of the laboratory work.

Funding

This study received no specific grant from any funding agency.

Author's contributions

SB designed the study, performed the analyses and drafted the manuscript. DL and SF provided project oversight, intellectual input and data interpretation. MC helped with analysis of the data and reviewed the manuscript; PB helped with the study design, prepared in-house Perl scripts, helped with the analyses and reviewed the manuscript. All authors read and approved the final manuscript.

Competing interests

The authors declare that they have no competing interests.

Author details

[1]School of Food and Nutrition, Massey University, Palmerston North, New Zealand. [2]Fonterra Research Institute, Palmerston North, New Zealand. [3]Statistics and Bioinformatics Group, Institute of Fundamental Sciences, Massey University, Palmerston North, New Zealand. [4]Infectious Disease Research Centre, Institute of Veterinary, Animal and Biomedical Sciences, Massey University, Palmerston North, New Zealand.

References

1. Nazina TN, Tourova TP, Poltaraus AB. Taxonomic study of aerobic thermophilic bacilli: descriptions of *Geobacillus subterraneus* gen. nov., sp. nov. and *Geobacillus uzenensis* sp. nov. from petroleum reservoirs and transfer of *Bacillus stearothermophilus*, *Bacillus thermocatenulatus*, *Bacillus thermoleovorans*, *Bacillus kaustophilus*, *Bacillus thermoglucosidasius* and *Bacillus thermodenitrificans* to *Geobacillus* as the new combinations *Geobacillus stearothermophilus*, *Geobacillus thermocatenulatus*, *Geobacillus thermoleovorans*, *Geobacillus kaustophilus*, *Geobacillus thermoglucosidasius* and *Geobacillus thermodenitrificans*. Int J Syst Evol Microbiol. 2001;51:433–446.
2. Ash C, Farrow J, Wallbanks S, Collins M. Phylogenetic heterogeneity of the genus *Bacillus* revealed by comparative analysis of small subunit ribosomal RNA sequences. Lett Appl Microbiol. 1991;13:202–6.
3. Aliyu H, Lebre P, Blom J, Cowan D, De Maayer P. Phylogenomic re-assessment of the thermophilic genus *Geobacillus*. Syst Appl Microbiol. 2016;39:527–33.
4. Bhalla A, Kainth AS, Sani RK. Draft genome sequence of lignocellulose-degrading thermophilic bacterium *Geobacillus* sp. strain WSUCF1. Genome Announc. 2013; doi:10.1128/genomeA.00595-13.
5. Boonmark C, Takahasi Y, Morikawa M. Draft genome sequence of *Geobacillus thermoleovorans* strain B23. Genome Announc. 2013; doi:10.1128/genomeA.00944-13.
6. Feng L, Wang W, Cheng J, Ren Y, Zhao G, et al. Genome and proteome of long-chain alkane degrading *Geobacillus thermodenitrificans* NG80-2 isolated from a deep-subsurface oil reservoir. Proc Natl Acad Sci U S A. 2007;104:5602–7.
7. Wiegand S, Rabausch U, Chow J, Daniel R, Streit WR, Liesegang H. Complete genome sequence of *Geobacillus* sp. strain GHH01, a thermophilic lipase-secreting bacterium. Genome Announc. 2013; doi:10.1128/genomeA.00092-13.
8. Donk PJ. A highly resistant thermophilic organism. J Bacteriol. 1920;5:373–4.
9. Zhao Y, Caspers MPM, Metselaar KI, de Boer P, Roeselers G, Moezelaar R, et al. Abiotic and microbiotic factors controlling biofilm formation by thermophilic sporeformers. Appl Environ Microbiol. 2013;79:5652–60.
10. Flint SH, Ward LJH, Walker KMR. Functional grouping of thermophilic *Bacillus* strains using amplification profiles of the 16S-23S internal spacer region. Syst Appl Microbiol. 2001;24:539–48.
11. Scott SA, Brooks JD, Rakonjac J, Walker KMR, Flint SH. The formation of thermophilic spores during the manufacture of whole milk powder. Int J Dairy Technol. 2007;60:109–17.
12. Tai SK, Lin HPP, Kuo J, Liu JK. Isolation and characterization of a cellulolytic *Geobacillus thermoleovorans* T4 strain from sugar refinery wastewater. Extremophiles. 2004;8:345–9.
13. Luecking G, Stoeckel M, Atamer Z, Hinrichs J, Ehling-Schulz M. Characterization of aerobic spore-forming bacteria associated with industrial dairy processing environments and product spoilage. Int J Food Microbiol. 2013;166:270–9.
14. Postollec F, Mathot A-G, Bernard M, Divanac'h M-L, Pavan S, Sohier D. Tracking spore-forming bacteria in food: from natural biodiversity to selection by processes. Int J Food Microbiol. 2012;158:1–8.
15. Suzuki Y, Kishigami T, Inoue K, Mizoguchi Y, Eto N, Takagi M, et al. *Bacillus thermoglucosidasius* sp.nov, a new species of obligately thermophilic bacilli. Syst Appl Microbiol. 1983;4:487–95.
16. Zarilla KA, Perry JJ. *Bacillus thermoleovorans*, sp. nov. a species of obligately thermophilic hydrocarbon utilizing endospore-forming bacteria. Syst Appl Microbiol. 1987;9:258–64.

17. LPSN:LPSN List of prokaryotic names with standing in nomenclature. http://www.bacterio.net/geobacillus.html. Accessed 22 April 2017.

18. Ahmad S, Scopes RK, Rees GN, Patel BKC. *Saccharococcus caldoxylosilyticus* sp nov., an obligately thermophilic, xylose-utilizing, endospore-forming bacterium. Int J Syst Evol Microbiol. 2000;50:517–23.

19. Bryanskaya AV, Rozanov AS, Slynko NM, Shekhovtsov SV, Peltek SE. *Geobacillus icigianus* sp nov., a thermophilic bacterium isolated from a hot spring. Int J Syst Evol Microbiol. 2015;65:864–9.

20. Coorevits A, Dinsdale AE, Halket G, Lebbe L, De Vos P, Van Landschoot A, et al. Taxonomic revision of the genus *Geobacillus*: emendation of *Geobacillus*, *G. stearothermophilus*, *G. jurassicus*, *G. toebii*, *G. thermodenitrificans* and *G. thermoglucosidans* (nom. corrig., formerly '*thermoglucosidasius*'); transfer of *Bacillus thermantarcticus* to the genus as *G. thermantarcticus* comb. nov.; proposal of *Caldibacillus debilis* gen. nov., comb. nov.; transfer of *G. tepidamans* to *Anoxybacillus* as *A. tepidamans* comb. nov.; and proposal of *Anoxybacillus caldiproteolyticus* sp nov. Int J Syst Evol Microbiol. 2012;62:1470–85.

21. Dinsdale AE, Halket G, Coorevits A, Van Landschoot A, Busse H-J, De Vos P, et al. Emended descriptions of *Geobacillus thermoleovorans* and *Geobacillus thermocatenulatus*. Int J Syst Evol Microbiol. 2011;61:1802–10.

22. Fortina MG, Mora D, Schumann P, Parini C, Manachini PL, Stackebrandt E. Reclassification of *Saccharococcus caldoxylosilyticus* as *Geobacillus caldoxylosilyticus* (Ahmad et al. 2000) comb. nov. Int J Syst Evol Microbiol. 2001;51:2063–71.

23. Kuisiene N, Raugalas J, Chitavichius D. *Geobacillus lituanicus* sp nov. Int J Syst Evol Microbiol. 2004;54:1991–5.

24. Nazina TN, Sokolova DS, Grigoryan AA, Shestakova NM, Mikhailova EM, Poltaraus AB, et al. *Geobacillus jurassicus* sp. nov., a new thermophilic bacterium isolated from a high-temperature petroleum reservoir, and the validation of the *Geobacillus* species. Syst Appl Microbiol. 2005;28:43–53.

25. Poli A, Laezza G, Gul-Guven R, Orlando P, Nicolaus B. *Geobacillus galactosidasius* sp nov., a new thermophilic galactosidase-producing bacterium isolated from compost. Syst Appl Microbiol. 2011;34:419–23.

26. Sung MH, Kim H, Bae JW, Rhee SK, Jeon CO, Kim K, et al. *Geobacillus toebii* sp nov., a novel thermophilic bacterium isolated from hay compost. Int J Syst Evol Microbiol. 2002;52:2251–5.

27. Nazina TN, Lebedeva EV, Poltaraus AB, Tourova TP, Grigoryan AA, Sokolova DS, et al. *Geobacillus gargensis* sp nov., a novel thermophile from a hot spring, and the reclassification of *Bacillus vulcani* as *Geobacillus vulcani* comb. nov. Int J Syst Evol Microbiol. 2004;54:2019–24.

28. Bryanskaya AV, Rozanov AS, Logacheva MD, Kotenko AV, Peltek SE. Draft genome sequence of *Geobacillus icigianus* strain G1w1^T isolated from hot springs in the valley of geysers, Kamchatka (Russian Federation). Genome Announc. 2014; doi:10.1128/genomeA.01098-14.

29. Yao N, Ren Y, Wang W. Genome sequence of a thermophilic bacillus, *Geobacillus thermodenitrificans* DSM465. Genome Announc. 2013; doi:10.1128/genomeA.01046-13.

30. Studholme DJ. Some (bacilli) like it hot: genomics of *Geobacillus* species. Microb Biotech. 2015;8:40–8.

31. Rossello-Mora R, Amann R. Past and future species definitions for bacteria and archaea. Syst Appl Microbiol. 2015;38:209–16.

32. Thompson CC, Amaral GR, Campeao M, Edwards RA, Polz MF, Dutilh BE, et al. Microbial taxonomy in the post-genomic era: rebuilding from scratch? Arch Microbiol. 2015;197:359–70.

33. Tindall BJ, Rossello-Mora R, Busse HJ, Ludwig W, Kaempfer P. Notes on the characterization of prokaryote strains for taxonomic purposes. Int J Syst Evol Microbiol. 2010;60:249–66.

34. Coorevits A, De Jonghe V, Vandroemme J, Reekemans R, Heyrman J, Messens W, et al. Comparative analysis of the diversity of aerobic spore-forming bacteria in raw milk from organic and conventional dairy farms. Syst Appl Microbiol. 2008;31:126–40.

35. Rainey FA, Fritze D, Stackebrandt E. The phylogenetic diversity of thermophilic members of the genus *Bacillus* as revealed by 16S rDNA analysis. FEMS Microbiol Lett. 1994;115:205–11.

36. Stackebrandt E, Goebel BM. A place for DNA-DNA reassociation and 16S ribosomal-RNA sequence-analysis in the present species definition in bacteriology. Int J Syst Bacteriol. 1994;44:846–9.

37. Weng FY, Chiou CS, Lin PHP, Yang SS. Application of *recA* and *rpoB* sequence analysis on phylogeny and molecular identification of *Geobacillus* species. J Appl Microbiol. 2009;107:452–64.

38. White D, Sharp RJ, Priest FG. A polyphasic taxonomic study of thermophilic bacilli from a wide geographical area. Antonie Van Leeuwenhoek. 1994;64:357–86.

39. Emerson D, Agulto L, Liu H, Liu L. Identifying and characterizing bacteria in an era of genomics and proteomics. Bioscience. 2008;58:925–36.

40. Goris J, Konstantinidis KT, Klappenbach JA, Coenye T, Vandamme P, Tiedje JM. DNA-DNA hybridization values and their relationship to whole-genome sequence similarities. Int J Syst Evol Microbiol. 2007;57:81–91.

41. Stackebrandt E. The richness of prokaryotic diversity: there must be a species somewhere. Food Technol Biotechnol. 2003;41:17–22.

42. Priest FG, Goodfellow M, Todd C. A numerical classification of the genus *Bacillus*. J Gen Microbiol. 1988;134:1847–82.

43. Caccamo D, Gugliandolo C, Stackebrandt E, Maugeri TL. *Bacillus vulcani* sp nov., a novel thermophilic species isolated from a shallow marine hydrothermal vent. Int J Syst Evol Microbiol. 2000;50:2009–12.

44. Kuisiene N, Raugalas J, Chitavichius D. Phylogenetic, inter, and intraspecific sequence analysis of *spo0A* gene of the genus *Geobacillus*. Curr Microbiol. 2009;58:547–53.

45. Tourova TP, Korshunova AV, Mikhailova EM, Sokolova DS, Poltaraus AB, Nazina TN. Application of *gyrB* and *parE* sequence similarity analyses for differentiation of species within the genus *Geobacillus*. Microbiology. 2010;79:356–69.

46. Zeigler DR. Application of a *recN* sequence similarity analysis to the identification of species within the bacterial genus *Geobacillus*. Int J Syst Evol Microbiol. 2005;55:1171–9.

47. Chun J, Rainey FA. Integrating genomics into the taxonomy and systematics of the bacteria and archaea. Int J Syst Evol Microbiol. 2014;64:316–24.

48. Whitman WB. Genome sequences as the type material for taxonomic descriptions of prokaryotes. Syst Appl Microbiol. 2015;38:217–22.

49. Thompson CC, Chimetto L, Edwards RA, Swings J, Stackebrandt E, Thompson FL. Microbial genomic taxonomy. BMC Genomics. 2013; doi:10.1186/1471-2164-14-913.

50. Kampfer P, Glaeser SP. Prokaryotic taxonomy in the sequencing era - the polyphasic approach revisited. Environ Microbiol. 2012;14:291–317.

51. Kim M, Oh H-S, Park S-C, Chun J. Towards a taxonomic coherence between average nucleotide identity and 16S rRNA gene sequence similarity for species demarcation of prokaryotes. Int J Syst Evol Microbiol. 2014;64:346–51.

52. Chan JZM, Halachev MR, Loman NJ, Constantinidou C, Pallen MJ. Defining bacterial species in the genomic era: insights from the genus *Acinetobacter*. BMC Microbiol. 2012; doi:10.1186/1471-2180-12-302.

53. Kumar N, Lad G, Giuntini E, Kaye ME, Udomwong P, Shamsani NJ, Young JPW, Bailly X. Bacterial genospecies that are not ecologically coherent: population genomics of *Rhizobium leguminosarum*. Open Biol. 2015;doi:10.1098/rsob.140133.

54. Burgess SA, Flint SH, Lindsay D. Characterization of thermophilic bacilli from a milk powder processing plant. J Appl Microbiol. 2014;11:350–9.

55. Burgess SA, Cox MP, Flint SH, Lindsay D, Biggs PJ. Draft genome sequences of three strains of *Geobacillus stearothermophilus* isolated from a milk powder manufacturing plant. Genome Announc. 2015; doi:10.1128/genomeA.00939-15.

56. Berendsen EM, Wells-Bennik MHJ, Krawczyk AO, de Jong A, van Heel A, Holsappel S, Eijlander RT, Kuipers OP. Draft Genome sequences of seven thermophilic spore-forming bacteria isolated from foods that produce highly heat-resistant spores, comprising *Geobacillus* spp., *Caldibacillus debilis*, and *Anoxybacillus flavithermus*. Genome Announc. 2016. doi: 10.1128/genomeA.00105-16.

57. Aleksandrzak-Piekarczyk T, Kok J, Renault P, Bardowski J. Alternative lactose catabolic pathway in *Lactococcus lactis* IL1403. Appl Environ Microbiol. 2005; 71:6060–9.

58. Hall BG. Predicting evolutionary potential. I. Predicting the evolution of a lactose-PTS system in *Escherichia coli*. Mol Biol Evol. 2001;18:1389–400.

59. Van der Heiden E, Delmarcelle M, Lebrun S, Freichels R, Brans A, Vastenavond CM, et al. A pathway closely related to the D-tagatose pathway of gram-negative Enterobacteria identified in the gram-positive bacterium *Bacillus licheniformis*. Appl Environ Microbiol. 2013;79:3511–5.

60. Stenesh J, Roe BA. DNA polymerase from mesophilic and thermophilic bacteria: I. Purification and properties of DNA polymerase from *Bacillus licheniformis* and *Bacillus stearothermophilus*. Biochim Biophys Acta. 1972; 272:156–66.

61. Golovacheva RS, Loginova LG, Salikhov TA, Kolesnikov AA, Zaitseva GN. New thermophilic species, *Bacillus thermocatenulatus* nov. sp. Microbiology. 1975; 44:230–3.

62. Sunna A, Tokajian S, Burghardt J, Rainey F, Antranikian G, Hashwa F. Identification of *Bacillus kaustophilus*, *Bacillus thermocatenulatus* and *Bacillus* strain HSR as members of *Bacillus thermoleovorans*. Syst Appl Microbiol. 1997;20:232–7.

63. Huson DH, Bryant D. Application of phylogenetic networks in evolutionary studies. Mol Biol Evol. 2006;23:254–267.58.

64. Richter M, Rossello-Mora R. Shifting the genomic gold standard for the prokaryotic species definition. Proc Natl Acad Sci U S A. 2009;106:19126–31.

65. Goh KM, Gan HM, Chan K-G, Chan GF, Shahar S, Chong CS, et al. Chai KP: Analysis of *Anoxybacillus* genomes from the aspects of lifestyle adaptations, prophage diversity, and carbohydrate metabolism. PLoS One. 2014;9(3). doi: 10.1371/journal.pone.0090549.

66. Sun ZH, Harris HMB, McCann A, Guo CY, Argimon S, Zhang WY, et al. Expanding the biotechnology potential of lactobacilli through comparative genomics of 213 strains and associated genera. Nat Commun. 2015; doi:10.1038/ncomms9322.

67. Wiedenbeck J, Cohan FM. Origins of bacterial diversity through horizontal genetic transfer and adaptation to new ecological niches. FEMS Microbiol Rev. 2011;35:957–76.

68. Vandamme P, Peeters C. Time to revisit polyphasic taxonomy. Anton Leeuw Int J Gen Mol Microbiol. 2014;106:57–65.

69. Claus D, Berkeley RCW. Genus *Bacillus* Cohn 1872. In: Sneath PHA, Mair NS, Sharpe ME, Holt JG, editors. Bergey's Manual of Systematic Bacteriology. 1986;2:1105–39.

70. Logan NA, De Vos P. Dinsdale AE genus *Geobacillus*. In: De Vos P, Garrity GM, Jones D, Krieg NR, Ludwig W, Rainey FA, Schleifer KH, Whitman WB, editors. Bergey's manual of systematic bacteriology, vol. 3. 2nd ed. New York, USA: Springer; 2009. p. 144–60.

71. Zeigler DR. The *Geobacillus* paradox: why is a thermophilic bacterial genus so prevalent on a mesophilic planet? Microbiology. 2014;160:1–11.

72. Rasko DA, Altherr MR, Han CS, Ravel J. Genomics of the *Bacillus cereus* group of organisms. FEMS Microbiol Rev. 2005;29:303–29.

73. Bezuidt OKI, Makhalanyane TP, Gomri MA, Kharroub K, Cowan DA. Draft genome sequence of thermophilic *Geobacillus* sp. strain Sah69, isolated from Saharan soil, Southeast Algeria. Genome Announc. 2015; doi:10.1128/genomeA.01447-15.

74. Blanchard K, Robic S, Matsumura I. Transformable facultative thermophile *Geobacillus stearothermophilus* NUB321 as a host strain for metabolic engineering. Appl Microbiol Biotechnol. 2014;98:6715–23.

75. Brumm P, Land ML, Hauser LJ, Jeffries CD, Chang YJ, Mead DA. Complete genome sequences of *Geobacillus* sp Y412MC52, a xylan-degrading strain isolated from obsidian hot spring in Yellowstone national park. Stand Genomic Sci. 2015; doi:10.1186/s40793-015-0075-0.

76. Brumm PJ, De Maayer P, Mead DA, Cowan DA. Genomic analysis of six new *Geobacillus* strains reveals highly conserved carbohydrate degradation architectures and strategies. Front Microbiol. 2015; doi:10.3389/fmicb.2015.00430.

77. De Maayer P, Williamson CE, Vennard CT, Danson MJ, Cowan DA. Draft genome sequences of *Geobacillus* sp. strains CAMR5420 and CAMR12739. Genome Announc. 2014; doi:10.1128/genomeA.00567-14.

78. Ortiz EM, Berretta MF, Navas LE, Benintende GB, Amadio AF, Zandomeni RO. Draft Genome Sequence of *Geobacillus* sp. Isolate T6, a thermophilic bacterium collected from a thermal spring in Argentina. Genome Announc. 2015; doi:10.1128/genomeA.00743-15.

79. Petkauskaite R, Blom J, Goesmann A, Kuisiene N. Draft genome sequence of pectic polysaccharide-degrading moderate thermophilic bacterium *Geobacillus thermodenitrificans* DSM 101594. Braz J Microbiol. 2017;48:7–8.

80. Pore SD, Arora P, Dhakephalkar PK. Draft genome sequence of *Geobacillus* sp. strain FW23, isolated from a formation water sample. Genome Announc. 2014; doi:10.1128/genomeA.00352-14.

81. Rozonov AS, Logacheva MD, Peltek SE. Draft genome sequences of *Geobacillus stearothermophilus* strains 22 and 53 isolated from the Garga hot spring in the Barguzin River valley of the Russian Federation. Genome Announc. 2014; doi:10.1128/genomeA.01205-14.

82. Sakaff MKLM, Rahman AYA, Saito JA, Hou S, Alam M. Complete genome sequence of the thermophilic bacterium *Geobacillus thermoleovorans* CCB_US3_UF5. J Bacteriol 2012;194:1239–1239.

83. Shintani M, Ohtsubo Y, Fukuda K, Hosoyama A, Ohji S, Yamazoe A, et al. Complete genome sequence of the thermophilic polychlorinated biphenyl degrader *Geobacillus* sp. strain JF8 (NBRC 109937). Genome Announc. 2014; doi:10.1128/genomeA.01213-13.

84. Siddiqui M, Rashid N, Ayyampalayam S, Whitman WB. Draft genome sequence of *Geobacillus thermopakistaniensis* strain MAS1. Genome Announc. 2014; doi:10.1128/genomeA.00559-1.

85. Takami H, Nishi S, Lu J, Shinamura S, Takaki Y. Genomic characterization of thermophilic *Geobacillus* species isolated from the deepest sea mud of the Mariana trench. Extremophiles. 2004;8:351–6.

86. Wissuwa J, Stokke R, Fedøy A-E, Lian K, Smalås AO, Steen IH. Isolation and complete genome sequence of the thermophilic *Geobacillus* sp. 12AMOR1 from an Arctic deep-sea hydrothermal vent site. Standards in genomic. Sciences. 2016;11:16.

87. Zhao Y, Caspers MP, Abee T, Siezen RJ, Kort R. Complete genome sequence of *Geobacillus thermoglucosidans* TNO-09.020, a thermophilic sporeformer associated with a dairy-processing environment. J Bacteriol 2012;194:4118–4118.

88. Zheng B, Zhang F, Chai L, Yu G, Shu F, Wang Z, et al. Permanent draft genome sequence of *Geobacillus thermocatenulatus* strain GS-1. Marine Genom. 2014;18:129–31.

89. Seemann T. Prokka: rapid prokaryotic genome annotation. Bioinformatics. 2014;30:2068–9.

90. Li L, Stoeckert CJ, Roos DS. OrthoMCL: identification of ortholog groups for eukaryotic genomes. Genome Res. 2003;13:2178–89.

91. Biggs PJ, Fearnhead P, Hotter G, Mohan V, Collins-Emerson J, Kwan E, et al. Whole-genome comparison of two *Campylobacter jejuni* isolates of the same sequence type reveals multiple loci of different ancestral lineage. PLoS One. 2011; doi:10.1371/journal.pone.0027121.

92. Edgar RC. MUSCLE: multiple sequence alignment with high accuracy and high throughput. Nucleic Acids Res. 2004;32:1792–7.

93. Bryant D, Moulton V. Neighbor-net: an agglomerative method for the construction of phylogenetic networks. Mol Biol Evol. 2004;21:255–65.

94. Harrigan WF Examination of cultures for motility by 'Hanging Drop' preparations. In: Harrigan WF, editor. Laboratory Methods in Food Microbiology. 3rd ed. San Diego, California, USA: Academic Press; 1998: 39–40.

95. Baldock JD. Heat resistance of rough and smooth variants of *Bacillus stearothermophilus*. Dissertation Abstracts International Section B The Sciences and Engineering. 1970;30:5088–9.

96. Humbert RD, Deguzman AN, Fields ML. Studies on variants of *Bacillus stearothermophilus* strain NCA 1518. Appl Microbiol. 1972;23:693–8.

97. Jung L, Jost R, Stoll E, Zuber H. Metabolic differences in *Bacillus stearothermophilus* grown at 55 degrees C and 37 degrees C. Arch Microbiol. 1974;95:125–38.

98. Walker PD. Wolf J taxonomy of *Bacillus stearothermophilus*. In: Barker AN, editor. Spore research. London and New York: Academic Press; 1971. p. 247–62.

99. Logan NA, Berkeley RCW. Identification of *Bacillus* strains using the API system. J Gen Microbiol. 1984;130:1871–82.

14

Initial nitrogen enrichment conditions determines variations in nitrogen substrate utilization by heterotrophic bacterial isolates

Suchismita Ghosh[1], Paul A. Ayayee[1*], Oscar J. Valverde-Barrantes[2], Christopher B. Blackwood[1], Todd V. Royer[3] and Laura G. Leff[1]

Abstract

Background: The nitrogen (N) cycle consists of complex microbe-mediated transformations driven by a variety of factors, including diversity and concentrations of N compounds. In this study, we examined taxonomic diversity and N substrate utilization by heterotrophic bacteria isolated from streams under complex and simple N-enrichment conditions.

Results: Diversity estimates differed among isolates from the enrichments, but no significant composition were detected. Substrate utilization and substrate range of bacterial assemblages differed within and among enrichments types, and not simply between simple and complex N-enrichments.

Conclusions: N substrate use patterns differed between isolates from some complex and simple N-enrichments while others were unexpectedly similar. Taxonomic composition of isolates did not differ among enrichments and was unrelated to N use suggesting strong functional redundancy. Ultimately, our results imply that the available N pool influences physiology and selects for bacteria with various abilities that are unrelated to their taxonomic affiliation.

Keywords: Dissolved organic nitrogen, Bacterial isolates, Nitrogen cycle

Background

Bacterial nitrogen (N) uptake and assimilation are influenced by availability and nature of dissolved organic and inorganic forms of N [1]. Simple N compounds are readily available to heterotrophic bacteria [2–4], whereas more complex N compounds require enzymatic degradation prior to uptake and assimilation [5, 6]. Heterotrophic bacterial communities use a variety of dissolved organic nitrogen (DON) compounds, including amino acids [7], nucleic acids [8], and proteins [9, 10], as carbon, N, and/or energy sources, or directly as specific compounds, such as via salvage pathways for amino acids [11]. In addition to DON, dissolved inorganic (DIN) species, such as nitrate [12] and ammonium [3], are also used to meet N requirements.

The ability of bacterial communities to utilize particular N types (simple vs. complex, labile vs. recalcitrant) depends on taxonomic composition [13], biochemical capacities, and competition with other bacteria for N [14]. Interactions under differing conditions result in varied N-utilization profiles [15, 16] among members of a bacterial community and may lead to ecological specialization [17, 18]. Ultimately, although N-utilization differs among heterotrophic bacterial communities, there is uncertainty regarding the scale at which common metabolic capabilities are shared regardless of the dominant forms of available N.

In this study, we investigated utilization of N substrates, ranging from labile to recalcitrant, by heterotrophic bacteria isolated from stream sediments under different N-enrichments (simple and complex). We sought to determine: 1) whether bacteria isolated from complex and simple N-enrichments would be taxonomically and compositionally different, and 2) if

* Correspondence: akwettey@gmail.com
[1]Department of Biological Sciences, Kent State University, Kent, OH 44242, USA
Full list of author information is available at the end of the article

N-substrate utilization by isolated bacteria was dependent on initial N-enrichment conditions.

Methods

Bacterial isolation

Stream sediment samples from three streams used in prior N studies: West Branch of Mahoning River near Ravenna, OH [19], Sycamore Creek located in Morgan County, IN [20], and Sugar Creek near Shirley, IN [20] were incubated in M9 minimal media, (amended with glucose as the carbon source) with 8 different N compounds. All final N concentrations were 94 mM. These included five single-source N treatments (nitrate in the form of NaNO₃, ammonium, urea, glycine, and tryptophan), an equimolar mixture of these compounds (ammonium + nitrate + urea + glycine + tryptophan), a bacterial protein (undefined cellular extract) and nutrient broth (complex medium; Difco BD nutrient broth [Becton, Dickinson and Company, Franklin Lakes, NJ, USA]). The bacterial protein was obtained as described in Ghosh et al. (2013) [21]. Briefly, soluble bacterial proteins were extracted from cultures of *Bacillus subtilis*, *Pseudomonas aeruginosa*, and *Staphylococcus aureus* incubated at 27 °C for 24 h and proteins were obtained using the Qproteome Bacterial Protein Prep Kit (Qiagen, MD, USA) and total DON quantified using a Shimadzu TNM-1(Shimadzu Corporation, Columbia, MD). Among the enrichments, ammonium, nitrate, and glycine were considered simple N-enrichments. Nutrient broth and the bacterial protein extract were considered complex enrichments, as were tryptophan and urea. In this study, urea was considered a complex enrichment due to low bacterial uptake compared to inorganic N species, amino acids and carbohydrates in a study of freshwater bacterial N turnover [22]. Tryptophan was considered complex due to its large molar mass, and chemical composition [23]. The defined-N-mixture (ammonium, nitrate, glycine, urea and tryptophan) was considered a simple enrichment for three reasons. First, the abundance of simple compounds relative to urea and tryptophan. Second, repression of the nitrogen assimilation control (*nac*) operon for urea uptake in the presence of ammonium and other simpler N compounds [1], as is the case in the defined-N-mix. Third, the high affinity for electrophilic substitutions in the indole ring of tryptophan renders it readily deoxidized in the presence of other compounds (including nitrate, carbon dioxide, and ammonia) leading to modifications into other compounds that could be utilized by bacterial cells [23].

Enrichments were incubated at 25°C for 24 h to isolate fast-growing bacteria or for 72 h to isolate slow-growing bacteria. Samples from each enrichment were used to inoculate plates of the same composition mixed with agar. Distinct colonies from respective plates were selected for isolation into pure cultures.

16S rRNA gene amplification and sequence analyses

Genomic DNA was extracted from bacteria isolates using the CTAB method followed by phenol: chloroform extraction and ethanol precipitation as in Moore et al. (2004) [24]. Polymerase chain reaction (PCR) was carried out with the universal primers 27F (5′-AGAGTTTGATCMTGGCTCAG-3′) and 1552R (5′-AAGGAGGTGATCCARCCGCA-3′) [25] in a PTC 200 DNA Engine Cycler (Biorad, Hercules, CA) with a thermal profile of 94 °C for 3 min and 35 cycles of 94 °C for 30 s, 58 °C for 30 s and 72 °C for 90 s followed by a final extension of 72 °C for 5 min. Each 25 µl PCR reaction mixture consisted of 2 µl of template DNA, 12.5 µl of water, 0.5 µl of both forward and reverse primers, and 12.5 µl of GoTaq Pre- Mixed Green Master Mix (Promega Corporation, Madison, WI). Amplified products were visualized on a 1% agarose via gel electrophoresis, purified and submitted for Sanger sequencing at the Advanced Genetic Technologies Center, at the University of Kentucky (Lexington, KY), using the same primer pair.

Resulting amplicon sequences were quality checked in Sequencher (Gene Codes Corporation, Ann Arbor, MI) using default settings. Sequences were classified using the Classifier tool in the Ribosomal Database Project (RDP) server [26]. Taxonomic affiliations of the isolates were determined at a cut-off threshold of 80% in RDP, and an operational taxonomic unit (OTU) table generated summarizing the taxa and abundance of isolates from each enrichment at the class level. This table was subsequently used to determine within-enrichment alpha diversity estimates (Chao1) [27] in QIIME (version 1.9.0) [28] following rarefaction. The reliance of Chao1 estimates on singletons, makes it a more robust estimate. A non-metric multidimensional scaling (NMDS) [29] analysis was performed on the Bray-Curtis distance matrix and axes used to visualize relatedness among the enrichments. Compositional difference between enrichments was assessed using the analysis of similarity (ANOSIM) multivariate test in QIIME.

Nitrogen substrate utilization assays

Substrate utilization by bacterial isolates was assessed spectrophotometrically in 96-well microtitre plates. 12 single-source N-substrates (94 mM each) ranging from labile to recalcitrant forms were used. The labile and recalcitrant designations are based on known resistance/refraction to degradation, bioavailability, and impacts on bacterial growth. The substrates were nitrate, ammonium, urea, glycine, proline, tryptophan, bacterial protein, peptidoglycan, nucleic acid (purified DNA), algal exudate, putrescine (polyamine), and humic matter. Humic matter, algal exudates and nucleic acids were obtained as described in Ghosh et al. (2013) [21]. Briefly, algal exudates were extracted from cultures of *Chlamydomonas*, *Chlorella*, and *Synedra* (Carolina Biological Supplies, Burlington, NC) grown in artificial stream water with 20 mg/L of $NaNO_3$, under constant light for 35 days. Humic matter was extracted from senescent red oak (*Quercus rubra*), witch hazel (*Hamamelis virginiana*), and corn leaves (*Zea mays*) in 0.027% NaCl and pooled. Nucleic acids were obtained following DNA extraction from cultures of *Bacillus subtilis*, *Pseudomonas aeruginosa*, and *Staphylococcus aureus* incubated at 27 °C for 24 h; extractions were performed using the Power-Soil DNA extraction kit (MoBio Laboratories, Carlsbad, CA) and nucleic acids were pooled among the three cultures. Following initial cell lysis and precipitation of bacterial cultures during DNA extraction, cell debris was collected and quantified to represent peptidoglycan. Putrescine was purchased from MP Biomedicals (MP Biomedicals, Santa Ana, CA, USA). Of N treatments, algal exudates, ammonium, nitrate, glycine, tryptophan, and urea were considered labile [21, 30] whereas, bacterial proteins, nucleic acid, and humic matter were considered recalcitrant [31, 32]. Peptidoglycan, polyamine (putrescine) and proline (Amresco Biochemicals and Life Science Research Products, Solon, OH, USA) were considered intermediate compounds. The rationale for these designation is that proline, as a N source in the presence of glucose, is suboptimal for *E. coli* growth [33], and disproportionately accumulates in particulate residues following microbial exposure, suggesting proline utilization following degradation of more bioavailable N sources [34]. In contrast, bacterial growth is positively correlated with tryptophan availability [35]. Peptidoglycan is designated an intermediate compound because the efficiency of peptidoglycan degradation by bacteria has ranges from 49% - 58% depending on whether they were from gram negative or positive bacterial sources, respectively [36]. Each of the 12 single-source N media had the same amount of nitrogen (94 mM) as the standard minimal media used in Maheswaran and Forchhammer 2003 [37] with glucose as the only carbon source.

Before beginning the assays, bacterial cultures were incubated in their respective broth media for 24–48 h depending on growth rate. After cultures reached an optical density (OD) of 0.4, they were centrifuged and washed five times with N-free minimal media and diluted 1:10 with the N-free minimal media to minimize transfer of N to the test plates. Washed cultures were subsequently transferred to plates, making up 10% of the final assay volume. Plates were incubated at room temperature for 6 days and OD determined at 600 nm every 12 h for the first 48 h, and every 24 h for the remaining 4 days. Treatments were carried out in triplicates for each isolate. Bacterial growth rates (day^{-1}) were calculated from OD_{600} values recorded at the different time points.

Assessment of substrate utilization and substrate range used by isolates was carried out by dividing the growth rates (day^{-1}) into ranges as: −1 for growth rates <0, 0 for rates between 0 and 29, 1 for rates between 30 and 39, 2 for rates between 40 and 99, and 3 for growth rates >100. Substrate range for each isolate was calculated by determining the mean score for each isolate across all 12 substrates. The score difference (Δ score = total isolate score − mean) for each isolate was determined and then used to categorize the substrate range of each isolate. Isolates with positive score differences were categorized as having broad substrate ranges and those with negative score differences were categorized as having narrow substrate ranges.

Statistical analysis

One-way analysis of variance (ANOVA) was used to examine differences among enrichments based on the Chao1 estimates without transformation. This was followed by visualization of the NMDS coordinates using the generated distance matrix, after the ANOSIM multivariate test of compositional differences. Differences in patterns of N-utilization by bacteria isolates were analyzed using a mixed-model analysis with actual growth rates as the dependent variable and N-enrichment and N-substrates as independent variables. Relationship between phylogenetic distance and substrate utilization (growth rates expressed as scores as described above) was examined using regression analysis, and the relationship between categorical bacterial N-utilization profiles (broad vs. narrow substrate ranges) and taxonomic affiliations was examined using contingency analysis followed by the Pearson's chi-square test. Statistical analyses were carried out in JMP 10 (SAS Institute Inc., Cary, NC, USA) and QIIME (version 1.9.0).

Results

Composition and diversity of bacterial isolates from N-enrichments

A total of 266 isolates representing 24 families were obtained (Additional file 1: Table S1). The highest number of isolates were from the nutrient broth enrichment (58), followed by tryptophan (34), ammonium (32), defined-N-mixture (31), glycine (29), nitrate (28) and urea (28), with the bacterial protein enrichment yielding the least number of isolates (26).

Taxonomically, four bacterial families, *Comamonadaceae*, *Enterobacteriaceae*, *Bacillaceae*, and *Pseudomonadaceae*, were most commonly isolated from complex N-enrichments (bacterial protein, nutrient broth, urea and tryptophan). In addition, 9 unique bacterial families (present in only one enrichment) were detected from these complex enrichments. Three families, *Alteromonadales incertae sedis* (relative abundance, 3.57%), *unassigned*

Sphaerobactereales (3.57%), *and Methylophilaceae* (3.57%), were only detected from the urea enrichment. *Planctomyce-taceae* (3.85%) was only detected from the bacterial protein enrichment and five bacterial families, *Burkholderiales incertae sedis* (1.69%), *Shewanellaceae* (6.78%), *Pseudoalter-omonadaceae* (1.69%), *Ferrimonadaceae* (1.69%), and *Rhodocyclaceae* (1.69%), were only detected only from the nutrient broth enrichment (Fig. 1). Partial 16S rRNA sequence data for each isolate are provided in Additional file 2: Document 1.

Three bacterial families (*Comamonadaceae, Entero-bacteriaceae,* and *Pseudomonadaceae*) were well represented in isolates from the simple N-enrichments (ammonium, glycine, nitrate and defined-N-mixture) and 4 unique bacterial isolates were detected. *Oceanos-pirillaceae* (3.45%) was only detected from the glycine enrichment. *Halobacteriaceae* (3.13%) and *Chitinopha-gaceae* (3.13%) were only detected from the ammonium enrichment and *Unassigned Calescibacterium (3.23%)* was only detected from the defined-N-mixture enrichments (Fig. 1).

Mean Chao1 diversity estimates at the family level differed significantly among N-enrichments (F = 2.22; df = 7, 136; *P* = 0.04), but there was no significant difference in alpha diversity between simple and complex N-enrichments when grouped together. Individually, the least diverse enrichments, tryptophan (1.83 ± 0.63, mean ± s.e.) and defined-N-mixture (1.94 ± 0.63), were significantly different from the most diverse enrichments,

glycine (4.35 ± 0.63) and nutrient broth (4.24 ± 0.63). Bacterial protein (3.58 ± 0.63), urea (3.2 ± 0.63), nitrate (3.14 ± 0.63) and ammonium (2.7 ± 0.63) enrichments had comparable richness estimates.

In spite of observed differences in Chao1 diversity estimates among N-enrichments and the presence of a few enrichment-specific isolates, overall community composition were very similar among N-enrichments (ANOSIM; Test statistic = −0.013, *P* = 0.55, number of permutations =1000, number of samples =24, number of groups =8). In the NMDS plot (Fig. 2, stress <0.01), the complex N-enrichments, tryptophan and nutrient broth grouped separately from each other and from the remaining two complex enrichments (urea and bacterial protein), and the simple N-enrichment glycine, was displaced from the other three simple N-enrichments, ammonium, nitrate, and defined-N mixture, which clustered closely together (Fig. 2). The observed displacements may be attributed to the presence of single unique isolates in several of the enrichments, but these were not sufficient to result in significant differences in overall community composition.

Bacterial isolate N-utilization

Substrate utilization by isolates differed significantly among the 8 initial source N-enrichments (F = 36.2; df = 7, 3184; *P* < 0.001). Overall, substrate utilization was lowest in bacteria obtained from the bacterial protein enrichment and highest in bacteria obtained from

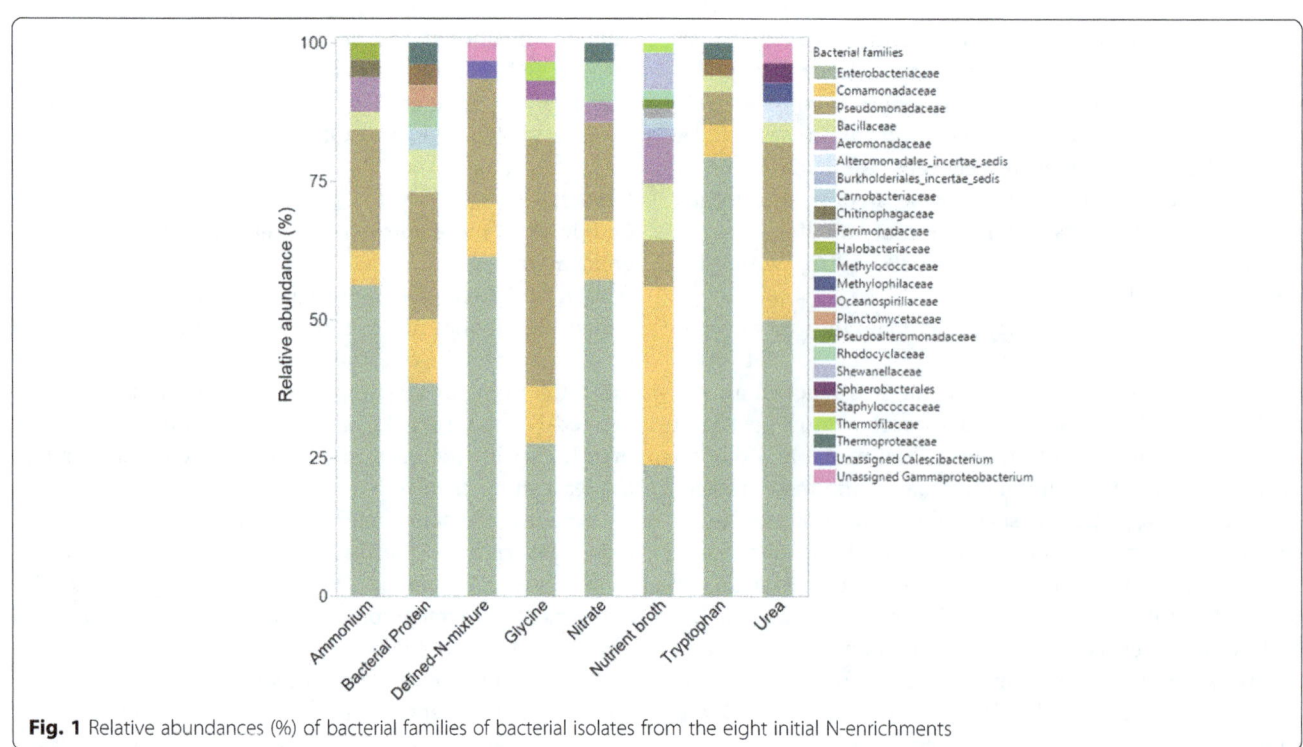

Fig. 1 Relative abundances (%) of bacterial families of bacterial isolates from the eight initial N-enrichments

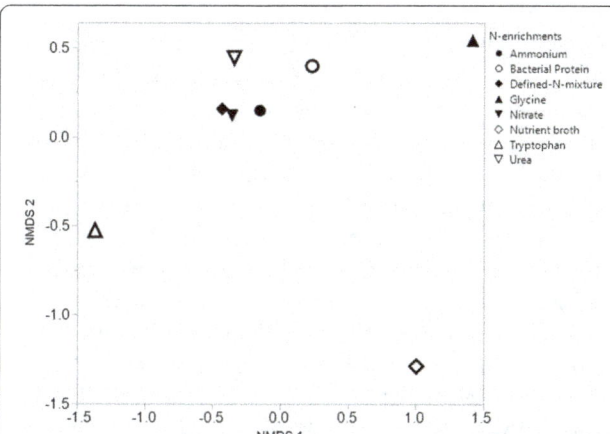

Fig. 2 Displacement of bacterial communities within the NMDS plot (Stress <0.01). Complex N-enrichments were tryptophan (*open triangle*), nutrient broth (open diamond), urea (*open inverted triangle*), and bacterial protein (*open circle*). Simple N-enrichments were glycine (filled triangle), ammonium (filled circle), nitrate (filled inverted triangle), and defined-N mixture (filled diamond)

Substrate utilizations were lowest on recalcitrant nucleic acid (6.02 ± 0.81) and humic matter substrates (11.74 ± 0.81) for bacteria from all enrichments, followed by peptidoglycan (17.9 ± 0.81) and bacterial protein (29.2 ± 0.81) substrates. On the other hand, all labile substrates, except for glycine and tryptophan were efficiently utilized by bacteria from all N-enrichments. Utilization of glycine, proline and tryptophan differed among bacteria in a N-enrichment driven manner; utilization of glycine and proline substrates were greater among bacteria from the simple enrichments, whereas utilization of tryptophan was greater among bacteria from the complex enrichments (Fig. 4). Growth rates for each of the 266 isolates are shown in Additional file 3: Table S2. The relationship between substrate range/utilization and N-enrichment was statistically significant (Pearson's test; Chi-square = 32.5, $P < 0.0001$), demonstrating that initial enrichment influenced subsequent substrate utilization and the range of substrates used. However, there was no significant linear correlation between average phylogenetic distance and average substrate utilization (R-statistic = 0, $P = 0.96$).

Finally, among enrichments there were differences in the range of substrates that were effectively utilized by bacteria. Bacterial isolates from the simple defined-N-mixture and ammonium N-enrichments had comparatively broader substrate ranges, followed by isolates from the complex tryptophan and urea N-enrichments (Fig. 5)

glycine, defined-N mix, and tryptophan (Fig. 3). There were significantly differences in substrate utilization by isolates among the 12 N-substrates used (F = 557.2; df = 11, 3180; $P < 0.001$), as well as significant N-enrichment by N-substrate differences in utilization by bacteria isolates (F = 3.9; df = 77, 3114; $P < 0.001$) (Fig. 4).

Fig. 3 Actual growth rates averaged across all N substrates (day^{-1}) (mean ± s.e.) for bacterial isolates from the eight initial N-enrichments (F $_{(7, 3184)}$ = 36.2, $P < 0.001$). The N-enrichments were: Nitrate, Ammonium, Glycine, Tryptophan, Urea, Defined-N-mixture, Bacterial Protein, and Nutrient Broth. Different letters represent significantly different growth rates on each N-enrichment at $P = 0.05$

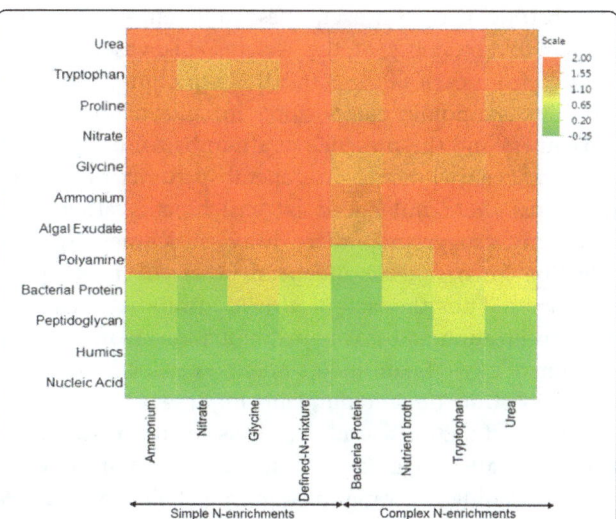

Fig. 4 Substrate utilization by groups of isolates from the initial enrichments on the twelve substrates used in the substrate assay (F$_{(77, 3114)}$ = 4, $P < 0.001$) depicted in a heat map. The color legend indicates the scaled scores from −0.25 to 2.00, with high and moderate substrate utilization shown as red and orange respectively, and the low and least substrate utilization shown as shades of yellow and green respectively

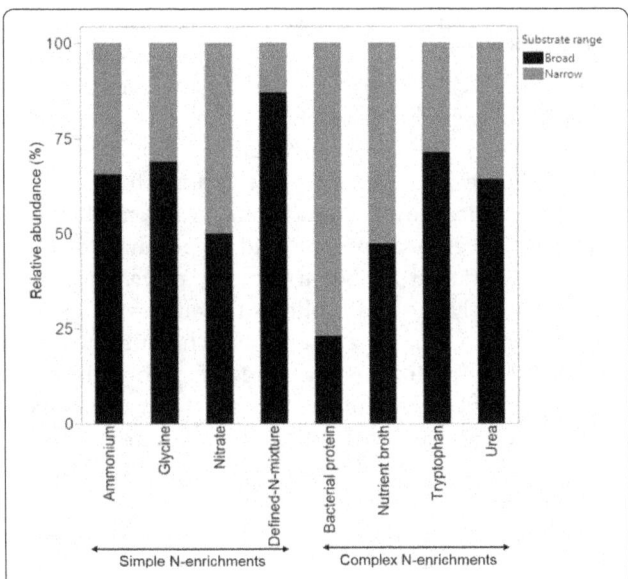

Fig. 5 Proportion (%) of isolates with broad and narrow substrate ranges from each enrichment following the substrate utilization assay

(Additional file 4: Table S3). The bacterial protein enrichment yielded isolates with the narrowest substrate range, whereas the proportions of isolates with narrow and broad substrate ranges were equivalent in the nitrate and nutrient broth enrichments.

Discussion

Initial N-enrichments significantly impacted subsequent N substrate use. However, these differences were not related to taxonomy of the isolates. Likewise, bacteria isolated from each of the 8 initial N-enrichments did not differ in taxonomic composition in spite of differences in richness and the presence of a number of unique taxa in specific enrichments. In general, patterns of N substrate use were influenced by enrichment rather than taxonomy, suggesting there was enrichment-specific selection for organisms independent of 16S rRNA gene sequences. Thus the lack of a relationship between substrate utilization and taxonomic affiliations is most likely explained by taxon-independent capacity for N-utilization (functional redundancy) [38].

Bacterial functional traits, such as nitrogen utilization and substrate ranges are influenced by environmental factors leading to variations in metabolic capabilities and, ultimately, ecological specialization within microbial assemblages and are taxon-independent [17, 18]. Additionally, substrate utilization patterns may be a function of acclimation and physiological change rather than reflective of genotypic differences. Isolates from two complex enrichments (tryptophan and urea) and three simple enrichments (ammonium, glycine and

defined-N-mixture) had similar substrate utilization profiles and greater proportions of broad substrate range isolates, suggestive perhaps of activated metabolic pathways enabling utilization of subsequent various N substrates regardless of the initial enrichments. The same explanation may be applied to the substrate utilization profiles of the defined-N-mixture enrichment, wherein a broad range of N compounds that can be utilized by bacteria is to be expected. As a result, the nitrogen-rich condition in this enrichment may have facilitated growth of metabolically versatile and broad substrate range. Utilization of other single N compounds and the production of intermediates, such as ammonium by bacteria isolates from these enrichments, may explain the breadth of N substrate use and similarities in N-profiles in the substrate assay [1].

Various operons within the bacterial nitrogen regulation system (*ntr*) enable degradation and/or uptake of diverse N sources [1, 5, 6]. Some of these operons are only activated by specific N sources leading to their rapid uptake, while others are repressed by certain N sources and only activated in their absence leading to instantaneous and primed N uptake pathways, respectively [1]. Priming may have contributed to the observed substrate ranges of isolates from complex enrichments. For example, one pathway for tryptophan use is non-oxidative degradation to ammonia, indole and pyruvate via the indole pathway [39]. The pyruvate and ammonia formed are then respectively used for respiration and amino acid biosynthesis [39]. Along these same lines, urea can be taken up by a variety of bacteria and hydrolyzed to ammonium and CO_2 by urease; the resulting ammonia is subsequently used for biosynthesis and growth [40]. Finally, glycine is oxidatively degraded into ammonium, CO_2 and a methylene group via the glycine cleavage system or glycine synthetase [41, 42]. Thus, among these bacterial communities, similar substrate utilization profiles may be attributed to shared/activated metabolic capacities by differently primed nitrogen utilization pathways selected by the various enrichments.

Although several bacteria families, including *Planctomycetaceae*, were obtained from the bacterial protein enrichment, isolates from this enrichment were predominantly narrow in their substrate ranges. The combined presence of refractory N compounds, such as membrane-bound proteins and histones in the protein extract [43], in addition to reported antimicrobial activity of histones [44] during the initial bacterial protein enrichment may have selected for bacteria with different traits. Thus, the reduced growth rate and subsequent narrow substrate range of these isolates during the substrate assay may be attributed to delayed or reduced activation of N scavenging enzymes in these bacteria.

Isolates from the nutrient broth enrichment had comparable proportions of members with broad and narrow substrate ranges. Some of the substrates effectively utilized by these isolates (labile free amino acids, algal exudate and ammonium) were also present in the nutrient broth enrichment (i.e. beef extract, labile and recalcitrant peptides and amino acids, nucleotide fractions, organic acids). Thus the subset of available and recalcitrant N compounds in the initial enrichment may have primed different nitrogen regulatory pathways in isolates from this enrichment, resulting in broad and narrow substrate ranges. Nitrate isolates also exhibited a similar profile as seen in nutrient broth. Lower growth rates of a variety of bacteria have been encountered when nitrate is provided as the only nitrogen source under aerobic conditions, due to lowered assimilatory nitrate reductase function [45, 46]. In addition, isolates under high nitrate conditions have been shown to reduce the production of N scavenging enzymes [32], and extracellular hydrolytic enzymes that degrade dissolved organic nitrogen species [47]. Thus the initial nitrate enrichment condition may have selected for isolates capable of effectively using some substrates but not others.

Nitrogen substrates examined inherently differed in their use regardless of the properties of the isolates. Specifically, the most complex, recalcitrant compounds (nucleic acids, peptidoglycan, and humics) were generally used poorly in comparison to other substrates. The crystalline and polymerized forms of these compounds makes them refractory to enzymatic degradation although degradation of humic matter [48], nucleic acids [49], and peptidoglycan [36] occurs under certain growth conditions. Thus, the observed low utilization of these recalcitrant substrates relative to the labile substrates may be a function of the minimal media conditions used in the substrate assay in this study.

Conclusions

We observed differences in N substrate use patterns of bacteria from some complex and simple N-enrichments while others were unexpectedly similar. This is attributed to priming and metabolic flexibility. Taxonomic composition of bacterial isolate groups from the N-enrichments did not differ and was unrelated to N use, suggesting breadths of function and strong functional redundancy. Given the considerable functional variations among bacterial isolates, further studies examining expression of functional gene markers (transcripts) related to N utilization, quantification of gene abundances, and direct quantification of substrate utilization via stable isotope techniques may provide insights into the metabolic processes responsible for observed similar N utilization profiles from different enrichment conditions.

Additional files

Additional file 1: Table S1. Taxonomical affiliations (to genus) of the 266 bacterial isolates from the 8 initial N-enrichments. Description of data: The name, and taxonomic identification to the genus level obtained for bacterial isolates obtained from this study using the Classifier tool in the Ribosomal Database Project (https://rdp.cme.msu.edu/classifier/classifier.jsp).

Additional file 2: Document 1. Partial bacterial 16S rRNA sequence data. Description of data: The partial 16S rRNA bacterial sequence information for all 266 bacterial isolates obtained in this study.

Additional file 3: Table S2. Title: Growth rates of all bacterial isolates from the initial N-enrichments on each of the 12 substrates. Description of data: The growth rates of each bacterial isolate from each N-enrichment on the 12 N-substrates used in the study.

Additional file 4: Table S3. Title: Score differences (total isolate score – mean) and substrate range classification of bacterial isolates from the initial N-enrichments across substrates. Description of data: The score difference between mean scaled growth rates and total scaled growth rates for each isolate on the 12 N-substrates. Positive score differences represent isolates with broad substrate range and negative score differences represent isolates with narrow substrate range.

Acknowledgement
We would like to thank Moumita Moitra and Erin Manis for their assistance in this study.

Funding
This work was supported by Kent State University via a Graduate Student Senate research grant, which had no roles in the design, execution, analysis and preparation of this manuscript.

Author contributions
SG, TR, and LL conceived and designed the study. CB helped design the study, assisted with statistical analysis and writing. SG collected the data. SG, OVB, CB, and PAA analyzed the data. SG, PAA, LL, CB, OVB, TR wrote the manuscript. All authors consent to the publication of the materials in this submission.

Competing interests
The authors declare no conflict of interest.

Author details
[1]Department of Biological Sciences, Kent State University, Kent, OH 44242, USA. [2]International Center for Tropical Botany (ICTB), Florida International University, Miami, FL 33199, USA. [3]School of Public and Environmental Affairs, Indiana University, Bloomington, Bloomington, IN 47405, USA.

References
1.   Merrick MJ, Edwards RA. Nitrogen control in bacteria. Microbiol Rev. 1995;59:604–22. http://www.ncbi.nlm.nih.gov/pmc/articles/PMC239390/.

2.   McCarthy JJ, Carpenter EJ. Nitrogen cycling in near-surface waters of the open ocean. Nitrogen Mar Environ. Elsevier BV; 1983. p. 487–512. http://dx.doi.org/10.1016/b978-0-12-160280-2.50022-5

3.   Wheeler PA, Kirchman DL. Utilization of inorganic and organic nitrogen by bacteria in marine systems. Limnol Oceanogr. 1986;31:998–1009. http://dx.doi.org/10.4319/lo.1986.31.5.0998

4.   Brookshire ENJ, Valett HM, Thomas SA, Webster JR. Coupled cycling of dissolved organic nitrogen and carbon in a forest stream. Ecology. 2005;86:2487–96. http://dx.doi.org/10.1890/04-1184

5.   Hoppe HG. Microbial extracellular enzyme cctivity: A new key parameter in aquatic ecology. Microb Enzym. Aquat Environ. 1991. p. 60–83. http://dx.doi.org/10.1007/978-1-4612-3090-8_4

6.   Sinsabaugh RL, Findlay S, Franchini P, Fischer D. Enzymatic analysis of riverine bacterioplankton production. Limnol Oceanogr. 1997;42:29–38. doi:10.4319/lo.1997.42.1.0029.

7.   Crawford CC, Hobbie JE, Webb KL. The utilization of dissolved free amino acids by estuarine microorganisms. Ecology. 1974;55:551–63. doi:10.2307/1935146.

8.   Paul JH, Cazares LH, David AW, DeFlaun MF, Jeffrey WH. The distribution of dissolved DNA in an oligotrophic and a eutrophic river of Southwest Florida. Hydrobiol. 1991;218:53–63. doi:10.1007/bf00006418.

9.   Hollibaugh JT, Azam F. Microbial degradation of dissolved proteins in seawater. Limnol Oceanogr. 1983;28:1104–16. doi:10.4319/lo.1983.28.6.1104.

10. Coffin RB. Bacterial uptake of dissolved free and combined amino acids in estuarine waters. Limnol Oceanogr. 1989;34:531–42. doi:10.4319/lo.1989.34.3.0531.

11. Liechti G, Goldberg JB. Helicobacter pylori relies primarily on the purine salvage pathway for purine nucleotide biosynthesis. J Bacteriol. 2011;194: 839–54. doi:10.1128/jb.05757-11.

12. Wheeler PA, Kokkinakis SA. Ammonium recycling limits nitrate use in the oceanic subarctic Pacific. Limnol Oceanogr. 1990;35:1267–78. doi:10.4319/lo.1990.35.6.1267.

13. Kassen R. The experimental evolution of specialists, generalists, and the maintenance of diversity. J Evol Biol. 2002;15:173–90. doi:10.1046/j.1420-9101.2002.00377.x.

14. Matias MG, Combe M, Barbera C, Mouquet N. Ecological strategies shape the insurance potential of biodiversity. Front Microbiol. 2013;3 doi:10.3389/fmicb.2012.00432.

15. Allison SD, Martiny JBH. Resistance, resilience, and redundancy in microbial communities. Proc Natl Acad Sci. 2008;105:11512–9. doi:10.1073/pnas.0801925105.

16. Carbonero F, Oakley BB, Purdy KJ. Metabolic flexibility as a major predictor of spatial distribution in microbial communities. PLoS One. 2014;9:e85105. doi:10.1371/journal.pone.0085105.

17. Futuyma D. The evolution of ecological specialization. Annu Rev Ecol Syst. 1988;19:207–33. doi:10.1146/annurev.ecolsys.19.1.207.

18. Devictor V, Clavel J, Julliard R, Lavergne S, Mouillot D, Thuiller W, et al. Defining and measuring ecological specialization. J Appl Ecol. 2010;47:15–25. doi:10.1111/j.1365-2664.2009.01744.x.

19. Olapade OA, Leff LG. Seasonal response of stream biofilm communities to dissolved organic matter and nutrient enrichments. Appl Environ Microbiol. 2005;71:2278–87.

20. Baxter AM, Johnson L, Edgerton J, Royer T, Leff LG. Structure and function of denitrifying bacterial assemblages in low-order Indiana streams. Freshw Sci. 2012;31:304–17. doi:10.1899/11-066.1.

21. Ghosh S, Leff LG. Impacts of labile organic carbon concentration on organic and inorganic nitrogen utilization by a stream biofilm bacterial community. Appl Environ Microbiol. 2013;79:7130–41. doi:10.1128/aem.01694-13.

22. Jørgensen NOG, Kroer N, Coffin RB, Hoch MP. Relations between bacterial nitrogen metabolism and growth efficiency in an estuarine and an open-water ecosystem. Aquat Microb Ecol. 1999;18:247–61. doi:10.3354/ame018247.

23. Alkhalaf LM, Ryan KS. Biosynthetic manipulation of tryptophan in bacteria: pathways and mechanisms. Chem Biol. 2015;22:317–28. doi:10.1016/j.chembiol.2015.02.005.

24. Moore E, Arnscheidt A, Krüger A, Strömpl C, Mau M. Section 1 update: Simplified protocols for the preparation of genomic DNA from bacterial cultures. Microb Ecol Man. 2008. p. 1905–19. http://dx.doi.org/10.1007/978-1-4020-2177-0_101

25. Lane DJ. 16S/23S rRNA sequencing. In: Stackebrandt E, Goodfellow M, editors. Nucleic acid tech. Bact. Syst. New York: John Wiley and Sons; 1991.

26. Wang Q, Garrity GM, Tiedje JM, Cole JR. Naive bayesian classifier for rapid assignment of rRNA sequences into the new bacterial taxonomy. Appl Environ Microbiol. 2007;73:5261–7. http://dx.doi.org/10.1128/aem.00062-07

27. Chao A. Non-parametric estimation of the number of classes in a population. Scand J Stat. 1984;11:265–70.

28. Caporaso JG, Kuczynski J, Stombaugh J, Bittinger K, Bushman FD, Costello EK, et al. QIIME allows analysis of high-throughput community sequencing data. Nat Methods. 2010;7:335–6. http://dx.doi.org/10.1038/nmeth.f.303

29. Mielke PW. Meteorological applications of permutation techniques based on distance functions. Handb Stat. Elsevier BV; 1984. p. 813–830. http://dx.doi.org/10.1016/s0169-7161(84)04036-0

30. Bronk DA. Dynamics of DON. Biogeochem Mar Dissolved Org Matter. Elsevier BV; 2002. p. 153–247. http://dx.doi.org/10.1016/b978-012323841-2/50007-5

31. Billen G. Protein degradation in aquatic environments. Microb Enzym Aquat Environ. 1991. p. 123–43. Available from: http://dx.doi.org/10.1007/978-1-4612-3090-8_7

32. Chróst RJ. Environmental control of the synthesis and activity of aquatic microbial ectoenzymes. Microb Enzym Aquat Environ. 1991. p. 29–59. http://dx.doi.org/10.1007/978-1-4612-3090-8_3

33. Bren A, Park JO, Towbin BD, Dekel E, Rabinowitz JD, Alon U. Glucose becomes one of the worst carbon sources for E.coli on poor nitrogen sources due to suboptimal levels of cAMP. Sci Rep. 2016;6:24834. 10.1038/srep24834

34. Takasu H, Nagata T. High proline content of bacteria-sized particles in the western North Pacific and its potential as a new biogeochemical indicator of organic matter diagenesis. Front Mar Sci. 2015;2 doi:10.3389/fmars.2015.00110.

35. Jaeger CHIII, Lindow SE, Miller W, Clark E, FM. Mapping of sugar and amino acid availability in soil around roots with bacterial sensors of sucrose and tryptophan. Appl Environ Microbiol. 1999;65:2685–90.

36. Jörgensen NOG, Stepanaukas R, Pedersen A-GU, Hansen M, Nybroe O. Occurrence and degradation of peptidoglycan in aquatic environments. FEMS Microbiol Ecol. 2003;46:269–80. doi:10.1016/s0168-6496(03)00194-6.

37. Maheswaran M. Carbon-source-dependent nitrogen regulation in Escherichia coli is mediated through glutamine-dependent GlnB signalling. Microbiol. 2003;149:2163–72. doi:10.1099/mic.0.26449-0.

38. Philippot L, Andersson SGE, Battin TJ, Prosser JI, Schimel JP, Whitman WB, et al. The ecological coherence of high bacterial taxonomic ranks. Nat Rev Microbiol. 2010;8:523–9. doi:10.1038/nrmicro2367.

39. Vederas JC, Floss HG. Stereochemistry of pyridoxal phosphate catalyzed enzyme reactions. Acc Chem Res. 1980;13:455–63. dx.doi.org/10.1021/ar50156a004

40. Solomon CM, Collier JL, Berg GM, Glibert PM. Role of urea in microbial metabolism in aquatic systems: a biochemical and molecular review. Aquat Microb Ecol. 2010;59:67–88. doi:10.3354/ame01390.

41. Jacob GS. Garbow JR SJ& KG. Solid-state NMR studies of regulation of N-(phosphonomethyl) glycine and glycine metabolism in Pseudomonas sp. strain PG2982. J Biol Chem. 1998;262:1552–7.

42. Kikuchi G, Motokawa Y, Yoshida T, Hiraga K. Glycine cleavage system: reaction mechanism, physiological significance, and hyperglycinemia. Proc Japan Acad Ser B. 2008;84:246–63. http://dx.doi.org/10.2183/pjab.84.246

43. Keil RG, Kirchman DL. Utilization of dissolved protein and amino acids in the northern Sargasso Sea. Aquat Microb Ecol. 1999;18:293–300. http://dx.doi.org/10.3354/ame018293

44. Sol A, Skvirsky Y, Blotnick E, Bachrach G, Muhlrad A. Actin and DNA protect histones from degradation by bacterial proteases but inhibit their antimicrobial activity. Front Microbiol. 2016;7 doi:10.3389/fmicb.2016.01248.

45. van 't Riet J, Knook DL, Planta RJ. The role of cytochrome b1 in nitrate assimilation and nitrate respiration in Klebsiella aerogenes.FEBS Lett 1972;23: 44–46. 10.1016/0014-5793(72)80280-1

46. Gates AJ, Luque-Almagro VM, Goddard AD, Ferguson SJ, Roldán MD, Richardson DJ. A composite biochemical system for bacterial nitrate and nitrite assimilation as exemplified by Paracoccus denitrificans. Biochem J. 2011;435:743–53. http://www.ncbi.nlm.nih.gov/pubmed/21348864.

47. Münster U, De Haan H. The role of microbial extracellular enzymes in the transformation of dissolved organic matter in humic waters. Ecol Stud. 1998. p. 199–257. http://dx.doi.org/10.1007/978-3-662-03736-2_10

48. Van Trump JI, Wrighton KC, Thrash JC, Weber KA, Andersen GL, Coates JD. Humic acid-oxidizing, nitrate-reducing bacteria in agricultural soils. MBio. 2011;2:e00044-11-e00044-11. http://dx.doi.org/10.1128/mbio.00044-11

49. Antheunisse J. Decomposition of nucleic acids and some of their degradation products by microorganisms. Antonie Van Leeuwenhoek. 1972;38:311–27. http://dx.doi.org/10.1007/bf02328101

15

Sequence-based identification of inositol monophosphatase-like histidinol-phosphate phosphatases (HisN) in *Corynebacterium glutamicum*, *Actinobacteria*, and beyond

Robert Kasimir Kulis-Horn, Christian Rückert, Jörn Kalinowski and Marcus Persicke[*] (ID)

Abstract

Background: The eighth step of L-histidine biosynthesis is carried out by an enzyme called histidinol-phosphate phosphatase (HolPase). Three unrelated HolPase families are known so far. Two of them are well studied: HAD-type HolPases known from *Gammaproteobacteria* like *Escherichia coli* or *Salmonella enterica* and PHP-type HolPases known from yeast and *Firmicutes* like *Bacillus subtilis*. However, the third family of HolPases, the inositol monophosphatase (IMPase)-like HolPases, present in *Actinobacteria* like *Corynebacterium glutamicum* (HisN) and plants, are poorly characterized. Moreover, there exist several IMPase-like proteins in bacteria (e.g. CysQ, ImpA, and SuhB) which are very similar to HisN but most likely do not participate in L-histidine biosynthesis.

Results: Deletion of *hisN*, the gene encoding the IMPase-like HolPase in *C. glutamicum*, does not result in complete L-histidine auxotrophy. Out of four *hisN* homologs present in the genome of *C. glutamicum* (*impA*, *suhB*, *cysQ*, and *cg0911*), only *cg0911* encodes an enzyme with HolPase activity. The enzymatic properties of HisN and Cg0911 were determined, delivering the first available kinetic data for IMPase-like HolPases.
Additionally, we analyzed the amino acid sequences of potential HisN, ImpA, SuhB, CysQ and Cg0911 orthologs from bacteria and identified six conserved sequence motifs for each group of orthologs. Mutational studies confirmed the importance of a highly conserved aspartate residue accompanied by several aromatic amino acid residues present in motif 5 for HolPase activity. Several bacterial proteins containing all identified HolPase motifs, but showing only moderate sequence similarity to HisN from *C. glutamicum*, were experimentally confirmed as IMPase-like HolPases, demonstrating the value of the identified motifs. Based on the confirmed IMPase-like HolPases two profile Hidden Markov Models (HMMs) were build using an iterative approach. These HMMs allow the fast, reliable detection and differentiation of the two paralog groups from each other and other IMPases.

Conclusion: The kinetic data obtained for HisN from *C. glutamicum*, as an example for an IMPase-like HolPases, shows remarkable differences in enzyme properties as compared to HAD- or PHP-type HolPases. The six sequence motifs and the HMMs presented in this study can be used to reliably differentiate between IMPase-like HolPases and IMPase-like proteins with no such activity, with the potential to enhance current and future genome annotations. A phylogenetic analysis reveals that IMPase-like HolPases are not only present in *Actinobacteria* and plant but can be found in further bacterial phyla, including, among others, *Proteobacteria*, *Chlorobi* and *Planctomycetes*.

Keywords: HisN, Cg0911, Histidinol-phosphate phosphatase (HolPase), Inositol monophosphatase (IMPase)-like, *Corynebacterium glutamicum*, Kinetic data, Sequence motifs, Phylogenetic analysis

---

* Correspondence: marcusp@cebitec.uni-bielefeld.de
Microbial Genomics and Biotechnology, Center for Biotechnology, Bielefeld University, Universitätsstraße 27, 33615 Bielefeld, Germany

Background

The gram-positive soil-bacterium *Corynebacterium gluta-micum*, a member of the order *Corynebacteriales* within the taxonomical class *Actinobacteria* [1], plays an important role in industrial amino acid fermentation, with annual production scales of more than 2.5 and 1.5 million tons L-glutamate and L-lysine, respectively [2]. Strains for the production of further amino acids including L-alanine, L-isoleucine, L-phenylalanine, L-serine, L-tryptophan, and L-valine are available [3]. It is obvious that the in-depth understanding of the amino acid biosynthesis pathways and their regulation in this organism is necessary not only for further improvement of existing production strains, but also facilitates the development of new production strains, like for the production of L-histidine [4].

The entire L-histidine biosynthesis pathway is present in *C. glutamicum* and has been reviewed recently [5]. So far, all organisms known to synthesize L-histidine, including archaea, bacteria, yeast, and plants, use the same pathway for the biosynthesis. Although there are differences in gene organization, including several gene fusion events, most of the enzymes seem to have a common ancestor [5, 6]. One interesting exception is the histidinol-phosphate phosphatase (HolPase) [EC 3.1.3.15] catalyzing the eighth step of L-histidine biosynthesis, the dephosphorylation of L-histidinol-phosphate (HolP) to L-histidinol. Three unrelated HolPase families are known so far. *C. glutamicum* possesses a HolPase belonging to the family of inositol monophosphatase (IMPase)-like proteins, a subgroup of the FIG (FBPase/IMPase/GlpX-like domain) superfamily encoded by *hisN* [7, 8]. IMPase-like HolPases are a characteristic of the *Actinobacteria* and genera possessing a HisN homolog can be found in almost all taxonomical orders of this bacterial class [5]. Additionally, IMPase-like HolPases have been discovered in plants [9]. Functional characterizations of IMPase-like HolPases have been conducted in *C. glutamicum* [7], *Streptomyces coelicolor* [10], and *Arabidopsis thaliana* [9]. The HolPase activity of the HisN homolog in *Mycobacterium tuberculosis* (gene *Rv3137*) is supported at least indirectly, since it is not possible to delete this gene if a L-histidine free medium is used during the required selection steps [11].

Outside the *Actinobacteria*, there exist at least two further major classes of HolPases. The first class belongs to the HAD (Haloacid dehalogenase-like hydrolase) superfamily of proteins. The HAD-type HolPase activity is in general present on a bifunctional His(NB) enzyme that catalyzes the eighth and additionally the sixth step of L-histidine biosynthesis, the dehydration of imidazole glycerol-phosphate (IGP) [12]. The two activities are independent of each other with the HolPase and IGP dehydratase activities being found in the N-terminal and C-terminal domain of the bifunctional protein, respectively [13]. Bifunctional HAD-type HolPases are in

general only found in *Gammaproteobacteria* [12], and have been extensively studied in *Salmonella enterica* serovar Typhimurium [14, 15] and *Escherichia coli* [13]. A monofunctional HAD-type HolPase has been discovered in the archaeon *Thermococcus onnurieneus* few years ago and homologs can be found in further archaeal genomes [16]. The second class of HolPases belongs to the PHP (polymerase and HolPase) subfamily of the metallo-dependent hydrolase (MDH) superfamily of proteins. The PHP-type HolPases are monofunctional and can be found in yeasts and in different bacterial lineages [12]. Examples for organisms with a well-studied PHP-type HolPase are *Saccharomyces cerevisiae* [17, 18], *Bacillus subtilis* [19], and *Lactococcus lactis* [20].

Our special interest in the corynebacterial HolPase arises from the observation that deletion of *hisN* in *C. glutamicum* results in pronounced L-histidine bradytrophy instead of complete auxotrophy [5]. A similar observation has been previously made with HolPase mutants of *S. cerevisiae*, resulting in the discovery of a second phosphatase with HolPase side activity [17]. Four HisN paralogs are encoded in the genome of *C. glutamicum* (Cg0911, SuhB, ImpA, and CysQ) [7, 21] and are therefore interesting candidates for alternative HolPases. The present study pursued three different aims: 1) The identification of an alternative HolPase in *C. glutamicum*; 2) The determination of the kinetic parameters of HisN in *C. glutamicum*, since up to our knowledge no such data has been reported for any IMPase-like HolPase so far; 3) The identification of one or more sequence motifs to reliably discriminate between IMPase-like HolPases and other IMPase-like proteins with no such activity.

Results

Genetic study on *hisN* and its four paralogs

During our previous investigation of different L-histidine gene deletion mutants of *C. glutamicum* [5], we observed that deletion of *hisN*, encoding the IMPase-like HolPase, does not result in L-histidine auxotrophy, but only in a pronounced bradytrophy of the mutant. Therefore, we started a closer investigation of the 8th step of L-histidine biosynthesis in *C. glutamicum* in general and the Δ*hisN* mutant in particular.

Growth of the Δ*hisN* mutant was visible after several days of incubation on minimal medium plates without L-histidine. Addition of L-histidine abolished the observed growth defect completely (Fig. 1). The residual growth of the Δ*hisN* mutant was not specific to one single mutant, but was observed with every confirmed *hisN* deletion mutant constructed during this study and was also confirmed for an independently constructed Δ*hisN* strain [7] that slightly differed in the extend of the *hisN* deletion (data not shown). The genome of *C. glutamicum* contains four genes encoding putative HisN

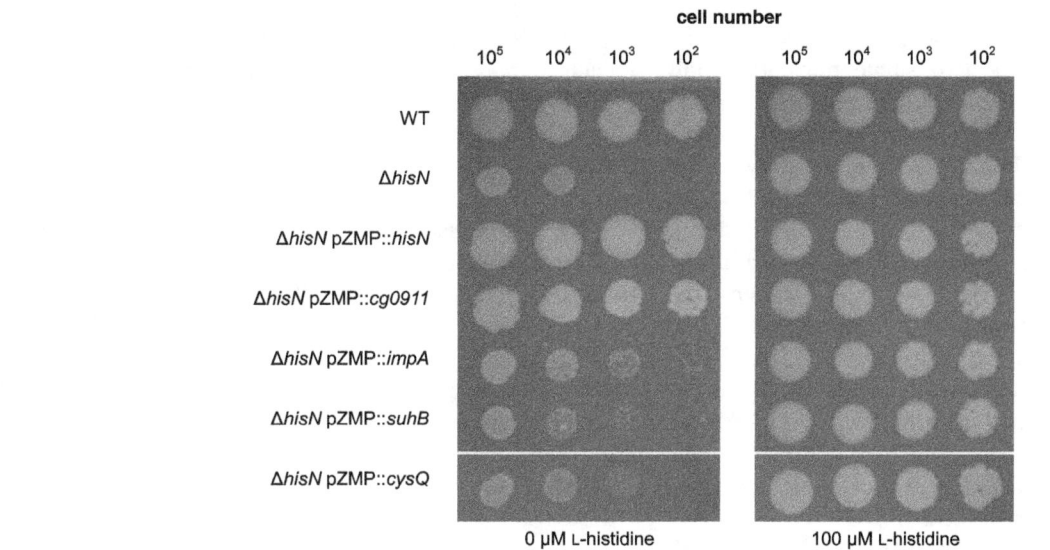

Fig. 1 Comparative growth test of *C. glutamicum* Δ*hisN* mutants with plasmid-based expression of *hisN* or one of its four paralogs. Equal amounts of cells were placed on MM1 minimal medium plates with or without L-histidine supplementation and incubated for 6 days at 30 °C

paralogs that have been already recognized in the original publication describing the HolPase activity of the *hisN* gene product [7]. All of them are grouped into the FIG superfamily of proteins according to their conserved domains and most of them are annotated as putative IMPases or fructose-1,6-bisphosphatases (Table 1). The degree of sequence similarity between HisN and one of its four putative paralogs is comparable in every case (24-26% identity, 37-41% similarity) with CysQ being least similar. In addition, all putative paralogs share the same degree of similarity if compared one to another. Since the four paralogs have so far not been analyzed for their function in *C. glutamicum*, we hypothesized that

one of them might be responsible for the residual growth of the Δ*hisN* mutant.

Moreover, two of the putative *hisN* paralogs, namely *cg0911* and *impA*, form operons with other L-histidine biosynthesis genes. The *cg0911* gene is transcribed together with *hisN* and *impA* is part of the larger *hisHA-impA-hisFI-cg2294* transcription unit [5].

In order to test if any of the four putative *hisN* paralogs encodes a gene with HolPase activity, the genes were cloned into the constitutive shuttle expression vector pZMP (approximately 15 copies per cell, *tac* promoter). Sequencing of the inserts revealed that the *cg0911* gene sequence from the *C. glutamicum* wild type

Table 1 HisN paralogs in *C. glutamicum*

HisN paralog (gene)	protein length [aa]	sequence similarity on protein level (identity, similarity, BLASTP score) [%], [%] [bits]					annotation	HolPase activity
		HisN	Cg0911	ImpA	SuhB	CysQ		
HisN (*cg0910*)	260	100, 100 515	26, 41 78	25, 41 86	27, 42 82	24, 37 52	HolPase [a]	yes [a, b]
Cg0911 (*cg0911*)	288 *	26, 41 77	100, 100 585	22, 35 58	27, 41 89	21, 30 84	putative IMPase [c] or archaeal FBPase [d]	yes (very low) [b]
ImpA (*cg2298*)	275	25, 41 81	22, 35 74	100, 100 561	22, 32 80	15, 26 35	putative IMPase [c] or archaeal FBPase [d]	not detected
SuhB (*cg2090*)	280	27, 42 82	27, 41 89	22, 32 80	100, 100 567	23, 34 48	putative IMPase [c] or archaeal FBPase [d]	not detected
CysQ (*cg0967*)	252	24, 37 52	21, 30 59	15, 26 35	23, 34 48	100, 100 521	putative PAPSPase [c, d]	not detected

* Based on RNAseq results, transcription of *cg0911* starts 9 nt downstream of the currently annotated *cg0911* CDS [5], resulting in a shorter Cg0911 protein (288 aa instead of 291 aa)
[a] [7]
[b] this study
[c] GenBank: BX927147.1 [21]
[d] GenBank: BA000036.3 [22]
PAPSPase = 3'-phosphoadenosine 5'-phosphosulfate (PAPS) 3'-phosphatase

strain used in this study is identical to that presented in the *C. glutamicum* ATCC 13032 reference sequence BA000036.3 [22] and has two single nucleotide polyphormisms as compared to reference sequence BX927147.1 [21] (one silent mutation and one resulting in a G50R mutation). The resulting plasmids were isolated from *E. coli* and subsequently transferred into the *C. glutamicum* Δ*hisN* strain. Since it was not possible to obtain an error free *impA* insert in *E. coli* (i.e. frame shift or promoter mutations; data not shown) the pZMP::*impA* assembly mix was directly used for transformation of *C. glutamicum* Δ*hisN* resulting in the correct Δ*hisN* pZMP::*impA* mutant (checked by sequencing of the *impA* insert and the promoter region). A comparative growth test was conducted on minimal medium plates to check if one of the genes is able to complement the Δ*hisN* growth defect *in trans* (Fig. 1).

Expression of *impA*, *suhB* or *cysQ* did not improve the growth of the Δ*hisN* strain on minimal medium. Beside the complementation by *hisN* itself, a complementation of the Δ*hisN* growth defect was only observed with *cg0911*. However, growth of the Δ*hisN* pZMP::*cg0911* strain was slower compared to the WT. Single colonies of this strain appeared 24 h later on the plates and remained smaller in size, even if the incubation was prolonged (data not shown). Supplementation with L-histidine resulted in the same growth phenotype of all tested mutants and did not differ from the WT. These results suggest that *cg0911* is encoding an enzyme with weak HolPase activity.

To obtain further insight into the function of the different *hisN* paralogs, deletion mutants were constructed. Each paralog was separately deleted in the WT. In addition, a Δ*hisN* Δ*cg0911* double mutant and a quintuple mutant, lacking *hisN* and all its paralogs, were generated. Growth of the different mutants was again monitored on minimal medium plates (Fig. 2).

The single deletion of one of the *hisN* paralogs in the WT had no effect on growth of the mutants. Unexpectedly, we did not observe a further reduction of growth of the Δ*hisN* Δ*cg0911* double or the Δ*hisN* Δ*cg0911* Δ*impA* Δ*suhB* Δ*cysQ* quintuple mutant as compared to the Δ*hisN* single mutant. Supplementation with L-histidine resulted in the same growth of all mutants. None of the *hisN* paralogs was needed for normal growth of *C. glutamicum* under the tested conditions. Moreover, although the complementation assay clearly demonstrated HolPase activity of the *cg0911* gene product in vivo if expressed on a multiple copy plasmid, this activity does not account for L-histidine biosynthesis in a measurable degree if present in single copy under control of the native promotor.

Enzymatic characterization of HisN and Cg0911
To the best of our knowledge, no kinetic data is available on the HolPase activity of HisN from *C. glutamicum* or any other organism possessing an IMPase-like HolPase. HolPase activity of the IMPase-like HolPases from *A. thaliana* and *S. coelicolor* has been deduced from complementation studies, and only the general phosphatase

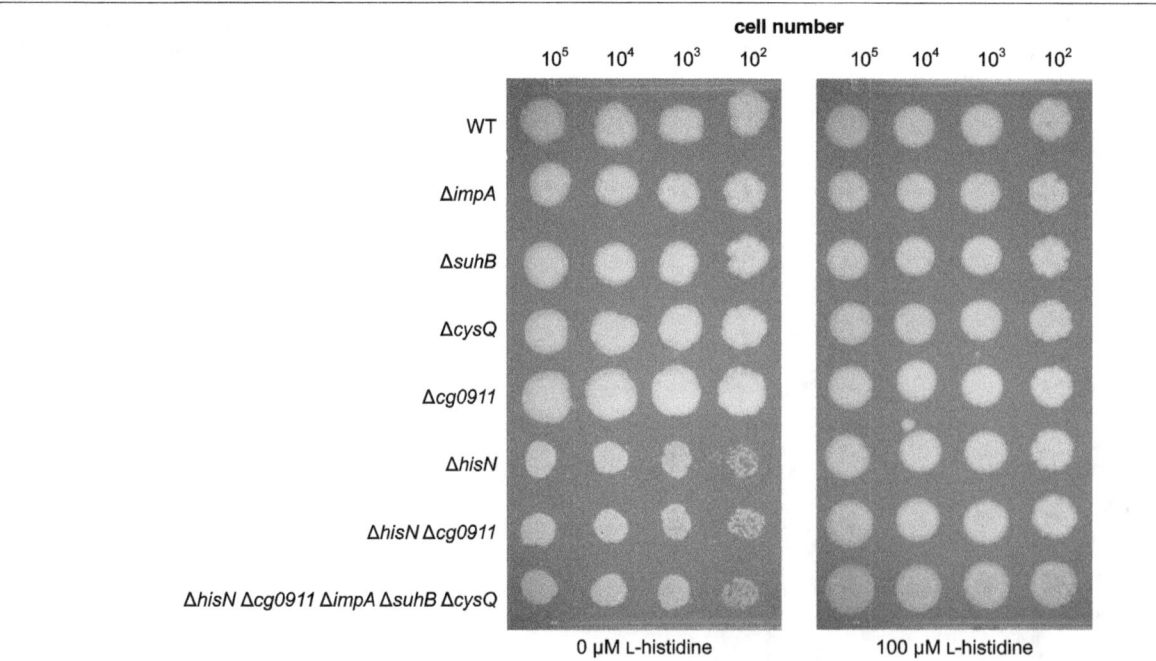

Fig. 2 Comparative growth test of mutants with deletion of *hisN* or paralogous genes. Equal amounts of cells were placed on MM1 minimal medium plates with or without L-histidine supplementation and incubated for 6 days at 30 °C

activity using the substrate *para*-nitrophenylphosphate (pNPP) has been demonstrated in vitro for the latter [9, 10]. Therefore, we determined the kinetic parameters of an IMPase-like HolPase with its natural substrate HolP using HisN$_{Cg}$ as an example and comparing it to the HolPase activity of Cg0911, the second IMPase-like protein in *C. glutamicum* possessing HolPase activity.

Both proteins were heterologously expressed in *E. coli* and purified tag-free using the commercial IMPACT™ system. Purity and molecular weight of the purified proteins were estimated by one-dimensional SDS-PAGE (Additional file 1: Figure S1) and identity was confirmed by MALDI-TOF-MS analysis (data not shown). The activity of HisN and Cg0911 was assayed by the release of inorganic phosphate (P$_i$) from HolP as described in Materials and Methods.

So far, all studied HolPases of the PHP- or HAD-type were shown to be dependent on divalent metal ions [13, 16, 20, 23]. The same holds true for eukaryotic and bacterial IMPases [11, 24, 25]. Therefore, in a first step, we evaluated the metal ion preference of HisN and Cg0911 as examples of IMPase-like HolPases (Fig. 3).

Both enzymes were inactive if metal ions were omitted from the reaction mixture. Presence of 10 mM EDTA also resulted in no activity (data not shown). In the presence of 5 mM Mg^{2+}, Mn^{2+}, or Co^{2+}, release of P$_i$ was detected. HisN showed a clear preference towards Mg^{2+} (100% activity) over Co^{2+} (20% activity) and Mn^{2+} (11% activity). The metal ion preference of Cg0911 was less pronounced. The enzyme still exhibited 78% of its maximal activity in the presence of Mn^{2+} and 47% in the

presence of Co^{2+}. No release of P$_i$ from HolP was detectable in the presence of Zn^{2+}, Cu^{2+}, Ca^{2+}, Fe^{2+}, or Ni^{2+} with either enzyme.

Next, activity of HisN and Cg0911 was assayed in response to the pH of the reaction buffer (Fig. 4a). The buffering substances were adapted to the intended pH values. HisN exhibited maximal activity at pH 7.35. HisN activity decreases almost uniformly beyond the optimal pH, with no activity present at around pH 6 and reduced to 10% at around pH 10. The pH profile of Cg0911 was shifted to the alkaline conditions by 0.5 to 1 pH units. Maximal Cg0911 activity was observed at around pH 8 and was only little reduced at pH 7.35, followed by a sharp loss in activity towards more acidic conditions. The drop in activity towards more alkaline conditions was less pronounced and the enzyme exhibited still 30% of its activity at around pH 10. Since both HisN and Cg0911 were highly active at pH = 7.35, and this pH value corresponded well to the internal pH value of 7.5 ± 0.5 in *C. glutamicum* [26], a pH of 7.35 was kept constant during all further measurements.

The activity of HisN and Cg0911 was determined in a temperature range from 20 to 50 °C (Fig. 4b). Maximal HisN activity was reached at 35–40 °C. No activity was observed at 50 °C, indicating heat denaturation of the protein. The HolPase activity of Cg0911 was even more heat sensitive. Maximal Cg0911 activity was reached at 30 °C and less than 5% of this activity remained at 40 °C. To retain comparability between the two enzymes, all following measurements were conducted at 30 °C, since both enzymes were active at this temperature and it reflects the optimal growth temperature of *C. glutamicum*.

The turnover number (k$_{cat}$) of HisN and Cg0911 in the presence of Mg^{2+}, as well as the HolP and Mg^{2+}-concentrations necessary for half maximal enzyme activity (K$_m$ values for HolP and Mg^{2+}) were determined (Table 2). The parameters were obtained by non-linear curve fitting of the data points to the Hill-equation [27].

HisN was very specific towards HolP with a K$_m$ value of about 25 μM. The k$_{cat}$ value was around 1 s^{-1} resulting in a catalytic efficiency of the enzyme of 4.41×10^4 s^{-1} M^{-1}. The Hill coefficient of HisN regarding HolP was around 1.5 indicating only a little cooperative effect.

The HolPase activity of Cg0911 was almost 80-times lower compared to HisN and the K$_m$ value for HolP was around 650 μM, resulting in a catalytic efficiency of 1.98×10^1 s^{-1} M^{-1}. The Hill coefficient of around 1.8 hints to some cooperative effect of HolP on Cg0911 HolPase activity. The kinetic parameters for Cg0911 indicate that HolP is not the preferred substrate of this protein and the ability to hydrolyze HolP might reflect only a side activity of the enzyme.

We also tested the affinity of the two enzymes towards bivalent magnesium ions. The K$_m$ values for Mg^{2+} were

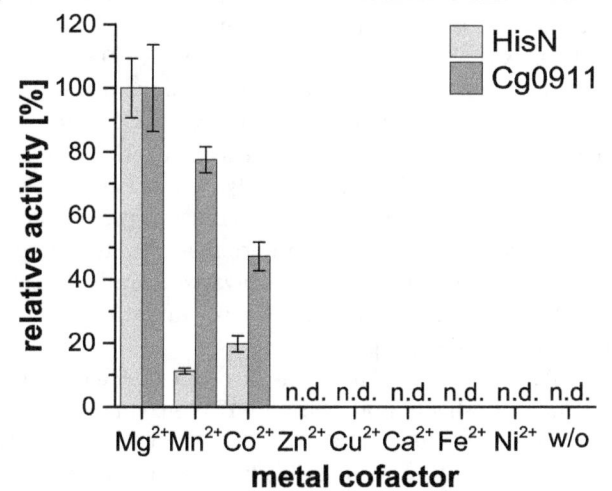

Fig. 3 Relative activity of HisN (light grey) and Cg0911 (dark grey) in the presence of different divalent metal ions. Activity with Mg^{2+} was set to 100% for each protein individually. Assay conditions: 30 °C, pH = 7.35, 5 mM divalent metal ions, 500 μM HolP. n.d. = not detectable, w/o = without, n = 4

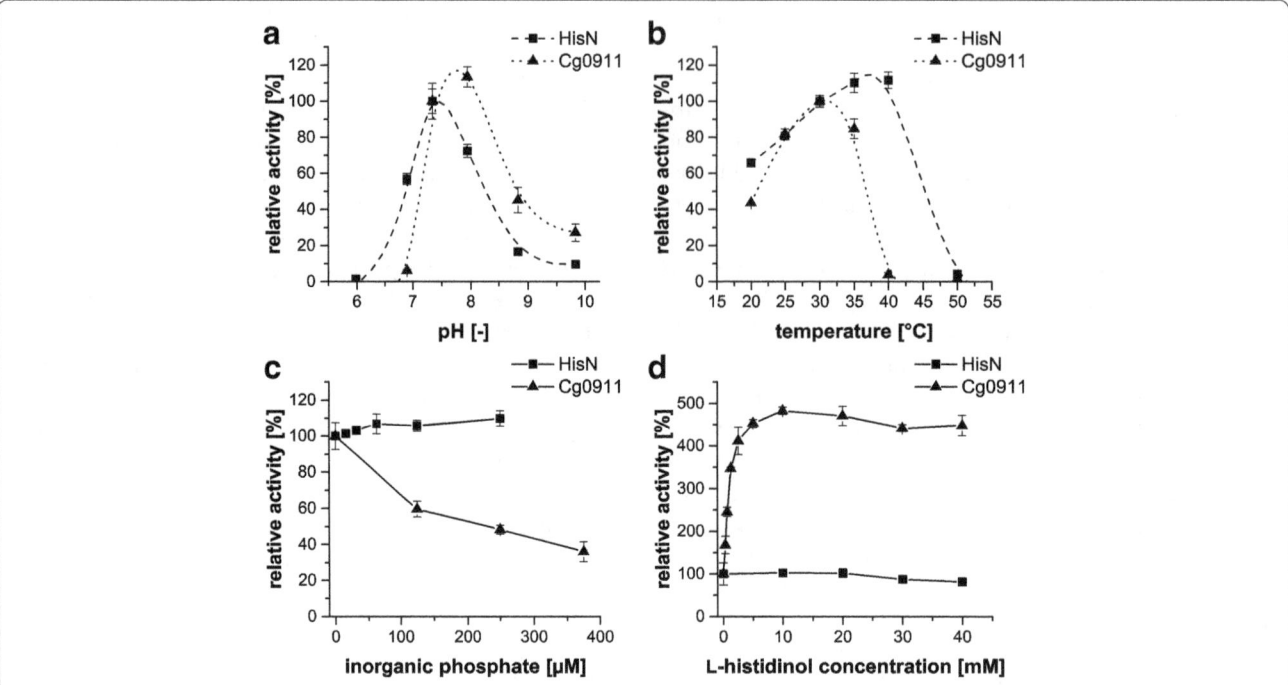

Fig. 4 Enzyme characteristics of HisN (■) and Cg0911 (▲). **a** Effect of different pH values. The activity at pH 7.35 was set to 100% for each protein individually. Assay conditions: 30 °C, 5 mM MgSO₄, 500 µM HolP, n = 4. **b** Effect of incubation temperature. The activity at 30 °C was set to 100% for each protein individually. Assay conditions: pH = 7.35, 5 mM MgSO₄, 500 µM HolP, n = 4. **c** Influence of inorganic phosphate (Pᵢ) and **d** influence of L-histidinol. The activity without the addition of Pᵢ or L-histidinol was set to 100% for each enzyme individually. Assay conditions: 30 °C, pH = 7.35, 5 mM MgSO₄, 250 µM (HisN) and 500 µM (Cg0911) HolP. 1 M TEA-puffer instead of 0.1 M was used in the L-histidinol experiments. n = 3

about 650 µM and 5000 µM for HisN and Cg0911, respectively. They were about 30-times and 10-times higher compared to the K_m values for HolP, respectively. The Hill coefficients regarding Mg^{2+} were around 2.4 for HisN and around 3.0 for Cg0911, indicating pronounced cooperativity of both enzymes in respect to the metal ion. This assumption is reinforced by the observation that no HisN or Cg0911 activity was measurable at Mg^{2+} concentrations ≤ 100 µM or ≤ 625 µM, respectively (data not shown).

Neither HisN nor Cg0911 showed any phosphatase activity against the general phosphatase substrate *para*-nitrophenyl phosphate (data not shown). The

ability to hydrolyze other natural phosphatase substrates was not tested.

Finally, the potential inhibition of HisN and Cg0911 by L-histidine or the two direct reaction products L-histidinol and Pᵢ was examined. No inhibitory effect of L-histidine was observed with concentrations up to 60 mM L-histidine (data not shown). A different effect of the addition of Pᵢ and L-histidinol to the reaction mixture was observed for HisN and Cg0911. While HisN was not inhibited by Pᵢ up to a concentration of 250 µM (higher concentrations were not tested for HisN), activity of Cg0911 decreased to 40% at 375 µM Pᵢ (Fig. 4c). Unfortunately, it was not possible to test the effect of higher Pᵢ concentrations, since the addition of external Pᵢ interferes with the detection of Pᵢ released during hydrolysis of HolP. It cannot be excluded, that HisN is inhibited by Pᵢ concentrations > 250 µM.

HolPase activity of HisN was also not affected by the presence of L-histidinol (Fig. 4d). The enzyme was fully active up to 20 mM L-histidinol. The slight reduction to 80% activity at 40 mM most likely reflects a pH artifact, since HisN activity is optimal at pH 7.35 and rising L-histidinol concentrations cause a drop in pH even in 1 M TEA buffer (pH 7.5 and pH 7.0 at 0 mM and 200 mM L-histidinol, respectively; estimated with pH indicator stripes at RT).

Table 2 Kinetic parameters of HisN and Cg0911 at 30 °C and pH 7.35, n = 6

		HisN	Cg0911
for HolP	k_{cat} [s⁻¹]	1.04 ± 0.06	0.013 ± 0.002
	K_m [µM]	23.6 ± 1.4	638.3 ± 70.5
	k_{cat}/K_m [s⁻¹ M⁻¹]	4.41×10^4	1.98×10^1
	Hill coefficient	1.47 ± 0.18	1.83 ± 0.13
for Mg^{2+}	K_m [µM]	644.8 ± 22.4	5155.8 ± 117.9
	k_{cat}/K_m [s⁻¹ M⁻¹]	1.61×10^3	2.46×10^0
	Hill coefficient	2.44 ± 0.20	3.04 ± 0.12

Surprisingly, we observed a stimulating effect of L-histidinol on the HolPase activity of Cg0911. The activity increased almost five-fold at L-histidinol concentrations ≥ 10 mM. Half maximal stimulation was reached at 0.86 ± 0.06 mM L-histidinol. Since no release of P_i was detectable if the substrate HolP was omitted from the assay (data not shown), any contamination of the L-histidinol reagent with P_i or other phosphorous substances can be excluded. It appears therefore, that Cg0911 is positively feedback regulated by L-histidinol.

Identification of sequence motifs for the discrimination of IMPase-like HolPases from other IMPase-like proteins in *C. glutamicum* and other bacteria

The presence of several IMPase-like proteins in one species (e.g. five in *C. glutamicum*) complicates the discrimination between an IMPase-like HolPase and IMPase-like proteins with different substrate specificities. Within the class *Actinobacteria*, it is rather easy to identify the HolPases due to a much higher sequence similarity to HisN$_{Cg}$ than to the other IMPase-like proteins. However, this becomes more difficult in other bacterial phyla or even in different kingdoms. Therefore, we were interested in the identification of amino acid motifs that allow the unambiguous discrimination of IMPase-like HolPases purely based on the protein sequence.

For each of the five IMPase-like proteins in *C. glutamicum* (HisN, Cg0911, SuhB, ImpA, and CysQ) we performed a multiple sequence alignment of potential orthologs from a wide range of bacteria to identify highly conserved amino acid residues. The comparison of the highly conserved residues in each group of orthologs allowed the determination of six amino acid motifs distributed over the entire protein sequence that can be used for the discrimination of HisN orthologs from other IMPase-like proteins. Orthologs were identified by a BLASTP search using the respective protein sequence from *C. glutamicum* as query. A BLASTP score ≥ 125 was set as cut-off for identification. This cut-off was chosen, because it was sufficient to reliably distinguish between HisN and the other IMPase-like proteins in *C. glutamicum*, *M. tuberculosis*, and *S. coelicolor* (data not shown). With very few exceptions, maximum one (HisN, SuhB, and CysQ) or three sequences (Cg0911 and ImpA) per genus were randomly chosen for the multiple sequence alignment (see Additional file 2 for a complete list of used sequences).

Since we were most interested in motifs for the identification of IMPase-like HolPase, only the HisN motifs will be described in detail below.

Motif 1 consist of a strictly conserved lysine (Lys36), a highly conserved aspartate (Asp38), a threonine or serine at position 40, followed by a highly conserved proline (Pro41), a strictly conserved valine (Val42) and

threonine or serine at position 43. An aspartate at position 46 is strictly conserved in all analyzed IMPase-like proteins and can be used for positioning of motif 1. Interestingly, motif 1 is completely absent in some of the HisN orthologs (11 out of 147 analyzed sequences mostly from *Alpha-* or *Gammaproteobacteria*). However, a different conserved motif is present in these cases consisting of lysine at position 34 or 35, an aromatic amino acid at position 40 and aspartate at position 41 followed by valine (Val42) and threonine (Thr43) (not shown).

Motif 2 consist of a highly conserved glycine (Gly68) followed by two strictly conserved glutamate residues (Glu69 and Glu70). Motif 2 is very similar in all analyzed groups of orthologs, with the exception of CysQ, where the conserved glycine is replaced by a strictly conserved serine (Ser64) and preceded by a highly conserved leucine (Leu63). Therefore, motif 2 is most suitable for the discrimination of HisN homologs from CysQ orthologs.

Motif 3 contains four of the active site key residues typical of all IMPase-like proteins (HisN$_{Cg}$: Asp85, Ile87, Asp88, and Thr90) [28, 29]. But not only these four residues are strongly conserved in HisN and the other IMPase-like proteins, but all residues ranging from positions 82 to 90. Therefore motif 3 is most suited for the identification of IMPase-like proteins in general. Striking differences between the different ortholog-groups within motif 3 appear only at position 91. There is a preference for a lysine at this position in HisN orthologs.

A highly conserved motif 4 is only present in CysQ orthologs. However, there exists a motif 4 in HisN orthologs, too. It consist of a moderately conserved arginine (Arg95), a strongly conserved glycine (Gly96) and proline (Pro98), followed by an aromatic amino acid at position 100, a strongly conserved threonine (Thr102) and a strongly conserved leucine (Leu103). Especially the combination of the aromatic amino acid at position 100 followed by Thr102 and Leu103 is very typical of HisN orthologs.

Motif 5 is the most characteristic motif of HisN orthologs. It consists of a highly conserved arginine (Arg187; replaced by Val or Leu in many alphaproteobacteria), a highly conserved glycine (Gly190) and an almost strictly conserved aspartate (Asp191). Only in some sequences of *Gammaproteobacteria* Asp191 is replaced with glutamate. Neither aspartate nor glutamate was found at this position in any other of the analyzed IMPase-like protein sequences. Moreover, several aromatic amino acids are present in motif 5 of HisN orthologs. One of these aromatic amino acids is present at position 188 or more likely 189, with the respective other position being occupied by a small residue (mostly glycine or alanine). Two more aromatic residues are usually present at position 193 and 195. Especially in actinobacterial

HisN orthologs, there is also an additional aromatic amino acid at position 192. Whereas usually only phenylalanine, tyrosine or tryptophan residues are present at positions 188, 189, 192 and 193 the aromatic amino acid histidine might be present at position 195. No aromatic amino acids are present at the positions 192–195 in the corresponding motifs of the other IMPase-like proteins. Therefore, this motif is very specific for HisN orthologs. Next to the already described characteristics, the HisN-specific motif 5 is lacking a highly conserved aspartate followed by a leucine residue that are present in Cg0911, SuhB and ImpA orthologs (Asp203 and Leu204 in SuhB$_{Cg}$). A specific motif 5 can also be identified in the other analyzed groups of IMPase-like proteins. Two consecutive arginine residues (Arg195 and Arg196 in SuhB$_{Cg}$), followed by the sequence GSAAL, are very typical of SuhB orthologs. On the other hand, two arginine residues interspaced by a non-conserved amino acid (Arg179 and Arg181 in ImpA$_{Cg}$) are very typical of ImpA orthologs.

The last motif, motif 6, is very similar in all analyzed IMPase-like ortholog groups. It contains the strictly conserved aspartate residue (Asp215) involved in coordination of the metal ions in the active site [28, 29]. Most interesting for discrimination between HisN and the other groups is position 219. Whereas a very highly conserved glycine (Cg0911, SuhB or ImpA) or a proline (CysQ) is usually present at this position in the other groups of orthologs, no glycine was present at position 219 in any of the analyzed HisN orthologs.

Although we included only sequences of bacterial IMPase-like HolPases in our motif search, all six identified HolPase motifs can also be found in the protein sequence of HISN7 from the plant A. thaliana (Additional file 1: Figure S2). HISN7$_{At}$ has been previously experimentally confirmed as IMPase-like HolPase [9], despite its low overall sequence similarity to HisN$_{Cg}$ (24% identity, 36% similarity, BLASTP-score: 103 bits).

Identification of IMPase-like HolPases based on the described sequence motifs and experimental validation of HolPase activity by complementation experiments

In order to prove the value of the identified HolPase motifs, different potential HisN orthologs were tested for their ability to complement a C. glutamicum ΔhisN strain, thus demonstrating HolPase activity of the respective gene products (Table 3). The potential HolPase genes from the actinobacterium Dietzia sp. strain Chol2 (genome announcement in preparation; preliminary locus tag Dietzia_sp.-Draft_1801, here referred to as hisN$_{Dz}$) and the alphaproteobacterium Zymomonas mobilis ZM4 ([30]; locus tag ZMO_RS06805, here referred to as hisN$_{Zm}$) were chosen, because the HolPase motifs are conserved in the respective gene products despite a relatively low overall sequence similarity to HisN$_{Cg}$.

In addition, we investigated potential HisN orthologs from Actinoplanes utahensis NRRL 12052 [31]. This actinobacterium possesses two genes encoding IMPase-like proteins that are most similar to HisN$_{Cg}$. The first gene product (locus tag MB27_13025, referred to as HisN$_{Au}$) is characterized by a high sequence similarity to HisN$_{Cg}$ and the presence of all six HolPase motifs. The second gene product (locus tag KHD72131.1, for convenience reasons referred to as HisN2$_{Au}$) is also more similar to HisN$_{Cg}$ then to another IMPase-like proteins in C. glutamicum and five of the six identified motifs are at least moderately conserved. However, motif 5 is absent in this protein (Table 3, Additional file 1: Figure S3).

The above described genes were cloned into the constitutive pZMP vector and tested for their ability to complement the L-histidine bradytrophic growth phenotype of the C. glutamicum ΔhisN mutant. As expected, hisN$_{Dz}$, hisN$_{Zm}$, and hisN$_{Au}$ were able to fully complement the C. glutamicum ΔhisN mutant (data not shown). In contrast, the expression of hisN2$_{Au}$ failed to complement the C. glutamicum ΔhisN mutant, even though the overall similarity of the gene product to

Table 3 List of potential HisN orthologs. Characteristic amino acid residues of HolPase motif 5 are underlined. HolPase activity was inferred from the ability of the corresponding gene to complement a C. glutamicum ΔhisN mutant

| potential HisN ortholog | protein length [aa] | sequence similarity on protein level (identity, similarity, BLASTP score) [%], [%] [bits] | | | | | HolPase motif 5 | HolPase activity |
		HisN	Cg0911	ImpA	SuhB	CysQ		
Dietzia sp. strain Chol2 HisN$_{Dz}$	278	29, 47 105	25, 40 76	23, 33 69	26, 38 49	23, 35 46	yes R**F**G**GDC**Y**A**Y	yes
Zymomonas mobilis HisN$_{Zm}$	259	32, 48 101	22, 35 69	22, 36 58	26, 40 58	24, 37 62	yes LLG**GDC**Y**N**Y	yes
Actinoplanes utahensis HisN$_{Au}$	266	55, 68 261	27, 40 76	26, 40 88	26, 40 71	26,32 56	yes RA**YGDFY**GY	yes
Actinoplanes utahensis HisN2$_{Au}$	259	36, 51 119	24, 35 70	24, 39 61	27, 36 67	24, 37 58	no ————QPS	no

HisN$_{Cg}$ is higher than that of HisN$_{Dz}$ and HisN$_{Zm}$. These results underline the importance of the motif 5 (Fig. 5) for HolPase activity of IMPase-like proteins.

Our continuing analyses revealed that many IMPase-like proteins within NCBI's non-redundant protein database are either not classified in more detail (mostly only as IMPases or IMPase-like proteins) or are even wrongly classified. Some examples of misclassified IMPases are

given in Table 4. By comparing the amino acid sequence of these IMPase-like proteins to the five IMPase-like proteins in *C. glutamicum* and by checking for the presence of the expected motifs, we were able to assign a more accurate function to these proteins.

This list demonstrates that many IMPase-like Hol-Pases are not recognized as such in the databases (class 1). By checking for the presence of the HolPase motifs it

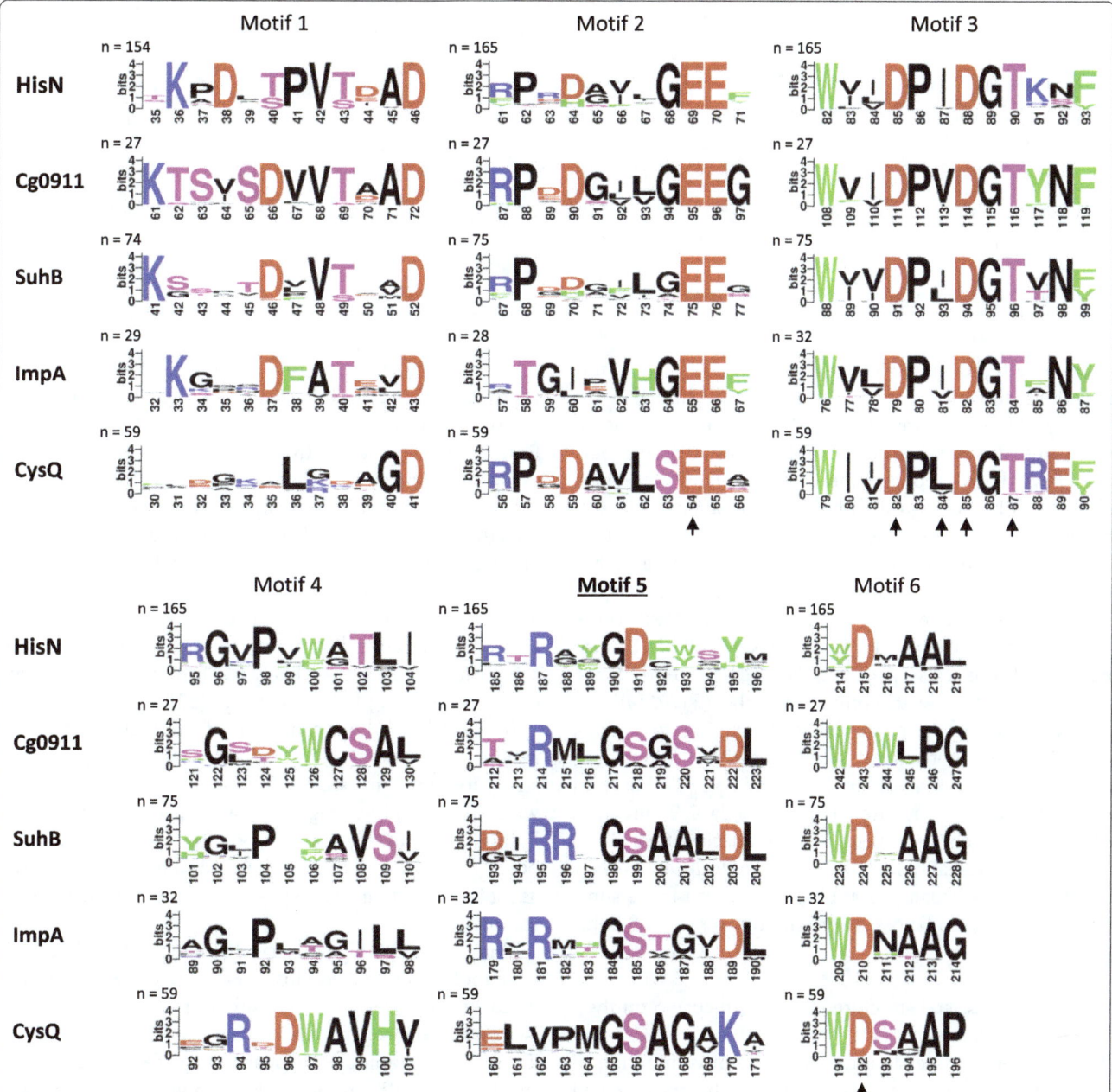

Fig. 5 Conserved amino acid residues in groups of potential HisN, Cg0911, SuhB, ImpA, and CysQ orthologs that can be used for the discrimination of the different protein groups. Numbering of the amino acids corresponds to the five proteins from *C. glutamicum*. The arrows indicate active site key residues that are involved in binding of the three catalytic metal ions, the phosphate group of the substrate and/or the activation of the water molecule for hydrolysis of the phosphate ester bond as derived from the solved crystal structures of various IMPases [28, 29]. Coloring: blue = Arg, Lys; red = Asp, Glu; pink = Ser, Thr; green = His, Phe, Trp, Tyr; black = all remaining

Table 4 List of randomly chosen bacterial IMPase-like proteins that are most likely wrongly annotated with a focus on HolPases. All proteins have been compared to the five IMPase-like proteins in *C. glutamicum* and reannotated according to the detected motifs

Organism	GI-number	current annotation	protein with highest similarity in *C. glutamicum*	presence of expected motifs	similarity to $HisN_{Cg}$ (BLASTP score) [bits]	new functional assignment
Class 1						
Pseudomonas pseudoalcaligenes	489546409	IMPase	$HisN_{Cg}$	yes	117	HisN
Rudaea cellulosilytica	648601632	IMPase	$HisN_{Cg}$	yes	108	HisN
Marinobacter psychrophilus	860341079	IMPase	$HisN_{Cg}$	yes	110	HisN
Rubrobacter radiotolerans	627778235	FBPase- or IMPase	$HisN_{Cg}$	yes	123	HisN
Class 2						
Streptosporangium roseum	502655523	IMPase	$HisN_{Cg}$	no	120	-
Rubrobacter xylanophilus	499884516	IMPase	$HisN_{Cg}$	no	120	-
Class 3						
Sediminibacterium salmoneum	739492005	HolPase	$SuhB_{Cg}$	yes	102	SuhB
Flavihumibacter solisilvae	743029970	HolPase	$SuhB_{Cg}$	yes	91	SuhB
Desulfatibacillum aliphaticivorans	654864062	HolPase	$SuhB_{Cg}$	yes	102	SuhB
Leptospirillum ferriphilum	738123182	HolPase	$SuhB_{Cg}$	yes	89	SuhB

is possible to accurately classify even those HisN homologs that show only a moderate overall similarity to $HisN_{Cg}$ (BLASTP score < 125 bits). On the other hand, there are also many examples of IMPase-like proteins that have been wrongly annotated as HolPases. Two classes can be distinguished here. The first class (class 2) consists of proteins which indeed are most similar to $HisN_{Cg}$, however the overall sequence similarity is rather low (BLASTP scores usually < 125 bits). Most importantly, HolPase motif 5 is missing in these proteins. Next to the two examples given in Table 4, $HisN2_{Au}$ (Table 3) also belongs to this class 2 of misclassified proteins. Since $hisN2_{Au}$ was unable to complement the *C. glutamicum* $\Delta hisN$ strain, all HisN homologs belonging to class 2 most likely do not exhibit HolPase activity. Their substrate specificity remains to be elucidated. The second class of wrongly annotated HolPases (class 3) includes sequences which have been simply misclassified. They exhibit a comparably low sequence similarity to $HisN_{Cg}$ (BLASTP scores usually < 100 bits), are indeed more similar to $SuhB_{Cg}$, and possess the motifs typical of SuhB orthologs.

Survey of the crystal structure of HisN$_{Zm}$ focusing on the conserved HolPase motifs

To get a better understanding of the putative function of some of the conserved residues within the six detected HolPase motifs, we had a closer look on the IMPase-like HolPase from *Z. mobilis* ($HisN_{Zm}$). $HisN_{Zm}$ is only moderately similar to $HisN_{Cg}$ but all HolPase motifs are present (Additional file 1: Figure S2) and we were able to experimentally verify its HolPase activity (see above).

The crystal structure of this protein has been solved recently by Hwang et al. in 2014. Up to date, it represents the only solved crystal structure of an IMPase-like HolPase. There is evidence from the crystal structure as well as from gel filtration experiments that native $HisN_{Zm}$ is a homodimer [32]. Notably, Hwang et al. did not recognize $HisN_{Zm}$, which they refer to as CbbF, being a HolPase. The protein has been crystallized by Hwang and coworkers in its *apo* form without metal ions, which are needed for enzymatic activity, or any substrate [32]. However, the crystal structure contains a sulfate ion at the position which most likely resembles the binding site of the substrate's phosphate group [32].

We examined the localization of the highly conserved residues of the six HolPase motifs in the $HisN_{Zm}$ crystal structure and investigated their putative interactions with other residues. Fig. 6 shows a part of the $HisN_{Zm}$ homodimer, centered on one of the two identical supposed active sites depicted as space-filling model (**a**) and as ribbon diagram with stick representation of selected residues (**b**).

Many of the highly conserved residues within the HolPase motifs are located close to the active center as indicated by the location of the sulfate ion. This sulfate ion is forming hydrogen bonds with Asp86, Asp89, Gly90, Thr91 (corresponding to position 85, 88, 89 and 90 of the HisN-specific motif 3; Fig. 5) and Asp210 (motif 6; position 215). A part of these residues, as well as Ile88 (motif 3; position 87) and Glu70 (motif 2; position 69), are supposed to be involved in coordination of three catalytic Mg^{2+} ions according to the known structures of different IMPases [28, 29].

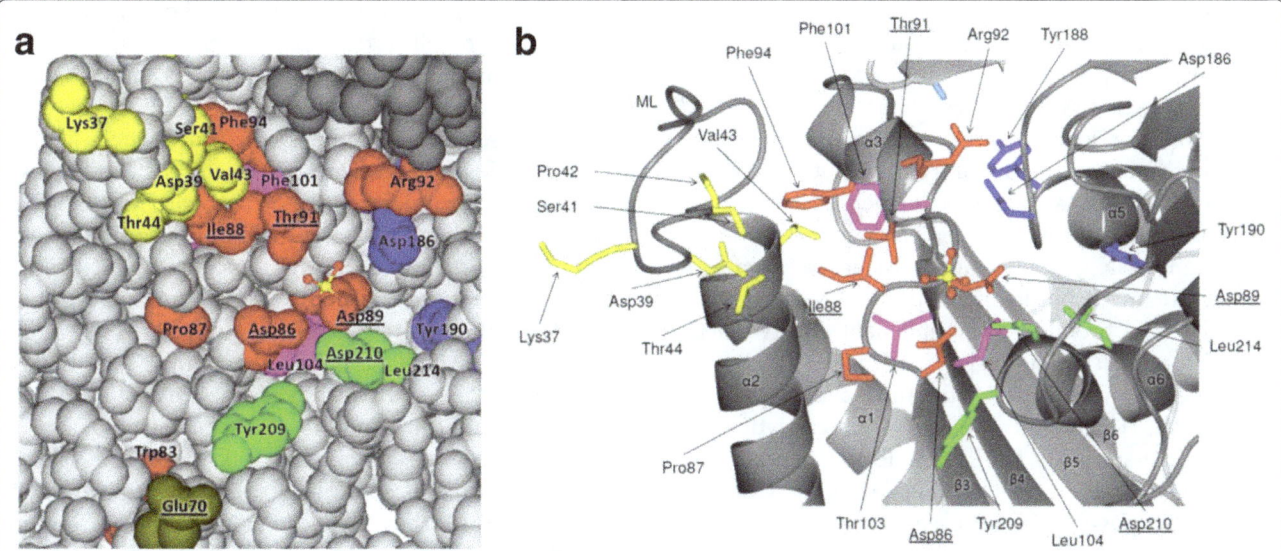

Fig. 6 Probable active site of HisN$_{Zm}$ depicted as space-filling model (**a**), illustrating surface exposed residues, and as ribbon diagram with stick representation of selected residues (**b**) based on the crystal structure of the homodimer as determined by Hwang et al. (2014) [32]. The two monomers are depicted in different gray shades. The side chains of highly conserved residues typical of IMPase-like HolPases are highlighted: yellow = motif 1, olive = motif 2, red = motif 3, pink = motif 4, blue = motif 5, green = motif 6 (compare Fig. 5). The location of the sulfate ion (ball-and-stick model) represents the most likely binding site of the phosphate moiety of the substrate HolP. Key active site residues involved in binding of metal ions (not present in the HisN$_{Zm}$ crystal), the substrates phosphate moiety and activation of the water molecule for ester hydrolysis, as derived from the structures of different IMPases [28, 29], are underlined

The side chains of most of the highly conserved residues of motif 1 (Lys37, Asp39, Ser41, Val43 and Thr44; corresponding to positions 36, 38, 40, 42, and 43 in Fig. 5) point to the side of the enzyme where the active site is located. This part of the enzyme (residues 29–41) has been recognized as a mobile catalytic loop in different IMPases-like proteins which changes its spatial position in response to binding of metal ions or the substrate [33]. Therefore, these residues might play a crucial role in recognition of the substrate HolP.

Several of the conserved aromatic amino acids within the different motifs seem to play an important role for the formation of the tertiary and quaternary structure of HisN$_{Zm}$ and IMPase-like HolPases in general. For example, Phe94, corresponding to the conserved aromatic amino acid at position 93 in the HisN-specific motif 3 (Fig. 5), has hydrophobic interactions with nine other amino acids (including residues from the motifs 1, 3 and 4). Two of these interactions are additionally stabilized by aromatic-aromatic interactions. All these interactions connect the α-helices 1, 2, and 3 with the active site key residue Ile88 and thereby contribute to the formation of the active site. A similar structural function might be attributed to some of the conserved branched chain amino acids. Leu214, for instance, which is very typical of IMPase-like HolPases (motif 6, position 219), has hydrophobic interactions with the likewise conserved Leu103 (motif 4, position 103) and three additional residues. One of the residues interacting with Leu214 is Tyr190,

one of the aromatic amino acids highly conserved within motif 5 of IMPase-like HolPases (motif 5, position 195). Therefore, this aromatic amino acid might primarily have a structural function. However, Tyr190 is the only of the typically three conserved aromatic amino acids within motif 5 that is at least partially exposed to the surface and located close to the supposed substrate binding site (Fig. 6a). It is therefore possible that Tyr190 is additionally involved in substrate recognition, possibly by aromatic interaction with the likewise aromatic substrate HolP. The two other highly conserved aromatic amino acids within motif 5 of IMPase-like HolPases correspond to Tyr188 (motif 5, position 193) and Leu183 (motif 5, position 188) in HisN$_{Zm}$. The side chains of both residues are located on the "back side" of the enzyme and distant from the active site. Leu183 is interacting with several other hydrophobic amino acids from the second subunit of the HisN$_{Zm}$ homodimer (among others with Leu183 itself). Tyr188 has some intramolecular hydrophobic and aromatic interactions stabilizing the tertiary structure, but it additionally forms a hydrogen bond with Arg29 from the second subunit. An important role of Leu183 and Tyr188 might therefore be the stabilization of the quaternary structure of the HisN$_{Zm}$ homodimer.

Most important, the analysis of the HisN$_{Zm}$ crystal structure suggest a direct involvement of Asp186 (motif 5, position 191) in substrate recognition. The Asp186 side chain is accessible to the solvent, points towards the supposed substrate binding site (Fig. 6a), and there is no indication that the carboxylic group is involved in the formation of any H-bonds or salt bridges. Consequently, Asp186 would be available for interaction with the substrate HolP, for instance by the formation of H-bonds between the amino group of HolP and the carboxylic group of the aspartate. This is in good agreement with our observation that replacement of the conserved Asp191 in $HisN_{Cg}$ with alanine, serine, or asparagine, but not glutamate, results in a considerably reduced ability of the gene products to complement a *hisN* deletion in *C. glutamicum* (Additional file 1: Figure S4). However, since there is no crystal structure available of any IMPase-like HolPase in complex with catalytic metal ions and the substrate HolP or at least the products L-histidinol or P_i, any interaction between HolP and Asp186 (and possibly Tyr190) remains speculative.

Distribution of HisN and Cg0911 orthologs within *bacteria*
The presence of an IMPase-like HolPases has so far only been experimentally proven in *C. glutamicum* and *S. coelicolor* [7, 10], but there is evidence, that this type of HolPase is a general feature of the *Actinobacteria* [5]. According to this assumption, we were able to prove the in vivo HolPase activity of HisN homologs in the actinobacterial genera *Actinoplanes* and *Dietzia* in the present study. However, the recent identification of the IMPase-like HolPase in the plant *Arabidopsis thaliana* [9], the results of our extensive BLAST-analysis in order to identify the HolPase motifs, and finally our experimental confirmation of a functional IMPase-like HolPase in the alphaproteobacterium *Z. mobilis* suggests that this type of HolPases might be more widespread than initially assumed. Therefore, we systematically examined the distribution of HisN orthologs within the bacterial kingdom and additionally extended the analysis to Cg0911 orthologs. The results are depicted in Fig. 7.

In this analysis, HisN and Cg0911 homologs were identified by a protein BLAST search (BLASTP) within NCBI's non-redundant protein sequences database. A BLASTP score ≥ 125 was set as cut-off for identification. This cut-off was chosen, because it was sufficient to reliably distinguish between HisN, Cg0911, and the other IMPase-like homologs ImpA, SuhB, and CysQ in *C. glutamicum, M. tuberculosis,* and *S. coelicolor* (data not shown). In addition, all putative HisN orthologs were checked for the presence of the HisN-specific motif 5. A HisN or Cg0911 ortholog was regarded a general feature of the genus if it was present in at least one species belonging to this genus. It was regarded a general feature of the family, if it was present in at least three or half of all genera, and the same criteria applied to the higher taxonomic levels.

According to this analysis, HisN orthologs are a general feature of all orders of the class *Actinobacteria* (BLASTP scores > 250 bits). The only exception are the *Kineosporiales*, however this is most likely attributed to the lack of sequence data for some of the genera. Indeed, HisN orthologs are present in *Kineococcus* and *Angustibacter*. HisN orthologs with an unusually low similarity to $HisN_{Cg}$ within the class *Actinobacteria* are present within the *Dietziaceae* (BLASTP scores ≤ 108 bits). However, despite the overall low sequence similarity we could identify all HolPase motifs (alternative motif 1) in all potential HisN orthologs within *Dietziaceae*. Additionally we proved the in vivo HolPase activity of the HisN homolog from *Dietzia* sp. strain Chol2 (see above). A BLASTP query revealed highest sequence similarity of these HisN orthologs to HisN orthologs from different species of the order *Rhizobiales* (max. BLASTP score: 210 bits), indicating a recent horizontal gene transfer event.

Although widely distributed within the class *Actinobacteria*, HisN orthologs are not generally present in all classes of the phylum *Actinobacteria*. They can be identified in *Acidimicrobiia* (genus *Ilumatobacter*; max. BLASTP score: 179 bits), *Nitriliruptoria* (genus *Nitriliruptor*; max. BLASTP score: 174 bits), and *Rubrobacteria* (genus *Rubrobacter*, max. BLASTP score: 123 bits), however with considerably lower BLASTP scores as compared to the class *Actinobacteria*. In contrast, no HisN orthologs were identified in the classes *Coriobacteriia* and *Thermoleophilia*, despite the availability of complete genome sequences.

IMPase-like HolPases were also identified outside the phylum *Actinobacteria*. The presence of HisN orthologs seems to be a general feature of the phyla *Chlorobi* (green sulfur bacteria), *Fibrobacteres* (cellulose-degrading bacteria), and *Nitrospinae* (marine nitrite oxidizing bacteria). It is also generally found in the class *Chloroflexia* within the phylum *Chloroflexi* (green non-sulfur bacteria) and the class *Planctomycetia* within the phylum *Planctomycetes* (aquatic bacteria). The HisN orthologs from *Planctomycetaceae* exhibit particularly high similarity to $HisN_{Cg}$ (BlastP scores: 162–194 bits). HisN orthologs were also identified in some members of the family *Chitinophagaceae*, phylum *Bacteroidetes*, and the order *Spirochaetales*, phylum *Spirochaetes*.

We also identified HisN orthologs within the phylum *Proteobacteria*, with the exception of *Epsilonproteobacteria*. They are generally present in the alphaproteobacterial order *Rhizobiales*, in the deltaproteobacterial order *Myxococcales*, and in the betaproteobacterial family *Burkholderiaceae*. Interestingly, HisN orthologs are also present in many *Gammaproteobacteria*, which are known for the presence of a bifunctional HAD-type HolPase [12]. Five families with a general occurrence of

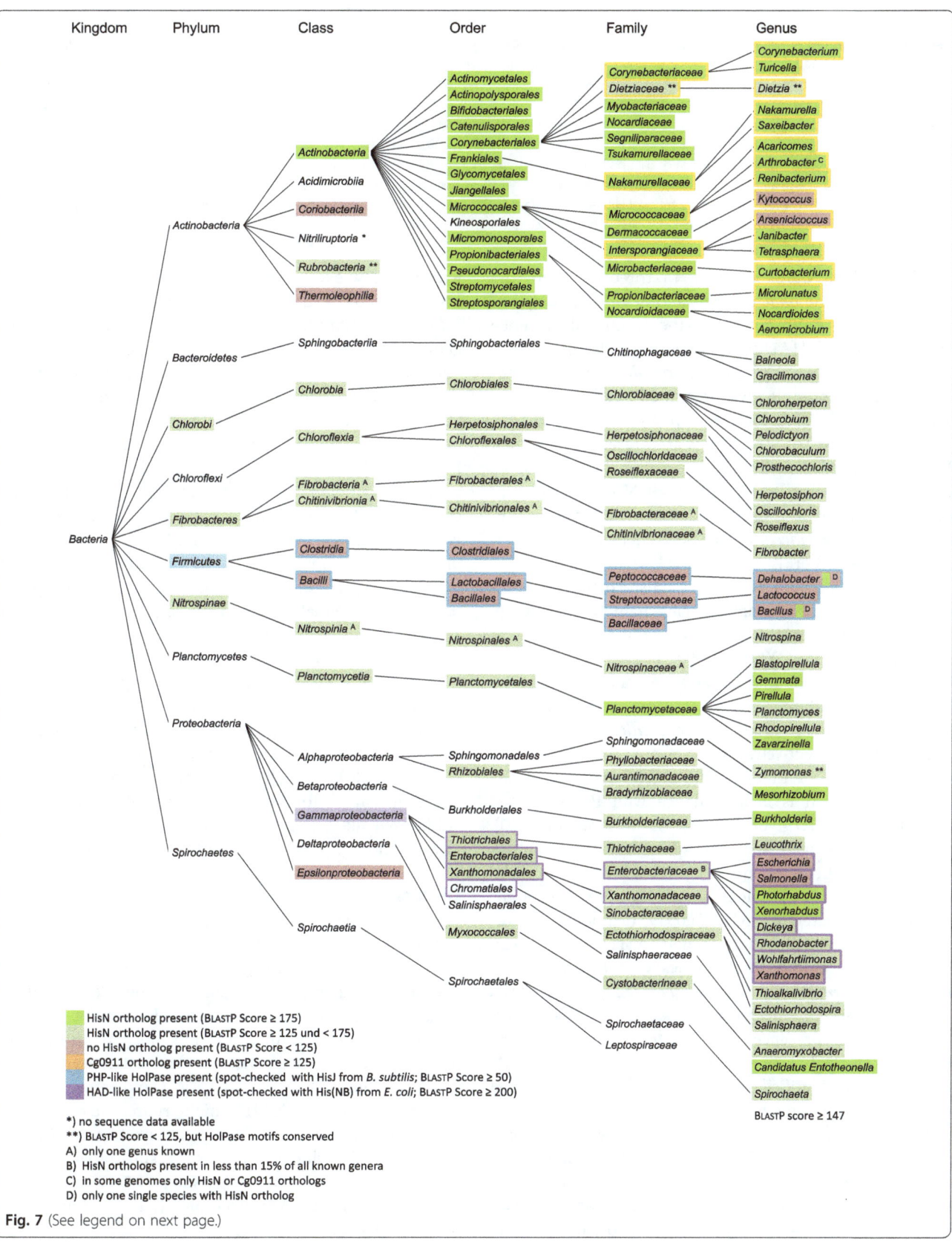

Fig. 7 (See legend on next page.)

(See figure on previous page.)
Fig. 7 Distribution of putative HisN and Cg0911 orthologs within *Bacteria*. Orthologs were identified by BlastP using the *C. glutamicum* ATCC 13032 HisN and Cg0911 protein sequences as query within NCBI's non-redundant protein sequences database (nr). A BlastP score ≥ 125 and a sequence coverage ≥ 80% were set as cut-off for identification. All hits were additionally checked for the HolPase motif 5 identified in this study (Fig. 5). Presence of a HisN or Cg0911 ortholog in at least one species is indicated on genus level by a *green* background or an *orange* surrounding, respectively. Since the actinobacterial branch focuses on the distribution of Cg0911 orthologs, display of HisN orthologs is reduced to family level and above. The same colors were applied on the family level and above if HisN or Cg0911 orthologs were identified in at least three or half of the entities of the lower level. A red background indicates that no HisN ortholog was found with the applied cut-off or did not possess the HolPase motif. The expected presence of a PHP-type HolPase [20] is marked with a *blue* surrounding and was spot-checked using the *B. subtilis* HisJ protein sequence via BlastP. The expected presence of a bifunctional HAD-type HolPase [12] is marked with a *purple* surrounding and was spot-checked using the *E. coli* His(NB) protein sequence via BlastP. Other kingdoms were not included in the analysis

HisN orthologs were observed. In three of them, the *Thiotrichaceae*, *Sinobacteraceae*, and *Ectothiorhodospiracea*, our analysis did not reveal the additional presence of a bifunctional HAD-type HolPase. In the other two, *Enterobacteriaceae* and *Xanthomonadaceae*, a bifunctional His(NB) homolog was identified in all genera with a putative HisN ortholog. However, in the case of *Enterobacteriaceae*, HisN orthologs are present in less than 15% of all genera listed in NCBI taxonomy (11 of 76, including "candidatus" genera). No HisN orthologs were identified in *Escherichia* and *Salmonella*, two genera from *Enterobacteriaceae* with a well characterized HAD-type HolPase [13–15].

Within the phylum *Firmicutes*, HisN orthologs were only identified in two single species, namely *Bacillus* sp. EGD-AK10 (draft; AVPM00000000.1) and *Dehalobacter sp.* FTH1 (draft; AQYY00000000.1). This might be attributed to a recent horizontal gene transfer, according to sequence similarity most likely from an actinobacterial species from the orders *Micrococcales* and *Propionibacteriales*, respectively. Apart from that, no HisN orthologs were identified in any other member of the *Firmicutes*, which is in accordance with the supposed presence of a PHP-type HolPase in this phylum [20], which was positively spot-checked during our analysis using the HisJ protein sequence of *B. subtilis*.

Unlike the HisN orthologs, which are spread throughout various bacterial phyla, Cg0911 orthologs are restricted to a few actinobacterial genera (Fig. 7). They can be generally found in the families *Corynebacteriaceae*, both in *Corynebacterium* and *Turicella*, *Dietziaceae*, *Nakamurellaceae*, *Micrococcaceae*, and *Intersporangiaceae*, but their presence is not restricted to these families. Interestingly, only a Cg0911 but not a HisN ortholog was identified in *Kytococcus* for which at least one complete genome is available (CP001686.1). In those species that contain both orthologs and where genome data was available (finished and draft genomes), we checked for the gene organization. In *C. glutamicum* the *cg0911* gene is directly followed by *hisN* and the two genes form an operon [5]. The same organization in such a *cg0911-hisN* homolog tandem was also found in almost all other available *Corynebacterium* genomes

with very few exceptions (data not shown). The *cg0911-hisN* homolog tandem was also present in *Turicella otitidis* (draft; CAJZ00000000.1), however not in any other genus possessing both homologs, and can be therefore considered a characteristic of *Corynebacteriaceae* only.

In order to validate the BlastP results, the tool JACKHMMER was used to create and refine HMMs based on all functionally validated orthologs of HisN (HisN$_{Cg}$, HisN$_{Au}$, HisN$_{Dz}$, and HisN$_{Zm}$) respectively Cg0911$_{Cg}$ using a E-value cutoff of 1e-65. In both cases the results corroborate the BlastP results. In addition, the searches with the HisN HMM revealed HisN to be generally present in the alphaproteobacterial orders *Rhodobacterales*, *Rhodospirillales*, *Caulobacterales*, and *Sphingomonadales*. Additionally, HisN orthologs were identified in several species within the phyla *Cyanobacteria* and *Verrucomicrobia*. While below the BlastP cutoff, motif 5 (as well as the others) was found to be present in 1687 out of 1695 sequences identified by the HMM, with the exeption of several sequences from "Candidatus Curtissbacterium" species and a few others (close to the gathering threshold). The HMMs obtained after the final iteration can be found in Additional file 3 (HisN) and Additional file 4 (Cg0911) and can be used for an easy classification of these two groups in the future.

Discussion
Of the four genes encoding HisN paralogs within the genome of *C. glutamicum* (namely *cg0911*, *impA*, *suhB*, and *cysQ*) only *cg0911* is capable of at least partially complementing the growth defect of the ΔhisN strain in L-histidine free medium in vivo. The results with the purified Cg0911 enzyme confirmed its HolPase activity also in vitro. However, the very low catalytic efficiency k_{cat}/K_m of only 1.98×10^1 s^{-1} M^{-1} indicates, that the HolPase activity of Cg0911 might represent only a side activity of this enzyme. The actual substrate of Cg0911 remains to be elucidated. Known substrates of other IMPase-like proteins are, e.g., inositol-1-P, inositol-2-P, inositol-3-P, glucitol-6-P, glycerol-2-P, 2'-AMP, and L-galactose-1-phosphate [25, 34, 35].

Particular surprising was the fact that *impA* does not encode a protein with HolPase activity. This gene is part of an operon with other L-histidine biosynthesis genes in *C. glutamicum* [5] and a similar gene arrangement is also observed in many other species of different genera including *Corynebacterium*, *Dietzia*, *Gordonia*, *Mycobacterium*, and *Nocardia* (data not shown). The substrate of ImpA remains to be elucidated, but its involvement in mycobacterial cell wall biosynthesis is discussed [36].

The concurrent deletion of *hisN* and all its four paralogs in the *C. glutamicum* quintuple mutant demonstrates two things: Firstly, the absence of all five IMPase-like proteins does not result in complete L-histidine auxotrophy. Thus, at least one additional protein with HolPase activity must exist in *C. glutamicum*. Such an alternative non-IMPase-like HolPase has been identified in *S. coelicolor* [10], however a homolog is not present in *C. glutamicum* (data not shown). A HolPase side activity has been demonstrated for alkaline phosphatases in *S. cerevisiae* [17] and *Neurospora crassa* [37] and has also been reported in *E. coli* [38]. Such a side activity might also be present in *C. glutamicum*. Eventually, one should also consider the possibility of non-enzymatic dephosphorylation of HolP within the cell.

Secondly, the activity of all this five IMPase-like proteins is totally dispensable for growth of *C. glutamicum* on minimal medium under the tested conditions. IMPases are thought to synthesize *myo*-inositol from IMP. *Myo*-inositol is supposed to be mainly used for the synthesis of the coryne- and mycobacterial cell envelope phospholipids phosphatidylinositol and phosphatidylinositol dimannoside [11, 36, 39–41]. The results of the *C. glutamicum* quintuple mutant suggest that the two phospholipids mentioned above are dispensable for *C. glutamicum* or that *myo*-inositol synthesis is carried out by a yet unknown enzyme. The IMPase-like proteins might be additionally involved in other reactions than the synthesis of *myo*-inositol. A high in vitro activity with the substrates sorbitol-6-phosphate, next to IMP itself, has been demonstrated for SuhB from *M. tuberculosis* [25]. CysQ from *M. tuberculosis* exhibits a more than tenfold higher turnover number with the substrate 3'-phosphoadenosine-5'-phosphate (PAP) as compared to IMP and accepts also 3'-phosphoadenoside-5'-phosphosulfate (PAPS) as a substrate [42]. Therefore, it has been suggested that CysQ primarily functions as regulator of the sulfur assimilation in *M. tuberculosis* [42]. Based on high sequence similarity of $CysQ_{Mt}$ to $CysQ_{Cg}$ (47% identity, 58% similarity) the same enzyme function can be assumed. However, since *C. glutamicum* uses a PAPS independent sulfur assimilation route [43], the function of $CysQ_{Cg}$ remains uncertain.

Our results with the purified HisN prove the in vitro HolPase activity of an IMPase-like HolPase (Table 2) for the first time. The catalytic efficiency k_{cat}/K_m of 4.41×10^4 s^{-1} M^{-1} is four orders of magnitude lower compared to that of the HAD-type HolPase from *E. coli* [13] or *T.*

onnurineus [16], but it is in good agreement with the values of several PHP-type HolPases [20]. The k_{cat} of HisN fits very well to the k_{cat} reported for $HisG_{Cg}$, the ATP-PR transferase catalyzing the first step of L-histidine biosynthesis in *C. glutamicum* [44], demonstrating an equal catalytic rate for at least two of the nine enzymes involved in L-histidine biosynthesis.

The low k_{cat} of $HisN_{Cg}$, especially as compared to that of the HolPase from *E. coli*, might be partially compensated for by the high affinity of the enzyme to its substrate. The K_m-value of $HisN_{Cg}$ for HolP of only roughly 25 μM is the lowest value reported for any HolPase so far. The absence of inhibition of $HisN_{Cg}$ by L-histidine and L-histidinol reflects another strategy to deal with the low turnover number. A resistance to inhibition by L-histidine and L-histidinol can also by observed with the PHP-type HolPase of *S. cerevisiae* (K_i for L-histidinol: 5–10 mM [18]), which, based on kinetic data of other PHP-type HolPases [20], have a rather low k_{cat}. HAD-type HolPases on the other hand exhibit a high k_{cat} and are strongly inhibited by these two substances (e.g. K_i for L-histidinol = 52 μM in *S. enterica*) [14]. $HisN_{Cg}$ is also not inhibited by P_i at least not up to a concentration of 250 μM P_i (a concentration that cannot be easily exceeded with the applied HolPase activity assay). However it was demonstrated for the HAD-type HolPase from *S. enterica* that it is not affected by P_i up to a concentration of 25 mM P_i [14].

Both $HisN_{Cg}$ and Cg0911 are strictly dependent on addition of bivalent metal ions to the reaction buffer for HolPase activity, with Mg^{2+} being the preferred ion and reduced activity with Mn^{2+} and Co^{2+} (Fig. 3). This is in accordance with results from a general study on IMPases in *Mycobacterium smegmatis*. This study demonstrated that IMPase activity is maximal with Mg^{2+}, is inhibited by Zn^{2+} and about 25% of activity can be obtained with Mn^{2+} [34]. The need for metal ion addition to the in vitro assay has also been observed for HAD-type HolPases that exhibit a binuclear metal cluster in the active center [13, 16]. In contrast to $HisN_{Cg}$ and Cg0911, some HAD-type HolPases are also active with Zn^{2+}, Cu^{2+} or Ni^{2+} [13, 16]. Interestingly, PHP-type HolPases, although exhibiting a trinuclear metal cluster in the active center, do not rely on addition of external metal ions for activity [20, 23]. This suggests a very tight binding of the metal ions in the metal cluster, resulting in a retention during the protein purification process. In contrast, binding of metal ions in the active site of HAD-type and IMPase-like HolPases seems to be much weaker, resulting in the need of metal ion addition after the protein purification process. This weak binding is supported by the relatively high K_m values of Mg^{2+} for HisN and Cg0911 (Table 2) and actinobacterial IMPases in general [34].

The tertiary structure of IMPase-like HolPases, as shown using the example of HisN$_{Zm}$, is very similar to that of various mammalian IMPases (data not shown) including the IMPases of *Homo sapiens* [29] and *Bos taurus* [28]. Three Mg^{2+} ions have been identified in the active site of these two intensively investigated proteins coordinated by five highly conserved amino acid residues [24, 28, 29]. These five residues are conserved in HisN$_{Zm}$, HisN$_{Cg}$, and all analyzed HisN orthologs. It is therefore very likely that the proposed three-metal mechanism for hydrolysis of inositol monophosphate in eukaryotic IMPases might be also employed for hydrolysis of HolP in IMPase-like HolPases. In that case, binding of the second Mg^{2+} ion would be cooperative [28], fitting well to the determined Hill-coefficients of 2.5-3 that indicate a cooperative effect of Mg^{2+} on the HolPase activity of HisN$_{Cg}$ and Cg0911.

The optimal pH for the HolPase activity of HisN$_{Cg}$ (pH ~7.5) and Cg0911 (pH ~8) reflects the internal pH of *C. glutamicum* (7.5 ± 0.5 [26]). The unusually high optimal pH for HisG activity of around 10 [44] is therefore no general attribute of enzymes involved in L-histidine biosynthesis in *C. glutamicum*. Moreover, the pH optima of Cg0911 and HisN$_{Cg}$ differ significantly from the pH optima of HAD- or PHP-type HolPases. HolPases of the PHP-family are most active at pH 8.5-9 [18, 20]. In contrast, HolPases from the HAD-family exhibit their maximal activity at a slightly acidic pH [14, 16].

Overall, there are significant differences in regard to K$_m$-values, turnover numbers, inhibition behavior, metal ion preference and pH-optima between HisN$_{Cg}$ (as one example of an IMPase-like HolPase) and HolPases of the HAD- or PHP-type. Therefore, IMPase-like HolPases do not only differ in protein sequence and tertiary structure from the two other HolPase families, but their differing enzymatic properties might reflect an adaptation to their host organism.

There are also interesting differences in some aspects of HolPase activity between HisN$_{Cg}$ and Cg0911. Next to the very obvious differences in k$_{cat}$ and K$_m$, the two enzymes also differ in their pH and temperature profiles. The HolPase activity of Cg0911 does not account significantly for the in vivo L-histidine biosynthesis in *C. glutamicum* under the tested conditions. However, since the catalytic properties of HisN$_{Cg}$ and Cg0911 are not identical, their might exist some growth conditions, where the HolPase activity of Cg0911 becomes relevant for the cell, for instance under alkaline stress conditions.

The most interesting observation concerning Cg0911 is the almost five-fold stimulation of HolPase activity by L-histidinol. This kind of a positive feedback by the direct reaction product on enzyme activity has been recently described for the RelA protein of *E. coli*, which synthesizes guanosine tetraphosphate (ppGpp) during the stringent response and is activated by ppGpp via positive allosteric feedback regulation [45].

The analysis of 165 potential bacterial HisN orthologs resulted in the formulation of six sequence motifs (Fig. 5) that can be used for the discrimination of IMPase-like HolPases from other IMPase-like proteins. This is of special interest, since there exist several IMPase-like protein families in bacteria (five are present in *C. glutamicum*) and there are substantial differences in their substrate specificity. The preferred substrates of the ImpA and Cg0911 orthologs still remain to be elucidated. However, at least in *Actinobacteria*, IMP is supposed to be the main substrate of SuhB orthologs [25], PAP and PAPS that of CysQ orthologs [42] and HolP that of HisN orthologs. The different substrate specificities illustrate that each of these IMPase-like proteins is involved in very different metabolic processes and underlines the need to clearly distinguish between the different paralogs. The motifs presented in this study ideally serve this purpose. Moreover, they can also be used for the identification of IMPase-like HolPases in plants.

Within the *Actinobacteria*, it is rather easy to distinguish the different IMPase-like proteins by comparing the amino acid sequences to the corresponding orthologs in *C. glutamicum*. However, since the amino acid sequences of all IMPase-like proteins share a big degree of similarity (approximately 20-30% sequence identity between the different IMPase-like proteins in *C. glutamicum*), it gets harder to classify IMPase-like proteins in more distantly related bacteria based on the overall sequence similarity alone. The motifs described in the present study are therefore of great help in assigning a specific function to a not yet characterized IMPase-like protein. We demonstrated this by proving HolPase activity of HisN orthologs from the genera *Dietzia* and *Zymomonas* that are only moderately similar to HisN$_{Cg}$ but possess the expected motifs. We could also demonstrate that a potential HisN$_{Cg}$ homolog from *A. utahensis*, with rather high overall sequence similarity but entirely lacking the HolPase motif 5, is not a functional HolPase.

The last result underlines the importance of motif 5 for HolPase activity of HisN orthologs. The detailed examination of the recently solved structure of the IMPase-like HolPase from *Z. mobilis*, suggests that the carboxylic group of the aspartate present in HolPase motif 5 might be involved in substrate recognition. A similar function might be also attributed to at least one of the aromatic amino acids present in motif 5 (Fig. 5: HisN motif 5, position 195). However, since no IMPase-like HolPase has been crystallized in the presence of the substrate HolP or the products L-histidinol and P$_i$, any interaction of the conserved residues with the substrates or the products remains speculative.

The application of the here presented motifs for the classification of IMPase-like proteins can help to improve the annotation of IMPase-like proteins. If a suspected HisN ortholog exhibits all HolPase motifs, and especially motif 5, it is very likely that this protein is a HolPase. Vice versa, if the motifs typical of one of the over IMPase-like proteins are identified in a suspected HisN ortholog, it is very likely that this protein is exactly this IMPase-like protein and not a HolPase. Consideration of the overall sequence similarity to $HisN_{Cg}$ is not sufficient to identify IMPase-like HolPase. $HisN_{Zm}$ is only moderately similar to $HisN_{Cg}$ (BLASTP score: 101 bits) but possesses all HolPase motifs and we were able to experimentally prove its HolPase activity. In contrast, the SuhB ortholog from *Sediminibacterium salmoneum* has an overall equal degree of similarity to $HisN_{Cg}$ (BLASTP score: 102 bits) but it is very unlikely a HolPase, since all the HolPase motifs are absent. This demonstrates the benefit of the presented motifs for the functional classification of IMPase-like proteins. The HMMs supplied as Additional file 3 (HisN) and Additional file 4 (Cg0911) should greatly facilitate the correct classification of IMPase-like HolPases.

Our analysis of the distribution of HisN homologs within the bacterial kingdom revealed, that IMPase-like HolPases are not restricted to the phylum *Actinobacteria* (Fig. 7). They appear to be generally present in different phyla like *Chlorobi, Fibrobacteres* and *Nitrospinae*. However, HolPase activity of these potential HisN orthologs has not been demonstrated so far. Interestingly, we also detected many putative IMPase-like HolPases in *Gammaproteobacteria*. This is surprising as many gammaproteobacteria are known to possess a bifunctional HAD-type HolPase homologous to the His(NB) protein from *E. coli* [12]. In those gammaproteobacterial families where we did not detect such an additional His(NB) homolog, it is likely that the IMPase-like proteins are the main HolPases. In contrast, the role of the putative HisN orthologs within those gammaproteobacterial families with concurrent occurrence of His(NB) and HisN homologs is less evident. It is possible, that these genera simply possess two different HolPases or that the HisN homologs from these families do not possess HolPase activity. It is also possible that a second, yet unknown activity of HisN orthologs is required in these genera. It has been demonstrated for the IMPase-like HolPase HISN7 from *A. thaliana*, that it additionally catalyzes the dephosphorylation of D-inositol-1-phosphate, D-inositol-3-phosphate, and L-galactose-1-phosphate [35].

Since we used a rather restrictive cut-off for the identification of HisN orthologs (BLASTP score ≥ 125 bits with $HisN_{Cg}$ as query), Fig. 7 does not give the exhaustive picture of the distribution of IMPase-like HolPases within bacteria. Indeed, the HMM for HisN reveals additional species that possess such an ortholog, however with a BLASTP score below the cut-off.

The distribution of Cg0911 orthologs is restricted to the four actinobacterial orders *Corynebacteriales, Frankiales, Micrococcales,* and *Propionibacteriales* and even within these orders only a few genera possess such a homolog. We identified only a Cg0911 ortholog but no HisN ortholog in *Kytococcus*. It is possible, that Cg0911 is the main HolPase in *Kytococcus*, since presence of other *his* genes indicates the general possibility of L-histidine biosynthesis (data not shown).

The occurrence of the genes encoding HisN and Cg0911 orthologs as a tandem within *Corynebacteriaceae* implies a recent gene duplication event. However, HisN and Cg0911 share only the same degree of similarity as do HisN or Cg0911 with one of the three other paralogs in *C. glutamicum* (Table 1). This indicates that all five paralogs evolved more or less simultaneously and are not a result of a recent duplication event. Although we currently do not know which phosphorylated substance might be the preferred substrate of Cg0911, one could assume that both enzymes are needed in the same context most of the time. However since the genes encoding HisN and Cg0911 orthologs are not clustered in all other genera besides *Corynebacterium* and *Turicella*, this clustering might be simply a result of chance or indicate a special, yet unknown function of Cg0911 orthologs in *Corynebacteriaceae*.

Conclusions

Here, using the example of the histidinol-phosphate phosphatase HisN from *C. glutamicum*, we present for the first time kinetic data on an IMPase-like HolPases with its natural substrate L-histidinol-phosphate. Based on this data, IMPase-like HolPases show remarkable differences in enzyme properties as compared with HAD- or PHP-type HolPases. Moreover, six sequence motifs have been presented in this study that can be used to reliably differentiate between IMPase-like HolPases and IMPase-like proteins with no such activity (like SuhB or CysQ), with the potential to enhance current and future genome annotations. A phylogenetic analysis reveals that IMPase-like HolPases are not only present in *Actinobacteria* and plants but can be found in further bacterial phyla, including *Chlorobi, Fibrobacteres,* and *Proteobacteria*.

Methods

Bacterial strains and cultivation conditions

All strains and plasmids used in this study are given in Additional file 1: Table S1 and Table S2, respectively. *Escherichia coli* DH5α MCR [46] was used for general cloning works and plasmid maintenance. *E. coli* ER2566 (New England Biolabs, Ipswich, MA) was used for heterologous gene expression in the context of protein purification. *E. coli* strains were grown in lysogeny broth (LB) Lennox medium. Solid medium contained 1.7% agar. Kanamycin (50 µg ml^{-1}) and ampicillin (200 µg ml^{-1}) were added where appropriate. *E. coli* strains were incubated at 37 °C if not stated otherwise.

Corynebacterium glutamicum ATCC 13032 [47, 48] and all thereof derived mutants (this work) were incubated at 30 °C. Caso broth (Carl Roth, Karlsruhe, Germany) or MM1 minimal medium was used for solid cultivations, supplemented with 1.7% agar. Nalidixic acid (50 µg ml^{-1}), kanamycin (25 µg ml^{-1}), L-histidine (100 µM), and sucrose (10%) were added where appropriate. The MM1 minimal medium (MMYE medium without yeast extract [49]) was constituted as follows: glucose (20 g l^{-1}), $(NH_4)_2SO_4$ (10 g l^{-1}), urea (3 g l^{-1}), $K_2HPO_4 \cdot 3\ H_2O$ (1.3 g l^{-1}), $MgSO_4 \cdot 7\ H_2O$ (400 mg l^{-1}), thiamine (500 µg l^{-1}), biotin (50 µg l^{-1}), $FeSO_4 \cdot 7\ H_2O$ (2 mg l^{-1}), $MnSO_4 \cdot 7\ H_2O$ (2 mg l^{-1}), NaCl (50 mg l^{-1}).

DNA of *Dietzia* sp. strain Chol2 [50], *Zymomonas mobilis* ZM4 [30], and *Actinoplanes utahensis* NRRL 12052 [31] was used for the amplification of *hisN* homologs from these bacteria.

Recombinant DNA work

A complete list of primers used in this study is given in Additional file 1: Table S3. Phusion high-fidelity DNA polymerase (Thermo Scientific, Dreieich, Germany) was used to amplify DNA fragments for cloning or for sequencing. To improve the amplification of GC-rich DNA from *Dietzia* and *A. utahensis* the GC-buffer provided by the supplier was used and dimethyl sulfoxide (DMSO) was added to the PCR mixture to a final concentration of 8.3%. Plasmids were constructed in two different ways: In the first method, vector and insert were cut with restriction enzymes and joined by a DNA ligase according to standard cloning procedures [51]. Restriction sites needed for cloning of the insert were included in the 5' overhang of the primers used for the amplification. All restriction and DNA-modifying enzymes were purchased from Thermo Scientific. In the second method, vector and insert were assembled in an isothermal enzymatic reaction by taking advantage of complementary DNA sequences at the end of the DNA fragments as described by Gibson (2011) [52]. For this purpose, vector DNA was linearized in a PCR reaction using the KOD hot start DNA polymerase (Novagen, San Diego, CA). Overlapping DNA-sequences (20–30 bp) were generated by including sequences complementary to the ends of the linearized vector within the 5' ends of the primers used for amplification of the insert. Specific mutations in an insert were introduced by including the desired mutation in the primers used for amplification of the DNA. All plasmids were constructed

and propagated in *E. coli* prior to transfer to *C. glutamicum*.

Competent *E. coli* cells were prepared according to an optimized $CaCl_2$ method and transformed with plasmid DNA by applying a heat-pulse [53]. Competent *C. glutamicum* cells were prepared as described previously and transformed with plasmid DNA via electroporation [54] at 2.5 kV, 200 Ω and 25 µF.

Gene deletion in *C. glutamicum*

Gene deletion in *C. glutamicum* relied on homologues recombination and a double cross-over event using the non-replicating pK18*mobsacB* vector [55] as described before [4]. The genomic regions flanking the deletion of interest were amplified from genomic DNA of the *C. glutamicum* wild type. These fragments (approximately 500 bp) were either fused via the gene splicing by overlap extension (gene SOEing) technique [56, 57] and used for ligation into the pK18*mobsacB* vector or they were directly used for the isothermal enzymatic assembly with the vector (Gibson assembly, [52]). The deletion plasmids constructed in either way were used for transformation of *C. glutamicum*. After selection for the double cross-over event, desired deletions were confirmed by PCR using primers binding to genomic sequences up- and down-stream of the deletion and that were not part of deletion plasmid. All deletion mutants generated in this study are listed in Additional file 1: Table S1.

C. glutamicum Δ*hisN* complementation experiments

The constitutive expression vector pZMP [58] was used for expression of putative HolPase genes in the *C. glutamicum* (plasmids listed in Additional file 1: Table S2). A SD-sequence exactly matching the 3' end of the 16S-rRNA in *C. glutamicum* was included within the 5' extension of the primers used for gene amplification. Gene expression from pZMP is under control of the *tac* promoter. Approximately 15 copies of the plasmid are present per *C. glutamicum* cell (unpublished observation). The DNA sequence of the inserts was confirmed by sequencing. Plasmid DNA was isolated from *E. coli* and used for transformation of *C. glutamicum* Δ*hisN*. Successful transformants were identified by selection for the plasmid-encoded kanamycin resistance. Presence of the insert was additionally confirmed by amplification of the insert using vector-specific primers and comparing the size of the PCR product with the expectation.

The *C. glutamicum* Δ*hisN* complementation experiments were conducted on MM1 minimal medium plates either supplemented or unsupplemented with L-histidine. The different mutants were diluted in liquid MM1 medium and drops containing the same amount of cells were applied to the plates. The plates were incubated for several days at 30 °C and pictures were taken in 24 h intervals. The ability of a putative HolPase gene to complement the genomic *hisN* deletion of *C. glutamicum* was assessed by comparing the growth of the expression mutants to the *C. glutamicum* Δ*hisN* or wild type strain.

Purification of the HisN$_{Cg}$ and Cg0911 enzymes

The commercial IMPACT^{TM} system (New England Biolabs, Ipswich, MA) was used for the tag-free purification of HisN and Cg0911 according to the manufacturer's instructions. In this system, a self-cleavable intein tag is translationally fused to the protein of interest. The tag binds specifically to chitin beads and its self cleavage activity is induced by thiol reagents, allowing the elution of unmodified protein.

The coding DNA sequences (CDS) of *hisN$_{Cg}$* and *cg0911* were amplified from genomic DNA without the start and stop codon (the start codon ATG is present on the vector) and inserted into the pXTB1 (New England Biolabs) vector by isothermal enzymatic assembly. This resulted in the translational fusion of the intein tag to the C-terminus of HisN$_{Cg}$ and Cg0911, respectively. The inserts were sequenced to exclude undesired mutations. The plasmids were transferred to *E. coli* ER2566 for heterologous gene expression. Details regarding cultivation, induction of protein expression, cell lyses, protein purification, concentration, and quality control have been described previously [4]. Diverging from this description, column and cleavage buffers used during affinity purification and on column cleavage contained 20 mM TRIS as buffering substance instead of Na$_2$HPO$_4$. Additionally, a different storage buffer was used (10 mM NaCl, 20 mM TRIS, pH 7.5). The purified proteins were mixed with glycerol (final concentration: 50% (w/v)) and stored at -80 °C.

HolPase activity assay

HolPase activity can be assayed by the release of inorganic phosphate (P$_i$) from the substrate HolP. However, conditions needed for the colorimetric detection of P$_i$ are not compatible with the standard conditions for enzymatic catalysis. Therefore, catalysis and P$_i$ detection had to be separated.

The reaction conditions for the catalysis step of the HolPase activity were based on the conditions described for the HolPase of *S. enterica* [59]. If not stated otherwise, reactions were conducted in 1.5 ml polypropylene microcentrifuge tubes in a water bath tempered to 30 °C. The standard reaction volume was 100 µl. 80 µl of reaction buffer (100 mM triethanolamine (TEA), pH 7.5 at 22 °C (corresponding to pH 7.35 at 30 °C)) were supplemented with MgSO$_4$ resulting in the desired final metal ion concentration (5 mM in standard settings) and were mixed with 10 µl protein solution (freshly diluted with reaction buffer to 0.1 µg µl^{-1} (HisN) or 1 µg µl^{-1} (Cg0911),

respectively) and preincubated for 10 min. The reaction was started by addition of 10 µl HolP solution (5 mM; HolP purchased from Paragos, Herdecke, Germany). Samples (30 µl) were taken in appropriate time intervals (e.g. after 1, 2, and 3 min). The reaction was stopped by immediately mixing the withdrawn sample with 30 µl of ice-cold EDTA solution (20 mM). Each enzymatic reaction was performed in at least three replicates.

Some modifications of the catalytic part of the assay were made to access individual enzyme parameters. (1) Temperature optimum: Variation of incubation temperature from 20 °C to 50 °C. The pH of the reaction buffer was adjusted to be 7.35 at every tested temperature. (2) pH-optimum: The pH of the reaction buffer was varied (pH 6–10). The buffer substances (always 100 mM) were changed according to their optimal buffering range (pH 6 [22 °C]/5.98 [30 °C]: 2-(N-morpholino)ethanesulfonic acid (MES); pH 7/6.90: 3-(N-morpholino)propanesulfonic acid (MOPS); pH 8/7.94: 4-(2-hydroxyethyl)-1-piperazineethanesulfonic acid (HEPES); pH 9/8.83: 2-(cyclohexylamino)ethanesulfonic acid (CHES); pH 10/9.83: CHES). (3) Metal ion preference: The metal salts CaCl$_2$, CoCl$_2$, CuSO$_4$, FeSO$_4$, MnSO$_4$, NiCl$_2$, or ZnCl$_2$ (5 mM final concentration) were used instead of MgSO$_4$. (4) Determination of K$_m$ and k$_{cat}$: The HolP concentration was kept constant at 500 µM in the measurements for the determination of the K$_m$ value for Mg^{2+}. At least nine different Mg^{2+} concentrations were tested ranging from 312.5 µM to 5 mM (HisN$_{Cg}$) or from 312.5 µM to 10 mM (Cg0911). The Mg^{2+} concentration was kept constant at 5 mM for the determination of the K$_m$ value for HolP. At least seven different HolP concentrations were tested ranging from 17.5 µM to 200 µM (HisN$_{Cg}$) or from 125 µM to 1 mM (Cg0911). Every concentration was tested in at least six replicates.

The detection of released P$_i$ was based on the complex formation of malachite green with phosphomolybdate under acidic conditions [60]. The method was adopted for measurement in transparent flat bottom 96-well plates in an Infinite M200 plate reader (Tecan, Männedorf, Switzerland). 90 µl H$_2$O, 20 µl reagent A (1.75% (w/v) ammonium heptamolybdate tetrahydrate in 6.3 N sulfuric acid), and 10 µl of a P$_i$-containing sample were mixed directly in a 96-well plate and incubated at RT for 10 min. Then, 20 µl of reagent C (0.35% (w/v) polyvinyl alcohol (MW ≈ 30 kDa, fully hydrolyzed), 0.035% (w/v) malachite green) were added. The mixture was incubated for additional 45 min at RT. The absorption at 610 nm was measured three times in 3 min intervals. Every P$_i$ containing sample was measured in three technical replicates. Samples with known P$_i$ concentrations (0.1 - 3.13 mM KH$_2$PO$_4$) were used to record a calibration curve.

Linearity of the measurement was ensured in the range from 0.07 - 0.85 absorption units at 610 nm.

The enzyme activity corresponds to the slope of the linear regression curve through the data points in a plot of released P_i against the time. Values for K_m, k_{cat} and the Hill-coefficients were determined by non-linear curve fitting of the data points in a plot of enzyme activity against the varied substrate concentration to the Hill equation [27] using OriginPro9.0 (function "Hill"; Origin Lab Corporation, Northampton, MA). The molecular masses of $HisN_{Cg}$ (27893.16 g mol^{-1}) and Cg0911 (30695.27 g mol^{-1}) were used for the calculation of the turnover numbers (k_{cat}) and were based on the pure amino acid sequence of the monomers.

Bioinformatics analyses

The online BLASTP suite of the National Center for Biotechnology Information (NCBI) was used for the identification of homologous protein sequences. If not stated otherwise, NCBI's non-redundant (nr) protein sequence database was queried using the default parameters [61].

The online tool EMBOSS NEEDLE, provided by the European Bioinformatics Institute (EMBL-EBI, [62]), was used in its default settings in order to determine the degree of identity [%] and similarity [%] between two homologous proteins (pairwise sequence alignment). Multiple sequence alignments were conducted with EMBL-EBI's online tool CLUSTAL OMEGA in the default settings [62].

The JACKHMMR tool was used online on the HMMER webserver, provided by the European Bioinformatics Institute (EMBL-EBI, [63]), to create HMMs for HisN and Cg0911. The UNIPROT REFPROT database was used for an iterative search with either a multiple alignment of all validated HisN orthologs ($HisN_{Cg}$, $HisN_{Au}$, $HisN_{Dz}$, and $HisN_{Zm}$) or the single Cg0911 sequence, respectively. Significance E-value cutoffs for sequence andd hit were set to 1e-65 and the the search was iterated until no new hits were found.

The graphical representations of amino acid motifs were created with the online tool WEBLOGO [64]. The homologous sequences used for these logos were identified by a BLASTP search and aligned using CLUSTAL OMEGA. The multiple alignments were improved manually if appropriate (local rearrangement of gaps or removal of whole sequences from the alignment).

The CCP4MG molecular-graphics software (version 2.10.5) was used for visualization and analysis of protein structures [65]. Additionally, the online PROTEIN INTERACTIONS CALCULATOR (PIC) was used to determine intra- and intermolecular interaction within the protein structures including hydrogen bonds, salt bridges, hydrophobic interactions, as well as aromatic interactions [66].

Additional files

Additional file 1: Supplementary Figures and Tables. The supplementary figures comprise the SDS-PAGE of purified HisN$_{Cg}$ and Cg0911, several alignments of IMPase-like proteins in *C. glutamicum* and other species, and the complementation assay for various *hisN$_{Cg}$* gene mutants. Furthermore, supplementary tables carry the used strains, plasmids and primers.

Additional file 2: Data collection for motif analysis. This data collection contains all sequence data of IMPase-like proteins belonging to the group Cg0911, CysQ, HisN, ImpA, and SuhB used for motif analysis. Furthermore, HisN orthologues not used for the motif analysis were listed.

Additional file 3: HisN-HMM. Hidden Markov model of 1681 HisN sequences based on an iterative approach using JACKHMMER.

Additional file 4: Cg0911-HMM. Hidden Markov model of 103 Cg0911 sequences based on an iterative approach using JACKHMMER.

Abbreviations

HolP: L-histidinol-phosphate; HolPase: Histidinol-phosphate phosphatase; IMPase: Inositol monophosphatase

Acknowledgements

Not applicable.

Funding

R. K. Kulis-Horn was supported by a CLIB-GC (Graduate Cluster Industrial Biotechnology) Phd grant co-funded by the Ministry of Innovation, Science and Research of the federal state of North Rhine-Westphalia (MIWF). This work was part of the SysEnCor research project (grant 0315598E) funded by the German Federal Ministry of Education and Research (BMBF). Besides financing the funding body played no role, active or passive, in study design, data collection and interpretation, and manuscript writing.

Authors' contributions

RKKH: Study Design and realization of all experiments (i.a. genetic studies, protein purification and characterization, database analyses, motif identification). Interpretation of the results. Preparation of the manuscript. JK: Guidance in study design and interpretation of the results. Review of the manuscript. CR: Additional bioinformatics analyses. Review of the manuscript. MP: Guidance in study design and interpretation of the results. Review of the manuscript. All authors read and approved the final manuscript.

Competing interests

The authors declare that they have no competing interest.

References

1. Gao B, Gupta RS. Phylogenetic framework and molecular signatures for the main clades of the phylum *Actinobacteria*. Microbiol Mol Biol Rev. 2012;76: 66–112.

2. Becker J, Wittmann C. Systems and synthetic metabolic engineering for amino acid production – the heartbeat of industrial strain development. Curr Opin Biotechnol. 2011;23:1–9.

3. Becker J, Wittmann C. Bio-based production of chemicals, materials and fuels – *Corynebacterium glutamicum* as versatile cell factory. Curr Opin Biotechnol. 2012;23:631–40.

4. Kulis-Horn RK, Persicke M, Kalinowski J. *Corynebacterium glutamicum* ATP-phosphoribosyl transferases suitable for L-histidine production – Strategies for the elimination of feedback inhibition. J Biotechnol. 2015;206:26–37.

5. Kulis-Horn RK, Persicke M, Kalinowski J. Histidine biosynthesis, its regulation and biotechnological application in *Corynebacterium glutamicum*. Microb Biotechnol. 2014;7:5–25.

6. Alifano P, Fani R, Lió P, Lazcano A, Bazzicalupo M, Carlomagno MS, Bruni CB. Histidine biosynthetic pathway and genes. structure, regulation, and evolution. Microbiol Rev. 1996;60:44–69.

7. Mormann S, Lömker A, Rückert C, Gaigalat L, Tauch A, Pühler A, Kalinowski J. Random mutagenesis in *Corynebacterium glutamicum* ATCC 13032 using an IS6100-based transposon vector identified the last unknown gene in the histidine biosynthesis pathway. BMC Genomics. 2006;7:205.

8. Marchler-Bauer A, Zheng C, Chitsaz F, Derbyshire MK, Geer LY, Geer RC, Gonzales NR, Gwadz M, Hurwitz DI, Lanczycki CJ, Lu F, Lu S, Marchler GH, Song JS, Thanki N, Yamashita RA, Zhang D, Bryant SH. CDD: conserved domains and protein three-dimensional structure. Nucleic Acids Res. 2012;41:D348.

9. Petersen LN, Marineo S, Mandala S, Davids F, Sewell BT, Ingle RA. The missing link in plant histidine biosynthesis: *Arabidopsis* myoinositol monophosphatase-like2 encodes a functional histidinol-phosphate phosphatase. Plant Physiol. 2010;152:1186–96.

10. Marineo S, Cusimano MG, Limauro D, Coticchio G, Puglia AM. The histidinol phosphate phosphatase involved in histidine biosynthetic pathway is encoded by SCO5208 (*hisN*) in *Streptomyces coelicolor* A3(2). Curr Microbiol. 2008;56:6–13.

11. Movahedzadeh F, Wheeler PR, Dinadayala P, Av-Gay Y, Parish T, Daffé M, Stoker NG. Inositol monophosphate phosphatase genes of *Mycobacterium tuberculosis*. BMC Microbiol. 2010;10:50.

12. Brilli M, Fani R. Molecular evolution of *hisB* genes. J Mol Evol. 2004;58:225–37.

13. Rangarajan ES, Proteau A, Wagner J, Hung M, Matte A, Cygler M. Structural snapshots of *Escherichia coli* histidinol phosphate phosphatase along the reaction pathway. J Biol Chem. 2006;281:37930–41.

14. Brady DR, Houston LL. Some properties of the catalytic sites of imidazoleglycerol phosphate dehydratase-histidinol phosphate phosphatase, a bifunctional enzyme from *Salmonella typhimurium*. J Biol Chem. 1973;248:2588–92.

15. Staples MA, Houston LL. Purification of the bifunctional enzyme, imidazoleglycerolphosphate dehydratase-histidinol phosphatase, of *Salmonella typhimurium*. Biochim Biophys Acta. 1980;613:210–9.

16. Lee HS, Cho Y, Lee J, Kang SG. Novel monofunctional histidinol-phosphate phosphatase of the DDDD superfamily of phosphohydrolases. J Bacteriol. 2008;190:2629–32.

17. Gorman JA, Hu AS. The separation and partial characterization of L-histidinol phosphatase and an alkaline phosphatase of *Saccharomyces cerevisiae*. J Biol Chem. 1969;244:1645–50.

18. Millay Jr RH, Houston LL. Purification and properties of yeast histidinol phosphate phosphatase. Biochemistry. 1973;12:2591–6.

19. le Coq D, Fillinger S, Aymerich S. Histidinol phosphate phosphatase, catalyzing the penultimate step of the histidine biosynthesis pathway, is encoded by *ytvP* (*hisJ*) in *Bacillus subtilis*. J Bacteriol. 1999;181:3277–80.

20. Ghodge SV, Fedorov AA, Fedorov EV, Hillerich B, Seidel R, Almo SC, Raushel FM. Structural and mechanistic characterization of L-histidinol phosphate phosphatase from the polymerase and histidinol phosphatase family of proteins. Biochemistry. 2013;52:1101–12.

21. Kalinowski J, Bathe B, Bartels D, Bischoff N, Bott M, Burkovski A, Dusch N, Eggeling L, Eikmanns BJ, Gaigalat L, Goesmann A, Hartmann M, Huthmacher K, Krämer R, Linke B, McHardy AC, Meyer F, Möckel B, Pfefferle W, Pühler A, Rey DA, Rückert C, Rupp O, Sahm H, Wendisch VF, Wiegräbe I, Tauch A. The complete *Corynebacterium glutamicum* ATCC 13032 genome sequence and its impact on the production of L-aspartate-derived amino acids and vitamins. J Biotechnol. 2003;104:5–25.

22. Ikeda M, Nakagawa S. The *Corynebacterium glutamicum* genome. features and impacts on biotechnological processes. Appl Microbiol Biotechnol. 2003;62:99–109.

23. Omi R, Goto M, Miyahara I, Manzoku M, Ebihara A, Hirotsu K. Crystal structure of monofunctional histidinol phosphate phosphatase from *Thermus thermophilus* HB8. Biochemistry. 2007;46:12618–27.

24. Strasser F, Pelton PD, Ganzhorn AJ. Kinetic characterization of enzyme forms involved in metal ion activation and inhibition of myo-inositol monophosphatase. Biochem J. 1995;307(Pt 2):585–93.

25. Nigou J, Dover LG, Besra GS. Purification and biochemical characterization of *Mycobacterium tuberculosis* SuhB, an inositol monophosphatase involved in inositol biosynthesis. Biochemistry. 2002;41:4392–8.

26. Follmann M, Ochrombel I, Krämer R, Trötschel C, Poetsch A, Rückert C, Hüser A, Persicke M, Seiferling D, Kalinowski J, Marin K. Functional genomics of pH homeostasis in *Corynebacterium glutamicum* revealed novel links between pH response, oxidative stress, iron homeostasis and methionine synthesis. BMC Genomics. 2009;10:621.

27. Weiss JN. The Hill equation revisited: uses and misuses. FASEB J. 1997;11:835–41.

28. Gill R, Mohammed F, Badyal R, Coates L, Erskine P, Thompson D, Cooper J, Gore M, Wood S. High-resolution structure of myo-inositol monophosphatase, the putative target of lithium therapy. Acta Crystallogr D Biol Crystallogr. 2005;61:545–55.

29. Singh N, Halliday AC, Knight M, Lack NA, Lowe E, Churchill GC. Cloning, expression, purification, crystallization and X-ray analysis of inositol monophosphatase from *Mus musculus* and *Homo sapiens*. Acta Crystallogr F Struct Biol Cryst Commun. 2012;68:1149–52.

30. Seo J, Chong H, Park HS, Yoon K, Jung C, Kim JJ, Hong JH, Kim H, Kim J, Kil J, Park CJ, Oh H, Lee J, Jin S, Um H, Lee H, Oh S, Kim JY, Kang HL, Lee SY, Lee KJ, Kang HS. The genome sequence of the ethanologenic bacterium *Zymomonas mobilis* ZM4. Nat Biotechnol. 2005;23:63–8.

31. Velasco-Bucheli R, del Cerro C, Hormigo D, Acebal C, Arroyo M, García JL, de La Mata I. Draft genome sequence of *Actinoplanes utahensis* NRRL 12052, a microorganism involved in industrial production of pharmaceutical intermediates. Genome Announc. 2015;3:e01411–14.

32. Hwang H, Park S, Kim J. Crystal structure of *cbbF* from *Zymomonas mobilis* and its functional implication. Biochem Biophys Res Commun. 2014;445:78–83.

33. Stieglitz KA, Johnson KA, Yang H, Roberts MF, Seaton BA, Head JF, Stec B. Crystal structure of a dual activity IMPase/FBPase (AF2372) from *Archaeoglobus fulgidus*. The story of a mobile loop. J Biol Chem. 2002;277:22863–74.

34. Nigou J, Besra GS. Characterization and regulation of inositol monophosphatase activity in *Mycobacterium smegmatis*. Biochem J. 2002;361:385–90.

35. Torabinejad J, Donahue JL, Gunesekera BN, Allen-Daniels MJ, Gillaspy GE. VTC4 Is a bifunctional enzyme that affects myoinositol and ascorbate biosynthesis in plants. Plant Physiol. 2009;150:951–61.

36. Parish T, Liu J, Nikaido H, Stoker NG. A *Mycobacterium smegmatis* mutant with a defective inositol monophosphate phosphatase gene homolog has altered cell envelope permeability. J Bacteriol. 1997;179:7827–33.

37. Morales M, Nozawa S, Thedei Jr G, Maccheroni Jr W, Rossi A. Properties of a constitutive alkaline phosphatase from strain 74A of the mold *Neurospora crassa*. Braz J Med Biol Res. 2000;33:905–12.

38. Garen A, Levinthal C. A fine-structure genetic and chemical study of the enzyme alkaline phosphatase of *E. coli* I. Purification and characterization of alkaline phosphatase. Biochim Biophys Acta. 1960;38:470–83.

39. Dinev Z, Gannon CT, Egan C, Watt JA, McConville MJ, Williams SJ. Galactose-derived phosphonate analogues as potential inhibitors of phosphatidylinositol biosynthesis in mycobacteria. Org Biomol Chem. 2007;5:952.

40. Jackson M, Crick DC, Brennan PJ. Phosphatidylinositol is an essential phospholipid of mycobacteria. J Biol Chem. 2000;275:30092–9.

41. Burkovski A. Cell envelope of *Corynebacteria*: Structure and influence on pathogenicity. ISRN Microbiol. 2013;2013:1–11.

42. Hatzios SK, Iavarone AT, Bertozzi CR. Rv2131c from *Mycobacterium tuberculosis* is a CysQ 3'-phosphoadenosine-5'-phosphatase. Biochemistry. 2008;47:5823–31.

43. Lee H. Sulfur metabolism and its regulation. In: Eggeling L, Bott M, editors. Handbook of *Corynebacterium glutamicum*. Boca Raton: Taylor & Francis; 2005. p. 351–76.

44. Zhang Y, Shang X, Deng A, Chai X, Lai S, Zhang G, Wen T. Genetic and biochemical characterization of *Corynebacterium glutamicum* ATP phosphoribosyltransferase and its three mutants resistant to feedback inhibition by histidine. Biochimie. 2012;94:829–38.

45. Shyp V, Tankov S, Ermakov A, Kudrin P, English BP, Ehrenberg M, Tenson T, Elf J, Hauryliuk V. Positive allosteric feedback regulation of the stringent response enzyme RelA by its product. EMBO Rep. 2012;13:835–9.

46. Grant SGN, Jessee J, Bloom FR, Hanahan D. Differential plasmid rescue from transgenic mouse DNAs into *Escherichia coli* methylation-restriction mutants. Proc Natl Acad Sci U S A. 1990;87:4645–9.

47. Kinoshita S, Nakayama S, Akita S. Taxonomic study of glutamic acid accumulating bacteria, *Micrococcus glutamicus,* nov. sp. Bull Agr Chem Soc Jpn. 1958;22:176–85.

48. Abe S, Takayama K, Kinoshita S. Taxonomical studies on glutamic acid-producing bacteria. J Gen Appl Microbiol. 1967;13:279–301.

49. Katsumata R, Ozaki A, Oka T, Furuya A. Protoplast transformation of glutamate-producing bacteria with plasmid DNA. J Bacteriol. 1984;159:306–11.

50. Holert J, Yücel O, Suvekbala V, Kulić Ž, Möller H, Philipp B. Evidence of distinct pathways for bacterial degradation of the steroid compound cholate suggests the potential for metabolic interactions by interspecies cross-feeding. Environ Microbiol. 2014;16:1424–40.

51. Schrimpf G. Gentechnische Methoden: Eine Sammlung von Arbeitsanleitungen für das molekularbiologische Labor. 3rd ed. Heidelberg: Spektrum Akad. Verl; 2002.

52. Gibson DG. Enzymatic assembly of overlapping DNA fragments. Methods Enzymol. 2011;498:349–61.

53. Inoue H, Nojima H, Okayama H. High efficiency transformation of *Escherichia coli* with plasmids. Gene. 1990;96:23–8.

54. Tauch A, Kirchner O, Löffler B, Götker S, Pühler A, Kalinowski J. Efficient electrotransformation of *Corynebacterium diphtheriae* with a mini-replicon derived from the *Corynebacterium glutamicum* plasmid pGA1. Curr Microbiol. 2002;45:362–7.

55. Schäfer A, Tauch A, Jäger W, Kalinowski J, Thierbach G, Pühler A. Small mobilizable multi-purpose cloning vectors derived from the *Escherichia coli* plasmids pK18 and pK19. Selection of defined deletions in the chromosome of *Corynebacterium glutamicum*. Gene. 1994;145:69–73.

56. Vallejo AN, Pogulis RJ, Pease LR. In vitro synthesis of novel genes. Mutagenesis and recombination by PCR. PCR Method Appl. 1994;4:S123–30.

57. Wehmeier L, Schäfer A, Burkovski A, Krämer R, Mechold U, Malke H, Pühler A, Kalinowski J. The role of the *Corynebacterium glutamicum* rel gene in (p)ppGpp metabolism. Microbiology. 1998;144:1853–62.

58. Walter F, Grenz S, Ortseifen V, Persicke M, Kalinowski J. *Corynebacterium glutamicum ggtB* encodes a functional γ-glutamyl transpeptidase with γ-glutamyl dipeptide synthetic and hydrolytic activity. J Biotechnol. 2015;232: 99–109.

59. Martin RG, Berberich MA, Ames BN, Davis WW, Goldberger RF, Yourno JD. [147] Enzymes and intermediates of histidine biosynthesis in *Salmonella typhimurium*. In: Tabor H, Tabor CW, editors. Methods in enzymology, Metabolism of amino acids and amines part B. New York: Academic; 1971. p. 3–44.

60. van Veldhoven PP, Mannaerts GP. Inorganic and organic phosphate measurements in the nanomolar range. Anal Biochem. 1987;161:45–8.

61. Camacho C, Coulouris G, Avagyan V, Ma N, Papadopoulos J, Bealer K, Madden TL. BLAST+: architecture and applications. BMC Bioinf. 2009;10:421.

62. Li W, Cowley A, Uludag M, Gur T, McWilliam H, Squizzato S, Park YM, Buso N, Lopez R. The EMBL-EBI bioinformatics web and programmatic tools framework. Nucleic Acids Res. 2015;43:W580–4.

63. Finn RD, Clements J, Arndt W, Miller BL, Wheeler TJ, Schreiber F, Bateman A, Eddy SR. HMMER web server: 2015 update. Nucleic Acids Res. 2015;43:W30–8.

64. Crooks GE, Hon G, Chandonia J, Brenner SE. WebLogo: a sequence logo generator. Genome Res. 2004;14:1188–90.

65. McNicholas S, Potterton E, Wilson KS, Noble MEM. Presenting your structures: the CCP4mg molecular-graphics software. Acta Crystallogr D Biol Crystallogr. 2011;67:386–94.

66. Tina KG, Bhadra R, Srinivasan N. PIC: Protein interactions calculator. Nucleic Acids Res. 2007;35:W473.

16

Application of hierarchical oligonucleotide primer extension (HOPE) to assess relative abundances of ammonia- and nitrite-oxidizing bacteria

Giantommaso Scarascia, Hong Cheng, Moustapha Harb and Pei-Ying Hong[*]

Abstract

Background: Establishing an optimal proportion of nitrifying microbial populations, including ammonia-oxidizing bacteria (AOB), nitrite-oxidizing bacteria (NOB), complete nitrite oxidizers (comammox) and ammonia-oxidizing archaea (AOA), is important for ensuring the efficiency of nitrification in water treatment systems. Hierarchical oligonucleotide primer extension (HOPE), previously developed to rapidly quantify relative abundances of specific microbial groups of interest, was applied in this study to track the abundances of the important nitrifying bacterial populations.

Results: The method was tested against biomass obtained from a laboratory-scale biofilm-based trickling reactor, and the findings were validated against those obtained by 16S rRNA gene-based amplicon sequencing. Our findings indicated a good correlation between the relative abundance of nitrifying bacterial populations obtained using both HOPE and amplicon sequencing. HOPE showed a significant increase in the relative abundance of AOB, specifically *Nitrosomonas*, with increasing ammonium content and shock loading ($p < 0.001$). In contrast, *Nitrosospira* remained stable in its relative abundance against the total community throughout the operational phases. There was a corresponding significant decrease in the relative abundance of NOB, specifically *Nitrospira* and those affiliated to comammox, during the shock loading. Based on the relative abundance of AOB and NOB (including comammox) obtained from HOPE, it was determined that the optimal ratio of AOB against NOB ranged from 0.2 to 2.5 during stable reactor performance.

Conclusions: Overall, the HOPE method was developed and validated against 16S rRNA gene-based amplicon sequencing for the purpose of performing simultaneous monitoring of relative abundance of nitrifying populations. Quantitative measurements of these nitrifying populations obtained via HOPE would be indicative of reactor performance and nitrification functionality.

Keywords: Single nucleotide primer extension, Quantitative monitoring, 16S rRNA gene-based amplicon sequencing, AOB/NOB ratio, Shock loading event

Background

The conventional process of removing nitrogen from municipal wastewater streams involves nitrification followed by denitrification. Nitrification in the wastewater treatment process is generally performed in two consecutive steps. The first step is ammonia oxidation by either ammonia-oxidizing bacteria (AOB) or ammonia-oxidizing archaea (AOA), and the second step is nitrite oxidation by nitrite-oxidizing bacteria (NOB). Representative AOB genera include *Nitrosomonas* and *Nitrosospira* while *Nitrososphaera* and *Nitrosopumilus* would account as two examples of AOA. Representative NOB include *Nitrospira* and *Nitrobacter*. In recent years, three members within the genus *Nitrospira*, namely Candidatus *Nitrospira inopinata*, Candidatus *Nitrospira nitrosa* and Candidatus *Nitrospira nitrificans*, were shown to perform complete nitrification in a single step and are referred to as comammox [1, 2]. Although comammox are not yet isolated from municipal

* Correspondence: peiying.hong@kaust.edu.sa
Biological and Environmental Science & Engineering Division (BESE), King Abdullah University of Science and Technology (KAUST), Water Desalination and Reuse Center (WDRC), Thuwal 23955-6900, Saudi Arabia

wastewater treatment plants (WWTP) and their contribution in terms of full nitrification in such systems remain unknown, comammox had been shown to use urea as ammonium source for nitrification and that this trait should in theory enable them to thrive in WWTP environments where urea is often present [1].

The commensal interaction between AOB/AOA and NOB, and the potential importance of Nitrospira-affiliated comammox suggest that the nitrification process can be prone to failure when there is a suboptimal ratio of AOB/AOA to NOB, particularly that of Nitrospira. This is primarily due to the effect of hydroxylamine, an intermediate of the conversion of ammonia to nitrite, which has a strong inhibitory effect on nitrite oxidizers when released by AOB/AOA [3]. Failure of the nitrification system in a WWTP can result in discharge of effluent that contains total nitrogen (TN) higher than the permissible limit (< 15 mg/L in Saudi Arabia). This can result in potentially detrimental environmental impacts in the form of eutrophication of the receiving water bodies. Hence, establishing an optimal ratio and abundance of AOB/AOA to NOB is important in ensuring the functionality of the nitrification process so that treated effluent will meet the discharge limits for TN.

In such instance, a quantitative method capable of efficiently targeting AOB/AOA and NOB relative abundances would be particularly useful in tracking the proportional ratio between these two groups. Various methods like quantitative PCR (qPCR) and fluorescent in-situ hybridization (FISH) were used to directly quantify the abundances of AOB and NOB. Through these approaches, it was determined that Nitrosomonas likely accounts for the largest fraction of AOB in conventional activated sludge systems. Additionally, in conventional municipal wastewater streams in which ammonium concentrations ranged from 12 to 50 mg/L, Nitrospira was the predominant NOB while Nitrobacter was generally in low abundance [3–5].

Although qPCR and FISH have been routinely used in earlier studies, both approaches have substantial limitations. For example, qPCR requires standard curves to be established for individual targets. New generation of PCR (i.e., digital PCR, dPCR) does not require standard curves to be generated but similar to conventional qPCR, remains limited by the number of fluorophores available in the ultraviolet spectrum. Hence, both quantitative PCR approaches have limited multiplexing throughput capability. Furthermore, both qPCR approaches can be prone to non-specific amplification when SYBR® green fluorescence reporters are utilized, and would require an additional melting curve analysis or verification step to differentiate between true-positive and false-positive amplifications. Although Taqman® probes can be developed for use in qPCR and dPCR, such probes are relatively expensive and remain challenging to design for a specific microbial target

within the internal region of two flanking qPCR primers. Digital PCR being a considerably new technology, also remains rather costly till this date compared to traditional qPCR. FISH, on the other hand, can be affected by the in-situ accessibility of labeled probes through the cell membranes and the copy numbers of rRNA genes inside AOB and NOB. For example, species of Nitrobacter, Nitrosomonas, Nitrosospira and Nitrospira are reported to have one copy of 16S rRNA gene per cell. This copy number of 16S rRNA genes per genome is lower than the mean copy number of 4.12 among the 2865 microbial genomes currently available in the same Ribosomal RNA Operon Copy Number database [6].

Hierarchical oligonucleotide primer extension (HOPE) was developed as a method to complement other quantitative methods like qPCR and FISH by providing a high-throughput multiplexing platform capable of addressing some of the previous methods' limitations. In this method, oligonucleotide primers are designed to target genes, usually 16S rRNA genes, at different hierarchical taxonomical levels (from domain to species). In the presence of a DNA template, the primers anneal at the designed targeting position of the 16S rRNA gene and extend with a single fluorescently labeled dideoxynucleoside triphosphate (ddNTP). The extended primers can then be identified on a DNA sequencer based on their anticipated fragment size and fluorescence. The peak areas of the individual fragments can then be used to determine the relative abundances of bacterial targets of interest [7].

The HOPE method has been developed to detect specific bacterial populations of interest including host-associated Bacteroidales and Bifidobacterium, cyanobacteria, and methanogens [8–11]. In these studies, HOPE has been demonstrated to multiplex up to 10 primers in a single reaction, allowing for the evaluation of a range of relevant bacterial indicators in a time-efficient manner [12]. The high-throughput capability of HOPE stems from the use of different fragment sizes of primers and fluorescence colors to distinguish between extended primers. To exemplify, most primers are synthesized with a length of 18 to 25 nucleotides (nt) while most DNA sequencer platforms are able to detect up to four different fluorescence colors. Hence, in theory, the combination of these two differentiating factors would mean that HOPE is capable of multiplex up to 32 primers in one single reaction tube. Aside from its multiplexing capability, it was demonstrated that HOPE has a detection limit of 0.1% of the total PCR-amplified bacterial targets, and that the entire duration required to carry out a single HOPE reaction and to analyze for the extended primer on a DNA sequencer is <1 h, which would be comparable to the time needed to carry out qPCR [12].

Given the advantages of HOPE, the method is a potentially useful tool in tracking the relative abundance of nitrifying populations. In the present study, we applied the HOPE approach to target groups of AOB and NOB (including *Nitrospira*-affiliated comammox) in wastewater environments. The method was tested against samples collected from a lab-scale biofilm-based trickling bioreactor. The total microbial community, including the nitrifying groups, was analyzed by 16S rRNA gene-based high-throughput sequencing for comparison against the HOPE data. To further demonstrate the use of HOPE in monitoring these relative abundances, the trickling bioreactor was subsequently challenged with high concentrations of incoming ammonium content so as to simulate a toxic shock event that could potentially crash the nitrification system. The changes in AOB and NOB proportions were then tracked by the HOPE approach and correlated to reactor performance. Wastewater samples from a fully operational wastewater treatment plant (WWTP) were also analyzed in order to further validate the method.

Methods

HOPE primer layout

HOPE was performed with a total of 13 primers targeting ammonia-oxidizing genera *Nitrosomonas* and *Nitrosospira* as well as nitrite-oxidizing genera *Nitrospira* and *Nitrobacter* (Table 1). Primers were sourced from the probeBase database [13]. The mismatch positions of non-targets were moved to the 3′-end of the HOPE primers as it was previously found that mismatches located at the 3′-end of the HOPE primers would not facilitate nucleotide base extension of primers, and would hence lead to improved specificity [14]. Re-positioning of the mismatches result in minimal change to the original coverage of the targeted bacterial groups. The coverage of the primers was then verified in-silico against the Ribosomal Database Project (RDP) database [15], and shown in Additional file 1. Primers were arranged into four tubes, with either 27F or 338F primer included in each tube to provide a normalization of

Table 1 HOPE primers used to target the different ammonia- and nitrite-oxidizing bacterial groups

	Target	Primer sequence (5'-3')	Extended ddNTP	PolyA-tail (nt)	Reference
NOB tube 1: *Nitrospira*					
27F	Most Bacteria	AGAGTTTGATCCTGGCTCAG	A	0	[38]
Ntspa712	4451 out of 8171 members of Nitrospirae; also targets comammox bacteria Candidatus *N. inopinata*	CGCCTTCGCCACCGGCCTTCC	T	0	[39]
Ntspa572	1835 out of 8171 members of Nitrospirae; also targets comammox bacteria Candidatus *N. inopinata*	AACCGCCTACGCTCCCTG	T	0	This study
Ntspa1429	107 out of 8171 members of Nitrospirae; *Nitrospira* cluster II	TGGCTTGGGCGACTTCAG	G	6	Modified from [37]
NOB tube 2: *Nitrobacter*					
Eub338Ia	Most Bacteria	GCTGCCTCCCGTAGGAG	T	8	[12]
Nit1017	107 out of 195 genus *Nitrobacter*	TGC TCC GAA GAG AAG GTC ACA	T	6	Modified from [40]
Nit1000	117 out of 195 genus *Nitrobacter*	TGC GAC CGG TCA TGG	A	0	[41]
AOB tube 1: *Nitrosomonas* and *Nitrosospira*					
27F	Most Bacteria	AGAGTTTGATCCTGGCTCAG	A	0	[38]
NmoCL6a	379 out of 7530 members of order Nitrosomonadales, *Nitrosomonas* cluster 6a; mainly targeting unclassified Nitrosomonadaceae	AAGCATAAGGTCTTTCGATCCCCT	G	3	Modified from [42]
NmoCL6b	1749 out of 7530 members of order Nitrosomonadales, *Nitrosomonas* cluster 6b; Major coverage also includes other non-AOB and NOB groups	GGATCAGGCTTGCGCCC	A	0	[42]
AOB tube 2: *Nitrosomonas* and *Nitrosospira*					
Eub338Ia	Most Bacteria	GCTGCCTCCCGTAGGAG	T	8	[12]
Nso190	1295 out of 7530 members of order Nitrosomonadales	CGATCCCCTGCTTTTCTC	C	9	[41]
Nmo218	321 out of 897 members of unclassified Nitrosomonadaceae	CGGCCGCTCCAAAAGCAT	A	0	[43]
Nse1472	208 out of 911 members of genus *Nitrosomonas*	ACCCCAGTCATGACCCCC	A	6	[44]
Nsv443	934 out of 5714 members of genus *Nitrosospira*	CCGTGACCGTTTCGTTCCGGC	T	0	[41]

Target coverage for each primer was identified using the RDP Probe match function the abundance of each nitrifying bacterial target against the total bacteria. To differentiate primers that extend with the same ddNTP and hence color during capillary electrophoresis, some primers were modified with a polyA tail at the 5′-end of the primer (Table 1). Primers in each tube were prepared to form a final concentration of 10 µM.

HOPE reference template preparation

To prepare the HOPE reference templates for nitrifying bacterial populations, extracted DNA from a separate trickling nitrification biofilter that was set up in year 2007 was first amplified with 27F (5′-AGA GTT TGA TCC TGG CTC AG-3′) to 1492R (5′-GGY TAC CTT GTT ACG ACT T-3′) primers and purified for use in 16S rRNA gene clone library construction. Approximately 100 clones were picked and sequenced to identify nitrifying bacterial species. Two clones associated with Candidatus *Nitrospira defluvii* and *Nitrosomonas eutropha* were obtained and used as the reference template for HOPE primers Ntspa712, Ntspa1429, Nso1225b and Nse1472. The reference standards for Ntspa572, Nit1017, NmoCL6a, NmoCL6b and NspCL3 were obtained by first amplifying the 16S rRNA gene portion targeted by that respective HOPE primer coupled with 27F. The reference standards for primers Nmo218 and NspCL1 were obtained by amplifying the 16S rRNA gene portion with a sense-strand version of that respective HOPE primer and 1492R. All amplicons were gel-excised, purified and submitted for Sanger sequencing at the KAUST Genomics Core lab. Sequences were checked to ensure perfect matches in regions targeted by the specific primers. Reference templates for Nit1000 and Nso190 were obtained by checking the above-mentioned reference templates for perfect match regions. Amplicons with regions that perfectly matched the targeted primers were then individually ligated into the pCR2.1 vector using the TA cloning kit (Thermo Fisher Scientific, Carlsbad, CA). The ligated vectors were individually transformed into TOP10 competent cells, and transformed cells were then extracted for the plasmid using PureYield plasmid miniprep system (Promega, Madison, WI). The insert gene was resequenced by the Sanger approach to verify for perfect match regions. Subsequently, plasmid DNA concentrations were measured using an Invitrogen Qubit 2.0 fluorometer (Thermo Fisher Scientific, Carlsbad, CA) and diluted to 10 ng/µL to be used as HOPE standards in determination of calibration factors. Calibration factors are normalization factors to account for differences in peak areas between two extended HOPE primers [8], and are calculated based on Eq. (1) below:

$$CF_{A-B} = \frac{\text{Peak area of extended primer A, } P_a}{\text{Peak area of extended primer B, } P_b} \quad (1)$$

where primer B is targeting at a higher hierarchical level compared to primer A.

HOPE reactions and capillary electrophoresis

For each sample that was to be tested with HOPE, the bacterial 16S rRNA genes were first amplified using 27F coupled with 1492R primer in PCR amplifications comprising of 5 ng DNA, 25 µL of Epicentre Biotechnologies FailSafe Premix F (Illumina, Madison, WI), 200 nM (each) of forward and reverse primers, 0.5 U of ExTaq DNA polymerase (Takara, Japan), and molecular-biology grade water as necessary to reach a total volume of 50 µL. PCR was performed with 1 cycle of 95 °C for 30 s followed by 25 cycles of thermal cycling (denaturation at 95 °C for 30 s; annealing at 55 °C for 45 s; extension at 72 °C for 60 s) and a final cycle of 72 °C for 10 min. All amplicons were purified using Wizard SV Gel and PCR Clean-up kit (Promega, Madison, WI), measured for their concentrations, and diluted to 10 ng/µL to be used as templates in HOPE reactions. Each HOPE standard and sample reaction was carried out in a reaction volume that was comprised of 2.5 µL of Applied Biosystems SNaPshot multiplex kit (Thermo Fisher Scientific, CA), 0.5 µL of 10 µM primer mix, 0.5 µL of DNA template and 1.5 µL of molecular-biology grade water. The HOPE thermal cycling program consisted of 20 cycles of denaturation (96 °C for 10 s), annealing (60 °C for 30 s) and extension (72 °C for 15 s). After the primer extension reaction, 5 µL of 200 U recombinant shrimp alkaline phosphatase, rSAP (New England Biolabs, Ipswich, MA), was added to each reaction and the mixture was incubated at 37 °C for 60 min prior to denaturation at 80 °C for 10 min. The rSAP-digested samples were prepared for capillary electrophoresis by first adding 1 µL of the sample to 12 µL of Applied Biosystems Hi-Di formamide and 0.3 µL of GeneScan 120 LIZ size standard, and the mixed samples were denatured at 96 °C for 5 min prior to capillary injection in the Applied Biosystems 3500 Series genetic analyzer. Extended primers were identified on GeneMapper v 4.1 based on the extended nucleotide base (i.e., color) and fragment size. Peak areas of the primers extended in the presence of reference standards were recorded for determination of calibration factors as described previously [8] and in Eq. (1). The same was performed for the primers extended in the presence of template obtained from tested samples. The relative abundance of 16S rRNA gene amplicons targeted by primer A with respect to those targeted by primer B can then be calculated based on Eq. (2) below:

Relative abundance of target (%)

$$= \frac{\text{Peak area of primer A}}{\text{Peak area of primer B} \times CF_{A-B}} \times 100\% \quad (2)$$

where primer B is targeting at a higher hierarchical level compared to primer A.

16S rRNA gene-based high-throughput sequencing and data analysis

For all samples quantified by HOPE, Illumina MiSeq amplicon sequencing was performed to validate the HOPE data and to provide information on the total microbial community. To prepare the 16S rRNA gene amplicon libraries, 515F (5′- Illumina overhang- GTG YCA GCM GCC GCG GTA A- 3′) and 907R (5′- Illumina overhang- CCC CGY CAA TTC MTT TRA GT- 3′) primers were modified to encode the overhang adaptor sequences, and used to amplify the 16S rRNA genes. The thermal cycling program included an initial denaturation stage at 95 °C for 3 min, followed by 25 cycles of denaturation at 95 °C for 30 s, annealing at 55 °C for 30 s and extension at 72 °C for 30 s, and then a final extension stage at 72 °C for 5 min. PCR amplicons were then purified by AMPure XP beads (Beckman Coulter, CA) prior to the index PCR. Nextera XT Index (Illumina, San Diego, CA) was incorporated into each of the individual samples during PCR. The thermal cycling program included a denaturation stage at 95 °C for 3 min, followed by 8 cycles of denaturation at 95 °C for 30 s, annealing at 55 °C for 30 s and extension at 72 °C for 30 s, and then a final extension stage at 72 °C for 5 min. The final indexed PCR amplicons were again purified by AMPure XP beads and quantified for the concentrations using an Invitrogen Qubit 2.0 fluorometer. The controls for all PCR reactions were negative for amplification. Purified amplicons were submitted to the KAUST Genomics Core lab for unidirectional sequencing on an Illumina MiSeq platform. Raw sequence reads were handled using procedures described previously [16]. To annotate the 16S rRNA gene sequences obtained from high-throughput sequencing, RDP Classifier was used for taxonomical assignments at a 95% confidence level [17].

Primer-E analysis and statistical tests

The taxonomical assignment for each sample obtained by high-throughput amplicon sequencing was calculated for relative abundances of the individual bacterial and archaeal genera. Special emphasis was made to collate the relative abundance of the *Nitrospira*, *Nitrosomonas*, *Nitrobacter* and *Nitrosospira* so as to provide validation against the relative abundances of these genera obtained by HOPE method. In addition, the relative abundance of all taxonomical assignments obtained by MiSeq amplicon sequencing were collated alongside the relative abundances of the bacterial targets identified by HOPE. Both datasets were individually input into Primer E version 7 [18], square-root transformed, and then computed for their Bray-Curtis similarities. The Bray-Curtis distance matrix was used for multivariate analysis on a non-metric threshold multidimensional scaling (nMDS) plot. The nMDS plots utilizes a distance matrix applied to all samples by which each is represented as a point in the two-dimensional space. The x

and y-axes of the nMDS plot do not represent any parameters. Instead, the distance between the positioning of two samples would denote the extent of similarity between these two samples. To illustrate, samples with higher similarity are closer in proximity in the nMDS and vice versa. The stress value measures the scatter in the nMDS plot, and ranges from 0 to 1, with 0 denoting a good representation of the positioning of the samples in the two-dimensional space shown in the nMDS plot and vice versa. Bacterial targets that exhibited >0.7 correlation with the multivariate patterns on the nMDS were overlaid as vectors. Vectors are directional lines emanating from a common origin, extending in the directions in which the marked bacterial targets increase in relative abundance, and hence play a dominant role in the subclustering of samples on the nMDS plot. All other significance tests were conducted using two-tailed t-tests on Microsoft Excel 2013.

Biofilm reactor setup and operation

The samples that were quantified for the respective nitrifying populations by HOPE and 16S rRNA gene-based amplicon sequencing were obtained from a trickling biofilm nitrification reactor. The trickling biofilm reactor was first packed with sponge cubes, each of dimension 1 by 1.5 cm, to a total volume of 4.5 L (Fig. 1). 500 mL of sludge collected from the aerobic activated sludge tank of the full scale WWTP was inoculated upon commencement of operation and another 500 mL of aerobic sludge was added 2 weeks later. During the startup phase (Phase A, day 0 to 39), basal substrate containing 40 mg/L ammonium (NH_4^+-N) was fed into the reactor. The components per L of basal substrate were modified from a previous study [19] and include: 153 mg NH_4Cl, 480 mg $NaHCO_3$, 24 mg $MgSO_4.7H_2O$, 20 mg K_2HPO_4, 11.3 mg $CaSO_4.2H_2O$ and 7.1 mg $FeCl_3.6H_2O$, 100 μL of trace elements that contained 0.274 g/L $Na_2MoO_4.2H_2O$, 2.74 g/L $ZnSO_4.7H_2O$, 0.036 g/L $CuSO_4.5H_2O$, 0.03 g/L $CoCl_2.6H_2O$, 7 g/L $FeSO_4.7H_2O$ and 1.55 g/L $MnCl_2.2H_2O$. Upon reaching stabilization and from day 40 onwards, the basal substrate was prepared by diluting the anaerobic effluent that was high in ammonium content [20]. The anaerobic effluent contained on average 160 mg/L NH_4^+-N and 40 mg/L of chemical oxygen demand (COD), and the ammonium content in the basal feed was diluted accordingly to provide 80 mg/L NH_4^+-N from day 40 to 83 (Phase B), 110 mg/L NH_4^+-N from day 84 to 120 (Phase C), and 150 mg/L NH_4^+-N from day 121 to 150 (Phase D1). Ammonium concentration was subsequently increased to 200 mg/L NH_4^+-N from day 151 to 179 (Phase E), and diluted to 150 mg/L NH_4^+-N from day 180 to 270 (Phase D2). Throughout operation, the pH of trickling biofilm reactor was maintained between 7.0 and 7.5 and aeration was provided at a sparging rate of 2 L/min.

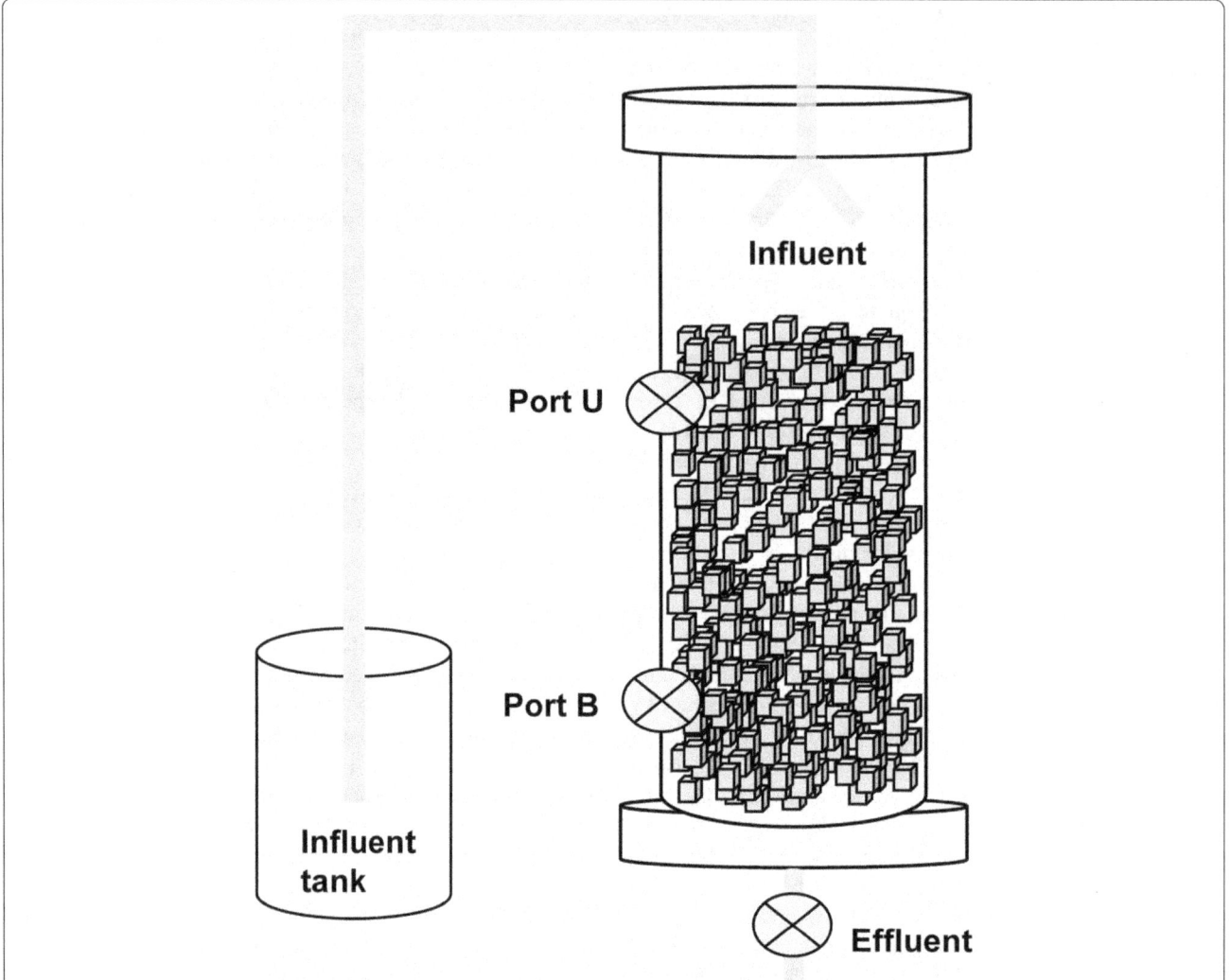

Fig. 1 Trickling nitrification biofilm reactor column. The reactor was operated through six phases A through E, with increasing ammonium (NH$_4^+$) concentration in the influent during phase A through E and a decrease to 150 mg/L NH$_4^+$ in the influent at phase D2. Samples were collected at the top (U) and bottom (B) locations from Port U and B, respectively. Influent and effluent were also collected from the marked sampling ports

Biofilm reactor sampling and DNA extraction

Three sponge cubes were sampled individually from both the top and bottom sampling ports (Fig. 1), and referred to as U and B samples subsequently in this study. New replacement sponges were placed back into the reactor after each biomass sampling so as to provide new substrata for the adherent bacteria. Subsequent samplings were carried out such as to avoid sampling the replacement sponges. Sponges with the adherent biomass were placed in 50 mL centrifuge tubes, each containing 20 mL of 1X phosphate buffer saline (PBS). The samples were then vortexed at maximum speed for 10 min and the sponge cubes removed aseptically by sterile forceps. The biomass suspensions in the 50 mL tubes were then centrifuged at 6800×g for 10 min to collect cell pellets. 0.2 g of biomass from cell pellet was weighed and extracted for its DNA. DNA was extracted using the UltraClean Soil DNA Isolation Kit

(MoBio Laboratories, Carlsbad, USA) with slight modifications to the protocol by adding lysozyme and achromopeptidase to the lysis buffer [21].

Full-scale WWTP sampling and DNA extraction

Another set of samples that were quantified for the respective nitrifying populations by HOPE and 16S rRNA gene-based amplicon sequencing were sampled from a full-scale wastewater treatment plant located in KAUST. The WWTP is a full-scale membrane bioreactor (MBR) equipped with the following process units: (i) grid mesh screen, (ii) primary clarifier, (iii) anoxic-oxic activated sludge tanks, (iv) submerged membrane tank, and (v) holding tank for chlorination. Influent was collected after the primary clarifier, effluent was collected at the MBR discharge point and chlorinated effluent was collected from the holding tank. Activated sludge was collected in the oxic zone of the sludge tank. Sampling was performed on a monthly basis from July to December 2015, with the exception of October 2015 during which two sample sets were collected. Influent samples were prepared for DNA extraction by centrifuging 50 to 100 mL of influent at 10000×g for 20 min to obtain the biomass pellet. 2 L each of effluent and chlorinated effluent were individually filtered through 0.4 μm polycarbonate membranes to retain the biomass. Biomass samples were extracted for their DNA using the UltraClean Soil DNA Isolation Kit (MoBio Laboratories, Carlsbad, USA) with slight modifications to the protocol by adding lysozyme and achromopeptidase to the lysis buffer [21].

Water quality measurement for biofilm reactor samplesand wastewater streams

Samples from the fresh basal substrate fed into the trickling biofilm reactor were collected as influent, and samples collected from the discharge port were collected as effluent (Fig. 1). Ammonium (NH$_4^+$-N), nitrite (NO$_2^-$-N), nitrate (NO$_3^-$-N) concentrations in influent and effluent samples (n = 75 each) collected every 2–3 d throughout the reactor operational period were measured using Test 'N Tube high range ammonia kit, TNTplus 839, and TNTplus 835, respectively. COD was measured using either LCK 314 (15–150 mg/L) or LCK 514 COD (100–2000 mg/L) cuvette test vials depending on the concentration to be measured. All measurements were conducted based on protocols specified by the manufacturer (Hach-Lange, Manchester, UK). The water quality data obtained from the trickling biofilm reactor were collated, log-transformed and normalized within Primer-E version 7 [18]. The normalized data were then used to generate a principal component analysis (PCA) plot. Water quality parameters that exhibited >0.7 correlation with the multivariate pattern on the PCA were overlaid as vectors. For the wastewater samples obtained from the WWTP, nitrate concentrations in all three streams were measured in lab using the TNTplus 835. Total nitrogen in all streams as well as ammonia in the influents were measured and provided by the plant operator.

Results

AOB abundance in biofilm reactor quantified by HOPE throughout the operational phases

The relative abundances of nitrifying bacterial populations quantified by the HOPE approach were collated for multivariate analysis on a multidimensional scaling (nMDS) plot (Fig. 2a). The positioning of the samples on the nMDS plot showed that operational phase was the dominant factor accounting for the subclustering of samples. Vector-based analysis showed an increase in the abundance of AOB in Phases E and D2 compared to the early operational phases. To illustrate, the relative abundance of AOB groups, specifically *Nitrosomonas* targeted by Nse1472 and NmoCL6a, increased from 2.2 ± 1.8% and 3.4 ± 1.3% of total bacteria, respectively, in Phase B to 18.9 ± 10.3% and 26.5 ± 9.5% of total bacteria, respectively, in Phase D2 (Table 2). The relative abundances of these AOB groups were significantly higher in the latter phases (i.e., Phases E and D2) than in earlier phases (i.e., Phases B and C) (t-test, $p < 0.001$). Both *Nitrosomonas* and *Nitrosospira* targeted by Nso190 also increased from 8.1 ± 5.2% in Phase B to 26.4 ± 8.2% in Phase E and 28.8 ± 8.7% in Phase D2 (Table 2).

However, the increase in the relative abundance of Nso190-targets is mainly due to the increase in the relative abundance of *Nitrosomonas*. This is because *Nitrosospira* detected by primer Nsv443 remained stable in its relative abundance against the total community throughout the operational phases (Phase A: not detected; Phase B: 2.1 ± 1.8%; Phase C: 4.3 ± 2.9%; Phase D1: 4.6 ± 3.0%; Phase E: 3.3 ± 1.5%; Phase D2: 5.2 ± 1.5%). The changes in the relative abundance of Nsv443-targeted *Nitrosospira* were not significantly different across Phases B to D2 (t-test, $p > 0.1$).

Abundance of NOB and *Nitrospira*-affiliated comammox in biofilm reactor quantified by HOPE throughout the operational phases

Vector analysis performed on the same nMDS plot also showed that samples collected from the early operational phases A and B were higher in relative abundance of NOB targeted by either Ntspa712 or Ntspa572. Both primers targeted *Nitrospirae* at the same hierarchical level but with different coverage (Additional file 1). Both primers also target at least one of the three recently identified comammox bacteria (i.e., Candidatus *N. inopinata*). The average relative abundance of *Nitrospira* targeted by Ntspa572 was 3.9 ± 0.3% in Phase A and 4.3 ± 1.2% in Phase B, and was significantly higher ($p < 0.05$) than that detected in Phases C (1.2 ± 1.6%), D1 and E (not detected), and D2 (0.01 ± 0.03%). Similarly, the relative abundance of *Nitrospira* targeted by Ntspa712 was 6.3 ± 2.3% in Phase A and 9.6 ± 3.1% in Phase B, and was significantly higher ($p < 0.05$) than that detected in Phases C (3.7 ± 4.6%), D1 (0.9 ± 1.8%), E (2.3 ± 2.9%) and D2 (2.7 ± 1.9%).

In contrast to *Nitrospira*, *Nitrobacter* detected by the Nit1017 primer increased in its relative abundance from 4.8 ± 3.2% in Phase B to 8.4 ± 3.6% in Phase D2 (Table 2). However, the Nit1000 primer did not perform in a similar manner as the Nit1017 primer despite both primers having similar detection coverage of most *Nitrobacter* spp. (Table 1). The Nit1000 primer consistently failed to detect any *Nitrobacter* in the latter operational phases, and in samples that were detected with *Nitrobacter*, the relative abundance was more than 2-fold lower than that detected by the Nit1017 primer.

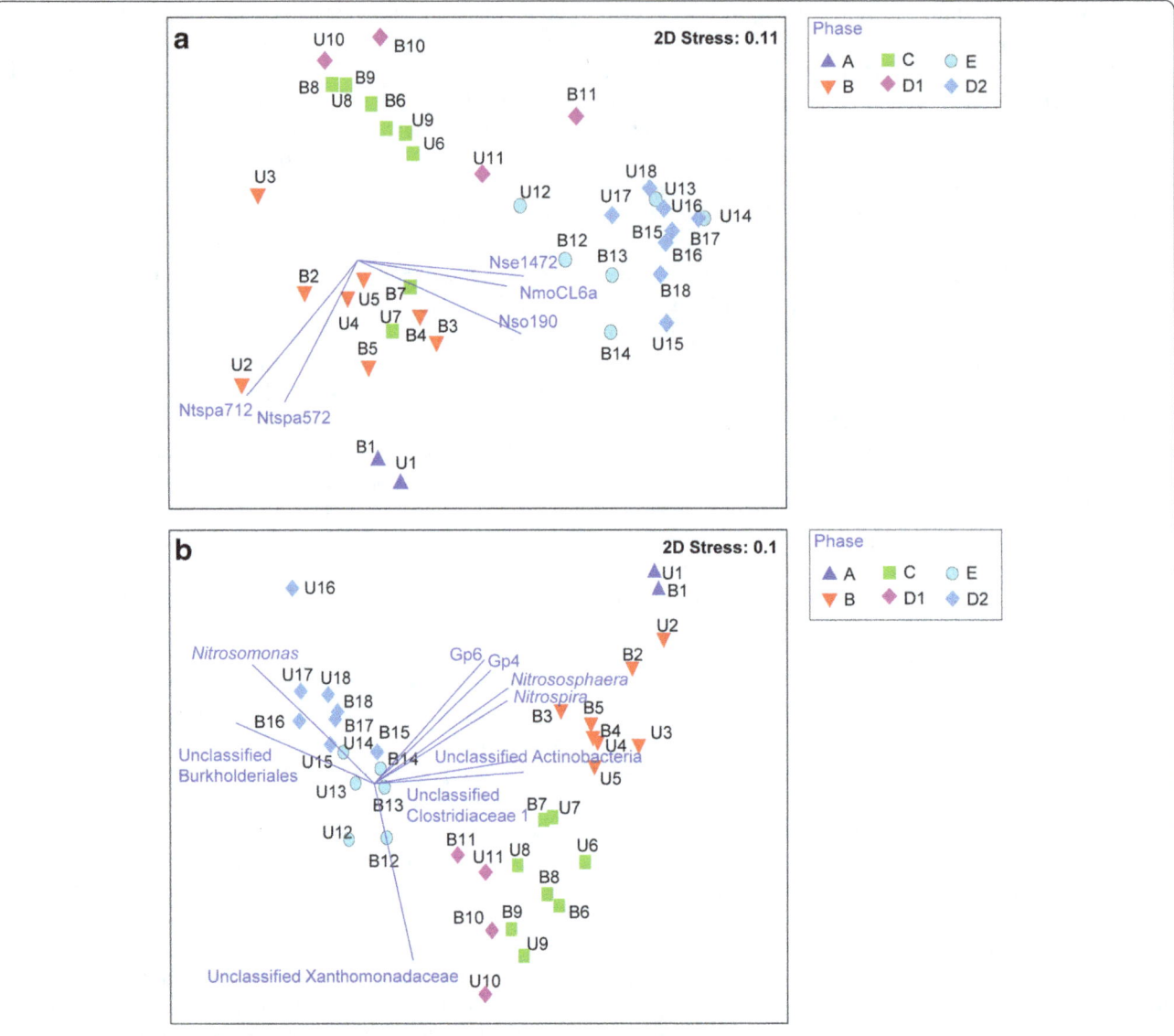

Fig. 2 Multivariate analysis of the microbial data. The relative abundance of AOB and NOB targeted by HOPE primers revealed differences throughout the operational phases. **a**. The relative abundances of the total microbial community obtained from high-throughput sequencing also showed similar clustering based on the operational phases (**b**)

Validation of HOPE dataset obtained for trickling biofilm reactor with 16S rRNA gene-based amplicon sequencing

Based on the HOPE primer coverage shown in Additional file 1, the relative abundance of genera *Nitrosomonas*, *Nitrosospira*, *Nitrobacter* and *Nitrospira* obtained by 16S rRNA gene-based amplicon sequencing was respectively compared against the relative abundance detected by Nse1472 (*Nitrosomonas*), Nsv443 (*Nitrosospira*), Nit1000 or Nit1017 (*Nitrobacter*), and Ntsp572 or Ntspa712 (*Nitrospira*) HOPE primers. The relative abundance of *Nitrosomonas* and *Nitrospira* obtained by 16S rRNA gene-based amplicon sequencing was also compared against that detected by Nso190 HOPE primer.

Generally, the relative abundance obtained by HOPE differed from that detected by amplicon sequencing but both datasets exhibited good Spearman's rank correlation (Table 2). To illustrate, the average relative abundance of *Nitrosomonas* reported by 16S rRNA gene-based amplicon sequencing was 2.0%, 1.4%, 0.1%, 0.1%, 6.8% and 11.6% in Phase A, B, C, D1, E and D2, respectively (Table 2). The average relative abundance of *Nitrosomonas* targeted by HOPE primer Nse1472 was 1.6%, 2.2%, 0.3%, 1.5%, 18.9% and 26.4% in Phase A, B, C, D1, E and D2, respectively. Similarly, the summation of relative abundance of *Nitrosomonas* and *Nitrosospira* obtained by amplicon sequencing differed from that detected by Nso190 HOPE

Table 2 Relative abundances of the different ammonia-oxidizing bacteria and nitrite-oxidizing bacteria obtained using HOPE

Method used		Phase A	Phase B	Phase C	Phase D1	Phase E	Phase D2	Spearman correlation between amplicon sequencing and HOPE data from primer with nearest coverage match
		Average relative abundance against the total bacteria ± standard deviation (%)						
		Ammonia-oxidizing bacteria (AOB)						
HOPE	NmoCl6a	6.7 ± 3.1	3.4 ± 1.3	2.9 ± 0.9	4.2 ± 2.6	23.2 ± 15.6	26.5 ± 9.5	
	Nmo218	3.7 ± 1.1	3.33 ± 2.6	Not detected	Not detected	0.01 ± 0.03	Not detected	
	Nse1472	1.6 ± 1.1	2.2 ± 1.8	0.3 ± 0.2	1.5 ± 1.5	18.9 ± 10.3	26.4 ± 5.0	
Amplicon sequencing	Nitrosomonas	2.0 ± 0.01	1.4 ± 1.1	0.1 ± 0.1	0.1 ± 0.2	6.8 ± 4.4	11.6 ± 3.0	Nitrosomonas with Nse1472: $\rho = 0.91$, $p = 2.0 \times 10^{-14}$
HOPE	Nsv443	Not detected	2.1 ± 1.8	4.3 ± 2.9	4.6 ± 3.0	3.3 ± 1.5	5.2 ± 1.5	
	Nso190	8.9 ± 1.4	8.1 ± 5.2	9.0 ± 4.8	11.5 ± 12.3	26.4 ± 8.2	28.8 ± 8.7	
Amplicon sequencing	Nitrosospira	0.03 ± 0.04	1.3 ± 1.1	2.8 ± 3.6	0.2 ± 0.8	0.5 ± 0.2	1.4 ± 0.6	Nitrosospira with Nsv443: $\rho = 0.62$, $p = 5.9 \times 10^{-5}$; Nitrosomonas and Nitrosospira with Nso190: $\rho = 0.82$, $p = 6.7 \times 10^{-10}$
		Nitrite-oxidizing bacteria (NOB)						
HOPE	Ntspa572	3.9 ± 0.3	4.3 ± 1.2	1.2 ± 1.6	Not detected	Not detected	0.01 ± 0.03	
	Ntspa712	6.3 ± 2.3	9.6 ± 3.1	3.7 ± 4.6	0.9 ± 1.8	2.3 ± 2.9	2.7 ± 1.9	
Amplicon sequencing	Nitrospira	10.5 ± 1.1	10.3 ± 2.7	2.2 ± 2.5	0.03 ± 0.02	0.003 ± 0.005	0.01 ± 0.02	Nitrospira with Ntspa712: $\rho = 0.56$, $p = 3.4 \times 10^{-4}$; Nitrospira with Ntspa572: $\rho = 0.86$, $p = 3.1 \times 10^{-11}$
HOPE	Nit1017	8.8 ± 3.5	4.8 ± 3.1	4.6 ± 0.8	4.7 ± 3.4	9.0 ± 3.9	8.4 ± 3.6	
	Nit1000	Not detected	1.0 ± 1.3	1.1 ± 0.3	0.5 ± 0.3	0.1 ± 0.3	Not detected	
Amplicon sequencing	Nitrobacter	Not detected	0.022 ± 0.05	0.01 ± 0.01	0.002 ± 0.002	Not detected	Not detected	Nitrobacter with Nit1017: $\rho = -0.40$, $p = 0.02$; Nitrobacter with Nit1000: $\rho = 0.68$, $p = 6.1 \times 10^{-6}$

The relative abundance of selected targets further correlated with the amplicon sequencing data primer (Table 2), although the datasets showed a Spearman's rank correlation coefficient of $\rho = 0.82$.

For NOB, although Nit1017 detected higher relative abundance of *Nitrobacter* than Nit1000 HOPE primer, the results obtained from Nit1017 did not correlate positively with those detected by amplicon sequencing ($\rho = -0.40$). Instead, *Nitrobacter* spp. were detected in a very low abundance of <0.1% throughout the operational phases by amplicon sequencing and coincides with the range of relative abundance detected by HOPE using Nit1000 primer ($\rho = 0.68$). In addition, the average relative abundance of *Nitrospira* reported by amplicon sequencing decreased from 10.5% to 0.01% from Phase A to D2, and the same decrease in relative abundance targeted by Ntspa572 was observed, albeit with different relative abundance that decreased from 3.9% to 0.01%.

Changes in biofilm reactor microbial community characterized by 16S rRNA gene-based amplicon sequencing

The total microbial communities of the trickling biofilm reactor were characterized by 16S rRNA gene-based amplicon sequencing (Fig. 2b). The nMDS plot generated by this data was similar to that which was obtained by HOPE in that the multivariate clusters were differentiated according to the operational phases of the reactor (Fig. 2b). Additional information was also obtained from the amplicon sequencing analysis. To illustrate, *Nitrososphaera*, which represent ammonia-oxidizing archaea, experienced a significant decrease in the relative abundance from 0.1% during early operational phases (Fig. 2b) to negligible levels during the late operational phases ($p = 0.004$). Other microbial populations that significantly decreased in their relative abundances across the different operational phases included unclassified Actinobacteria, Clostridiaceae I, Rhodospirillales and Deltaproteobacteria (t-test, $p < 0.005$). Unclassified Xanthomonadaceae had the highest relative abundance of 15.0 ± 5.9% during Phases C and D1, and averaged a relative abundance of 4.6 ± 3.1% in the remaining phases. Unclassified Burkholderiales increased significantly from 2.0 ± 1.2% in Phases A through D1 to 23.0 ± 10.8% in Phases D2 and E ($p = 5.4 \times 10^{-6}$).

Performance of trickling nitrification biofilm reactor

Based on the collated water quality data, effluent from the trickling nitrification biofilm reactor clustered into the six operational phases, which corresponded with the influent ammonium content (Fig. 3). As the influent ammonium content was increased stepwise from 40 mg/L to 200 mg/L, the COD in the effluent also increased and was significantly higher in the effluent collected during Phase D1, D2 and E compared to those collected during Phase A and C (t-test, $p = 2.1 \times 10^{-7}$, Fig. 4a). Ammonium removal was optimally achieved at an average of 44% reduction per day during

days 59 through 109 of operation (Phases B and C), which corresponded with an influent concentration of 80–110 mg/L (i.e., U3-U8). During these phases, ammonium was primarily converted to nitrate (average concentration 71.9 ± 21.5 mg/L) with minimal nitrite accumulation in the effluent (0.41 ± 0.45 mg/L). Subsequently, the trickling biofilm reactor experienced deterioration in its nitrification efficacy as influent ammonium content increased to 150 mg/L and 200 mg/L (Phases D and E). Ammonium was only partially converted to nitrate and an accumulation of nitrite was observed in the effluent collected during these phases (i.e., U11-U18). Nitrite levels in the effluent during the two latter phases were significantly higher than during Phases A through C (t-test, $p = 9.9 \times 10^{-9}$, Fig. 4a).

Correlative trends of AOB and NOB proportions with reactor performance

The ratios in the relative abundance of these AOB (i.e., Nse1472, Nsv443 and Nso190) and NOB targeted by Ntspa712 HOPE primer were further evaluated and compared to the reactor performance based on its water quality parameters (Fig. 4a and b). It was determined that the average ratio of AOB targeted by Nse1472 and Nso190 normalized against Ntspa712-targeted NOB was 0.2 and 2.5, respectively, during the phases that corresponded to stable reactor performance as evidenced by low nitrite content in the effluent (i.e., Phases A through C). During Phases E and D2, which corresponded with deterioration in the reactor performance, the ratio of AOB targeted by Nse1472 and Nso190 normalized against Ntspa712-targeted NOB increased to 13.0 and 14.0, respectively. There was, however, a less significant increase in the ratio of AOB targeted by Nsv443 against Nit1017-targeted NOB (Fig. 4b). These ratios obtained using HOPE were compared against those generated from 16S rRNA-based amplicon sequencing data (Fig. 4c). The ratio of the relative abundance of *Nitrosomonas* compared against *Nitrospira* (including *Nitrospira*-affiliated comammox) as determined by amplicon sequencing was 0.12 during the stable operational phases. This ratio markedly increased to 948 when the reactor deteriorated in performance. Likewise, the ratio of *Nitrosospira* normalized against *Nitrospira* was observed to be 0.66 during Phase A to C, and increased to 72.7 in the latter phases.

Detection of AOB and NOB in WWTP samples using HOPE and amplicon sequencing

Both HOPE and amplicon sequencing revealed that the activated sludge from the local full-scale WWTP had lower relative abundances of *Nitrosomonas*, *Nitrosospira* and *Nitrobacter* compared to Ntspa712-targeted *Nitrospira* (Table 3). Nitrate concentrations were consequentially higher in the effluent ($p < 0.0001$) than in the influent and chlorinated effluent. HOPE and amplicon sequencing were further used to evaluate the nitrifying bacterial populations

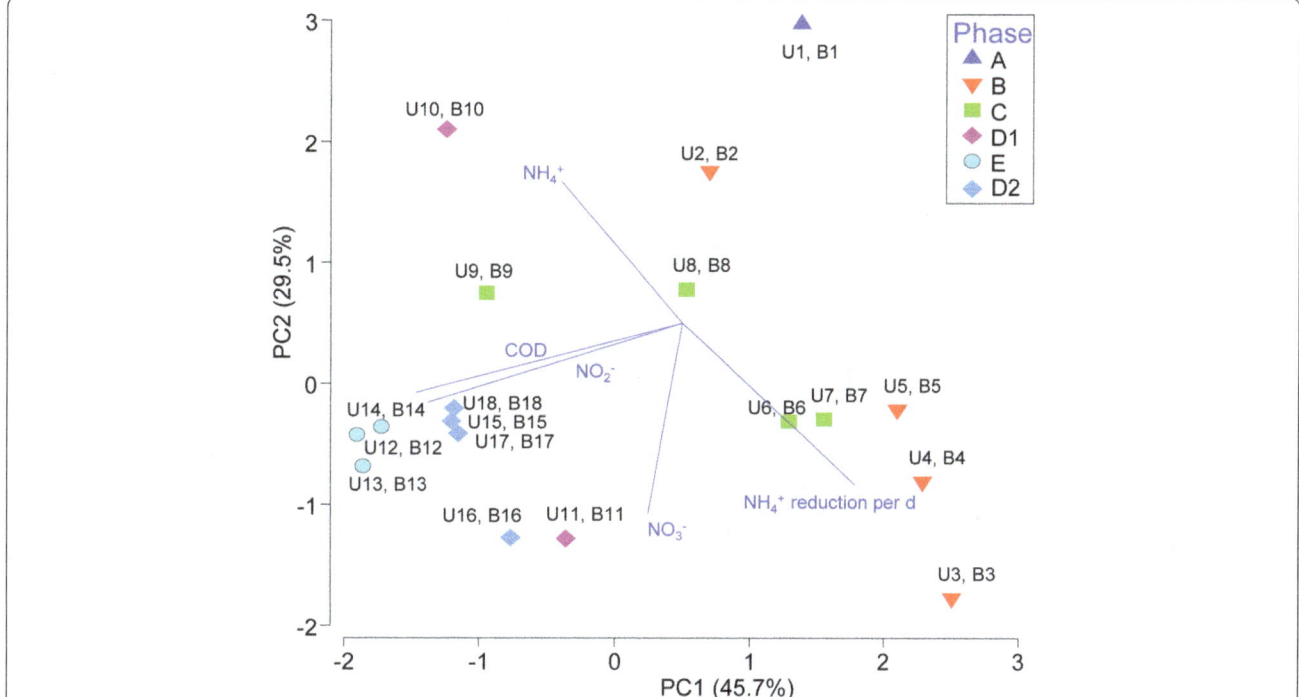

Fig. 3 Principal component analysis (PCA) of the measured water quality parameters in the effluent stream throughout the operational phases. The PCA plot suggests that there was high ammonium NH_4^+ oxidation during Phases B and C but an accumulation of nitrite and an increase in the chemical oxygen demand (COD) during the late operational phases D2 and E

present in these wastewater samples. Similar to the activated sludge, the wastewaters did not have any detectable *Nitrobacter* but instead, had relatively higher abundances of *Nitrospira* compared to *Nitrosomonas* and *Nitrosospira*. To illustrate, the relative abundance of Ntspa712-targeted NOB ranged from 3.1% in the influent to 1.0% in the chlorinated effluent. This relative abundance in *Nitrospira* was at least 2 times higher than the AOB targeted by Nse1472 and Nsv443 (Table 3). Amplicon sequencing also showed a relatively higher abundances of *Nitrospira* compared to both AOB (i.e., *Nitrosomonas* and *Nitrosospira*) in the three types of wastewater streams.

Discussion

This study demonstrates the development of HOPE as a high-throughput method to quantify for the nitrifying bacterial populations including AOB, NOB and *Nitrospira*-affiliated comammox. Along with the AOA, which was not targeted by the HOPE approach in this study, these nitrifying populations play an important role in the activated sludge process to convert ammonia in the untreated wastewater to nitrite and nitrate. The nitrate is subsequently converted to dinitrogen and nitrogen by denitrifiers. Collectively, the nitrifying and denitrifying populations reduce the total nitrogenous content in wastewater to a level permissible for discharge or reuse. It has previously been hypothesized that the theoretical

AOB/NOB ratio in a nitrification process would be 2 due to the differences in electron generation and biomass yield between AOB and NOB [22–25]. Simultaneous nitrification/denitrification can possibly increase this ratio to 3 (i.e., more NOB than AOB) [26]. However, the inverse may have a potentially detrimental impact on overall functionality of the nitrification system, especially since excess AOB can lead to accumulation of the hydroxylamine intermediate which is inhibitory towards NOB [3]. As such, the HOPE method is developed with the intention to serve as a practically applicable tool for monitoring the AOB and NOB ratio and hence used to infer the efficiency of the nitrification process in wastewater treatment systems.

For this purpose, the HOPE primers were designed to target *Nitrosomonas*, *Nitrosospira*, *Nitrobacter* and *Nitrospira* at the order, subcluster and/or genus level, and the relative abundance at each hierarchical taxonomical level was normalized against that of the total bacteria. Primers were arranged into multiple reaction tubes that involve multiplexing of up to four targets per reaction, and tested against samples collected from a trickling nitrification biofilm reactor. It was observed from the HOPE data that the relative abundance of *Nitrosomonas* increased with increasing NH_4^+ concentrations while abundances of *Nitrosospira* did not. In contrast to the AOB, the NOB genus *Nitrospira* was detrimentally affected by the shock loading

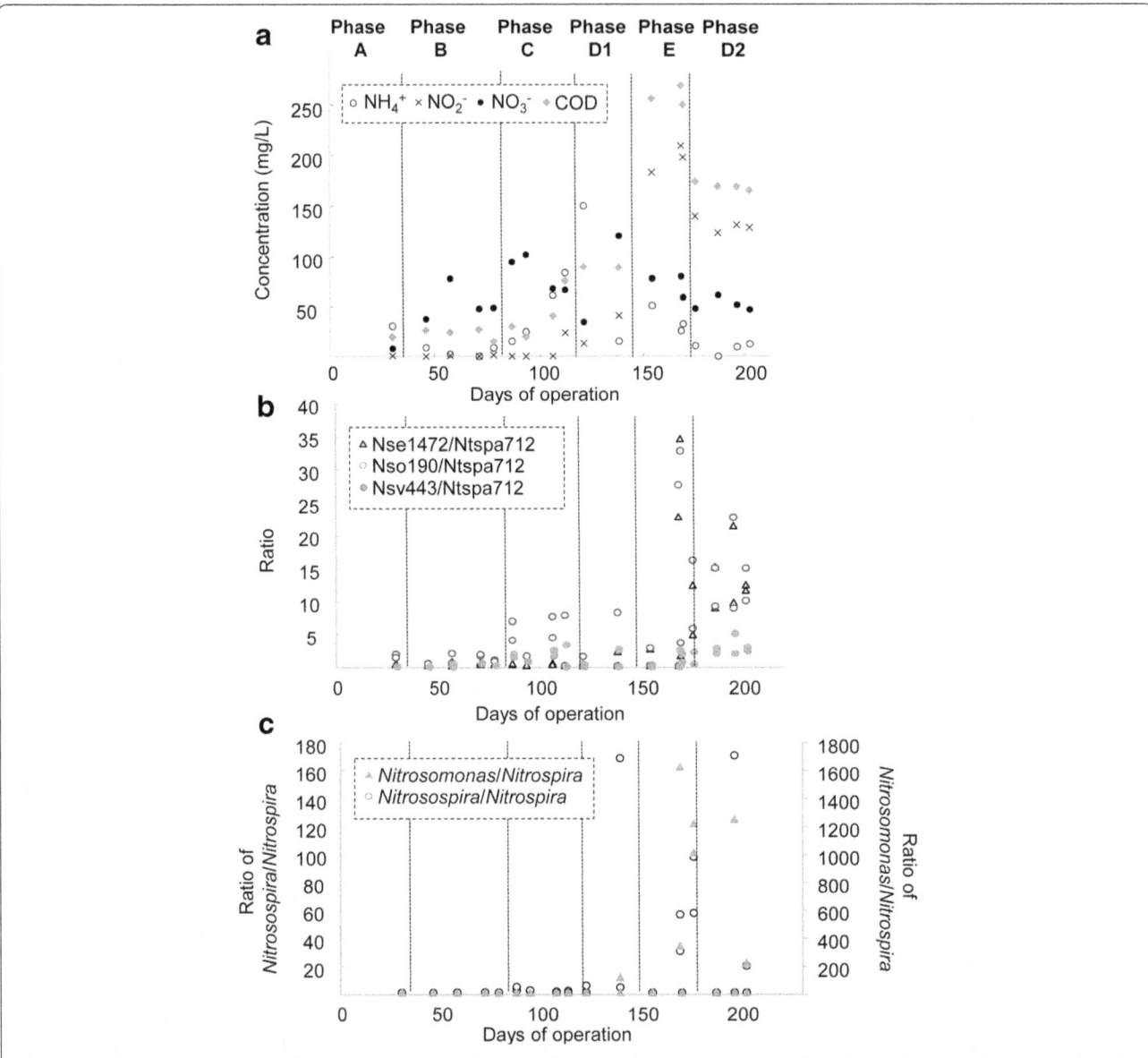

Fig. 4 Correlative trends between effluent quality of the trickling nitrification biofilm reactor and the abundance ratio of predominant AOB and NOB. Concentrations of ammonium (NH_4^+), nitrite (NO_2^-), nitrate (NO_3^-) and chemical oxygen demand (COD) in the effluent stream of the trickling nitrification biofilm reactor column throughout the different operational phases (**a**). The ratio of Nse1472-, Nso190- and Nsv443-targeted AOB against Ntspa712-targeted NOB, obtained using HOPE (**b**). The ratio of *Nitrosospira* and *Nitrosomonas* against *Nitrospira*, obtained using amplicon sequencing (**c**)

of the reactor. During the stable phases B and C, *Nitrospira* targeted by the Ntspa572-primer was more predominant than *Nitrobacter*. However, with the accumulation of nitrite in the late operational phases, *Nitrospira* was depleted while *Nitrobacter*, targeted by the Nit1000-primer, remained relatively stable in its relative abundance against the total bacteria. This observation coincides with results reported by Knapp and Graham [27], in which *Nitrospira* was observed to be more perturbed than *Nitrobacter* during a destabilizing experiment [27]. In addition, given that both Ntspa572 and Ntspa712 HOPE primers target at

least one of the three recently identified comammox (Table 1), the higher relative abundance of HOPE-targeted *Nitrospira* than *Nitrobacter* is likely to suggest that comammox is contributing to an equally important nitrification role in this reactor. Similarly, the depletion in the collective relative abundance of Ntspa572-targeted *Nitrospira* suggests the susceptibility of comammox to shock loading events.

The high NH_4^+ content exposure appeared to affect specific groups of AOB and NOB differently. It is generally thought that *Nitrospira*-like bacteria are k-strategists

Table 3 Water quality for the influent, effluent and chlorinated effluent samples collected from the wastewater treatment plant and the corresponding relative abundances of AOB and NOB as quantified using the HOPE and amplicon sequencing approach

	Total nitrogen, TN (mg/L)	NH_4^+-N (mg/L)	NO_3^--N (mg/L)	Relative abundance (%) with respect to Eub338la (HOPE data)					Relative abundance (%) with respect to total bacteria (Amplicon sequencing data)			
				Nse1472	Nsv443	Ntspa712	Ntspa572	Nit1000	Nitrosomonas	Nitrosospira	Nitrobacter	Nitrospira
Influent	17 ± 1.3	12.3 ± 2.4	2.9 ± 0.8	0.28 ± 0.43	Not detected	3.1 ± 1.0	0.6 ± 0.5	Not detected	Not detected	Not detected	Not detected	0.02 ± 0.04
Effluent	17 ± 1.4	Not applicable	14 ± 2.3	0.61 ± 0.73	1.1 ± 2.9	2.3 ± 1.7	2.0 ± 1.6	Not detected	0.004 ± 0.007	0.008 ± 0.014	Not detected	1.8 ± 2.6
Chlorinated effluent	14 ± 2.4		8.6 ± 0.9	0.03 ± 0.07	0.2 ± 0.5	1.0 ± 1.3	1.6 ± 2.2	Not detected	0.01 ± 0.02	0.003 ± 0.008	Not detected	1.0 ± 1.5
Oxic sludge	Not applicable			0.23 ± 0.13	0.21 ± 0.28	8.1 ± 3.4	4.8 ± 2.4	0.17 ± 0.21	Not detected	Not detected	Not detected	6.4 ± 1.7

with a high affinity for NO_2^- and oxygen, and reach high densities under substrate-limited conditions. This is in contrast to the *Nitrobacter* species, which are r-strategists with a lower NO_2^- and oxygen affinity and outcompete *Nitrospira* only at higher substrate concentrations [28–30]. An additional postulation to account for the better survival of *Nitrobacter* during shock loading events is its greater metabolic diversity under stressed conditions [27, 31]. This genus exhibits versatile metabolism and is able to grow either mixotrophically or chemoorganotrophically [32]. Although *Nitrobacter* was able to withstand shock loading and increase in its relative abundance as quantified by HOPE, it was observed that the increase in its relative abundance did not correlate with any decrease in nitrite content of the effluent. Furthermore, the increase in *Nitrobacter* did not correlate with a subsequent conversion of nitrite to nitrate in the effluent. This suggests that *Nitrobacter* may not be playing as important of a role as *Nitrospira* and comammox in nitrite oxidation in the wastewater treatment process.

Given that both *Nitrosomonas* and *Nitrospira* seem to correlate with nitrification functionality, the HOPE-obtained relative abundances of both groups were further expressed as a proportional ratio. The ratio of *Nitrosomonas* targeted by either Nse1472 or Nso190 HOPE primer against Ntspa572-targeted *Nitrospira* ranged from 0.2 to 2.5 in the nitrification biofilm reactor during the phases that corresponded to stable reactor performance as evidenced by the low nitrite content in the effluent. However, the exceedingly high ammonium content of the later phases of operation (i.e., Phase E and D2) resulted in a toxic shock response by the nitrifying bacterial populations, which correlated with an observed increase in the AOB/NOB ratio. This ratio increased to a range of 13.0 to 13.7 when the bioreactor performance deteriorated. Based on energetic calculations, it was proposed that a theoretical AOB/NOB ratio of 2 should be obtained during a stable and functional nitrification process (i.e., Phase A and B) [26]. This theoretical ratio was in agreement with the observation made in this study, and reiterates the need to maintain an optimal proportion of AOB to NOB in a well-performing nitrification process.

The HOPE method had been previously validated against other conventional methods such as qPCR and FISH [7, 8, 11]. In this study, to validate the data obtained from HOPE, the relative abundances of nitrifying bacterial populations in the trickling biofilter and in samples collected from a full-scale WWTP were further validated against results from amplicon sequencing (Tables 2 and 3). The observed difference in the abundance values between HOPE and amplicon sequencing is likely due to differences in the primer coverage for the individual AOB and NOB groups (Additional file 1). For example, Nso190 targets 27.2% of the *Nitrosomonas* and 17.4% of the *Nitrosospira* sequences

collated in the RDP Classifier database. In contrast, the universal primers 515F and 907R used for amplicon sequencing target a higher percentage of *Nitrosomonas* and *Nitrosospira*. Furthermore, the relative abundances reported by HOPE were expressed by normalizing against total bacteria as detected by universal primers that differed from those used in amplicon sequencing. Although the relative abundances determined by both methods varied in terms of the absolute values, both datasets showed good Spearman's rank correlation. Furthermore, multivariate nMDS plots from both HOPE and amplicon sequencing also showed good correlation at a significant confidence level (ρ = 0.79, p = 0.001). This indicates that the observed trends in the nitrifying bacterial population dynamics revealed by both methods were well aligned.

Although similar conclusions were obtained from both HOPE and amplicon sequencing with regards to the nitrifying population dynamics in the presence of shock loading, a limitation of the HOPE approach is that it can only detect bacterial targeted by the designed primers. For instance, through the use of amplicon sequencing and not HOPE, it was determined that the detrimental effect of the shock loading event on *Nitrospira* can possibly be due to its competition for substrates (e.g. oxygen) with heterotrophic bacteria. Such heterotrophic bacteria included unclassified Burkholderiales and unclassified Xanthomonadaceae, which increased in relative abundance throughout all operational phase. Furthermore, the HOPE technique developed in this study does not contain primers that target the AOA. Although AOA had been previously found abundant in wastewater treatment systems [33, 34], both genus *Nitrososphaera* and *Nitrosopumilus* were found to be in low abundance (< 0.1%) in the samples when evaluated by high-throughput amplicon sequencing. Further optimization on the HOPE technique would be required to design a range of AOA-targeting primers, as well as to improve its current detection limits of 0.1% so as to facilitate the quantification of potentially low abundance species.

Regardless, an advantage of the HOPE approach is that new primers, for example those designed to target the different sublineages of *Nitrospira* and *Nitrosomonas* that were selected for by varying nitrite and ammonium concentrations [35–37], can be easily added to the multiplexing tube arrangement when these primers become available. The additional depth offered by high-throughput amplicon sequencing would be useful as a tool for periodically benchmarking HOPE-based results obtained on a more frequent basis. Although this study did not attempt to challenge the limits of high-throughput multiplexing capability of HOPE, the method can in theory perform up to 32-plexes per reaction tube, and modifications to the HOPE approach can be efficiently made to facilitate the simultaneous detection of the expanding list of nitrifying bacteria. Furthermore, this study demonstrated a good correlation between the results obtained from the HOPE approach and high-throughput sequencing. Thus, depending on the intended purpose of an experimental or practical study, HOPE may be particularly useful for the frequent tracking of the occurrence and abundance of selected microbial populations without the need to perform high-throughput amplicon sequencing. This in turn reduces the need to sieve through a full range of microbial population data for every sampling event.

Conclusion

In summary, this study demonstrates the applicability and adaptability of HOPE for assessing abundances of predominant AOB and NOB groups that have been identified as integral to wastewater treatment systems. This method allows for the simultaneous monitoring of relative abundances of AOB, NOB and comammox groups, which can be used to provide indicative data of nitrification performance and efficiency. Coupled with previous application of HOPE to a variety of target bacteria and sample types [7–11], this study demonstrates the versatility and applicability of HOPE for a wide range of microbial ecology studies.

Abbreviations

AOA: Ammonia-oxidizing archaea; AOB: Ammonia-oxidizing bacteria; COD: Chemical oxygen demand; ddNTP: Dideoxynucleoside triphosphate; dPCR: Digital polymerase chain reaction; FISH: Fluorescent in-situ hybridization; HOPE: Hierarchical oligonucleotide primer extension; MBR: Membrane bioreactor; NH_4^+-N: Ammonium; nMDS: Non-metric threshold multidimensional scaling; NO_2^--N: Nitrite; NO_3^--N: Nitrate; NOB: Nitrite-oxidizing bacteria; nt: Nucleotides; PBS: Phosphate buffer saline; PCA: Principal component analysis; qPCR: Quantitative polymerase chain reaction; RDP: Ribosomal database project; rSAP: Recombinant shrimp alkaline phosphatase; TN: Total nitrogen; WWTP: Wastewater treatment plant

Acknowledgements
The authors would like to thank Mr. George Princeton Dunsford for granting access to KAUST wastewater treatment plant, and Dr. Muhammad Raihan Jumat for sampling assistance.

Funding
This study was supported by funding from King Abdullah University of Science and Technology (KAUST Center Competitive Funding Program grant FCC/1/1971–06-01) awarded to P.-Y. Hong. The funders had no role in the study design, data collection, analysis and interpretation of the data, decision to publish, or preparation of the manuscript.

Authors' contributions
GS and PYH designed the experiments, GS and PYH performed the experiments and analyses, HC and MH provided assistance in reactor operation. GS, MH and PYH drafted the manuscript. All authors read and approved the final manuscript.

Competing interests
The authors declare that they have no competing interest.

References
1. van Kessel MA, Speth DR, Albertsen M, Nielsen PH, Op den Camp HJ, Kartal B, Jetten MS, Lucker S. Complete nitrification by a single microorganism. Nature. 2015;528(7583):555–9.
2. Daims H, Lebedeva EV, Pjevac P, Han P, Herbold C, Albertsen M, Jehmlich N, Palatinszky M, Vierheilig J, Bulaev A, et al. Complete nitrification by Nitrospira bacteria. Nature. 2015;528(7583):504–9.
3. Stuven R, Vollmer M, Bock E. The Impact of Organic-Matter on Nitric-Oxide Formation by Nitrosomonas-Europaea. Arch Microbiol. 1992;158(6):439–43.
4. Wagner M, Rath G, Koops H-P, Flood J, Amann R. In situ analysis of nitrifying bacteria in sewage treatment plants. Water Sci Technol. 1996;34(1–2):237–44.
5. Hallin S, Lydmark P, Kokalj S, Hermansson M, Sorensson F, Jarvis A, Lindgren PE. Community survey of ammonia-oxidizing bacteria in full-scale activated sludge processes with different solids retention time. J Appl Microbiol. 2005; 99(3):629–40.
6. Stoddard SF, Smith BJ, Hein R, Roller BR, Schmidt TM. rrnDB: improved tools for interpreting rRNA gene abundance in bacteria and archaea and a new foundation for future development. Nucleic Acids Res. 2015;43(Database issue):D593–8.
7. Hong PY, Wu JH, Liu WT. Relative abundance of Bacteroides spp. in stools and wastewaters as determined by hierarchical oligonucleotide primer extension. Appl Environ Microbiol. 2008;74(9):2882–93.
8. Hong PY, Wu JH, Liu WT. A high-throughput and quantitative hierarchical oligonucleotide primer extension (HOPE)-based approach to identify sources of faecal contamination in water bodies. Environ Microbiol. 2009; 11(7):1672–81.
9. Wu J-H, Hsu M-H, Hung C-H, Tseng I-C, Lin T-F. Development of a hierarchical oligonucleotide primer extension assay for the qualitative and quantitative analysis of Cylindrospermopsis raciborskii subspecies in freshwater. Microbes Environ. 2010;25(2):103–10.
10. Wu JH, Chuang HP, Hsu MH, Chen WY. Use of a hierarchical oligonucleotide primer extension approach for multiplexed relative abundance analysis of methanogens in anaerobic digestion systems. Appl Environ Microbiol. 2013; 79(24):7598–609.
11. Hong PY, Yap GC, Lee BW, Chua KY, Liu WT. Hierarchical oligonucleotide primer extension as a time- and cost-effective approach for quantitative determination of Bifidobacterium spp. in infant feces. Appl Environ Microbiol. 2009;75(8):2573–6.
12. Wu JH, Liu WT. Quantitative multiplexing analysis of PCR-amplified ribosomal RNA genes by hierarchical oligonucleotide primer extension reaction. Nucleic Acids Res. 2007;35(11):e82.
13. Loy A, Horn M, Wagner M. probeBase: an online resource for rRNA-targeted oligonucleotide probes. Nucleic Acids Res. 2003;31(1):514–6.
14. Wu JH, Hong PY, Liu WT. Quantitative effects of position and type of single mismatch on single base primer extension. J Microbiol Methods. 2009;77(3): 267–75.
15. Cole JR, Wang Q, Fish JA, Chai B, McGarrell DM, Sun Y, Brown CT, Porras-Alfaro A, Kuske CR, Tiedje JM. Ribosomal Database Project: data and tools for high throughput rRNA analysis. Nucleic Acids Res. 2014;42(Database issue):D633–42.
16. Ansari MI, Harb M, Jones B, Hong PY. Molecular-based approaches to characterize coastal microbial community and their potential relation to the trophic state of Red Sea. Sci Rep. 2015;5:9001.
17. Cole JR, Wang Q, Cardenas E, Fish J, Chai B, Farris RJ, Kulam-Syed-Mohideen AS, McGarrell DM, Marsh T, Garrity GM, et al. The Ribosomal Database

Project: improved alignments and new tools for rRNA analysis. Nucleic Acids Res. 2009;37(Database issue):D141–5.

18. Clarke K, Gorley R. PRIMER version 7: user manual/tutorial. Plymouth: PRIMER-E; 2015. p. 296.

19. Hanaki K, Wantawin C, Ohgaki S. Effects of the Activity of Heterotrophs on Nitrification in a Suspended-Growth Reactor. Water Res. 1990;24(3):289–96.

20. Harb M, Xiong Y, Guest J, Amy G, Hong P-Y. Differences in microbial communities and performance between suspended and attached growth anaerobic membrane bioreactors treating synthetic municipal wastewater. Environ Sci Water Res Technol. 2015;1(6):800–13.

21. Hong PY, Wheeler E, Cann IK, Mackie RI. Phylogenetic analysis of the fecal microbial community in herbivorous land and marine iguanas of the Galapagos Islands using 16S rRNA-based pyrosequencing. ISME J. 2011;5(9):1461–70.

22. Aleem MIH. Generation of reducing power in chemosynthesis II. Energy-linked reduction of pyridine nucleotides in the chemoautotroph, *Nitrosomonas europaea*. Biochimica et Biophysica Acta (BBA) Enzymol Biol Oxidation. 1966;113(2):216–24.

23. Ferguson SJ. Is a proton-pumping cytochrome oxidase essential for energy conservation in Nitrobacter? FEBS Lett. 1982;146(2):239–43.

24. Hagopian DS, Riley JG. A closer look at the bacteriology of nitrification. Aquac Eng. 1998;18(4):223–44.

25. Hooper AB, Vannelli T, Bergmann DJ, Arciero DM. Enzymology of the oxidation of ammonia to nitrite by bacteria. Antonie Van Leeuwenhoek. 1997;71(1):59–67.

26. Winkler MKH, Bassin JP, Kleerebezem R, Sorokin DY, van Loosdrecht MCM. Unravelling the reasons for disproportion in the ratio of AOB and NOB in aerobic granular sludge. Appl Microbiol Biot. 2012;94(6):1657–66.

27. Knapp CW, Graham DW. Nitrite-oxidizing bacteria guild ecology associated with nitrification failure in a continuous-flow reactor. FEMS Microbiol Ecol. 2007;62(2):195–201.

28. Schramm A, de Beer D, Wagner M, Amann R. Identification and activities in situ of Nitrosospira and Nitrospira spp. as dominant populations in a nitrifying fluidized bed reactor. Appl Environ Microb. 1998;64(9):3480–5.

29. Schramm A, de Beer D, van den Heuvel JC, Ottengraf S, Amann R. Microscale distribution of populations and activities of Nitrosospira and Nitrospira spp. along a macroscale gradient in a nitrifying bioreactor: Quantification by in situ hybridization and the use of microsensors. Appl Environ Microb. 1999;65(8):3690–6.

30. Kim DJ, Kim SH. Effect of nitrite concentration on the distribution and competition of nitrite-oxidizing bacteria in nitratation reactor systems and their kinetic characteristics. Water Res. 2006;40(5):887–94.

31. Abeliovich A. The Nitrite Oxidizing Bacteria. In: Dworkin M, Falkow S, Rosenberg E, Schleifer K-H, Stackebrandt E, editors. The Prokaryotes: Volume 5: Proteobacteria: Alpha and Beta Subclasses. New York, NY: Springer New York; 2006. p. 861–72.

32. Nowka B, Daims H, Spieck E. Comparison of oxidation kinetics of nitrite-oxidizing bacteria: nitrite availability as a key factor in niche differentiation. Appl Environ Microbiol. 2015;81(2):745–53.

33. Zhang T, Jin T, Yan Q, Shao M, Wells G, Criddle C, Fang HH P. Occurrence of ammonia-oxidizing Archaea in activated sludges of a laboratory scale reactor and two wastewater treatment plants. J Appl Microbiol. 2009;107(3):970–7.

34. Park HD, Wells GF, Bae H, Criddle CS, Francis CA. Occurrence of ammonia-oxidizing archaea in wastewater treatment plant bioreactors. Appl Environ Microbiol. 2006;72(8):5643–7.

35. Lydmark P, Almstrand R, Samuelsson K, Mattsson A, Sorensson F, Lindgren PE, Hermansson M. Effects of environmental conditions on the nitrifying population dynamics in a pilot wastewater treatment plant. Environ Microbiol. 2007;9(9):2220–33.

36. Gruber-Dorninger C, Pester M, Kitzinger K, Savio DF, Loy A, Rattei T, Wagner M, Daims H. Functionally relevant diversity of closely related Nitrospira in activated sludge. ISME J. 2015;9(3):643–55.

37. Maixner F, Noguera DR, Anneser B, Stoecker K, Wegl G, Wagner M, Daims H. Nitrite concentration influences the population structure of Nitrospira-like bacteria. Environ Microbiol. 2006;8(8):1487–95.

38. Hongoh Y, Ohkuma M, Kudo T. Molecular analysis of bacterial microbiota in the gut of the termite Reticulitermes speratus (Isoptera; Rhinotermitidae). FEMS Microbiol Ecol. 2003;44(2):231–42.

39. Daims H, Nielsen JL, Nielsen PH, Schleifer KH, Wagner M. In situ characterization of Nitrospira-like nitrite-oxidizing bacteria active in wastewater treatment plants. Appl Environ Microbiol. 2001;67(11):5273–84.

40. Hovanec TA, DeLong EF. Comparative analysis of nitrifying bacteria associated with freshwater and marine aquaria. Appl Environ Microbiol. 1996;62(8):2888–96.

41. Mobarry BK, Wagner M, Urbain V, Rittmann BE, Stahl DA. Phylogenetic probes for analyzing abundance and spatial organization of nitrifying bacteria. Appl Environ Microbiol. 1996;62(6):2156–62.

42. Stephen JR, Kowalchuk GA, Bruns MAV, McCaig AE, Phillips CJ, Embley TM, Prosser JI. Analysis of beta-subgroup proteobacterial ammonia oxidizer populations in soil by denaturing gradient gel electrophoresis analysis and hierarchical phylogenetic probing. Appl Environ Microbiol. 1998;64(8):2958–65.

43. Gieseke A, Purkhold U, Wagner M, Amann R, Schramm A. Community structure and activity dynamics of nitrifying bacteria in a phosphate-removing biofilm. Appl Environ Microbiol. 2001;67(3):1351–62.

44. Juretschko S, Timmermann G, Schmid M, Schleifer KH, Pommerening-Roser A, Koops HP, Wagner M. Combined molecular and conventional analyses of nitrifying bacterium diversity in activated sludge: Nitrosococcus mobilis and Nitrospira-like bacteria as dominant populations. Appl Environ Microbiol. 1998;64(8):3042–51.

17

The role of the two-component systems Cpx and Arc in protein alterations upon gentamicin treatment in *Escherichia coli*

Emina Ćudić[†], Kristin Surmann[†], Gianna Panasia[1,3], Elke Hammer[2] and Sabine Hunke[1*] [ID]

Abstract

Background: The aminoglycoside antibiotic gentamicin was supposed to induce a crosstalk between the Cpx- and the Arc-two-component systems (TCS). Here, we investigated the physical interaction of the respective TCS components and compared the results with their respective gene expression and protein abundance. The findings were interpreted in relation to the global proteome profile upon gentamicin treatment.

Results: We observed specific interaction between CpxA and ArcA upon treatment with the aminoglycoside gentamicin using Membrane-Strep-tagged protein interaction experiments (mSPINE). This interaction was neither accompanied by detectable phosphorylation of ArcA nor by activation of the Arc system via CpxA. Furthermore, no changes in absolute amounts of the Cpx- and Arc-TCS could be determined with the sensitive single reaction monitoring (SRM) in presence of gentamicin. Nevertheless, upon applying shotgun mass spectrometry analysis after treatment with gentamicin, we observed a reduction of ArcA ~ P-dependent protein synthesis and a significant Cpx-dependent alteration in the global proteome profile of *E. coli*.

Conclusions: This study points to the importance of the Cpx-TCS within the complex regulatory network in the *E. coli* response to aminoglycoside-caused stress.

Keywords: *E. coli* , Two - component system , Cpx, Arc, Aminoglycosides, Protein interactions, Absolute quantification, SRM, Proteome profiling

Background

Aminoglycoside antibiotics are toxic to bacterial cells by targeting the 30S subunit of the ribosome and by subsequent inhibition of the translation process [1]. However, treatment with aminoglycosides is commonly used to understand cellular processes that enable bacteria to resist these antibiotics. It was discovered that sub-inhibitory concentrations (1 μg ml^{-1}) of the aminoglycoside gentamicin induce the biofilm formation of *Escherichia coli* (*E. coli*) and *Pseudomonas aeruginosa* (*P. aeruginosa*) causing aminoglycoside-resistance of these strains [2]. Both, biofilm formation and two-component system (TCS) signaling were found to increase the resistance towards aminoglycosides [2, 3]. TCSs comprise a sensor kinase (SK) and a

response regulator (RR) [4–6]. After sensing a specific stimulus, the SK autophosphorylates and transfers the phosphoryl group to the respective RR. The response is terminated by dephosphorylation of the RR either by the SK if it possesses phosphatase activity, or by intrinsic activity of the RR [6, 7]. Mahoney and Silhavy (2013) found that hyper-activation of the Cpx (conjugative pilus expression)-TCS pathway in *E. coli* enhances the viability of cells grown in presence of gentamicin [3]. In contrast to this, Kohanski et al. (2008) hypothesized that treatment of *E. coli* cells with gentamicin induces a crosstalk between the Cpx-TCS and the Arc (anoxic redox control)-TCS of *E. coli* resulting in bacterial cell death [8]. This hypothesis was therefore particularly significant, as crosstalk between different TCSs is commonly circumvented by distinct mechanisms to ensure specific responses [9]. Nevertheless, TCSs are able to crosstalk under certain environmental conditions allowing for a diversified response by integration of multiple signals [10]. Crosstalk can occur by different mechanisms, such as:

* Correspondence: Sabine.Hunke@UOS.de
[†]Equal contributors
[1]FB 5 Microbiology, Department of Biology/Chemistry, University Osnabrück, Barbarastraße 11, 49076 Osnabrück, Germany
Full list of author information is available at the end of the article

i) integration of additional signals by an accessory protein regulated by another TCS, ii) phosphorylation of another, non-cognate RR by the SK, and iii) target promotor recognition by a non-cognate RR [11]. Here, we refer to 'crosstalk' at RR-phosphorylation-level. Such crosstalk between TCS in *E. coli* has been already reported between the Cpx-TCS and the EnvZ-OmpR-TCS [12, 13], or the quorum-sensing QseBC-TCS [14].

A functional interaction between the Cpx- and Arc-TCS has been in discussion since 1980. Mutations in *cpxA* and *arcA* resulted in reduced synthesis of isoleucine and valine [15], reduced conjugation [16], and reduced synthesis of porins [17]. That time, the response regulator of the Arc-TCS, ArcA, was named CpxB as the supposed cognate response regulator of CpxA [18]. Later, the genuine cognate response regulator of CpxA, CpxR, was identified [19]. Further studies identified ArcB as the specific SK of ArcA [20, 21].

Today it is known that the Cpx-TCS consists of the SK CpxA, the RR CpxR, and the periplasmic accessory molecule CpxP [22]. The SK CpxA exhibits kinase and phosphatase activity. Its autophosphorylation is inhibited by direct interaction with a CpxP homodimer under non-activating conditions [23, 24]. The Cpx-TCS is responsible for detection and response to conditions disturbing the integrity of the cell envelope which includes a large variety of signals, e.g. elevated pH [25, 26], or changes in the membrane composition [27]. Moreover, defective lipopolysaccharide (LPS) synthesis and assembly were found to induce the Cpx-TCS [28, 29]. Activation of the Cpx-TCS results in positive or negative regulation of several genes on the transcript level [30–33] and is associated with alterations of protein abundance [34]. Additionally, the Cpx-TCS can be regulated independently of CpxP [35], for example by the outer membrane lipoprotein NlpE which is known to specifically induce Cpx-TCS activity upon surface attachment or upon its overexpression [36–38].

The Arc-TCS monitors changes during respiratory growth. ArcB is autophosphorylated under reducing conditions followed by transphosphorylation of the RR ArcA by a 3-step phosphorelay reaction in order to induce the expression of genes e.g. involved in fermentative metabolism [39–44].

Kohanski et al. (2008) proposed within their study that treatment with 5 µg ml^{-1} gentamicin causes accumulation of misfolded proteins in the inner membrane and the periplasm resulting in hyper-activation of CpxA [8]. They concluded from their results that gentamicin treatment results in the accumulation of misfolded proteins in the inner membrane leading to radical formation and membrane depolarization. Kohanski et al. (2008) suggested that the Cpx-TCS is a key player in this process and that the SK CpxA cross-phosphorylates the RR ArcA [8]. Given that ArcA ~ P regulates several metabolic genes leading to free radical formation, Kohanski et al. (2008) concluded that this

crosstalk between CpxA and ArcA finally results in cell death [8] (Fig. 1). In contrast to this, Mahoney & Silhavy (2013) showed that a dominant mutation of *cpxA* resulting in constitutive activation of the Cpx-TCS pathway in *E. coli* enhances the viability of cells grown in presence of 5 µg ml^{-1} gentamicin [3]. Increased viability was found to be CpxR-dependent although none of the tested Cpx regulon members was responsible for the monitored increased resistance against aminoglycoside antibiotics. However, increased viability in presence of aminoglycosides may be caused by the CpxR-dependent prevention from membrane damage induced by the misfolded proteins generated due to gentamicin treatment. In addition, it was found that CpxR is not required for cell death caused by bactericidal antibiotics [3]. Recently, it was demonstrated that cell death due to bactericidal antibiotics does not require oxygen or reactive oxygen species, thus further calling the hypothesis of Kohanski et al.'s (2008) model into question [45, 46].

To decipher these contrary findings, we aimed to investigate the postulated crosstalk between the Cpx- and the Arc-TCS at the molecular level in *E. coli* M1G655 wild type upon gentamicin treatment in comparison to bacteria grown in absence of gentamicin. Besides protein interaction experiments, relative quantification (gene expression) and absolute quantification (protein amounts) of the components of the Cpx- and the Arc-TCS were performed. For a deeper insight into alterations of the protein composition upon aminoglycoside treatment and the role of CpxA-dependent phosphorylation, global protein profiling experiments were performed on the *E. coli* MG1655 wild type and an isogenic *cpxAR* mutant.

Results

Cpx and ArcA interact specifically upon aminoglycoside treatment In order to investigate the postulated crosstalk between the Cpx- and Arc-TCS (CpxA with ArcA or ArcB with CpxR) after aminoglycoside treatment in *E. coli* ([8]; Fig. 1), we first analyzed the physical interactions between the sensor kinases (SK) CpxA and ArcB with the response regulators (RR) CpxR and ArcA of *E. coli* strain MG1655 in vivo. For this purpose, bacterial two-hybrid (BACTH) and Membrane-Strep-tagged protein interaction experiments (mSPINE) were performed (Fig. 2a,b). BACTH makes use of fusions of two putatively interacting proteins with two complementary fragments of *Bordetella pertussis* adenylate cyclase, namely T18 and T25. Interactions of proteins fused to T18 and T25 restore the abolished function of the adenylate cyclase, allowing detection and quantification of protein interactions by ß-galactosidase activity assay [47]. In this study, the SKs CpxA and ArcB were C-terminally fused to T18, whereas the RRs CpxR and ArcA were N-terminally fused to T25.

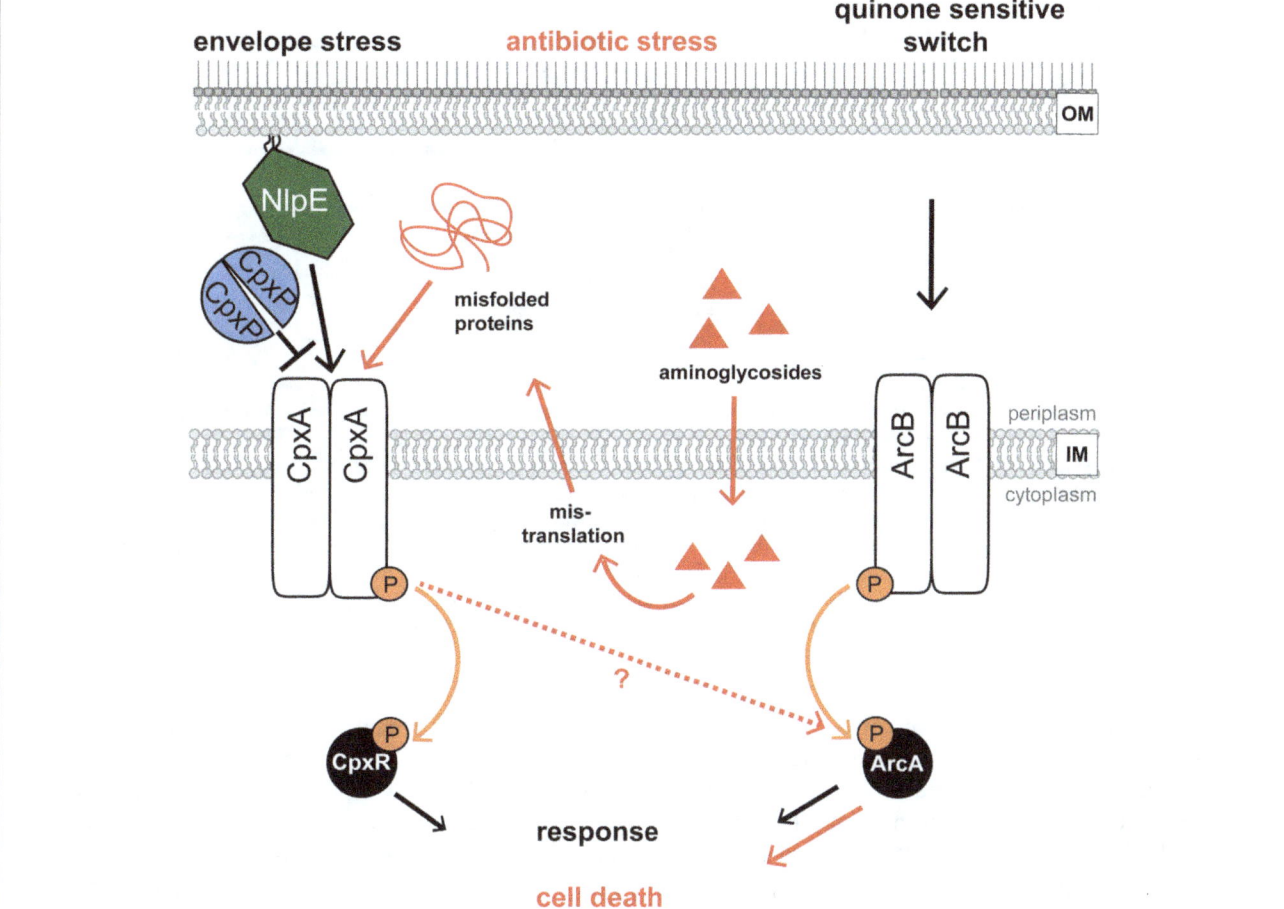

Fig. 1 Organization and regulation of the Arc- and Cpx-TCS in *E. coli*. The Arc system is commonly induced under conditions provoking a reduced state of the quinone pool of the electron transport chain. The autophosphorylation of the SK ArcB is followed by a phosphoryl group transfer onto its cognate RR ArcA resulting in a specific response. The phosphorelay reaction for the phosphoryl group transfer from ArcB to ArcA is simplified for better understanding. The envelope stress sensing Cpx system is induced by e.g. misfolded proteins or *nlpE* overexpression, followed by autophosphorylation of CpxA, phosphotransfer onto its cognate RR CpxR and response of the cells. The Cpx system is inhibited by interaction of the periplasmic accessory protein CpxP with CpxA. As postulated by Kohanski et al. (2008) addition and uptake of aminoglycosides lead to mistranslation of proteins in the cytoplasm, and result in misfolded, periplasmic proteins activating the Cpx system [8]. According to this controversially discussed hypothesis, crosstalk between CpxA and ArcA takes place under this condition leading to cell death (red marked pathway). IM = inner membrane; OM = outer membrane

We co-overexpressed different combinations of SK and RR in an adenylate-cyclase-deficient strain and measured the ß-galactosidase activity for three biological replicates each (Fig. 2a). The ß-galactosidase activities for the natural SK/RR pairs CpxA/CpxR and ArcB/ArcA amounted to about 1100 Miller units (MU) and 800 MU, respectively (Fig. 2a). Interactions between CpxA and its non-cognate response regulator ArcA resulted in higher ß-galactosidase activity (about 1600 MU) than observed for the positive control (+) using fusions of T18 and T25 with leucine zipper domains of the transcription factor GCN4 derived from *Saccharomyces cerevisiae* (about 1200 MU; Fig. 2a). In contrast, interactions between ArcB and CpxR yielded lowest ß-galactosidase activity (about 250 MU) below the estimated background using empty vectors (about 400

MU) and thus no interaction could be identified for this SK/RR-pair (Fig. 2a). In sum, BACTH strongly suggested that physical interaction between CpxA and ArcA after overexpression of both interaction partners is possible without the need of any specific stimulus.

Next, we investigated whether CpxA and ArcA interact specifically due to aminoglycoside-caused stress. We therefore applied the Membrane-SPINE method under conditions almost identical to the experiments performed by Kohanski et al. (2008) [8]. Kohanski et al. (2008) used 5 µg ml^{-1} of the aminoglycoside gentamicin for aminoglycoside treatment and incubated the cells for 30 min [8]. In our study, cells were incubated for 40 min. The cross linker formaldehyde requires at least 20 min incubation time for reliable interaction data (Fig. 2b). The Membrane-SPINE

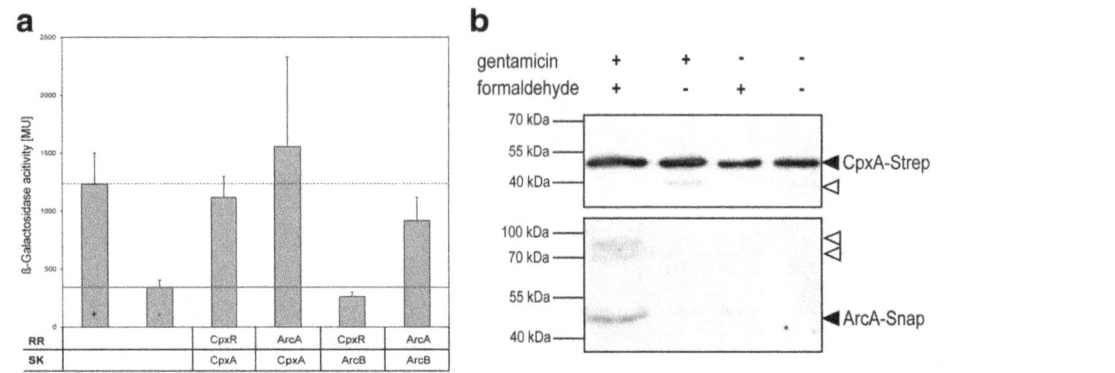

Fig. 2 CpxA and ArcA interact in vivo. **a** Bacterial two-hybrid assay (BACTH) was used to identify interactions between CpxA and ArcA. The sensor kinases CpxA and ArcB were C-terminally fused to the T18-fragment, whereas the response regulators CpxR and ArcA were N-terminally fused to the T25-fragment of the *Bordetella pertussis* adenylate cyclase. Different combinations of sensor kinase and response regulator were co-overexpressed and the ß-galactosidase activity in Miller units [MU] was determined. CpxA/CpxR- and ArcB/ArcA-interactions serve as controls. Plasmids expressing fusions of T18 and T25 with leucine zipper domains of the transcription factor GCN4 derived from *Saccharomyces cerevisiae* (+) serve as a positive control. Empty vectors non-expressing the T18- and T25-fragment (–) serve as a negative control. The average of three biological replicates is shown. **b** The chromosomally-encoded fusion protein ArcA-Snap and plasmid-encoded CpxA-Strep were used to analyze interactions between CpxA and ArcA using Membrane-SPINE. Interactions were determined for cells treated with or without of 5 µg ml^{-1} gentamicin for 40 min. Additionally, samples without addition of the crosslinker formaldehyde or gentamicin served as controls. Out of two biological replicates, one representative blot is shown. Black triangles show specific bands of CpxA-Strep and ArcA-Snap, whereas white triangles represent unspecific bands

approach combines the purification of a specific Strep-tagged membrane protein with the reversible fixation of protein complexes by adding the aforementioned cross linker formaldehyde, allowing snapshots of interactions in living cells [48]. Here we used a MG1655-derivate *E. coli* strain harboring a chromosomal fusion of the *arcA* gene with a Snap-Tag and a fusion of *cpxA* with a Strep-Tag on the medium copy plasmid pMal-p2X (~20 copies per cell; [49]). It was demonstrated in a previous study that CpxA-Strep is correctly localized in the inner membrane and fully functional for transphosphorylation of purified CpxR [49].

The interaction between CpxA-Strep and ArcA-Snap could be specifically observed after treatment with gentamicin and after addition of formaldehyde (Fig. 2b). Moreover, in samples treated with gentamicin but without formaldehyde or vice versa, only weak bands of ArcA-Snap were observed (Fig. 2b). The quantification of band densities revealed the highest band intensity for ArcA-Snap for samples treated with gentamicin and formaldehyde (Fig. 2b). The amount of CpxA-Strep remained constant in all four samples (Additional file 1: Figure S1).

We proved that CpxA and ArcA interact physically and specifically after aminoglycoside treatment of the cells, fulfilling the first requirement for a crosstalk between the Cpx- and the Arc-TCS.

Gentamicin treatment alters the expression levels of arcB and arcA to different extent

After verifying physical interaction of CpxA and ArcA in vivo, we hypothesized that aminoglycoside treatment could lead to a shift in ratios between CpxA, ArcB, and

ArcA, thereby enhancing the probability of a crosstalk between CpxA and ArcA under physiological conditions. Kohanski et al. (2008) postulated Cpx activation after gentamicin treatment (Fig. 1, [8]) and it is known that activation of the Cpx-TCS leads to increased amounts of CpxA [34]. Hence, crosstalk between CpxA and ArcA might be forced by a significantly increased amount of CpxA compared to that of ArcB. Moreover, the experiments by Kohanski et al. (2008) [8] as well as the experiments within this study were performed under aerobe, and thus Arc non-inducing conditions [39–44]. Therefore, an additional, significant deficiency of ArcB compared to CpxA could make crosstalk between CpxA and ArcA more likely. To address these aspects as well as to analyze the hypothesis of aminoglycoside driven Cpx activation, we estimated the expression levels of *cpxA*, *arcB*, and *arcA* first. For this purpose, we used identical growth conditions as described by Kohanski et al. (2008) [8].

We performed quantitative reverse transcription-PCR (qRT-PCR) for *cpxA*, *arcB*, and *arcA* in *E. coli* MG1655 WT cells 30 min after addition of 5 µg ml^{-1} gentamicin and compared the expression levels with those from cells (i) grown under WT conditions (LB-medium), and (ii) upon activation of the Cpx system (overexpression of the outer membrane protein *nlpE*; Cpx-ON). The *nlpE* gene encodes the outer membrane lipoprotein NlpE and the overexpression of *nlpE* represents a well-known, Cpx-activating condition [34, 38]. The expression of *cpxA* increased (1.4-fold) after Cpx activation (Fig. 3). The expression of *arcB* and *arcA* behaved in opposite manners by decreasing after Cpx activation (Fig. 3). The

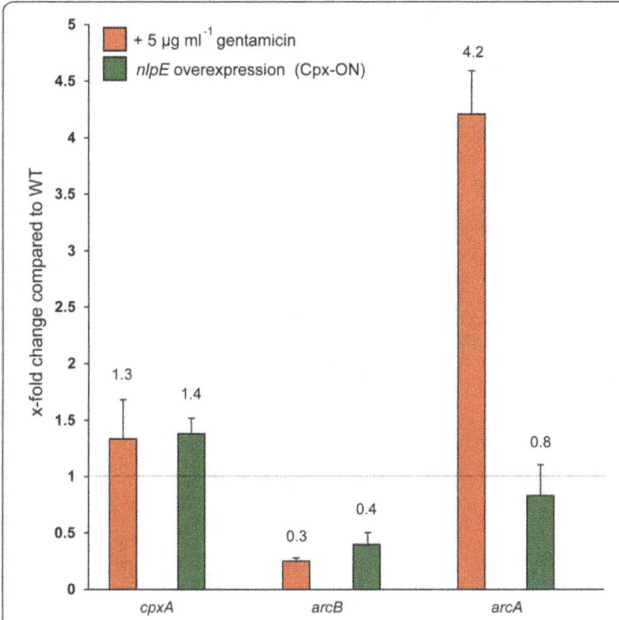

Fig. 3 Impact of different stressors on the expression level of *cpxA*, *arcB*, and *arcA*. Changes in expression levels for the genes *cpxA*, *arcB*, and *arcA* were determined by q-RT-PCR after addition of 5 µg ml^{-1} gentamicin (red bars) and after *nlpE* overexpression (Cpx-ON; green bars). All depicted values represent ratios between the condition of interest and WT cells. Shown are mean data and standard deviations of five biological replicates measured in three technical replicates each

expression of *arcB* decreased significantly after addition of gentamicin (0.3-fold), whereas the expression of *arcA* significantly increased in comparison to WT conditions (4.2-fold; Fig. 3). The significant changes in *arcB* expression match our initial hypothesis suggesting a deficiency of ArcB compared to CpxA after gentamicin treatment. However, the expression of *cpxA* changed only slightly after aminoglycoside treatment (1.3-fold). In awareness of the fact that expression levels of genes are not directly linked with activation and do not display the absolute amounts of proteins within the cell, we went one step further and quantified the absolute numbers of the Cpx- and the Arc-TCS proteins under different conditions.

Absolute quantification of the Arc- and the Cpx-TCS

To further investigate if a possible crosstalk between the Arc- and the Cpx-TCS is indicated by altered amounts of respective molecules, we monitored the absolute number of proteins per cell upon gentamicin treatment. As controls, we also quantified the Arc- and the Cpx-TCS under WT conditions (LB-medium) and during Cpx-activating conditions (Cpx-ON by *nlpE* overexpression).

It is important to consider that protein translation is not fully blocked within 30 min of growth in presence of 5 µg ml^{-1} gentamicin as performed for the experiments

in this study. This is reasoned in previous data demonstrating that cells still grow for at least 60 min in presence of sub-lethal gentamicin concentrations of 5 µg ml^{-1} [8]. Accordingly, the employed cells continued growing after addition of 5 µg ml^{-1} gentamicin in the present study (Additional file 1: Figure S2). Moreover, ribosome binding by aminoglycosides as gentamicin does not elicit immediate hold-up of translation, but mistranslation of proteins by false-incorporation of amino acids into the peptide strands [50].

We employed the single reaction monitoring (SRM) method which combines highly sensitive detection of selected molecules with exact and absolute quantification by spiking in heavy isotope labeled standard peptides of targeted proteotypic peptides from a desired protein [51, 52]. Recently, the SRM method was used to establish a global approach to absolutely quantify the *E. coli* proteome under different conditions [53]. These data also include absolute amounts of the Cpx- and the Arc-TCS. However, gentamicin treatment and Cpx activating conditions were not tested. The data for the present study are provided as Table 1, in which values labeled with * were recently published [34]. Even though differences were expected and existent between our study and Schmidt et al. (2016), the ranges of the determined copy numbers were comparable for bacteria grown in LB-medium [Schmidt et al. (2016); e.g. CpxA 41 molecules/cell our study, 50 molecules/cell Schmidt et al. (2016); ArcA 1088 molecules/cell our study, 5464 molecules/cell Schmidt et al. (2016)] [53].

We provide values for ArcB and ArcA as well as for CpxA, CpxR, and the accessory protein CpxP under each tested condition. Similar to the qRT-PCR data, we observed changes in absolute protein amounts of the Cpx and Arc-TCS components upon Cpx activation and gentamicin treatment in comparison to native conditions (Fig. 4a). The absolute amounts of CpxA, ArcB and ArcA decreased after gentamicin treatment (Fig. 4a), disproving the initially raised hypothesis of a significant increase of CpxA and/or a significant decrease of ArcB supporting crosstalk between CpxA and ArcA.

Regarding cells grown under WT conditions, we observed a stoichiometry of SK to RR being CpxA:CpxR = 1:10, ArcB:ArcA = 1:13, and CpxA:ArcA = 1:27 (Fig. 4b). The determined ratios after addition of gentamicin were nearly the same (Fig. 4b), demonstrating a high robustness for both systems with regard to the stoichiometry. In contrast, activation of the Cpx-TCS (*nlpE* overexpression) significantly changed the ratio between CpxA and ArcA to 1:6 in comparison to 1:27 under native conditions.

Overall, the interaction between CpxA and ArcA distinctly observed in the Membrane-SPINE experiment was not caused by a shift in ratios between CpxA, ArcB, and ArcA in vivo.

Table 1 Absolute quantification of CpxAR, CpxP, and ArcBA by selected reaction monitoring

Peptide	Protein	Replicate	Average amount [molecules/cell]	SD [molecules/cell]	CV [%]
AEDSPLGGLR	CpxA	wild type conditions	41*	11	28
		with 5 µg ml^{-1} gentamicin	32	6	19
		nlpE overexpression (Cpx-ON)	78*	4	5
EHLSQEVLGK	CpxR	wild type conditions	393*	93	24
		with 5 µg ml^{-1} gentamicin	348	43	12
		nlpE overexpression (Cpx-ON)	738*	69	9
DVTQWQK	CpxP	wild type conditions	36*	15	42
		with 5 µg ml^{-1} gentamicin	41	17	42
		nlpE overexpression (Cpx-ON)	133*	28	21
FTQQGQVTVR	ArcB	wild type conditions	85	20	24
		with 5 µg ml^{-1} gentamicin	66	11	16
		nlpE overexpression (Cpx-ON)	82	8	10
SLIGPDGEQYK	ArcA	wild type conditions	1088	162	15
		with 5 µg ml^{-1} gentamicin	796	150	19
		nlpE overexpression (Cpx-ON)	484	41	8

The peptides used for absolute determinations of protein molecules per cell are given as average values from five biological replicates with their standard deviations and coefficient of variance (CV) in percent for each of the four applied conditions

Molecular amounts labeled with * represent data that were already published previously [32] but are mentioned here to provide complete information

Alterations in the proteome profile of *E. coli* MG1655 upon gentamicin treatment

On the one hand, our results demonstrated gentamicin-dependent interaction between CpxA and ArcA. On the other hand, quantification of absolute protein amounts did not explain why these two proteins interact specifically upon gentamicin treatment. Kohanski et al. (2008) postulated a functional interaction between CpxA and ArcA, and Cpx activation upon gentamicin treatment on the basis of expression profiles of selected ArcA ~ P-dependent genes [8]. Here, we wanted to add information on the impact of the Cpx-TCS upon gentamicin treatment. Thus, we monitored the proteome profiles of *E. coli* MG1655 WT and an isogenic *cpxAR* mutant under control conditions (LB-media only) and upon gentamicin treatment for 30 min in four biological replicates (Additional file 2: Table S1). Avoiding the possibility of CpxA-mediated ArcA-phosphorylation in the absence of CpxAR allowed us to investigate if ArcA ~ P targets are differently affected in comparison to WT cells, pointing to Cpx-specific Arc-activation in the WT. Moreover, implementing the *cpxAR* mutant provides information about changes in protein levels being dependent on the presence of CpxAR and corresponding alterations in transcription, thereby rejecting background effects that are only caused by gentamicin treatment. Finally, phosphorylation of ArcA by ArcB can be neglected under the analyzed conditions as the Arc-TCS is commonly inactive under aerobic growth conditions [39–44].

Applying shotgun proteomics, we detected and quantified in total 1467 proteins in the dataset of the WT strain and the respective *cpxAR* mutant treated with or without gentamicin. Almost no alterations in ArcA or ArcB amounts were visible when comparing WT conditions and gentamicin treatment in both strains. In contrast, the Cpx-TCS seems to play an important role in the cell response upon gentamicin treatment.

In WT cells, 115 proteins were significantly altered in their level after gentamicin treatment compared to non-treated WT. In the *cpxAR* mutant this was only true for 39 proteins when comparing gentamicin to control conditions (Additional file 2: Table S1). A selection of proteins already affected by aminoglycoside addition after 30 min is provided in the heatmap in Fig. 5 visualizing relative protein intensities in all tested conditions as well as the impact of gentamicin on the WT (WT$_G$/WT) or the *cpxAR* mutant (*cpxAR*$_G$/*cpxAR*).

An initiated inhibition of protein translation by gentamicin is partly visualized by lower intensities of some proteins upon gentamicin treatment (Fig. 5). Nevertheless, a CpxAR-dependent adaptation to gentamicin could be observed. Only three of these proteins with different behavior in the WT and the mutant strain contain binding sites for CpxR ~ P at the DNA level. These were the two outer membrane proteins OmpF and OmpC with decreased levels in the mutant and slight alteration in the WT upon gentamicin treatment. The third protein with CpxR ~ P binding site at the DNA level, the RNA polymerase heat shock sigma factor RpoH, was strongly increased in WT cells (WT$_G$/WT 11.6) and not changed in level regarding the mutant. A CpxAR-dependent increase in protein amounts upon gentamicin treatment

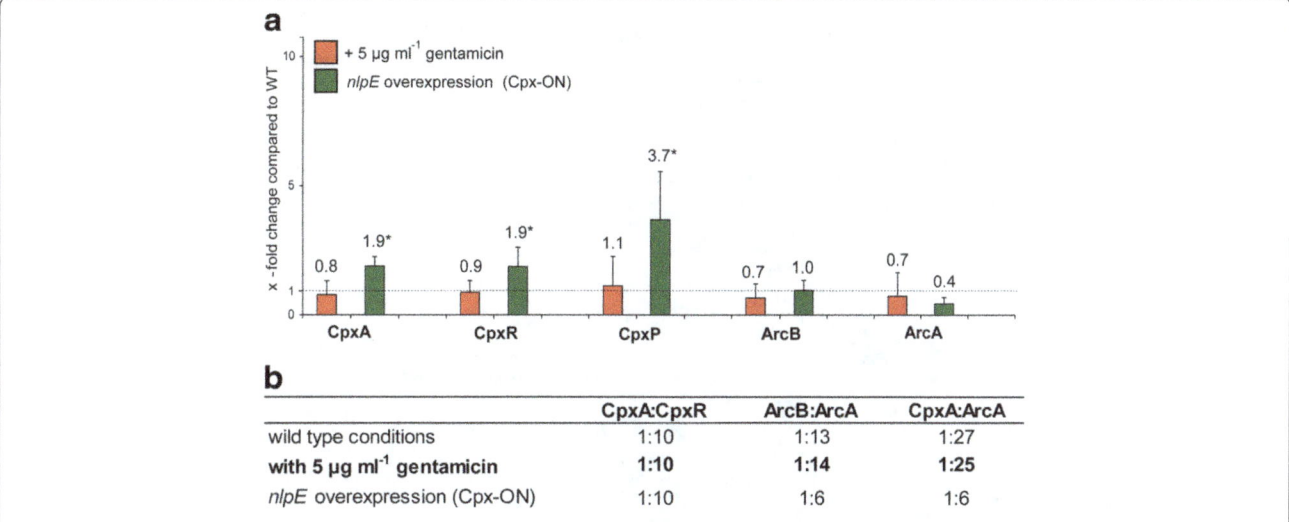

Fig. 4 Absolute quantification of CpxAR, CpxP, and ArcBA under various conditions. A) Selected reaction monitoring (SRM) was used to determine the absolute amounts of CpxA, CpxR, CpxP, ArcB, and ArcA after addition of 5 μg ml^{-1} gentamicin (red bars) and after *nlpE* overexpression (Cpx-ON; green bars). All depicted values represent ratios between the condition of interest and WT cells. Shown are mean data and standard deviations of five biological replicates measured. Values labeled with * represent molecular amounts published in a previous study [34] but are listed here for complete information. B) The absolute amounts of CpxA, CpxR, ArcA, and ArcB were used to calculate the SK:RR-ratio for each tested condition

was further shown for the two heat shock proteins IbpA and IbpB to high extent (WT$_G$/WT 20.7 and 20.5), and to a lesser extent for the two multidrug efflux pump proteins MdtE and MdtF (WT$_G$/WT 2.0 and 2.1), as well as for the Lon protease (WT$_G$/WT 2.2). Finally, proteins of the Nar family (respiratory nitrate reductases) were found to be decreased upon gentamicin treatment (WT$_G$/WT < 0.2). Hence, we found remarkable changes in the global proteome profiles comparing the WT- and the *cpxAR* strain (single comparison WT$_G$/WT vs. *cpxAR$_G$/cpxAR*).

Some proteins were altered in level independently of CpxAR presence after gentamicin treatment as indicated by similar ratios of WT$_G$/WT compared to *cpxAR$_G$/ cpxAR*. These displayed changes caused by gentamicin without any measurable Cpx-involvement. Importantly, monitoring not only decreased protein levels due to initiated translation inhibition triggered by gentamicin, but also increased protein levels indicate intact transcription and translation within the observed growth period and gentamicin concentration. Among these were also three proteins which were already identified as Cpx targets [34]: the enterobactin synthetase component E (EntE) and two iron-dependent transport proteins, the ferric iron-catecholate outer membrane transporter (CirA) and the ferrienterobactin receptor (FepA). These proteins were likewise increased in WT and *cpxAR cells* upon gentamicin treatment. Other Cpx-independent increased protein levels were found e.g. for the transcription regulator IscR and some metabolic enzymes such as the aspartate-ammonia ligase AsnA, the glucose dehydrogenase (Gcd), and the isochorismate synthase EntC as

well as the uncharacterized protein YacC. Lower levels upon aminoglycoside uptake in both strains were found for the galactitol-specific enzyme IIA component of PTS (GatA) and the transferase PurT.

Twelve of 15 detected ArcA ~ P-regulated proteins were found significantly decreased in level upon gentamicin treatment in the WT strain. In contrast, gentamicin had no effect on most ArcA ~ P targets in the *cpxAR* mutant (Fig. 5). Among the ArcA ~ P targets identified in the WT strain were for example three anaerobic glycerol-3-phosphate dehydrogenases GlpA, GlpB, and GlpC, the anaerobic C4-dicarboxylate transporter DcuC, and two trehalose metabolizing enzymes TreB and TreC. Only GatC and GatA were likewise decreased in the mutant, and LldD was doubled in amount in the strain lacking *cpxAR*. Although a Cpx-dependent effect on the ArcA ~ P targets can be clearly seen, this is most likely not a result of CpxA-mediated ArcA phosphorylation, as these targets were not altered according to known regulation (positive or negative) via ArcA ~ P.

Our results indicated that CpxA does not cross-phosphorylate ArcA. To investigate this further, we determined the ArcA isoform pattern by 2D gel electrophoresis for WT conditions, after Cpx activation (*nlpE* overexpression, Cpx-ON) and upon gentamicin addition. We did not observe any ArcA phosphorylation in these samples. Unfortunately, aspartate phosphorylations as described for ArcA ~ P [54] are known to be very unstable. Using 2DE gels we were able to detect ArcA as the prevailing protein in two spots under all conditions at the predicted mass range and separated from each other by a pH shift into

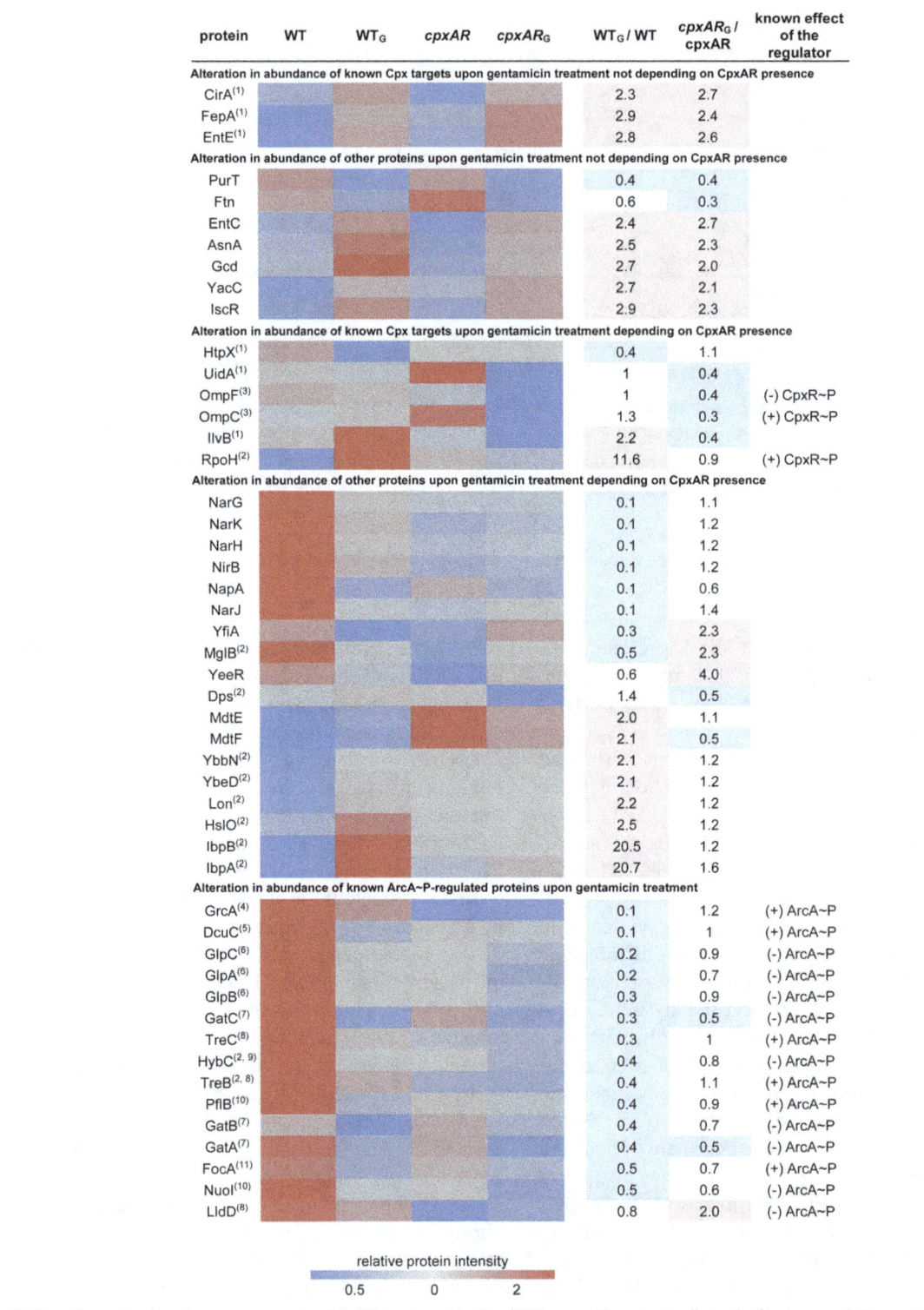

Fig. 5 Altered protein abundance upon gentamicin (G) treatment in *E. coli* WT or *cpxAR* mutant cells. In the heatmap, relative proteins intensities are visualized, thereby red fields indicate higher relative intensities, dark blue lower relative intensities and grey fields show relative intensities around one. In addition, gentamicin treatment depending ratios between the WT and the *cpxAR* mutant are presented as well, ratios to the control >2 are labeled red, those <0.5 in blue. Proteins are grouped by CpxAR or ArcA dependence. Some proteins were already described to be Cpx targets [(1) Surmann et al. (2016) [34]], found regulated upon gentamicin on transcript level [(2), Kohanski et al. (2008) [8]], and/or positively (+) or negatively (−) CpxR ~ P or ArcA ~ P-regulated [(3) Batchelor et al. (2005) [69], (4) Wyborn et al. (2002) [70], (5) Zientz et al. (1999) [71], (6) Iuchi et al. (1990) [20], (7) Salmon et al. (2005) [72], (8) Liu & de Wulf (2004) [44], (9) Richard et al. (1999) [73], (10) Shalel-Levanon et al. (2005) [74], (11) Kaiser & Sawers (1997) [75]]

the acid milieu pointing to the expected phosphorylation, which, however, could not be confirmed by standard nanoLC-MS/MS. Since the spot volume was comparable under WT conditions and upon gentamicin treatment (example 2D gel shown as Additional file 1: Figure S3) no hints for an alteration of the ArcA status could be found.

The proteome profile after classical Cpx activation differs significantly from the proteome composition after gentamicin treatment

To elucidate if gentamicin induces the Cpx system as hypothesized by Kohanski et al. (2008) [8] and demonstrated by Kashyap et al. (2011) on transcriptomic level [55], we compared proteome profiles of cells treated with gentamicin with a previously published proteome profile of cells grown under Cpx activating conditions via overexpression of *nlpE* [34]. Therefore, the proteome data acquired within this study was re-evaluated essentially as performed for the proteome profiles acquired after Cpx activation by *nlpE* overexpression [34]. Here, employment of the $(WT_G/WT)/$ $(cpxAR_G/WT)$-ratio emphasizing higher amounts of respective proteins WT_G cells compared to $cpxAR_G$ cells revealed only eleven proteins as Cpx-dependently increased after gentamicin treatment. Among these were proteins involved in metabolism: FruK [$(WT_G/WT)/(cpxAR_G/WT)$: 16.7], FruA (13.8), IlvB (8.1), or stress response: IbpB (10.1), IbpA (9.7), RpoH (5.3), YgiQ (5.1), YceA (4.1), YncD (3.1), YrbL (2.2), and DeaD (2.0) (Additional file 3: Table S2). In contrast, we found seven proteins with decreased levels after addition of gentamicin depicted by the $(cpxAR_G/WT)/$ (WT_G/WT)-ratio that underlines higher amounts of the respective protein in absence of CpxAR after gentamicin treatment. These proteins were YodD [$(cpxAR_G/WT)/$ (WT_G/WT): 8.7], GarR (3.2), YfiA (3.0), YoaC (2.9), PptA (2.5), GcvP (2.2), and PhnA (2.0). None of the depicted proteins strongly suggesteds a binding site for ArcA ~ P or CpxR ~ P, except for RpoH harboring a binding site for CpxR ~ P, as asserted previously. Furthermore, the protein IlvB is the only protein that was identified in a previous study being specifically increased after Cpx activation by *nlpE* overexpression [34]. Hence, missing or low Cpx activation could be verified by comparing these two data sets.

Further comparison of the two protein profiles (gentamicin addition versus *nlpE* overexpression) revealed that NlpE-mediated Cpx activation caused increase of the proteins AceA, GltA, and FadB, and decrease of the proteins GadA, CydA, and CydB [34]. All of these proteins are known to be regulated by ArcA ~ P at the transcriptional level, but in an opposite manner to Cpx, thereby being negatively regulated by ArcA ~ P as it holds true for e.g. *aceA*, *gltA*, and *fadB*. Hence, Cpx activation itself seems not to pronounce putative crosstalk to ArcA, since the Cpx response dominated in regulation of the aforementioned proteins.

Protein abundance correlation studies of the Cpx- and the arc-TCS revealed significant correlations to various chaperone proteins

Finally, we wanted to identify proteins whose protein levels correlate to the protein levels of CpxA and ArcA in order to investigate to what extent CpxA and ArcA affect the protein pattern of *E. coli* MG1655 WT cells. Therefore, we profiled the proteome under control conditions (WT), Cpx activation by *nlpE* overexpression (Cpx_{ON}) as well as after treatment with gentamicin (WT_G) and analyzed the correlation of protein abundances of all 1462 detected proteins with CpxA and ArcA amounts applying a linear model. Proteins whose levels increased likewise to those of CpxA or ArcA showed a high correlation with the maximum value of 1. In contrast, proteins whose levels decreased to the same extent as the target proteins increased obtained negative values (defined as significant for correlation values >0.8 or < −0.8). The results revealed a significant negative correlation between ArcA and CpxA/CpxR and a significant positive correlation between CpxA- and CpxR-levels (Additional file 4: Table S3). No significant correlation between the amounts of RR ArcA and its SK ArcB was detected, indicating that the ArcB-ArcA interaction might not be relevant under the applied aerobic conditions. Further, amounts of at least 15 annotated proteins with chaperone function were found to be significantly associated with the amounts of ArcA, ArcB, CpxA or CpxR. Four chaperons were significantly correlated with ArcA and CpxAR (Additional file 4: Table S3). In contrast to the proteins of the Cpx-TCS, these chaperones (Skp, SurA, GroL, and ClpA) showed specifically regulated protein levels during Cpx activation by *nlpE* overexpression, gentamicin treatment and WT conditions. The levels of Cpx- and Arc-TCS proteins did not change between gentamicin and WT conditions (Additional file 4: Table S3).

Discussion

The effect of the Cpx-TCS upon gentamicin treatment has been controversially discussed. Kohanski et al. (2008) postulated a gentamicin-induced crosstalk between the SK CpxA and the RR ArcA resulting in cell death [8]. This hypothesis was based on monitoring CpxA-dependent expression of ArcA ~ P targeted genes independently of CpxR. In contrast, Mahoney and Silhavy (2013) showed that the Cpx system protects cells from aminoglycoside antibiotics and hydroxyurea [3]. Mahoney and Silhavy (2013) were able to demonstrate that survival in presence of 5 µg ml^{-1} gentamicin strictly requires CpxR [3]. However, they were not able to identify the respective Cpx regulon member targeted by CpxR in presence of gentamicin.

We demonstrated in our study that physical interaction between CpxA and ArcA is possible without any kind of stimulus. This occurs under conditions overexpressing both interaction partners, as it was the case for the performed BACTH interaction analyses. However, expressing one interaction partner from the chromosome (ArcA-Snap) in combination with a mid-copy plasmid-derived CpxA-Strep variant using the mSPINE approach, we were even able to demonstrate that the interaction is dependent on the presence of 5 μg ml^{-1} gentamicin. One theory raised within this study was that a significant increase of CpxA and/or a significant decrease of ArcB in presence of gentamicin might promote interaction between CpxA and ArcA instead of the cognate ArcB-ArcA interaction. A significant decrease of *arcB* expression accompanied by constant expression levels of *cpxA* after gentamicin treatment supported this theory at least partially. However, with respect to the absolute quantification of the Cpx- and Arc-TCS on the protein level after gentamicin treatment, this hypothesis was rejected as the absolute amounts of CpxA, ArcB and ArcA decreased. These differences between the transcriptional and the protein level emphasize the need to combine expression and protein data to fully consider post-transcriptional and post-translational effects. Nevertheless, it is important to note that the CpxA:ArcA-ratios displayed 1:6 after Cpx-TCS activation by *nlpE* overexpression switching to 1:25 after gentamicin treatment. Since the CpxA:CpxR-ratio remained constant among all conditions being 1:10, RR-competition between CpxR and ArcA is higher upon gentamicin presence. It remains unclear whether these changes alone enable CpxA-ArcA interaction upon gentamicin treatment.

In the second part of our analyses, we focused on the output of the CpxA-ArcA interaction upon gentamicin treatment. Based on the fact that Kohanski et al. (2008) found a CpxA-dependent expression of ArcA ~ P targeted genes [8], we aimed to decipher such effects on the protein level. Furthermore, we addressed the question whether gentamicin treatment provokes Cpx activation.

Gentamicin treatment caused significant Cpx-dependent changes in the global proteome profile of *E. coli* MG1655. For a dozen proteins being regulated by ArcA ~ P at the DNA level, we could detect a Cpx- and Gentamicin specific alteration. However, this alteration did not correlate to the kind of ArcA ~ P regulation being positive, or negative, respectively. This finding was supported by the fact that ArcA is most likely not phosphorylated after gentamicin treatment as visualized by 2D gel electrophoresis.

Nevertheless, we could observe that the Cpx system is highly involved in modulating the global proteome profile in presence of gentamicin. This became evident in several proteins being significantly altered in a Cpx-dependent manner. These included the strongly increased protein RpoH. Its relation to the activation of CpxR during stress in *E. coli* was shown previously [56]. It has already been demonstrated that accumulation of unfolded proteins induces the production of heat shock proteins as RpoH [57]. Further, it has been discussed that RpoH may be additionally stabilized under this condition by a passive mechanism: The chaperones DnaK, DnaJ, and GrpE could be titrated away from RpoH in presence of unfolded proteins instead of interacting with RpoH promoting its degradation [58]. Nevertheless, observing changes in levels of some proteins as RpoH only in presence of CpxAR proves additional involvement of the Cpx-TCS by an unknown mechanism. Interestingly, two multidrug efflux pump proteins were also increased Cpx-dependently after gentamicin treatment. Multidrug efflux pumps contribute to the resistance of *E. coli* towards antibiotics via export of the respective substances [59]. Moreover, MdtE, MdtF and the outer membrane channel TolC are known to promote increased tolerance to different ß-lactam antibiotics [60]. Previous transcriptomic data identified the operons encoding the NADH dehydrogenase Nuo, the ferrous iron transporter EfeUOB, the succinate dehydrogenase SQR, and the cytochrome *bo* terminal oxidase complex as downregulated by CpxR ~ P [33]. Furthermore, knock-out mutants of the first gene corresponding to each of the named operons increased the cell viability in the presence of 3 μg ml^{-1} of the aminoglycoside amikacin [33]. Within the data presented here, proteins referring to these operons were not significantly altered after gentamicin treatment, except for NuoI (WT$_G$/WT 0.5). This might be due to employing another aminoglycoside in a different concentration and in a different *E. coli* strain background. However, monitoring the Cpx-dependent increase of MdtE and MdeF after gentamicin treatment might represent a protective mechanism against aminoglycoside antibiotics controlled by the Cpx-TCS. This is in contrast to the postulated Cpx- and Arc-mediated cell death under this condition [8], but in line with Mahoney & Silhavy (2013) proposing a protective role for the Cpx system in presence of aminoglycosides [3].

Moreover, our results strongly suggest that the Cpx system is not activated in presence of gentamicin. We found that the absolute amounts of CpxA, CpxR, and CpxP changed to a lower extent upon gentamicin addition compared to alterations upon well-described Cpx activating conditions (*nlpE* overexpression). Regarding the global proteome, only three proteins with altered protein abundance in dependency of CpxAR presence possess a binding site for CpxR ~ P at the DNA level. Cpx activation after gentamicin treatment became even more implausible after comparing the proteome profiles acquired after gentamicin treatment and *nlpE* overexpression which differed significantly. Here, the question whether and how these findings match previous data published by Kohanski et al. (2008) deserves further contemplation [8]. First, we can conclude that the putative interaction between CpxA and ArcA is most likely not dependent on Cpx activation, but most likely on gentamicin addition. This is reasonable considering that Cpx activation by *nlpE* overexpression evoked increased abundance of some proteins known to be negatively regulated by ArcA ~ P and vice versa. Second, response regulator competition is one well-known mechanism to prevent crosstalk [61]. Matching this mechanism, increase of CpxR with simultaneous decrease of ArcA upon *nlpE* overexpression implies that CpxR dominates regarding CpxA-mediated phosphorylation under this particular condition. This is of high importance taking into account that ArcA ~ P not only regulates some Cpx targets in an opposite manner, but also that a crosstalk between the Cpx- and the Arc-TCS is postulated to be lethal, and thus, should be avoided. Moreover, it is in general disputable how the Cpx system should be able to differ between activating conditions obviously avoiding crosstalk to the Arc-TCS (*nlpE* overexpression), and activating conditions enforcing crosstalk to the Arc-TCS (gentamicin treatment), further pointing to CpxA-ArcA interaction independently of the Cpx-active state. For future studies, it would be interesting to address whether activation of the Cpx system by e.g. *nlpE* overexpression promotes interactions between CpxA and ArcA.

Summarizing our overall results we suppose that the interaction between CpxA and ArcA monitored specifically after aminoglycoside treatment does not point to a cross-phosphorylation of ArcA via CpxA and subsequently to activation of the Arc system by CpxA. Nevertheless, our data support the hypothesis of Kohanski et al. (2008) [8], as an intersection of these two TCSs could be found by proteomic analyses. In contrast to Kohanski et al. (2008) [8], we identified protein levels of multidrug efflux pump proteins as Cpx-specifically increased in presence of gentamicin. This is more of an argument for protection against aminoglycosides than for Cpx- and Arc-mediated cell death, confirming previously recorded data [3, 33]. Altogether, our data point to an unknown, but distinct involvement of CpxAR upon gentamicin presence.

Conclusions

Gentamicin caused pronounced changes in the global proteome profile of *E. coli*, which were found to be partially dependent on the presence of CpxAR. These alterations occurred despite the fact that the Cpx-TCS seems to be not, or to a lesser extent, activated by gentamicin treatment compared to classical Cpx activating conditions like *nlpE* overexpression. Nevertheless, CpxA and ArcA were found to specifically interact upon gentamicin addition. However, this interaction seems not to be relevant for CpxA-mediated phosphorylation and activation of ArcA, although an intersection of these two pathways could be found by proteomic analyses. While the type of intersection between these two pathways

remains enigmatic, the results of the present study increase the understanding of a complex regulatory network in the *E. coli* response to aminoglycoside caused stress and on the complex network of TCSs in *E. coli*.

Methods
Bacterial strains and plasmids
All *E. coli* strains and plasmid used in this study are described in Table 2. Strains were grown in Luria-Bertani (LB) medium [62]. When necessary, antibiotics were included at the following concentration: 150 µg ml^{-1}ampicillin, 5 µg ml^{-1} gentamicin.

Harvest of bacterial culture
E. coli cells were diluted 1:100 from an overnight culture and grown aerobically at 37 °C in LB medium to an optical density at 600 nm (OD$_{600}$) of ~0.5. Subsequently, IPTG (isopropyl-ß-D-thiogalactopyranoside) was added to a final concentration of 1 mmol l^{-1} to induce the overexpression of *nlpE* (pT*nlpE*) in MG1655. For crosstalk-inducing conditions, gentamicin was added in

Table 2 *E. coli* strains and plasmids used in this study

Strain / plasmid	Relevant genotype	Reference or source
MG1655	F-lambda- *ilvG*- *rfb*-50 *rph*-1	Blattner et al. (1997) [68]
EMC07E	F-lambda- *ilvG*- *rfb*-50 *rph*-1 *cpxAR::kan*	Surmann et al. (2016) [34]
GP01E	F-lambda- *ilvG*- *rfb*-50 *rph*-1, *arcA*-Snap	this work
BTH-101	F-, *cya*-99, *araD139*, *galE15*, *galK16*, *rpsL1*, (Str r), *hsdR2*, *mcrA1*, *mcrB1*	Euromedex, Souffelweyersheim, France
pT*nlpE*	*nlpE* cloned in pTrc99A, Amp^R	Zhou et al. (2011) [24]
pKT01E	CpxA-Strep on pMal-p2X without MalE, Amp^R	Tschauner et al. (2014) [49]
pUT18	T18-fragment (amino acids 225–399 of *cyaA*), *lac* Promotor, MCS located at 5'-end of T18; in pUC19 (high copy number), Amp^R	Euromedex
pKT25	T25-fragment (amino acids 1–224 of *cyaA*), *lac* Promotor, MCS located at 3'-end of T25; in pSU40 (low copy number), Kan^R	Euromedex
pSH106E	*cpxA* in pUT18, Amp^R	this work
pSH105E	*arcB* in pUT18, Amp^R	this work
pEL16E	*cpxR* in pKT25, Kan^R	this work
pGP02E	*arcA* in pKT25, Kan^R	this work
pKT25-zip	leucine zipper of GCN4 from *Saccharomyces cerevisiae* in pKT25, Kan^R	Euromedex
pUT18C-zip	leucine zipper of GCN4 from *Saccharomyces cerevisiae* in pUT18C, Amp^R	Euromedex

a final concentration of 5 μg ml^{-1}. After additional growth for 30 min to an OD_{600} of ~1, cells were harvested by centrifugation at 5000 x g for 10 min and immediately frozen at −80 °C. The number of cells per milliliter cell culture was assigned using light microscopy and a Thoma chamber.

Analysis of CpxA- and ArcA-interaction in vivo by bacterial two-hybrid (BACTH)

The method of bacterial two-hybrid was performed as described by Karimova et al. (1998) [47]. Bacterial two-hybrid (BACTH) is based on fusions of two putative interacting proteins with two complementary fragments of *Bordetella pertussis* adenylate cyclase, namely T18 and T25. T18 and T25 build up the catalytic domain of the adenylate cyclase CyaA which converts ATP to cAMP. Separation of T18 from T25 leads to the loss of adenylate cyclase function and transcriptional activation of e.g. catabolic operons mediated by cAMP together with CAP (catabolite activator protein). Interactions of proteins fused to T18 and T25 can restore the function of the adenylate cyclase in a strain lacking the *cya* gene. Cyclic AMP can be generated again and the transcription of several genes as the *lac-* or *mal*-operon can be activated. In this study, *cpxA* and *arcB* derived from *E. coli* were cloned into the vector pUT18 containing the T18-fragment resulting in C-terminal fusions of CpxA (pSH106E) and ArcA (pSH105E) with T18. The *cpxR-* and *arcA* genes were cloned into pKT25 containing the T25-fragment leading to N-terminal fusions of CpxR (pEL16E) and ArcA (pGP02E) with T25. These combinations of N- and C-terminal fusions were chosen as the N-terminal domain of a SK contains the input domain [63]. Furthermore, it is not known whether N-terminal fusions to e.g. CpxA allow for correct integration of this fusion protein into the inner membrane as this failed for a N-terminal CpxA fusion protein variant [64]. Contrarily, the C-terminal domains of RRs often contain the output domain essential for binding to respective promoter regions [65]. *E. coli* cells of the strain BTH-101 lacking the adenylate cyclase were co-transformed with different combinations of the described plasmids. After transformation, cells were grown aerobically in LB-Medium (pH 7) to an OD_{600} of ~0.6. Putative protein interactions were quantified by determination of the ß-galactosidase activity measuring the color shift of the substrate homologue *ortho*-nitrophenyl-ß-galactoside (ONPG) from colorless to yellow after cleavage.

Analysis of CpxA- and ArcA-interaction in vivo by membrane-SPINE

Membrane-SPINE was performed as described previously with min minor modifications [48]. In brief, GP01E cells (chromosomal fusion of *arcA* with Snap-tag in MG1655) were transformed with pKT01E harboring a C-terminal fusion of *cpxA* with Strep-tag. Cells were grown in 520 ml LB (pH 7) supplemented with ampicillin at 37 °C until OD_{600} ~ 0.7. The expression of *cpxA*-Strep was induced by addition of IPTG (isopropyl-ß-D-thiogalactopyranoside) in a final concentration of 0.5 mmol l^{-1}. After growth until OD_{600} ~ 0.9, cells were split and either non-treated or treated with gentamicin in a final concentration of 5 μg ml^{-1}. After additional incubation for 20 min with or without gentamicin, formaldehyde was added in a final concentration of 0.6%, whereby samples without addition of formaldehyde served as an additional control. Cells were harvested after additional 20 min of incubation by centrifugation (3000 x g for 30 min at 4 °C) and resuspended in 16.6 ml buffer P1 (20 mmol l^{-1} Tris-HCl, 0.5 mol l^{-1} sucrose, pH 8.0). By addition of 2 ml buffer P2 (2 mg ml^{-1} lysozyme in 0.1 mol l^{-1} EDTA, pH 7.5) spheroplasts were generated and afterwards collected by centrifugation (10,000 x g, 30 min, 4 °C). Spheroplasts were resuspended in 6 ml buffer P3 (20 mmol l^{-1} Tris-HCl, 0.1 mmol l^{-1} PMSF, pH 8.0) and subsequently disrupted by ultrasonication (Branson Sonifier 250, Emerson Industrial Automation, Ferguson, MO, USA) on ice (five pulses, each 30 s). Cell debris and unbroken cells were removed by centrifugation (10,000 x g, 10 min, 4 °C). After ultracentrifugation of the supernatant (100,000 x g, 30 min, 4 °C), the resulting pellet containing the membrane fraction [5 mg ml^{-1} membrane proteins; measured using the Implen P330 Nanophotometer (Implen, München, Germany)] was resuspended in buffer P3 supplemented with 2% dodecyl maltoside (DDM, Glycon, Luckenwalde, Germany) and stirred on ice for 1 h. Non-solubilized proteins and solubilized proteins were separated by ultracentrifugation (100,000 x g, 30 min, 4 °C). CpxA-Strep was purified using a Strep-tactin column (1 ml Superflow-StrepTactin sepharose, IBA, Göttingen, Germany). The column was equilibrated with buffer W (100 mmol l^{-1} Tris-HCl, 150 mmol l^{-1} NaCl, 1 mmol l^{-1} EDTA, 0.05% DDM, pH 8.0) and washed with 20 column volumes buffer W. Afterwards, CpxA-Strep and chemically crosslinked proteins were eluted with buffer E (100 mmol l^{-1} Tris-HCl, 150 mmol l^{-1} NaCl, 1 mmol l^{-1} EDTA, 2.5 mmol l^{-1} Desthiobiotin, 0.05% DDM, pH 8.0). Using a centrifugal filter unit (Amicon YM10 filter device, Milipore) elution fractions were 10-fold concentrated. The samples were boiled at 95 °C for 20 min to remove crosslinks. Proteins were separated by SDS-PAGE and analyzed by immunoblotting. CpxA-Strep and ArcA-Snap were detected using the primary antibody α-Strep-MAB-classic (Iba, Göttingen, Germany) in a working dilution of 1:10,000 and the antibody α-SnapTag-rabbit IgG (NEB, Frankfurt am Main, Germany) in a working dilution of 1:1000, respectively. Protein band visualization was carried out using Super-Signal West Pico chemoluminescent Substrate ECL-kit (Thermo Scientific Pierce Protein Biology Products) with a peroxidase-conjugated anti-rabbit IgG antibody (GE Healthcare) in a working dilution of 1:5000 and the ChemiDoc™ MP imaging system with Image Lab™ software (BIO-RAD, München, Germany).

Quantitative reverse transcription-PCR (qRT-PCR)

The expression levels of the genes *cpxA*, *arcB*, and *arcA* were analyzed by quantitative reverse transcription-PCR (qRT-PCR) using five biological replicates for each condition. Every biological replicate was tested with three technical replicates. The extraction of total RNA was performed using the RNeasy minikit (Qiagen, Hilden, Germany) according to manufacturer's instructions. After dilution of the RNA to a final concentration of 20 ng μl^{-1} and digestion of residual DNA contaminations by DNase I, cDNA was synthesized via the RevertAid First Strand cDNA synthesis kit (Fermentas). The qRT-PCR experiments were performed according to the following protocol: The first cycle (95 °C/ 2 min) was followed by forty repeats of Cycle 2 including several heating steps (95 °C/15 s; X °C/30 s; 72 °C/30 s). Temperature X displays the primer-specific annealing temperature. The protocol was completed by three more cycles: 95 °C/1 min, 55 °C/1 min, and 55 °C/10 s. The last cycle includes a temperature increment of 0.5 °C after every 10 s to enable melt curve data collection and real-time analysis. The primer pairs used in this study are CpxA-qRT_fw2/CpxA-qRT_rev2 annealing in *cpxA* (TCT GTT CCG GGC GAT TGA TA and TTA TCT TCG CCA TCA CGC AC), ArcB-qRT_fw/ArcB-qRT_rev annealing in *arcB* (ACT GGA GGA GTC ACG ACA AC and TGT GTC TCT TCG CGC TCT TT) and ArcA-qRT_fw/ArcA-qRT_rev annealing in *arcA* (GGC GAA TGT TGC GTT GAT GT and AGC TTT CAA CGC TAC GAC GT). Since an internal standard is needed we used the primer pair GapA-qRT-fw/GapA-qRT-rev (CTC CAC TCA CGG CCG TTT CG and CTT CGC ACC AGC GGT GAT GTG) amplifying the *E. coli* housekeeping gene *gapA* (glyceraldehyde-3-phosphate dehydrogenase A). The levels of expression of *cpxA*, *arcB*, and *arcA* were normalized to the expression level of *gapA* and the x-fold change of the expression after Cpx activation compared to WT cells was calculated.

Preparation of protein extracts

As described before [34], bacterial cell pellets from 10 ml culture harvested at an OD_{600} of ~1 were reconstituted in 150 µl buffer containing 8 mol l^{-1} urea and 2 mol l^{-1} thiourea and disrupted by ultrasonication (50 W, 3 × 30 s, on ice, SonoPuls, Bandelin electronic, Berlin, Germany). After centrifugation (20,000 x g, 4 °C, 1 h) the protein concentration of the supernatant was determined using a Bradford assay (Biorad, Munich, Germany). Absolute amounts of protein per *E. coli* cell were determined by counting the number of bacteria per

ml cell culture in a Thoma chamber using light microscopy and afterwards the total protein amount was correlated to the bacterial counts. Applying this method revealed the cellular amount of protein of *E. coli* K12 to be 1.4×10^{-7} µg (average from four conditions and five biological replicates).

Heavy spike-in for absolute quantification by SRM

Two heavy labeled (^{13}C and ^{15}N arginine and lysine) proteotypic peptides for ArcA and three peptides for ArcB were obtained from JPT (JPT, Berlin, Germany). The peptide setup for the proteins of the Cpx-TCS was published recently [34]. In this work we refer to these data for the Cpx-TCS to compare the results to the newly acquired Arc-TCS data. The standard peptides possessed a C-terminal amino acid tag which was needed for quantification during manufacturing. It was eliminated by tryptic digestion during sample processing for mass spectrometry. The peptides were obtained as lyophilized powder and each one nmol was reconstituted in 100 µL buffer comprising 80% (v/v) aqueous ammonium bicarbonate solution (100 mmol l^{-1}) and 20% (v/v) ACN. Prior to usage the peptide solutions were stored in aliquots of 10 µl at –80 °C. Pure peptide solutions were tryptically digested and analyzed by shotgun MS and SRM to control incorporation rate and purity of the peptides. All peptides were fully labeled and the share of contamination (trypsin and keratin) amounted to <1%. For method optimization and absolute quantification of the proteins, standard peptides were spiked into sample background and digestion occurred as described below.

Protease digestion in solution

Four µg protein from each sample were diluted in 20 mmol l^{-1} aqueous ammonium bicarbonate solution to a final urea concentration below 2 mol l^{-1}. Prior to SRM analyses, heavy peptides were added to the sample in this step. Proteins were reduced with 2.5 mmol l^{-1} dithiothreitol at 60 °C for 1 h and subsequently alkylated with 10 mmol l^{-1} iodoacetamide at 37 °C for 30 min in the dark. The final urea concentration was decreased to 1 mol l^{-1} with 20 mmol l^{-1} aqueous ammonium bicarbonate solution to ensure the efficiency of trypsin digestion. Trypsin was added to the sample in a protease to protein ratio of 1:25 (w/w). After 16–18 h incubation at 37 °C, digestion was stopped with 1% (v/v) acetic acid. Peptides in the supernatant after centrifugation (10 min, 16,000 x g) were desalted and purified using µC18-ZipTip columns (Merck Millipore, Darmstadt, Germany). Elution buffer was evaporated in a vacuum centrifuge and peptides were dissolved in 20 µl 0.1% (v/v) aqueous acetic acid containing 2% (v/v) acetonitrile (ACN). Previously, we had tested digestion efficiency of this protocol for *E. coli* K12 cell pellets by 1D gel analysis with silver staining and revealed to be

>99.9% [66]. Samples were stored short-term at −20 °C before shotgun MS or SRM analysis.

Preparation and proteome analysis of 2D PAGE

In order to detect a possible phosphorylation of ArcA during Cpx activation on protein level, two dimensional polyacrylamide gel electrophoresis (2D PAGE) was conducted. Each 400 μg protein extracts (see above for preparation) of *E. coli* grown under WT conditions, gentamicin treatment or Cpx activation (*nlpE* overexpression) were utilized. In the first dimension, proteins were separated by their isoelectric point (pI) during isoelectric focusing on immobilized pH gradient (IPG) strips using a Multiphor™ II system (GE Healthcare Life Sciences, Freiburg, Germany) according to factory instructions in a pH range between 4.5 and 5.5 to ensure a high resolution in the expected range of ArcA (pI 5.2 according to ExPASy.org).

In the second dimension proteins were separated by their molecular mass. The IPG strips were first equilibrated for 15 min at room temperature in aqueous buffer consisting of 6 mol l^{-1} urea, 1.5 mol l^{-1} Tris HCl, pH 8.8, 87% (v/v) glycerol, 20% (m/v) sodium dodecyl sulfate (SDS) and 10 mg ml^{-1} dithiothreitol. Subsequently, a second equilibration took place for 15 min with a similar buffer only replacing dithiothreitol with 25 mg ml^{-1} iodoacetamide and a little bromophenol blue. Equilibrated IPG strips were transferred onto a 0.5% (m/v) agarose gel prepared in running buffer [72 g glycin, 15 g Tris, and 25 ml 20% (m/v) SDS filled up to 500 ml with *A. dest.*]. Proteins were separated at 20 mA in about 1.5 h. Coomassie Brilliant Blue G-250 (Merck Millipore, Darmstadt, Germany) staining allowed visualization of separated proteins using the program Delta 2D (DECODON, Greifwald, Germany) after scanning the gels. Spots from each two gels (WT and gentamicin treatment or WT and Cpx activation) were aligned allowing size comparison and determination of presence or absence in one of the two conditions.

For protein identification, selected single protein spots (Additional file 1: Figure S3) were cut out from the gel with the help of a pipette tip and transferred to a 1.7 ml reaction tube. To discolor the gel 200 μl 200 mmol l^{-1} aqueous ammonium bicarbonate were added to the gel spot and allowed for 15 min incubation at 37 °C. The supernatant was removed and the step was repeated until the dye was completely removed. Afterwards, the gel was dehydrated by adding 100 μl acetonitrile for 15 min at 37 °C. After removal of the supernatant, the dehydration was repeated once. To the dried gel piece 10 μl 10 ng $μl^{-1}$ trypsin (Promega, Madison, WI, USA, to detect tryptic peptides cleaved after lysine and arginine) dissolved or 10 ng $μl^{-1}$ proteinase K (Sigma-Aldrich, St. Louis, USA, to increase number of detected peptides due to less specific cleavage) in 20 mmol l^{-1} aqueous

ammonium bicarbonate were added. After 1 h incubation at room temperature, excessive protease solution was removed, if the gel was not fully reconstituted with liquid, the required volume of 20 mmol l^{-1} aqueous ammonium bicarbonate was added to the sample. Subsequent tryptic digestion was allowed for 14 h at 37 °C in horizontal position of the reaction tubes to avoid drying the gel. Peptides were extracted from the gel first, in 10 μl 0.1% (v/v) aqueous acetic acid for 30 min in a ultrasonication bath and, second, after separation of the first supernatant, by adding 10 μL 50% (v/v) acetonitrile in 0.05% (v/v) aqueous acetic acid under same conditions. Both supernatant containing peptides were combined and freeze-dried by lyophilization. Dried peptides were reconstituted in 40 μl 0.1% (v/v) aqueous acetic acid and purified with μC18-ZipTip columns (Merck Millipore, Darmstadt, Germany) and prepared for nanoLC-MS/MS as described above.

Data acquisition by mass spectrometry

Equal to our previously published study [34] for proteome profiling peptide separation was performed on a NanoAcquity BEH130 C18 column (10 cm length, 100 μM inner diameter and 1.7 μm particle size) using a nanoAcquity UPLC (Waters, Manchester, UK). Separated peptides were ionized using electrospray and analyzed with an LTQ Orbitrap Velos mass spectrometer (Thermo Electron, Bremen, Germany) operated in data-dependent mode. Up to 20 of the most intense ions were sequentially isolated for collision induced dissociation (CID) in the linear ion trap. Shotgun LC-MS/MS analysis was carried out for four or five [correlation study between CpxA or ArcA and other detected proteins under WT, Cpx activating conditions (*nlpE* overexpression) and gentamicin treatment of *E. coli* WT strain], Additional file 4: Table S3) or four (influence of gentamicin on the proteome of WT or *cpxAR* mutant, Additional file 2: Table S1; Additional file 3: Table S2) independent biological replicates (BR) per condition.

Protein digests from the 2D gel were investigated on a Q Exactive mass spectrometer (Thermo Fisher Scientific, Waltham, MA, USA) after separation of peptides using a Dionex UltiMate 3000 RSLC nano-LC system (Dionex/Thermo Fisher Scientific, Idstein, Germany), and ionization with a TriVersa NanoMate source (Advion, Ltd., Harlow, UK). Here, after the first analysis peptides were fragmented using higher energy collision dissociation (HCD) instead of CID. Further details on shotgun data acquisition are provided as Additional files.

For SRM analysis, a nano-HPLC (EASY-nLC, Proxeon Biosystems A/S, Odense, Denmark) with the help of an Acclaim PepMap 100 reverse phase column (3 μm, 75 μm i.d. × 150 mm, LC Packings, Dionex, Idstein, Germany) was coupled to a TSQ Vantage (Thermo Electron). First, separated and ionized targeted peptides (precursors) were analyzed in the first quadrupole. After fragmentation by CID the corresponding likewise targeted products were analyzed in a further quadrupole. Required collision energy (CE) was optimized at precursor level beginning from factory defaults (depending on the m/z ratio of the precursor) by applying different eV in steps of + or −2 eV. Settings for final SRM analyses are provided as Additional files. For each peptide, the doubly charged precursor and the four most abundant product ions (transitions) were chosen for SRM acquisition (Additional file 1: Table S4). SRM data were recorded for five independent biological replicates for each condition.

Analysis of proteome data

Mass spectrometric data from proteome profiling was investigated with the Rosetta Elucidator software (Ceiba Solutions, Boston, MA, USA). Protein identifications resulted from an automated database search against a Swiss-Prot database rel. 06–2014 limited to *E. coli* K12 entries using Sequest v. 2.7. Quantitative analysis was based on summed intensities of aligned single isotope features representing peptides (PeptideTeller probability >0.95). Only proteins that were identified by at least two peptides or by one peptide provided that the sequence coverage exceeded 10%, respectively, were subjected for further analysis. With the help of the Genedata Analyst software v8.2 (Genedata AG, Basel, Switzerland) protein intensity values were median normalized and statistically analyzed using two group t-test together with multiple testing corrections after to Benjamini-Hochberg.

The normalized intensity values of each protein from the complete dataset of four conditions and five biological replicates (for *nlpE* overexpression only four biological replicates were available) were used to determine correlations between the intensity behavior of CpxA and ArcA and all other quantified proteins. Therefore, a linear model was applied in Genedata Analyst and statistically tested. Correlations comprising a q-value (Benjamini-Hochberg corrected) < 0.05 were regarded as statistically significant. Positive correlation values indicated similar, negative values an opposite behavior of protein intensities. Values >0.8 were defined significantly positively correlated and values < −0.8 as significantly negatively correlated. A value of 1 indicated the same behavior as the compared protein.

Ratios in comparison to the WT or between control and gentamicin treated sample (four BR) were calculated from the mean of all BRs per condition. Ratios between two conditions with values <0.5 or >2, along with a multiple testing corrected q-value <0.05 were regarded as regulated in the corresponding condition.

Data from separated proteins from 2D gels were identified using the MASCOT search algorithm against the *E. coli* database (limited to tryptic peptides or non-limited when proteinase K was used for hydrolysis of

proteins) mentioned above allowing a false discovery rate of <1% and carbamidomethylation of cysteine set as static, and oxidation of methionine and phosphorylation of aspartate as variable modification.

For SRM analysis method development and optimization as well as quantification were accomplished using the open source program Skyline v2.5 [67]. The final transition list is provided as Additional file 1: Table S4. Quantification was performed as described recently [34]. Pairs of heavy and light peptides were identified by equal peak elution pattern and retention time. Dilution series of each heavy standard peptide (0, 0.1, 0.5, 1, 5, 10, 50, and 100 fmol µg^{-1} protein) mixed with background of WT bacteria were acquired as duplicates by SRM after tryptic digestion. From that the linear range (R^2 > 0.99), in which absolute quantification was possible, was determined for each peptide. The peptide per protein with the highest intensity and thereby best signal to noise ratio was chosen for quantification. Finally, each of the samples was spiked with 0.5 fmol µg^{-1} or 10 fmol µg^{-1} for each peptide. The ratio from the peak area of the heavy (synthetic peptide) to light (natural sample peptide) eluting at same retention time, which amounted closer to one (single point calibration) was utilized for absolute quantification. Average values and coefficient of variance were calculated over all replicates for each condition.

Additional files

Additional file 1: Supporting information and supporting figures.

Additional file 2: Table S1. Proteome profile of *E. coli* MG1655 and its isogenic *cpxAR* mutant with and without incubation with 5 µg ml^{-1} gentamicin.

Additional file 3: Table S2. Proteins regulated upon gentamicin treatment.

Additional file 4: Table S3. Correlation studies between intensities of CpxA or ArcA and other detected proteins including WT condition, Cpx induction, and gentamicin treatment.

Abbreviations
ArcA ~ P: Phosphorylated ArcA; BACTH: Bacterial two-hybrid; CpxR ~ P: Phosphorylated CpxR; SRM: Single reaction monitoring; TCS: Two - component system

Acknowledgements
We thank Eva Limpinsel, Ulrike Lissner, Katrin Schoknecht, Sophie Eisenlöffel, and Manuela Gesell Salazar for technical assistance and Heather Heizel for critically reading of the manuscript.

Funding
This work was supported by DFG grants Hu1011/2-1 and SFB944 to S.H.

Authors' contributions

EC, KS, EH, and SH designed the experiments. EC, KS, and GP performed the experiments. All authors analyzed the data. EC, KS, EH, and SH wrote the manuscript. All authors read and approved the final manuscript.

Competing interests

The authors declare that they have no competing interests.

Author details

[1]FB 5 Microbiology, Department of Biology/Chemistry, University Osnabrück, Barbarastraße 11, 49076 Osnabrück, Germany. [2]Department of Functional Genomics, Interfaculty Institute of Genetics and Functional Genomics, University Medicine Greifswald, Friedrich-Ludwig-Jahn-Straße 15A, 17475 Greifswald, Germany. [3]Department of Biology, Institute of Molecular Microbiology and Biotechnology, Universität Münster, Corrensstraße 3, 48149 Münster, Germany.

References

1. Magnet S, Blanchard JS. Molecular insights into aminoglycoside action and resistance. Chem Rev. 2005;105(2):477–98.
2. Hoffman LR, D'Argenio DA, MacCoss MJ, Zhang Z, Jones RA, Miller SI. Aminoglycoside antibiotics induce bacterial biofilm formation. Nature. 2005;436(7054):1171–5.
3. Mahoney TF, Silhavy TJ. The Cpx stress response confers resistance to some, but not all, bactericidal antibiotics. J Bacteriol. 2013;195(9):1869–74.
4. Chang C, Stewart RC. The two-component system. Regulation of diverse signaling pathways in prokaryotes and eukaryotes. Plant Physiol. 1998;117(3):723–31.
5. Hoch JA. Two-component and phosphorelay signal transduction. Curr Opin Microbiol. 2000;3(2):165–70.
6. Gao R, Stock AM. Biological insights from structures of two-component proteins. Annu Rev Microbiol. 2009;63:133–54.
7. Stock AM, Robinson VL, Goudreau PN. Two-component signal transduction. Annu Rev Biochem. 2000;69:183–215.
8. Kohanski MA, Dwyer DJ, Wierzbowski J, Cottarel G, Collins JJ. Mistranslation of membrane proteins and two-component system activation trigger antibiotic-mediated cell death. Cell. 2008;135(4):679–90.
9. Groban ES, Clarke EJ, Salis HM, Miller SM, Voigt CA. Kinetic buffering of cross talk between bacterial two-component sensors. J Mol Biol. 2009;390(3):380–93.
10. Wanner BL. Is cross regulation by phosphorylation of two-component response regulator proteins important in bacteria? J Bacteriol. 1992;174(7):2053–8.
11. Yoshida M, Ishihama A, Yamamoto K. Cross talk in promoter recognition between six NarL-family response regulators of Escherichia Coli two-component system. Genes Cells. 2015;20(7):601–12.
12. Siryaporn A, Goulian M. Cross-talk suppression between the CpxA-CpxR and EnvZ-OmpR two-component systems in E. Coli. Mol Microbiol. 2008;70(2):494–506.
13. Skerker JM, Perchuk BS, Siryaporn A, Lubin EA, Ashenberg O, Goulian M, Laub MT. Rewiring the specificity of two-component signal transduction systems. Cell. 2008;133(6):1043–54.
14. Guckes KR, Kostakioti M, Breland EJ, Gu AP, Shaffer CL, Martinez CR, Hultgren SJ, Hadjifrangiskou M. Strong cross-system interactions drive the activation of the QseB response regulator in the absence of its cognate sensor. Proc Natl Acad Sci U S A. 2013;110(41):16592–7.
15. McEwen J, Silverman P. Mutations in genes cpxA and cpxB of Escherichia Coli K-12 cause a defect in isoleucine and valine syntheses. J Bacteriol. 1980;144(1):68–73.
16. McEwen J, Silverman P. Chromosomal mutations of Escherichia Coli that alter expression of conjugative plasmid functions. Proc Natl Acad Sci U S A. 1980;77(1):513–7.
17. McEwen J, Sambucetti L, Silverman PM. Synthesis of outer membrane proteins in cpxA cpxB mutants of Escherichia Coli K-12. J Bacteriol. 1983;154(1):375–82.
18. McEwen J, Silverman P. Genetic analysis of Escherichia Coli K-12 chromosomal mutants defective in expression of F-plasmid functions: identification of genes cpxA and cpxB. J Bacteriol. 1980;144(1):60–7.
19. Dong J, Iuchi S, Kwan HS, Lu Z, Lin EC. The deduced amino-acid sequence of the cloned cpxR gene suggests the protein is the cognate regulator for the membrane sensor, CpxA, in a two-component signal transduction system of Escherichia Coli. Gene. 1993;136(1–2):227–30.
20. Iuchi S, Matsuda Z, Fujiwara T, Lin EC. The arcB gene of Escherichia Coli encodes a sensor-regulator protein for anaerobic repression of the arc modulon. Mol Microbiol. 1990;4(5):715–27.
21. Iuchi S. Lin EC: arcA (dye), a global regulatory gene in Escherichia Coli mediating repression of enzymes in aerobic pathways. Proc Natl Acad Sci U S A. 1988;85(6):1888–92.
22. Jones CH, Danese PN, Pinkner JS, Silhavy TJ, Hultgren SJ. The chaperone-assisted membrane release and folding pathway is sensed by two signal transduction systems. EMBO J. 1997;16(21):6394–406.
23. Fleischer R, Heermann R, Jung K, Hunke S. Purification, reconstitution, and characterization of the CpxRAP envelope stress system of Escherichia Coli. J Biol Chem. 2007;282(12):8583–93.
24. Zhou X, Keller R, Volkmer R, Krauss N, Scheerer P, Hunke S. Structural basis for two-component system inhibition and pilus sensing by the auxiliary CpxP protein. J Biol Chem. 2011;286(11):9805–14.
25. Danese PN, Silhavy TJ. CpxP, a stress-combative member of the Cpx regulon. J Bacteriol. 1998;180(4):831–9.
26. Nakayama S, Watanabe H. Involvement of cpxA, a sensor of a two-component regulatory system, in the pH-dependent regulation of expression of Shigella sonnei virF gene. J Bacteriol. 1995;177(17):5062–9.
27. Mileykovskaya E, Dowhan W. The Cpx two-component signal transduction pathway is activated in Escherichia Coli mutant strains lacking phosphatidylethanolamine. J Bacteriol. 1997;179(4):1029–34.
28. Klein G, Kobylak N, Lindner B, Stupak A, Raina S. Assembly of lipopolysaccharide in Escherichia Coli requires the essential LapB heat shock protein. J Biol Chem. 2014;289(21):14829–53.
29. Klein G, Lindner B, Brabetz W, Brade H, Raina S. Escherichia Coli K-12 suppressor-free mutants lacking early Glycosyltransferases and late Acyltransferases: minimal lipopolysaccharide structure and induction of envelope stress response. J Biol Chem. 2009;284(23):15369–89.
30. De Wulf P, McGuire AM, Liu X, Lin EC. Genome-wide profiling of promoter recognition by the two-component response regulator CpxR-P in Escherichia Coli. J Biol Chem. 2002;277(29):26652–61.
31. Shimohata N, Chiba S, Saikawa N, Ito K, Akiyama Y. The Cpx stress response system of Escherichia Coli senses plasma membrane proteins and controls HtpX, a membrane protease with a cytosolic active site. Genes Cells. 2002;7(7):653–62.
32. Gerken H, Charlson ES, Cicirelli EM, Kenney LJ, Misra R. MzrA: a novel modulator of the EnvZ/OmpR two-component regulon. Mol Microbiol. 2009;72(6):1408–22.
33. Raivio TL, Leblanc SK, Price NL. The Escherichia Coli Cpx envelope stress response regulates genes of diverse function that impact antibiotic resistance and membrane integrity. J Bacteriol. 2013;195(12):2755–67.
34. Surmann K, Ćudić E, Hammer E, Hunke S. Molecular and proteome analyses highlight the importance of the Cpx envelope stress system for acid stress and cell wall stability in Escherichia Coli. Microbiology. 2016;5(4):582–96.
35. Hunke S, Keller R, Müller VS. Signal integration by the Cpx-envelope stress system. FEMS Microbiol Lett. 2012;326(1):12–22.
36. DiGiuseppe PA, Silhavy TJ. Signal detection and target gene induction by the CpxRA two-component system. J Bacteriol. 2003;185(8):2432–40.
37. Otto K, Silhavy TJ. Surface sensing and adhesion of Escherichia Coli controlled by the Cpx-signaling pathway. Proc Natl Acad Sci U S A. 2002;99(4):2287–92.
38. Snyder WB, Davis LJ, Danese PN, Cosma CL, Silhavy TJ. Overproduction of NlpE, a new outer membrane lipoprotein, suppresses the toxicity of periplasmic LacZ by activation of the Cpx signal transduction pathway. J Bacteriol. 1995;177(15):4216–23.
39. Georgellis D, Kwon O, De Wulf P, Lin EC. Signal decay through a reverse phosphorelay in the arc two-component signal transduction system. J Biol Chem. 1998;273(49):32864–9.
40. Georgellis D, Kwon O, Lin EC, Wong SM, Akerley BJ. Redox signal transduction by the ArcB sensor kinase of Haemophilus influenzae lacking the PAS domain. J Bacteriol. 2001;183(24):7206–12.

41. Jung WS, Jung YR, Oh DB, Kang HA, Lee SY, Chavez-Canales M, Georgellis D, Kwon O. Characterization of the arc two-component signal transduction system of the capnophilic rumen bacterium Mannheimia succiniciproducens. FEMS Microbiol Lett. 2008;284(1):109–19.

42. Kwon O, Georgellis D, Lin EC. Phosphorelay as the sole physiological route of signal transmission by the arc two-component system of Escherichia Coli. J Bacteriol. 2000;182(13):3858–62.

43. Malpica R, Sandoval GR, Rodríguez C, Franco B, Georgellis D. Signaling by the arc two-component system provides a link between the redox state of the quinone pool and gene expression. Antioxid Redox Signal. 2006;8(5–6):781–95.

44. Liu X, De Wulf P. Probing the ArcA-P modulon of Escherichia Coli by whole genome transcriptional analysis and sequence recognition profiling. J Biol Chem. 2004;279(13):12588–97.

45. Keren I, Wu Y, Inocencio J, Mulcahy LR, Lewis K. Killing by bactericidal antibiotics does not depend on reactive oxygen species. Science. 2013;339(6124):1213–6.

46. Liu Y, Imlay JA. Cell death from antibiotics without the involvement of reactive oxygen species. Science. 2013;339(6124):1210–3.

47. Karimova G, Pidoux J, Ullmann A, Ladant D. A bacterial two-hybrid system based on a reconstituted signal transduction pathway. Proc Natl Acad Sci U S A. 1998;95(10):5752–6.

48. Müller VS, Tschauner K, Hunke S. Membrane-SPINE: a biochemical tool to identify protein-protein interactions of membrane proteins in vivo. J Vis Exp. 2013;81:e50810.

49. Tschauner K, Hörnschemeyer P, Müller VS, Hunke S. Dynamic interaction between the CpxA sensor kinase and the periplasmic accessory protein CpxP mediates signal recognition in E. Coli. PLoS One. 2014;9(9):e107383.

50. Kohanski MA, Dwyer DJ, Collins JJ. How antibiotics kill bacteria: from targets to networks. Nat Rev Microbiol. 2010;8(6):423–35.

51. Gallien S, Duriez E, Domon B. Selected reaction monitoring applied to proteomics. J Mass Spectrom. 2011;46(3):298–312.

52. Schmidt C, Lenz C, Grote M, Lührmann R, Urlaub H. Determination of protein stoichiometry within protein complexes using absolute quantification and multiple reaction monitoring. Anal Chem. 2010;82(7):2784–96.

53. Schmidt A, Kochanowski K, Vedelaar S, Ahrné E, Volkmer B, Callipo L, Knoops K, Bauer M, Aebersold R, Heinemann M. The quantitative and condition-dependent Escherichia Coli proteome. Nat Biotechnol. 2016;34(1):104–10.

54. Iuchi S. Phosphorylation/dephosphorylation of the receiver module at the conserved aspartate residue controls transphosphorylation activity of histidine kinase in sensor protein ArcB of Escherichia Coli. J Biol Chem. 1993;268(32):23972–80.

55. Kashyap DR, Wang M, Liu LH, Boons GJ, Gupta D, Dziarski R. Peptidoglycan recognition proteins kill bacteria by activating protein-sensing two-component systems. Nat Med. 2011;17(6):676–83.

56. Zahrl D, Wagner M, Bischof K, Koraimann G. Expression and assembly of a functional type IV secretion system elicit extracytoplasmic and cytoplasmic stress responses in Escherichia Coli. J Bacteriol. 2006;188(18):6611–21.

57. Parsell DA, Sauer RT. Induction of a heat shock-like response by unfolded protein in Escherichia Coli: dependence on protein level not protein degradation. Genes Dev. 1989;3(8):1226–32.

58. Kanemori M, Mori H, Yura T. Induction of heat shock proteins by abnormal proteins results from stabilization and not increased synthesis of sigma 32 in Escherichia Coli. J Bacteriol. 1994;176(18):5648–53.

59. Li XZ, Nikaido H. Efflux-mediated drug resistance in bacteria: an update. Drugs. 2009;69(12):1555–623.

60. Nishino K, Senda Y, Yamaguchi A. The AraC-family regulator GadX enhances multidrug resistance in Escherichia Coli by activating expression of mdtEF multidrug efflux genes. J Infect Chemother. 2008;14(1):23–9.

61. Laub MT, Goulian M. Specificity in two-component signal transduction pathways. Annu Rev Genet. 2007;41:121–45.

62. Miller JH. A short course in bacterial genetics : a laboratory manual and handbook for Escherichia coli and related bacteria. New York: Cold Spring Harbor Laboratory Press; 1992.

63. West AH, Stock AM. Histidine kinases and response regulator proteins in two-component signaling systems. Trends Biochem Sci. 2001;26(6):369–76.

64. Kefala G, Kwiatkowski W, Esquivies L, Maslennikov I, Choe S. Application of Mistic to improving the expression and membrane integration of histidine kinase receptors from Escherichia Coli. J Struct Funct Genom. 2007;8(4):167–72.

65. Galperin MY. Structural classification of bacterial response regulators: diversity of output domains and domain combinations. J Bacteriol. 2006;188(12):4169–82.

66. Surmann K, Laermann V, Zimmann P, Altendorf K, Hammer E. Absolute quantification of the Kdp subunits of Escherichia Coli by multiple reaction monitoring. Proteomics. 2014;14(13–14):1630–8.

67. Maclean B, Tomazela DM, Shulman N, Chambers M, Finney GL, Frewen B, Kern R, Tabb DL, Liebler DC, MacCoss MJ. Skyline: an open source document editor for creating and analyzing targeted proteomics experiments. Bioinformatics. 2010;26(7):966–8.

68. Blattner FR, Plunkett G, Bloch CA, Perna NT, Burland V, Riley M, Collado-Vides J, Glasner JD, Rode CK, Mayhew GF, et al. The complete genome sequence of Escherichia Coli K-12. Science. 1997;277(5331):1453–62.

69. Batchelor E, Walthers D, Kenney LJ, Goulian M. The Escherichia Coli CpxA-CpxR envelope stress response system regulates expression of the porins ompF and ompC. J Bacteriol. 2005;187(16):5723–31.

70. Wyborn NR, Messenger SL, Henderson RA, Sawers G, Roberts RE, Attwood MM, Green J. Expression of the Escherichia Coli yfiD gene responds to intracellular pH and reduces the accumulation of acidic metabolic end products. Microbiology. 2002;148(Pt 4):1015–26.

71. Zientz E, Janausch IG, Six S, Unden G. Functioning of DcuC as the C4-dicarboxylate carrier during glucose fermentation by Escherichia Coli. J Bacteriol. 1999;181(12):3716–20.

72. Salmon KA, Hung SP, Steffen NR, Krupp R, Baldi P, Hatfield GW, Gunsalus RP. Global gene expression profiling in Escherichia Coli K12: effects of oxygen availability and ArcA. J Biol Chem. 2005;280(15):15084–96.

73. Richard DJ, Sawers G, Sargent F, McWalter L, Boxer DH. Transcriptional regulation in response to oxygen and nitrate of the operons encoding the [NiFe] hydrogenases 1 and 2 of Escherichia Coli. Microbiology. 1999;145(Pt 10):2903–12.

74. Shalel-Levanon S, San KY, Bennett GN. Effect of ArcA and FNR on the expression of genes related to the oxygen regulation and the glycolysis pathway in Escherichia Coli under microaerobic growth conditions. Biotechnol Bioeng. 2005;92(2):147–59.

75. Kaiser M, Sawers G. Overlapping promoters modulate Fnr- and ArcA-dependent anaerobic transcriptional activation of the focApfl operon in Escherichia Coli. Microbiology. 1997;143(Pt 3):775–83.

18

Temporal dynamics in microbial soil communities at anthrax carcass sites

Karoline Valseth[1,2], Camilla L. Nesbø[1,3], W. Ryan Easterday[1], Wendy C. Turner[4], Jaran S. Olsen[2], Nils Chr. Stenseth[1] and Thomas H. A. Haverkamp[1*] (iD)

Abstract

Background: Anthrax is a globally distributed disease affecting primarily herbivorous mammals. It is caused by the soil-dwelling and spore-forming bacterium *Bacillus anthracis*. The dormant *B. anthracis* spores become vegetative after ingestion by grazing mammals. After killing the host, *B. anthracis* cells return to the soil where they sporulate, completing the lifecycle of the bacterium. Here we present the first study describing temporal microbial soil community changes in Etosha National Park, Namibia, after decomposition of two plains zebra (*Equus quagga*) anthrax carcasses. To circumvent state-associated-challenges (i.e. vegetative cells/spores) we monitored *B. anthracis* throughout the period using cultivation, qPCR and shotgun metagenomic sequencing.

Results: The combined results suggest that abundance estimation of spore-forming bacteria in their natural habitat by DNA-based approaches alone is insufficient due to poor recovery of DNA from spores. However, our combined approached allowed us to follow *B. anthracis* population dynamics (vegetative cells and spores) in the soil, along with closely related organisms from the *B. cereus* group, despite their high sequence similarity. Vegetative *B. anthracis* abundance peaked early in the time-series and then dropped when cells either sporulated or died. The time-series revealed that after carcass deposition, the typical semi-arid soil community (e.g. *Frankiales* and *Rhizobiales* species) becomes temporarily dominated by the orders *Bacillales* and *Pseudomonadales*, known to contain plant growth-promoting species.

Conclusion: Our work indicates that complementing DNA based approaches with cultivation may give a more complete picture of the ecology of spore forming pathogens. Furthermore, the results suggests that the increased vegetation biomass production found at carcass sites is due to both added nutrients and the proliferation of microbial taxa that can be beneficial for plant growth. Thus, future *B. anthracis* transmission events at carcass sites may be indirectly facilitated by the recruitment of plant-beneficial bacteria.

Keywords: *Bacillus anthracis*, Metabolism, Metagenomics, Semi-arid, Shotgun sequencing, Taphonomy, Time-series analysis, Sporulation, Microbial diversity

Background

The microbial composition of arid soils across the globe is distinct from other soil environments [1, 2]. The forces that shape arid soil microbial community composition include low water availability, temperature and UV radiation [3–8]. Especially, water restriction has a large influence since it affects several environmental factors, such as salinity, pH, and the availability of (in-) organic matter, which further

modulate soil microbial diversity and activity [4, 5, 9–11]. The combination of the above factors may explain why soils in arid environments, such as deserts and arid savannahs, are taxonomically distinct from other soil types [1].

Most research on arid soil microbial diversity is directed to understand the community dynamics under changing environmental conditions, e.g. precipitation changes. However, soil communities can also be affected by the deposition of animal carcasses and their decomposition. This may have a profound impact on the soil microbiome as the carcass influences both biotic (adding new microbes) and abiotic (adding nutrients and moisture) factors. Moreover, after the introduction of a

* Correspondence: thhaverk@ibv.uio.no
[1]Department of Biosciences, Centre for Ecological and Evolutionary Synthesis (CEES), University of Oslo, The Kristine Bonnevie Building, UiO, campus Blindern, Blindern, Oslo, Norway
Full list of author information is available at the end of the article

carcass, a succession within the microbial community will occur, changing abundances in accordance with nutrients being released during carcass decomposition [12]. For instance, bacteria belonging to *Proteobacteria* and *Acidobacteria* are the most common in soils during the initial stages of carcass decomposition, while *Firmicutes* are more prominent during active decomposition [13, 14]. The described succession is however, found under experimentally controlled conditions, where carcasses were secured at one location for the entire experiment. In contrast, in natural ecosystems carcasses are often consumed and/or dragged away from the site of death by scavengers [15]. Thus microbes and nutrients may only transiently enter the soil at the site of death [16], which might induce different microbial soil dynamics at natural carcass sites.

The present study investigates the effects of animal carcasses on soil microbial communities after an animal has died of an anthrax infection. The disease anthrax is caused by *Bacillus anthracis*, a gram-positive, rod shaped, sporulating bacterium [17]. This species belongs together with *Bacillus cereus*, *Bacillus thuringiensis* and several other *Bacillus spp.*, to the *B. cereus* group [18], which is commonly found in soil as vegetative cells or as spores [19]. The *B. cereus* group bacteria are indistinguishable from each other using 16S rRNA gene sequences and show high genetic identity (> 99.6%) for certain housekeeping genes [18, 20, 21]. However, *B. anthracis* can genetically be distinguished from the other 'species' based on single nucleotide polymorphisms (SNPs) (e.g. in the *plcR* gene [22, 23]) and in most cases by the presence of two virulence plasmids specific to *B. anthracis* pXO1 and pXO2 [24–26]. The lifecycle of *B. anthracis* is different from other *B. cereus* group bacteria in that it predominantly targets, infects and kills mammalian herbivores, instead of insects as found for *B. thuringiensis*. Grazing by herbivores seems to be the main route for transmission in natural settings such as found in the semi-arid savannah in Etosha National Park (ENP), Namibia [27]. After causing the death of its host, *B. anthracis* returns to the soil where it sporulates. Hence, *B. anthracis* spores will often be found in high densities in the top layer of soil where haemorrhagic fluids have leaked from an anthrax carcass [27–29].

Long term measurements at carcass sites in the ENP have identified several processes occurring in and above the soil. It was found that *B. anthracis* cell counts in rhizosphere soils increased in the second year after carcass deposition, but not in surface soils [27]. Those results contrast experimental work where no multiplication of cells was observed in the second year [30]. Furthermore, it was observed that grass biomass and quality increased at the localised area of anthrax carcass sites [27]. This increase in localised plant growth is mostly

likely due to nutrient release and/or rhizosphere plant-microbe interactions. The increase in above ground plant biomass results in attraction of grazers, which increases the potential exposure to the pathogen through ingestion of grasses and potentially roots / soil at carcass sites [16, 27]. It is thought that the bacterium sporulates shortly after entry into the soil and that it will stay dormant until favourable conditions arrive [31]. This suggests that there is only a short period, after the entry of *B. anthracis* into the soil, where the pathogen could interact with other microbes or plants to enhance its transmission [16, 30–33]. It is unclear if such interactions influence further transmission of the pathogen. In particular there is a lack of understanding of the dynamics of the (arid) soil microbial community after the influx of animal fluids (e.g. blood, gut contents) with high densities of *B. anthracis* vegetative cells. In order to address how interactions between microbes / plants and *B. anthracis* benefits pathogen transmission, it is important to understand the dynamics of the arid soil microbial community after carcass deposition and influx of *B. anthracis* vegetative cells into the soil.

There are several technical challenges in following both the *B. anthracis* and microbial community dynamics in soil. Currently, DNA-based methods such as 16S rRNA amplicon sequencing and shotgun metagenomics are established methods for the study of complete microbial communities [34]. As mentioned above *B. anthracis* is indistinguishable from other *B. cereus* group species on 16S rRNA gene sequences. Therefore, 16S rRNA amplicons are not suitable for studies aiming to distinguish members from this group. *B. anthracis* and many other microbes can be detected with shotgun metagenomics [35]. Again, *B. anthracis* is genetically highly similar to *B. cereus* group bacteria, which makes bioinformatic identification of these species challenging [18]. Therefore methods able to detect and distinguish various *B. cereus* group species are needed to complement and confirm the shotgun metagenomic results. Such methods include cultivation using selective media (e.g. PLET) and specific qPCR assays [23, 36]. The combination of these methods can then be used to monitor environmental *B. anthracis* populations and to generate hypotheses about possible species interactions, which subsequently can be tested with specifically designed experiments.

Here we present an analysis of soil microbial community composition following leaching of fluids into the soil from two spatially and temporally proximate zebra anthrax carcasses in ENP. The ENP is a semi-arid savannah environment, which for much of the year has sparse water availability, meagre vegetation and limited nutrient resources [37, 38]. Hence, carcass nutrients will likely have great influence on the localised soil microbial community [39]. Large herds of ungulates ensure that every year there is an abundance of carcasses in ENP, often through predation or diseases such as anthrax. It is estimated that up to 400 plains zebras (*Equus quagga*) per year die from anthrax infections in the ENP [40]. Due to the unpredictability of the presence of disease cases, and our interest in controlling for temporal and spatial variability in soil microbiota among study sites, our study was limited to two carcass sites situated proximately in space and time. Our aim was to describe the microbial community succession taking place in the soil after the influx of haemorrhagic fluids containing *B. anthracis* vegetative cells. Shotgun metagenomic sequencing of samples obtained through the first month of decomposition was employed to investigate temporal dynamics of the community structure and function. We investigate the temporal change of the taxonomic composition and the metabolic potential by identifying major metabolic pathways. Finally, we use a combination of techniques (qPCR, cultivation and metagenomic sequencing) to circumvent state (vegetative cell vs. spore) associated challenges to accurately track *B. anthracis* abundances.

Results

Carcass information and rainfall recording

On 03.03.2014 we identified two plains zebra (*Equus quagga*) carcasses less than 1 km apart in ENP. Hereafter referred to as Carcass 1 (Ca1) and Carcass 2 (Ca2). Ca1 was intact when sampled on day 0, while Ca2 was minimally scavenged (some intestine was dragged out the anus by vultures). The time of collection ($\approx$ 14:00), the state of both carcasses and only the presence of a few avian scavengers and not mammalian, indicates that both animals were likely dead for less than 12 h and certainly fewer than 24 h. Our study contrasts with other carcass decomposition experiments, since scavenging was not restricted [12–14, 39, 41]. As such, the carcass nutrients and fluids will only leak onto the soil for a short time before scavengers consume soft tissue and move the remains off the site.

After three days both carcasses were completely consumed by scavengers and the bones were found approximately 5 m away from each sampling site. The sampling sites were visited at days: 0, 3, 7, 14, 21 and 30, to collect material for cultivation and DNA extraction. At day zero an uncontaminated control sample (Ctrl0) was taken at both sites to function as a reference sample. The soil in the study area has a alkaline pH around 8.7 ± 0.4 (Additional file 1: Table S1), low moisture and dominance of bacteria (Additional file 2: Figure S1), which is characteristic of arid soils [42].

There was some rainfall at Okaukuejo in the days prior to day 0 followed by little rain until day 18 when heavy rainfalls occurred (day 18–21, Additional file 3: Figure S2).

Detection of B. anthracis

DNA extraction efficiency of bacterial gram-positive cells and/or spores can be poor [43]. To control for extraction efficiency we spiked soil samples with 3.4 X 10^6 *B. anthracis* vaccine strain Sterne 34F2 spores (Onderstepoort Biological Products).

The qPCR showed that soil samples spiked with *B. anthracis* spores (see Additional file 4: for details) had an estimated recovery of 347,183 (Fig. 1a) and 654,419 (Fig. 1d) genomes representing 10.2% and 19.2% DNA extraction efficiency, respectively (equation in Additional file 4). QPCR on the carcass samples revealed that Ca1 had high abundance of *B. anthracis* at day 0 (Fig. 1a), while a similar peak occurred at day 3 for Ca2 (Fig. 1d). At both sites we find low *B. anthracis* abundances in the Ctrl0 samples. The presence of *B. anthracis* in these samples could either be due to spill-over of *B. anthracis* cells between contaminated and uncontaminated soils (see methods), or *B. anthracis* spores were present in those samples before carcass deposition. With our data it is not possible to determine which explanation is correct.

In order to distinguish *B. anthracis* metagenome reads from those of *B. cereus* and *B. thuringiensis*, mapping of reads was performed using the aln algorithm (Burrows-Wheeler Aligner (BWA-aln)) [44] with very strict mapping parameters (Additional file 4: Methods). In addition, we added closely related *Bacillus spp.* reference strains as bait for sequences not unique to *B. anthracis* [45]. (Note that without the usage of closely related reference strains, mapping of metagenomic reads against *B. anthracis* becomes highly unreliable). Such mapping against the genomes of two *B. anthracis* strains, K1 and K2 isolated from the carcasses studied here [46], resulted in a similar pattern as observed in *B. anthracis* specific qPCR experiments (Fig. 1a, d) with abundances peaking at day 0 for Ca1 and at day 3 for Ca2. For both carcasses there are no significant differences between the mapping results when using the K1 (Fig. 1b, e) or K2 (Fig. 1c, f) genomes. There are few reads (2–94) mapping to *B. cereus*, *B. thuringiensis* or *B. subtilis* in the Ctrl0 samples (Fig. 1, Additional file 5: Table S2). The frequencies for these three species peak at day 3 at both carcass sites, but abundances are lower than for *B. anthracis* (Fig. 1b–f). After day 3 the numbers of reads mapping to all *Bacillus* spp. decrease gradually, except for a slight increase in Ca1 at day 30. A close inspection of the mapping process revealed that many reads with low mapping qualities, e.g. poor alignments due to mismatches, were removed in our final filtering step (Additional file 4: Methods). Removal of reads with mismatches did not change the abundance pattern of our time-series for *B.*

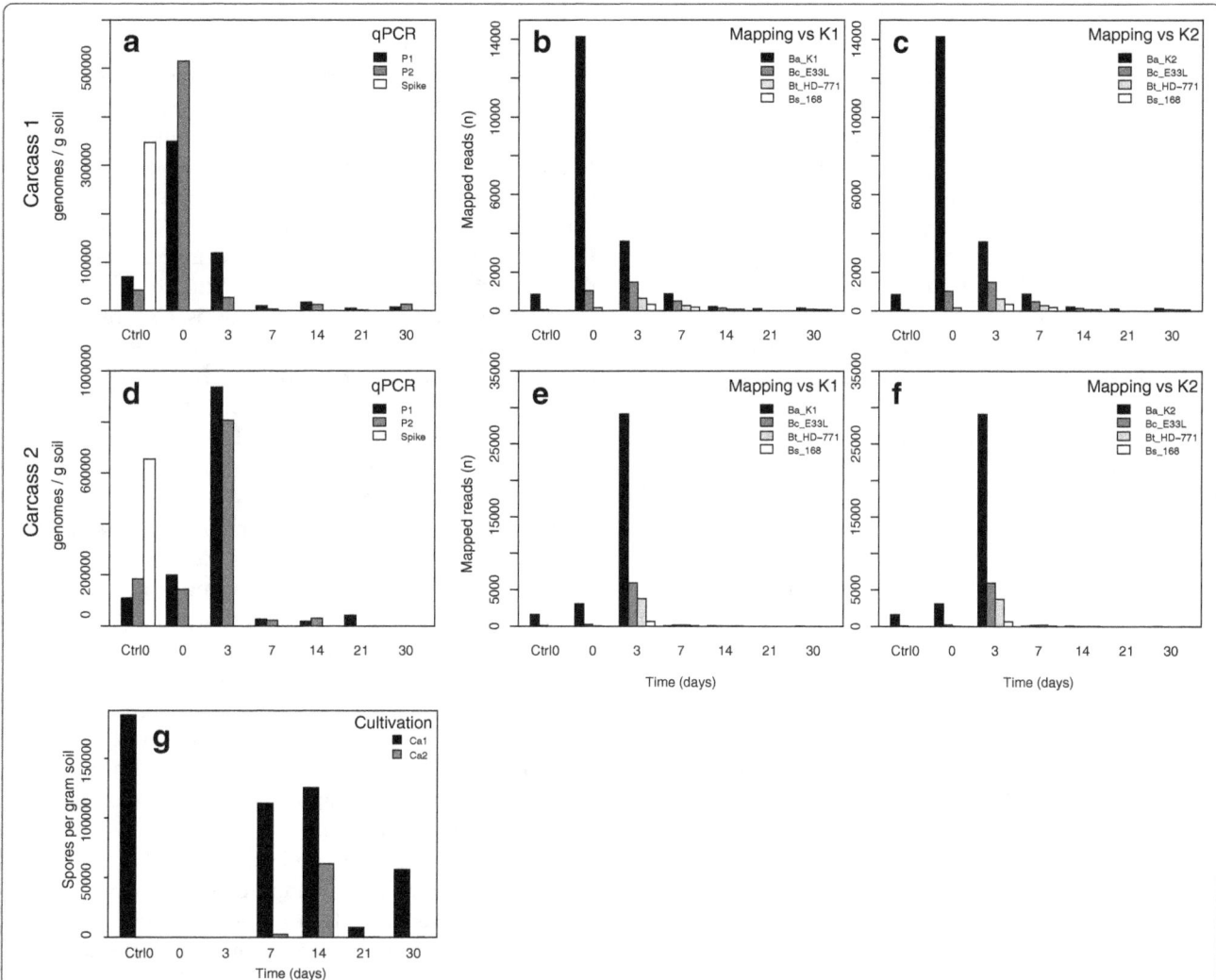

Fig. 1 Estimation of *B. anthracis* abundance in soil samples using qPCR, metagenomic reads mapping and cultivation. For all panels, sampling time-points are on the x-axis. **a** and **d** qPCR results with estimated number of *B. anthracis* genomes per gram soil along the y-axis at different time-points (x-axis) for Carcass 1 (Ca1) (**a**) and Carcass 2 (Ca2) (**d**). The spiked sample is control soil from day 0 with 3.4×10^6 Stern34F2 *B. anthracis* spores added. P1 and P2 represent the two replicates at each time-point. **b**, **c**, **e** and **f** shows the results from mapping the metagenomic reads against the *B. anthracis* isolate genomes (K1 /K2) isolated from Ca1 and Ca2 as well as the other reference strains; *B. cereus* E33L, *B. thuringiensis* HD-771 and *B. subtilis* 168. Reads matching the references are on the y-axis. **b** and **e** and (**c** and **f**) shows the mapping of the metagenomes against the K1 and K2 strain, respectively as well as the other reference strains. Ca1 in panes (**B** and **C**) and Ca2 in panes (**E** and **F**). (**G**) *B. anthracis* spore counts of soil samples from Ca1 and Ca2 in spores per gram dry soil. Spore counts were estimated by culturing of heat-shocked soil samples and adjusted to dry weight of soil

anthracis, while there was a significant change for the other species used (Additional file 6: Figure S3). This difference is likely due to the presence of DNA sequences in the metagenomes related, but not identical to the *B. cereus* group genomes available and used as references.

Culturing from soil samples that had been heated to kill vegetative cells, showed no *B. anthracis* spores present at days 0 and 3 (Fig. 1g) in contrast to the qPCR results. Ca1 had >100,000 spores on day 7 and 14, followed by a sharp drop in spores at day 21, and an increase to >50,000 spores at day 30. Ca2 had >2000 spores on day 7 and >60,000 spores on day 14; at the rest of the time-points Ca2 only had <130 spores. Some of these aberrations from the general trend in counts may be products of sampling biases. In addition, the reduction of *B. anthracis* levels on day 21 (Fig. 1g) could also be due to a response to the rainfall prior to day 21 (Additional file 3: Figure S2).

Temporal changes of GC content and average genome size

Changes in the average % guanine-cytosine (GC) content of metagenomics data indicate changes in microbial community composition [47]. The average GC-content of the time-series metagenomes ranged between 54.6

and 69.0% (Table 1). Interestingly, in both time-series there was a marked drop in average %GC content at day 3 and 7 before it increased again at day 14. This drop was more prevalent in Ca1 than Ca2. This change is due to addition of an extra %GC content peak (≈43%) in the %GC profile, which reduces the average GC-content, and the skewness of the GC-content (Table 1).

The drop in GC content corresponded with a drop in average genome size (AGS) and an increase in genome equivalents (Table 1) [48]. The AGS drop and the shift in average % GC suggests that the carcass soil community composition is changing due to actively growing microbes responding to the carcass nutrients. In addition, the drop in AGS may indicate that the metabolic capacity of the soils changes over the time-series, since AGS can be used as a measure for the metabolic complexity of an ecosystem [49].

Taxonomic classification of shotgun sequences

Many tools exist for the taxonomic classification of metagenomic shotgun sequences. Our aim was to use a tool that captures most of the diversity present in the samples. We therefore compared metaxa2 [50], which classifies rRNA sequences in the metagenomes using a modified Silva SSU /LSU database, with metaBIT (wrapper around Metaphlan2) [51, 52], Kraken [53] and MEGAN [54]. Metaphlan2 and Kraken uses a database derived from microbial whole genome sequences, with the difference that Metaphlan2 only uses signature sequences to identify taxa, instead of whole genomes. MEGAN uses the Non-Redundant protein database from NCBI (NR) for classification.

By comparing the relative abundance of reads classified by each tool we found that the tools differed in the amount of reads classified per sample (Additional file 7: Table S3; Additional file 8: Figure S4) [55]. Interestingly, Kraken showed different temporal dynamics with respect to the change in relative abundance of classified reads compared to the other tools. MetaBIT shows an especially large change between days 0 and 3 compared to the other tools (Additional file 8: Figure S4). We assume that in both cases these results are due to the databases used by metaBIT and Kraken. This is also reflected in the number of prokaryotic taxa identified, with MEGAN detecting a maximum of 180 prokaryotic orders, while the three other tools detected fewer taxa (metaBIT:18, Kraken: 121 and metaxa2: 159). This shows that detection of environmental prokaryotes is significantly influenced by database choice.

The results of the above comparison suggest that metaxa2 is the best available choice among the four tools tested here to describe the bacterial and eukaryotic composition of the soil communities in our study (Additional file 7: Table S3), since the database it relies on is both well curated and contains the widest taxonomic range available. Moreover, MEGAN is a good complement to metaxa2, since it uses most reads for taxonomic and functional classification and can provide diversity measures that support community comparisons (e.g. PCoA, clustering etc).

The metaxa2 data was used to analyse the community diversity using rarefaction and rank-abundance plots. The number of classified reads per sample varied from 19,661 to 110,996 corresponding to about 50% of the identified rRNA sequences (Additional file 2: Figure S1a). The

Table 1 Overview metagenome shotgun samples

Dataset	Time	Raw PE reads	Cleaned PE reads	Cleaned Singleton reads	Average GC-content (%)	GC-content skewness	Average Genome size (Mb)	Genome Equivalents
Carcass 1	Ctrl0[a]	18,736,078	17,119,526	138,893	69.0	−1.52	4.7	1679
	0	18,545,478	18,157,853	189,906	67.7	−1.34	4.2	1975
	3	18,270,652	17,901,288	155,300	60.7	−0.24	3.3	2487
	7	19,794,059	19,333,716	131,415	54.6	−0.08	3.1	2819
	14	18,686,371	18,157,423	227,754	66.4	−0.97	4.0	2034
	21	18,513,371	17,999,309	141,642	67.6	−1.32	4.2	1935
	30	17,526,182	16,999,968	116,608	66.4	−0.96	3.4	2237
Carcass 2	Ctrl0	20,731,918	19,940,019	180,050	68.8	−1.51	4.9	1871
	0	22,832,502	22,263,495	251,377	68.8	−1.67	4.7	2116
	3	20,693,407	20,053,092	167,471	65.2	−0.63	3.8	2386
	7	17,689,524	17,356,265	119,172	64.8	−0.85	4.1	1912
	14	18,078,980	17,079,305	168,140	67.5	−1.46	4.1	1894
	21	17,591,012	17,104,126	148,834	68.7	−1.44	4.8	1615
	30	16,364,557	15,672,082	228,390	68.1	−1.33	4.7	1485

[a]Ctrl0 is the control sample on day 0 without blood

classified reads were dominated by bacteria, with a relative abundance of 70% ± 0.13% and 69% ± 0.1% for Ca1 and Ca2, respectively (Additional file 2: Figure S1b), which stayed constant regardless of time-point. The rRNA reads clustered into 557 OTUs using 97% similarity cut-off. The two Ctrl0 samples have the lowest number of OTUs, while the highest number of OTUs was found in Ca1 day 7 (Ca1_7) (Fig. 2a). The rarefaction curves (Fig. 2a) indicate that we have captured the dominant taxa present, since the curves start levelling off at around 5400 sequences. Nonetheless, the curves do not become horizontal, which indicates that additional low abundant taxa can still be detected with additional sequencing effort. Ca2 had fewer total rRNA reads than Ca1 at all time-points, except the control samples (Additional file 9: Table S4), which results also in fewer OTUs for Ca2 time-points compared to Ca1. An OTU rank abundance plot (Fig. 2b) indicates that Ca1 has higher OTU abundances for the dominant OTUs at all the time-points compared to Ca2. For both carcasses, the soil microbial community is dominated by a few OTUs on days 3 and 7. At the other time-points the evenness of the community is higher as seen in the day 30

samples where almost all OTUs have similar abundances (Fig. 2b). The OTU abundance distribution for the different time-points suggests different soil communities for each of the samples. And indeed, when we analysed community diversity differences using PCoA, based on MEGAN shotgun read classification, we found a clear separation between the communities (Fig. 3). Here metagenomic sequences from both carcasses follow a similar trend, where they are most divergent from the Ctrl0 and day 0 samples at days 3 and 7, before clustering closer to the Ctrl0 and day 0 sample again on days 14 and 21. Ca1 however, has a deviation from the trend on day 30: this sample is close to the Ca1 day 14 sample. The day 30 sample for Ca2 is similar to both Ctrl0 samples.

Temporal dynamics of the soil microbial communities
The 50 most abundant orders from the metaxa2 analysis were extracted and visualised for Ca1 and Ca2 to track fluctuations of the soil community over a 30-day time-period (Fig. 4). For both carcasses there is a clear relative abundance change in part of the microbial community after the influx of body fluids, while

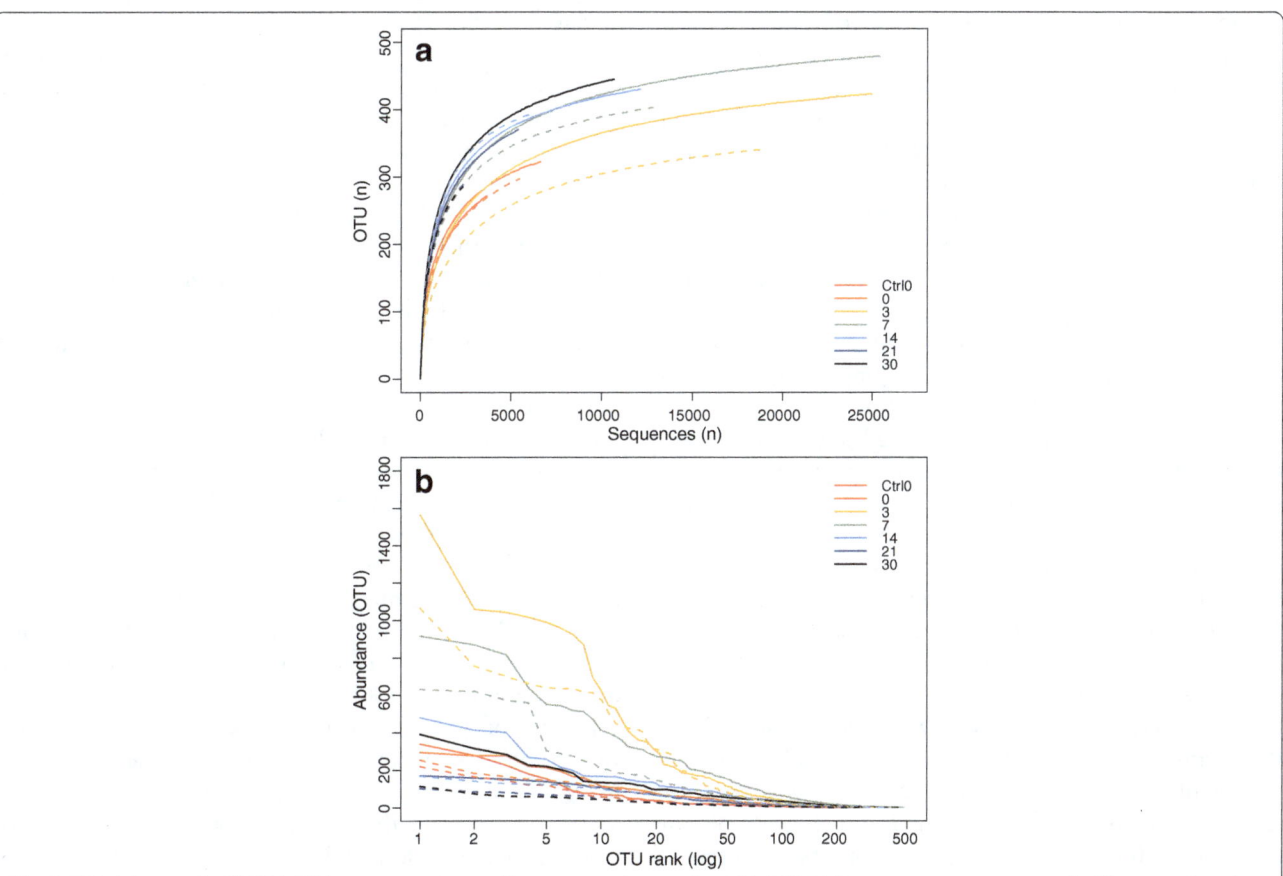

Fig. 2 OTU richness of 16S /18S rRNA sequences from soil metagenomic samples. 16S /18S rRNA sequences were extracted with metaxa2 and diversity was analysed with MetaAmp. **a** Rarefaction curves. X-axis indicates number of rRNA sequences, and y-axis shows OTU numbers **b** Rank abundance of OTUs, where OTU rank is on the x-axis and the abundance per OTU is on the y-axis. The solid lines are Ca1 samples and the dotted lines are Ca2 samples. Line colour indicates time-points as indicated in the legend

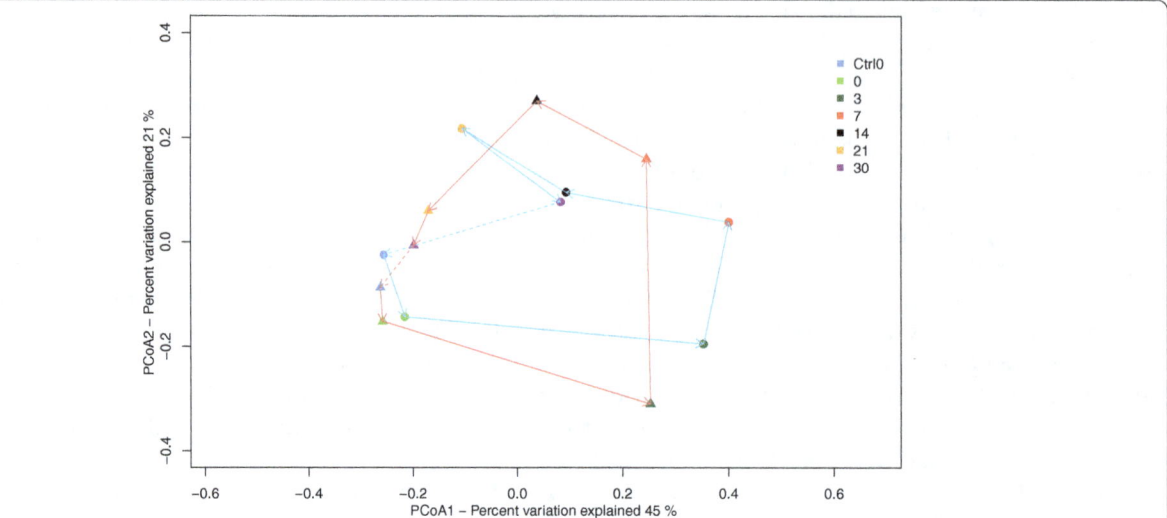

Fig. 3 Soil microbial community relationships by time-point and carcass site. Principal coordinate analysis of a distance matrix created by normalised total counts and using Bray-Curtis dissimilarity. Ca1 is represented by the circles and Ca2 by the triangles, the time-points are visualised in different colours and the arrows are pointing in the direction of increasing days, red arrows for Ca1 and light blue for Ca2. The dotted arrows show the relationship between the day 30 to the Ctrl0 sample

another part of the community does not seem to respond to this influx. Relative abundances may reflect an increase or decrease of a taxon due to changes in the abundance of another taxon, which can be misleading. We therefore show in Fig. 4b the log5 average fold changes between the time points of the raw counts for both carcass communities. This illustrates both the average direction of change over time and the average size of the read abundance change.

Several orders show a sharp abundance increase after day 0 (log5 average fold change >1) at both carcass sites. These include *Bacillales* and *Pseudomonodales* species. In contrast, bacteria belonging to orders such as *Frankiales*, *Rhizobiales* and *Solirubrobacteriales* do not show large changes in their read abundances after the blood/nutrient influx at either carcass site. This is despite the large increase in total reads classified (Additional file 2: Figure S1). Finally, there are also orders that show limited increases. The abundance increase can be early (T = 0 days) and then decline (e.g. *Clostridiales*, *Bacteriodales*), or increase only later ($T \geq 7$ days) in the time-series (e.g. *Burkholderiales*, *Corynebacteriales*).

The orders *Bacillales* and *Pseudomonadales* have the highest abundances at days 3 and 7 (Additional file 10: Figure S5), with genera such as *Acinetobacter*, *Lysinibacillus* and *Kurthia* being dominant (Additional file 9: Table S4). These orders are present at all time-points for both carcasses, but are especially abundant on days 3–14. Many of the genomes in the NCBI database from these genera are small (Size <3 Mbp) and their presence can explain the

drop in average genome size (AGS) that we observed at day 3 (Table 1).

Between days 14 and 21 the relative rRNA read abundance for *Bacillales* drops at Ca1 and Ca2 by 33 and 23%, respectively. For *Pseudomonadales*, we identify a drop of 17 and 13% for Ca1 and Ca2. In total 9 orders show a relative rRNA read abundance drop (>10%) by day 21 in Ca1 and 12 orders in Ca2. At day 30 the abundances of both *Bacillales* and *Pseudomondales* increased again in Ca1, but not in Ca2. Interestingly, for the "stable" orders there was an overall increase in relative rRNA read abundances at day 21 and a subsequent drop at day 30.

The abundance changes observed at days 21 and 30 are likely influenced by heavy rainfall (16–103 mm) recorded at all the weather stations in the ENP in the three days prior to sampling day 21 (Additional file 3: Figure S2). Thus, we assume that the abundance increase at day 30 for many orders could be a response to this heavy rainfall. Moreover, the precipitation could explain the species abundance profile changes since it can result in short-term changes in the microbial community [56].

In addition to bacterial orders that are normal for soils, we also observed representatives of several bacterial orders likely introduced from the zebra. For instance, *Fusobacteriales* are present in the first days for both Ca1 and Ca2, where they are most abundant on day 0 for both samples. They are completely absent on day 14 and 30 in Ca1 and from day 7 onwards in Ca2. The increase in abundance for this order on day 0 is most likely due to the introduction of *Fusobacterium* spp. such as *Fusobacterium equinum*, which

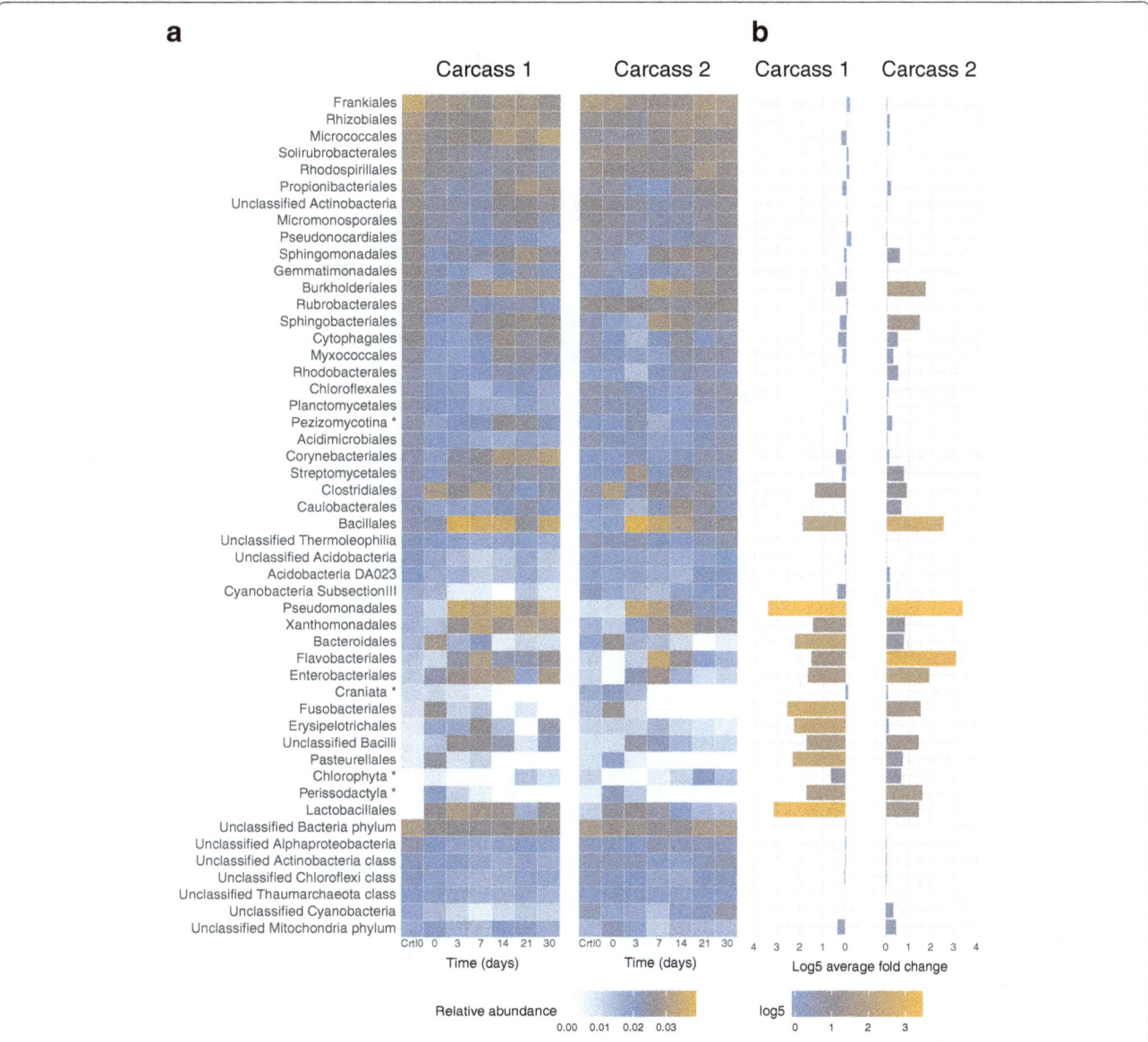

Fig. 4 Temporal dynamics of microbial order abundances at carcass 1 and 2. **a** Heatmap visualisation of the relative abundance for the 50 most abundant orders at each time-point for carcass 1 and 2. **b** Barplots that show the log5 average fold change of the raw reads counts across the time-series for the 50 taxa shown in figure (**a**). The taxa in the heatmaps (**a**) are sorted using the order abundances in the control sample of Carcass1 at day 0 (Control). In order to visualise the relative abundances (**a**) values were log5 normalised and then scaled so that the sum of each column equals 1. Eukaryotic orders are marked with *, the remaining orders are bacterial. The bottom seven entries have not been classified to lower phylogeny than Kingdom, Phylum or Class

is a known inhabitant of the oral cavity and lower respiratory tract of horses [57]. In Ca1 sequences classified to the order *Pasteurellales* are highly abundant on day 0, but completely disappear after day 7. Their abundance at day 0 is mainly caused by *Actinobacillus* spp. and *Pasteurella caballi* (Additional file 9: Table S4). The later species is a commensal of the upper respiratory tract of horses [58].

Microbial community metabolism

The soil metagenomes at the early time-points show a decline in AGS compared to the control sample and the later time-points (Fig. 5). AGS is ecologically informative, as in general, soil bacteria have larger genomes than specialised, opportunistic, parasitic and symbiotic bacteria [59]. Thus AGS change suggests taxonomic and metabolic changes taking place within the microbial community. We therefore correlated KEGG pathway abundances of 112 pathways with AGS to study the temporal change of metabolic potential in the metagenomic time-series. For Ca1 we identified 29 pathways that were significantly correlated with AGS ($p < 0.05$), with 23 being specific to Ca1 (Fig. 5, Additional file 11: Table S5,

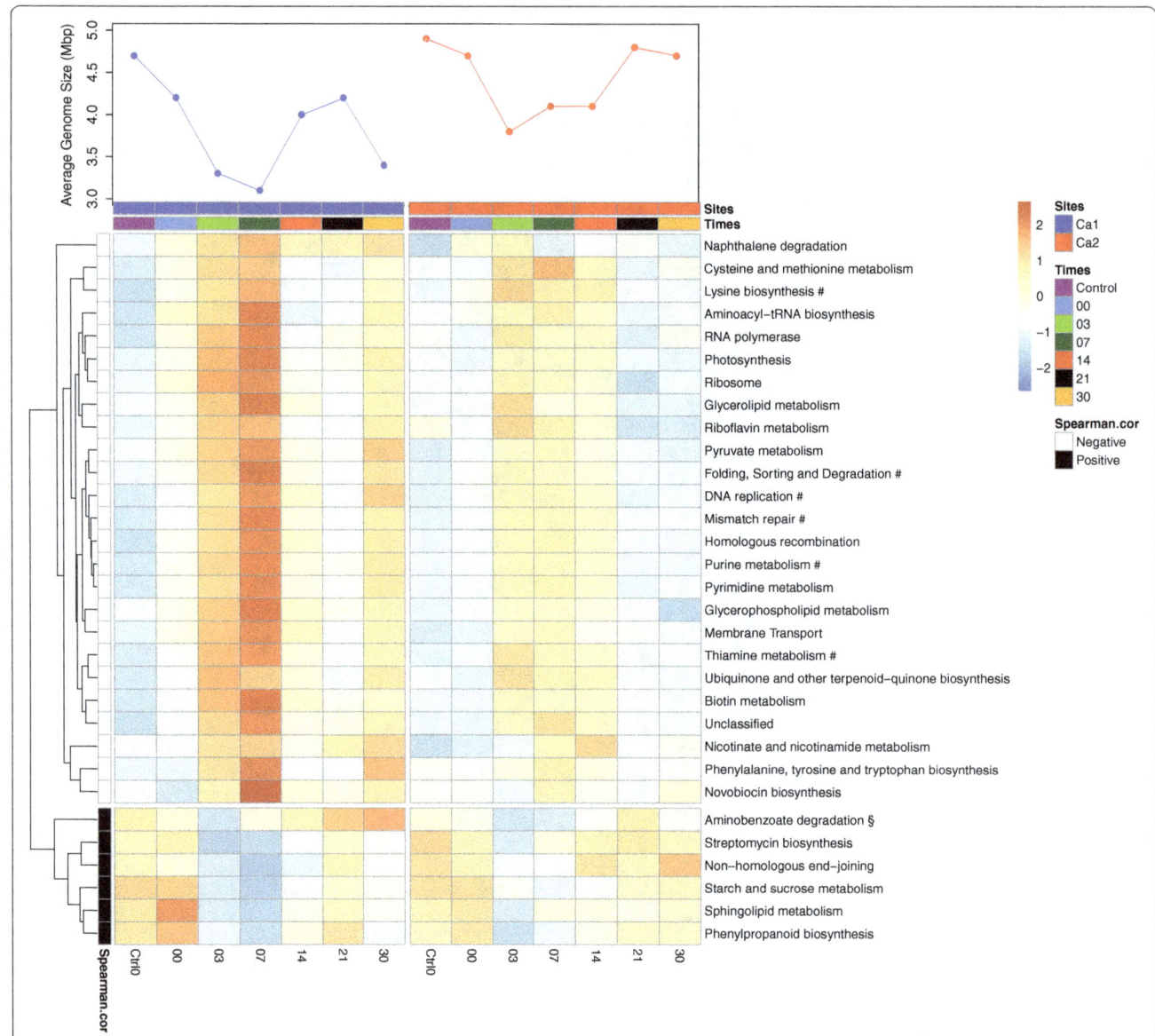

Fig. 5 Heatmap of significantly different metabolic pathways per time-point. KEGG metabolic pathways significantly correlating with AGS for microbial communities of Ca1 and Ca2. Pathways were determined by MEGAN classification. KEGG pathways abundances were normalised with DESEQ2 [96] and correlated to the AGS using the Spearman correlation method with the False Discovery Rate (FDR) test to calculate probabilities (FDR cut-off at 0.05). KEGG-pathways correlating positively or negatively with a *p*-value <0.05 are shown for Ca1 (blue) and Ca2 (red) for all the time-points, the AGS is shown in the top panel. # indicates pathways that are significant in both Ca1 and Ca2, § indicates pathways that are significant in Ca2, and the rest are only significant in Ca1. Pathway abundances were centred and scaled per row and positive and negative correlations were clustered based on the sign of the Spearman correlations

Additional file 12: Table S6). Six pathways showed a negative correlation with AGS for both Ca1 and Ca2. For the Ca2 metagenomes we identified 7 pathways correlated with AGS, with only one pathway (aminobenzoate degradation) positively correlated with a large average genome size in Ca2 but not Ca1. In Ca1 six pathways show a decrease in abundance that is correlated with the decrease of AGS (Fig. 5, top panel). These pathways show an abundance drop at days 3 and 7 and then a subsequent increase. In Ca2 these pathways show a similar pattern

(Additional file 11: Table S5, Additional file 12: Table S6). For the metabolic pathways: novobiocin metabolism (KEGG map00401), streptomycin metabolism (KEGG map00521) and naphthalene degradation (KEGG map00626), the genes identified are not unique for the pathways. For example, no reads were mapped to the genes involved in the final step of novobiocin production and only a few reads to the final step of streptomycin production. The later examples highlight the difficulty in functional annotation of shotgun metagenomic data and indicate the need for cautious interpretation of such results [60, 61].

Discussion

Shotgun metagenomic sequencing, which allows detection of seemingly all species in a community based on the presence of their DNA, is generally considered a superior method to culturing [1, 62, 63]. This is because it addresses "the great plate count anomaly", which claims that only about 1% of the microorganisms seen during microscopy can be cultured [64, 65]. However, DNA-based analyses of microbial communities also have technical pitfalls, particularly when studying spore-forming microorganisms such as *B. anthracis*, which can result in large discrepancies between actual and estimated abundances [43]. For *B. anthracis* this is further confounded by the high levels of DNA sequence identity shared with other members of the *B. cereus* group.

The technical pitfalls of DNA-based analyses are evident from our data, where *B. anthracis* spores could readily be cultured from our samples, but were not as easily detected in metagenomes without using a very strict mapping approach. The onset of *B. anthracis* sporulation occurs within the first 72 h after carcass deposition and can continue up to eight days post-mortem [29, 66]. The qPCR data (Fig. 1) indicates that over time *B. anthracis* abundance peaks before almost completely disappearing from the samples. These results, together with the cultivation data, suggest that the reduced quantities of *B. anthracis* observed in the qPCR experiments are due to reduced DNA extraction efficiency (because of sporulation) [67]. This also suggests that abundances of *B. anthracis* and similar organisms are likely underestimated in the metagenomes at several time-points due to low extraction efficiency. Nonetheless, the metagenomes still captured a large part of the microbial community present in the carcass site soils, and can therefore be used to study the temporal changes taking place.

Why does a carcass promote plant growth?

The data generated reflects the temporal dynamics of the soil microbial community in a natural ecosystem after inoculation of a pathogen (*B. anthracis*), other host-associated microbes and nutrient influx from zebra carcasses. Turner et al. [16] showed that carcass sites in the ENP have higher quality and more abundant vegetation, which could be related to higher levels of phosphate, nitrogen and lower pH than the surrounding soils.

The microbial communities at both carcass sites show a clear response to the influx of bodily fluids, with similar community composition (at order level) of the dominant taxa for the two carcasses with only minor variation in abundances (Fig. 4). Shifts in the community are indicated by several observations: changes in relative abundance of classified reads between the samples (Additional file 3: Figure S2), changes of the AGS of the community studied (Fig. 5) and by variation in OTU diversity and abundance for each of the time-points (Fig. 2). The AGS decreases in the first week of the sample period before increasing again, suggesting a community shift towards copiotrophic bacteria in the first week (Table 1). This is expected as copiotrophic bacteria are better adapted to local increases in nutrient availability than typical soil bacteria and therefore can increase drastically in a short amount of time as observed here [68].

In contrast to the copiotrophic bacteria, the typical soil microbial community consisted of orders like *Frankiales*, *Rhizobiales* and *Solirubrobacterales* (Fig. 4), which remain relatively stable throughout the sample period. This suggests that these bacteria do not need, or are unable to use, the nutrients from the carcass. Moreover, *Rhizobiales* are nitrogen-fixing bacteria living in symbiosis with plants that thrive under nitrogen poor conditions [69]. The extra nitrogen provided by an animal carcass would make them less competitive in soils. *Frankiales* spp. can be plant symbionts, but have also been found to grow on rock beds, which indicate a lifestyle specialized for oligotrophic or extreme conditions [70, 71]. In that respect, it is reasonable that these oligotrophic taxa do not react dramatically to the nutrients of a decomposing animal.

Among the copiotrophic bacteria we see an increase in genera like *Acinetobacter*, *Lysinibacillus* and *Kurthia* in the first week, which is similar to observations in other decomposition studies [12]. These genera belong to the orders *Pseudomonadales* and *Bacillales* and are known to contain species that can be either pathogenic or beneficial for both animals and plants [72–76].

Interestingly, many of the orders reacting quickly to the nutrient influx remain at a relatively high level throughout the sampling period with variations in abundance of genera within orders (Fig. 4). For example, *Xanthomonadales* is abundant at relatively stable levels from day 3 onwards, but the genus *Wohlfahrtiimonas* within this order is only abundant on day 7 at Ca1. *Wohlfahrtiimonas* is a known parasite of different *Wohlfahrtia* spp. (flesh flies) and other such insects. The abundance of *Wohlfahrtiimonas* seen on day 7 for Ca1 may be a result of an increased number of insect hosts appearing through the decomposition process. This example indicates that short-term species composition changes might be due to local conditions in the soil and are not necessarily due to direct competition or sample variation.

Nonetheless, after day 7 other bacterial orders, such as *Burkholderiales*, increase in abundance. This order is typically found in higher abundances in the rhizosphere of plants than in the surrounding soils because they actively feed on plant root exudates [77–79]. Furthermore, *Burkholderiales* are known to have members that are plant growth-promoting bacteria (PGPB), like the *Bacillales* and *Pseudomonadales* genera described above [80–82]. Turner et al. [16] showed that decomposition of carcasses improves soil fertility and enhances plant growth. Our results suggest that this process is promoted by the soil microbial community, which shows a sharp increase in the abundance of orders known to have PGPBs. The promoting functions of PGPBs are many, such as nitrogen fixation, siderophore and phytohormone production, phosphorus solubilisation, and the suppression of plant diseases [79]. *Bacillus anthracis* is an obligate-killer pathogen, able to transmit only by killing its host. Interestingly, by taking advantage of the nutrient- and microbial-driven stimulation of plant growth occurring at carcass sites, transmission of *B. anthracis* is enhanced since this plant growth is an attractive food source for herbivores [16].

Microbial metabolism in arid environments

To get a functional overview of the microbial communities in our study we investigated how metabolic pathways correlate with changes in AGS over time (Fig. 3). Several of the positively correlating pathways can be linked to environmental stressors typical for arid soils, such as desiccation, high temperatures and (UV) radiation. For instance, starch and sucrose (S&S) metabolism is essential for the synthesis of the disaccharides sucrose and trehalose, which can play a role in prokaryotic desiccation resistance [83]. The importance of DNA repair in the surface soil exposed to high radiation levels is reflected in the positive correlation of the non-homologous end joining (NHEJ) repair system, which is also needed for survival during quiescent states such as in the spore stage [84–86].

The presence of other positively correlating pathways can be explained by their role in nutrient acquisition. Arid soils have low organic matter content and most of it comes from decaying plant material [56]. The positive correlation of the aminobenzoate degradation pathway probably reflects its involvement in the degradation of recalcitrant compounds such as lignin.

Many of the pathways negatively correlated with AGS are essential pathways involved in membrane transport, DNA metabolism (transcription, translation, replication, repair and recombination), amino-acid, lipid, carbohydrate and cofactor metabolism (Additional file 13: Figure S6). This reflects the dominance of opportunistic bacteria, especially at day 3–7, with small genomes that lack the arid soil 'specialist' pathways discussed above.

Overall, the analysis of the functional metabolism shows a clear change between the different time-points.

The "normal" microbial soil community metabolism in ENP is represented by species that have genes that are involved in dealing with stress encountered in arid soils. The carcass fluids and nutrients activate many dormant or low abundant copiotrophic lineages present in the soil, which is reflected by the pathways needed for proliferation especially present in the first week of our experiment. These two states of the soil community, dormant vs. active, are typical for arid environments where water input stimulates in a pulse-like manner microbial activity and if sufficient even plant growth [56]. Rainfall gives enough moisture to activate both plant growth and the microbial decomposition compartment of the soil, providing plants with necessary nutrients [56, 87, 88]. However, in contrast to rainfall, animal carcasses not only provide moisture to the soil, but also additional nutrients. This pulse of nutrients and moisture allows the microbial community at very local scales to break down carcass nutrients, laying the foundation for enhanced plant growth directly and in the future [16, 56].

Conclusion

Our work is the first study to use shotgun metagenomics to describe the temporal changes in the soil microbial community after deposition of anthrax carcasses. We observed that the metagenomic approach is not the best way to study *B. anthracis* in a natural setting as it sporulates quickly making detection via DNA extraction difficult. In addition, it is too genetically similar to its close relatives such that most currently available tools to analyse (meta-) genomic sequences are not able to differentiate between such similar organisms. Nevertheless, the metagenomic data allows us to assess how this pathogen influences the microbial soil community.

The metagenomic time-series analysis indicated that carcass introduction stimulated the increase in abundance of part of the microbial soil community. This could play an indirect role in the future transmission of *B. anthracis* by stimulating the growth of taxa known to have PGPBs. The taxonomic shift in the microbial community was accompanied by shifts in the metabolic potential. The typical semi-arid soil community was involved in stress evasion, while the carcass introduction increased pathways involved in proliferation. Nonetheless, the time-series showed that within a month the semi-arid bacterial soil community was similar to the community at $T = 0$ with respect to taxonomic and metabolic composition.

Our work provides a background to study the ecology of spore-forming pathogens in a natural setting using culture-independent methods. It highlights the difficulty of using DNA based approaches to study spore-forming organisms such as *B. anthracis* in arid soils where they are dormant and poorly detectable.

Methods

Carcass and sample information

Sample collection in ENP was done after receiving permission from the Ministry of Environment and Tourism of Namibia (permit number: 1857/2013). Two fresh (likely <24 h since death) plains zebra (*Equus quagga*) anthrax carcasses (Ca1 & Ca2) were found on 03.03.2014 between Sprokieswoud and Charl Marais Dam, 835 m apart at coordinates S19.031/E015.548 (Ca1) and S19.037/E015.553 (Ca2), respectively. Information about ENP, carcass sites and sample collection can be found in the Additional file 4. Soil samples were taken from the area of blood-spill at six time-points (T), starting at T = 0 days (03.03.2014), T = 3, 7, 14, 21, 30 days. A control sample (Ctrl0) was also taken from the grids at T = 0 days, avoiding areas covered in blood (Additional file 8: Figure S4).

DNA isolation and purification

A total of 14 soil samples were collected (7 per carcass). DNA was isolated in two replicates from each sample (P1 and P2) resulting in a total of 14 DNA samples per carcass site. DNA was isolated using the FastDNA® spin kit for soil (MP Biomedicals, Santa Ana, California, USA), following the manufacturer's protocol with adjustments specified in Additional file 4. The DNA samples were filter sterilised using an Ultrafree® Durapore PVDF 0.1 μM spinfilter (Millipore, Darmstadt, Germany).

DNA was shipped dry, following an ethanol precipitation, from ENP to Norway (see Additional file 4: for details). Upon arrival DNA was re-suspended in DES buffer from FastDNA® spin kit for soil. Samples were concentrated and purified using Agencourt® AMPure® XP beads (Beckman Coulter, Beverly, Massachusetts, USA), and treated with a PowerClean® Pro DNA clean-Up Kit, (MO BIO Laboratories, Carlsbad, California, USA) to remove any remaining inhibitors.

Metagenome sequencing and quality control

Environmental DNA samples were sequenced at the Norwegian Sequencing Centre (NSC) using Illumina MiSeq® (Illumina Inc., San Diego, California, USA). A 250 bp paired-end sequencing library (400 bp insert length) was generated using a Regular TruSeq® adapter ligation kit (Illumina Inc., San Diego, California, USA). Fastq files were processed using cut-adapt (v1.8) [89] and prinseq-lite (v0.20.4) [90] (See Additional file 4: for settings). Metagenomic read GC-content was calculated using infoseq (EMBOSS v. 6.5.7 [91]). The infoseq output was used to calculate average GC-content and GC-content skewness (timeData package version: 3012.100) using R-Studio (v3.3.0).

The average genome size (AGS) of each metagenome was estimated using MicrobeCensus on clean paired-end files (Additional file 4: Methods). The likelihood of sampling a universal single copy is inversely correlated with the AGS of a metagenomes [48, 92]. Thus universal single copy abundance will show significant differences when AGS between metagenomes differ.

Bacillus anthracis quantification from soils

Two methods were used to quantify *B. anthracis* in the soil samples; serial dilution culturing and quantitative PCR (qPCR) (see Additional file 4: for details).

The genome coverages of the *B. anthracis* isolate K1 and K2 genomes (Accession numbers: LBBZ00000000 and LBCA00000000 [46]) isolated from the Ca1 and Ca2 sites respectively, were investigated by mapping metagenomic sequences using the Burrows-Wheeler Aligner (BWA v0.7.8) [44]. The genomes of *B. cereus* E33L, *B. thuringiensis* HD-771 and *Bacillus subtilis* 168 were also used as references (see Additional file 4: for details).

Taxonomic and functional classification of metagenomic reads

Taxonomic composition of the metagenomes was determined using metaxa2 (v2.0.1) which extracts SSU rRNA sequences [50], MEGAN (v5.10.15) [54], Kraken (0.10.5-beta) [53] and metaBIT [51]. For the details on the comparison of the taxonomic classifiers see Additional file 4. The metaxa2 taxonomic profiles were visualised with a heatmap using ggplot2 in R-studio (see Additional file 4: for settings). Metaxa2 extracted SSU reads were run through the MetaAmp 1.1 pipeline to obtain operational taxonomical unit (OTU) abundances.

Shotgun sequences from each dataset were compared to the NCBI nr database (accessed on 20.01.2016) using Diamond (v0.7.11) [93] and the output was used for taxonomic and functional classification in MEGAN (v5.10.15) [54]. Temporal species composition dynamics of metagenomic samples was examined by principle coordinate analysis (PCoA) in R-Studio (v3.3.0). Changes in the metabolic potential of the community over time were analysed by correlation of KEGG KO-terms and pathways [94, 95] abundances with AGS size (For details see Additional file 4).

Additional files

Additional file 1: Table S1. Soil nutrients and soil composition at carcass sites.

Additional file 2: Figure S1. Abundance and relative abundance of 16S/18S rRNA taxonomically classified into kingdoms. (a) Raw abundance and (b) relative abundance of metaxa2 classified reads. These are the 16S/18S reads extracted and classified by metaxa2. Roughly 50% of the extracted rRNA sequenced did not get assigned to any taxonomy and were assumed to be false positives.

Additional file 3: Figure S2. Rainfall at the Okaukuejo field station. The precipitation is measured in mm at Okaukuejo, which is approximately 40 km from the sampling site. Because of the very local rainfalls in the area, this can only be used as an indication of rainfall at our sample sites. The graph shows rainfall from the week before sampling started and to the end of the sampling period, the 2nd of April 2014. The blue arrow indicates the sampling period with sampling days marked in red.

Additional file 4: A text document contained supplementary methods, equations, results and figure (mentioned only in the supplementary information).

Additional file 5: Table S2. Counts of mapped metagenomic reads of each sample against reference genomes.

Additional file 6: Figure S3. Estimation of *B. anthracis* abundance in soil samples using qPCR, metagenomic reads mapping and cultivation. For all panels, sampling time-points are on the x-axis. (A and D) qPCR results with estimated number of *B. anthracis* genomes per gram soil along the y-axis at different time-points (x-axis) for Carcass 1 (Ca1) (A) and Carcass 2 (Ca2) (D). The spiked sample is control soil from day 0 with 3.4×10^6 Stern34F2 *B. anthracis* spores added. P1 and P2 represent the two replicates at each time-point. (B, C, E and F) shows the results from mapping the metagenomic reads against the *B. anthracis* isolate genomes (K1/K2) isolated from Ca1 and Ca2 as well as the other reference strains; *B. cereus E33L*, *B. thuringiensis HD-771* and *B. subtilis 168*. Reads with mismatches were not filtered away. Reads matching the references are on the y-axis. (B and E) and (C and F) shows the mapping of the metagenomes against the K1 and K2 strain, respectively as well as the other reference strains. Ca1 in panes (B and C) and Ca2 in panes (E and F). (G) *B. anthracis* spore counts of soil samples from Ca1 and Ca2 in spores per gram dry soil. Spore counts were estimated by culturing of heat-shocked soil samples and adjusted to dry weight of soil.

Additional file 7: Table S3. Comparison of taxonomic classification with four different methods: MetaBIT, Metaxa2, MEGAN, and Kraken.

Additional file 8: Figure S4. Comparison of the percentage of classified reads by four metagenomic classification tools. Note the very different scales on the y-axis. Barplots show the relative abundances of classified reads for each time point metagenome of Ca1 (blue) and Ca 2 (orange). The tools are Metaxa2 (A), metaBIT (B), MEGAN (C), Kraken (D).

Additional file 9: Table S4. Metaxa2 taxonomic classifications on pair1 16S /18S rRNA sequences.

Additional file 10: Figure S5. Soil microbial community relationships by time-point and carcass site. Principal coordinate analysis of a distance matrix created by normalised total counts and using Bray-Curtis dissimilarity. Ca1 is represented by the circles and Ca2 by the triangles, the time-points are visualised in different colours and the arrows are pointing in the direction of increasing days, red arrows for Ca1 and light blue for Ca2. The dotted arrows show the relationship between the day 30 to the Ctrl0 sample. The black arrows indicate the 10 most abundant orders and their effect on the soil microbial communities per time-point. Arrows were fitted using envfit with 1000 permutations using the vegan package in R-studio.

Additional file 11: Table S5. Overview of the Spearman correlation coefficients of all identified KEGG pathways and their FDR corrected *p*-values.

Additional file 12: Table S6. KEGG KO-term abundances for significantly correlating KEGG pathways. Abundances are normalized using DESEQ2 and sorted on carcass 1, time point $T = 0$.

Additional file 13: Figure S6. Sample area. (**a**) Ca1, soil samples were taken from within the 30×30 cm grid, (**b**) Ca2, samples were taken from within the red square (resembling the 30×30 cm metal grid shown in (**a**)).

Acknowledgements

We thank Namibia's Ministry of Environment and Tourism for permission to conduct this research (permit number: 1857/2013). In addition, we thank Dr. Mari Espelund for help optimizing DNA extractions, Claudine C. Cloete and Zoe Barandongo for help with sample collecting and DNA extractions, the Etosha Ecological Institute and the Norwegian Defence Research Institute for providing laboratory facilities and the Department of Biosciences student funding at University of Oslo for travel money. We also thank Eric de Muinck and Pål Trosvik for critical feedback on our manuscript.

Funding

This work was supported by the Research Council of Norway (Project No. 180444/V40) awarded to CLN and NSF OISE-1103054 awarded to WCT.

Authors' contributions

All authors contributed to the design of the experiment. KV, WRE, WCT conducted the environmental sampling and qPCR experiments, bacterial cultivation and DNA extractions. KV prepared samples for metagenomic sequencing and performed bioinformatic quality control. KV, CLN and THA executed bioinformatics analysis of shotgun data. KV prepared the manuscript. All authors contributed to critical revision of the manuscript. All authors read and approved the final manuscript.

Competing interests

The authors declare that they have no competing interests.

Author details

[1]Department of Biosciences, Centre for Ecological and Evolutionary Synthesis (CEES), University of Oslo, The Kristine Bonnevie Building, UiO, campus Blindern, Blindern, Oslo, Norway. [2]Norwegian Defence Research Establishment, Kjeller, Norway. [3]Department of Biological Sciences, University of Alberta, Edmonton, Alberta, Canada. [4]Department of Biological Sciences, University at Albany, State University of New York, Albany, New York, USA.

References

1. Fierer N, Leff JW, Adams BJ, Nielsen UN, Bates ST, Lauber CL, Owens S, Gilbert JA, Wall DH, Caporaso JG. Cross-biome metagenomic analyses of soil microbial communities and their functional attributes. Proc Natl Acad Sci U S A. 2012; 109(52):21390–5.
2. O'Brien SL, Gibbons SM, Owens SM, Hampton-Marcell J, Johnston ER, Jastrow JD, Gilbert JA, Meyer F, Antonopoulos DA. Spatial scale drives patterns in soil bacterial diversity. Environ Microbiol. 2016;18(6):2039–51.

3.   Rousk J, Bååth E, Brookes PC, Lauber CL, Lozupone C, Caporaso JG, Knight R, Fierer N. Soil bacterial and fungal communities across a pH gradient in an arable soil. ISME J. 2010;4(10):1340–51.

4.   Bell CW, Tissue DT, Loik ME, Wallenstein MD. Acosta - Martinez V, Erickson RA, Zak JC: soil microbial and nutrient responses to 7 years of seasonally altered precipitation in a Chihuahuan Desert grassland. Glob Chang Biol. 2014;20(5):1657–73.

5.   Johnson SL, Kuske CR, Carney TD, Housman DC, Gallegos-Graves LV, Belnap J. Increased temperature and altered summer precipitation have differential effects on biological soil crusts in a dryland ecosystem. Glob Chang Biol. 2012;18(8):2583–93.

6.   Garcia-Pichel F, Loza V, Marusenko Y, Mateo P, Potrafka RM. Temperature drives the continental-scale distribution of key microbes in topsoil communities. Science. 2013;340(6140):1574–7.

7.   Niu F, He J, Zhang G, Liu X, Liu W, Dong M, Wu F, Liu Y, Ma X, An L, et al. Effects of enhanced UV-B radiation on the diversity and activity of soil microorganism of alpine meadow ecosystem in Qinghai–Tibet plateau. Ecotoxicology. 2014;23(10):1833–41.

8.   Bell CW, Acosta-Martinez V, McIntyre NE, Cox S, Tissue DT, Zak JC. Linking microbial community structure and function to seasonal differences in soil moisture and temperature in a Chihuahuan desert grassland. Microb Ecol. 2009;58(4):827–42.

9.   Bi J, Zhang NL, Liang Y, Yang HJ, Ma KP. Interactive effects of water and nitrogen addition on soil microbial communities in a semiarid steppe. J Plant Ecol. 2012;5(3):320–9.

10. Rath KM, Maheshwari A, Rousk J. The impact of salinity on the microbial response to drying and rewetting in soil. Soil Biol Biochem. 2017;108:17–26.

11. Van Horn DJ, Okie JG, Buelow HN, Gooseff MN, Barrett JE, Takacs-Vesbach CD. Soil microbial responses to increased moisture and organic resources along a salinity gradient in a polar desert. Appl Environ Microbiol. 2014; 80(10):3034–43.

12. Howard GT, Duos B, Watson-Horzelski EJ. Characterization of the soil microbial community associated with the decomposition of a swine carcass. Int Biodeterior Biodegradation. 2010;64(4):300–4.

13. Cobaugh KL, Schaeffer SM, DeBruyn JM. Functional and structural succession of soil microbial communities below decomposing human cadavers. PLoS One. 2015;10(6):e0130201.

14. Weiss S, Carter DO, Metcalf JL, Knight R. Carcass mass has little influence on the structure of gravesoil microbial communities. Int J Legal Med. 2016; 130(1):253–63.

15. Bump JK, Peterson RO, Vucetich JA. Wolves modulate soil nutrient heterogeneity and foliar nitrogen by configuring the distribution of ungulate carcasses. Ecology. 2009;90(11):3159–67.

16. Turner WC, Kausrud KL, Krishnappa YS, Cromsigt JP, Ganz HH, Mapaure I, Cloete CC, Havarua Z, Küsters M, Getz WM. Fatal attraction: vegetation responses to nutrient inputs attract herbivores to infectious anthrax carcass sites. Proc R Soc Lond B Biol Sci. 2014;281(1795):20141785.

17. Sharp RJ, Roberts AG. Anthrax: the challenges for decontamination. J Chem Technol Biotechnol. 2006;81(10):1612–25.

18. Liu Y, Lai Q, Göker M, Meier-Kolthoff JP, Wang M, Sun Y, Wang L, Shao Z. Genomic insights into the taxonomic status of the Bacillus Cereus group. Sci Rep. 2015;5:14082.

19. Jensen GB, Hansen BM, Eilenberg J, Mahillon J. The hidden lifestyles of Bacillus cereus and relatives. Environ Microbiol. 2003;5(8):631–40.

20. Helgason E, Økstad OA, Caugant DA, Johansen HA, Fouet A, Mock M, Hegna I, Kolstø A-B. Bacillus anthracis, Bacillus cereus, and Bacillus thuringiensis—one species on the basis of genetic evidence. Appl Environ Microbiol. 2000;66(6):2627–30.

21. Zwick ME, Joseph SJ, Didelot X, Chen PE, Bishop-Lilly KA, Stewart AC, Willner K, Nolan N, Lentz S, Thomason MK. Genomic characterization of the Bacillus cereus sensu lato species: backdrop to the evolution of Bacillus anthracis. Genome Res. 2012;22(8):1512–24.

22. Mignot T, Mock M, Robichon D, Landier A, Lereclus D, Fouet A. The incompatibility between the PlcR-and AtxA-controlled regulons may have selected a nonsense mutation in Bacillus anthracis. Mol Microbiol. 2001; 42(5):1189–98.

23. Easterday WR, Ert MN, Simonson TS, Wagner DM, Kenefic LJ, Allender CJ, Keim P. Use of single nucleotide polymorphisms in the plcR gene for specific identification of Bacillus anthracis. J Clin Microbiol. 2005;43:1995–7.

24. Green BD, Battisti L, Koehler TM, Thorne CB, Ivins BE. Demonstration of a capsule plasmid in Bacillus anthracis. Infect Immun. 1985;49(2):291–7.

25. Mikesell P, Ivins BE, Ristroph JD, Dreier TM. Evidence for plasmid-mediated toxin production in Bacillus anthracis. Infect Immun. 1983; 39(1):371–6.

26. Toby IT, Widmer J, Dyer DW. Divergence of protein-coding capacity and regulation in the Bacillus cereus sensu lato group. BMC Bioinformatics. 2014; 15(11):1.

27. Turner WC, Kausrud KL, Beyer W, Easterday WR, Barandongo ZR, Blaschke E, Cloete CC, Lazak J, Van Ert MN, Ganz HH. Lethal exposure: an integrated approach to pathogen transmission via environmental reservoirs. Sci Rep. 2016;6:27311.

28. Dragon DC, Bader DE, Mitchell J, Woollen N. Natural dissemination of Bacillus anthracis spores in northern Canada. Appl Environ Microbiol. 2005; 71(3):1610–5.

29. Bellan SE, Turnbull PC, Beyer W, Getz WM. Effects of experimental exclusion of scavengers from carcasses of anthrax-infected herbivores on Bacillus anthracis sporulation, survival, and distribution. Appl Environ Microbiol. 2013;79(12):3756–61.

30. Ganz HH, Turner WC, Brodie EL, Kusters M, Shi Y, Sibanda H, Torok T, Getz WM. Interactions between Bacillus anthracis and plants may promote anthrax transmission. PLoS Negl Trop Dis. 2014;8(6):e2903.

31. Schuch R, Fischetti VA. The secret life of the anthrax agent Bacillus anthracis: bacteriophage-mediated ecological adaptations. PLoS One. 2009;4(8):e6532.

32. Saile E, Koehler TM. Bacillus anthracis multiplication, persistence, and genetic exchange in the rhizosphere of grass plants. Appl Environ Microbiol. 2006; 72(5):3168–74.

33. Dey R, Hoffman PS, Glomski IJ. Germination and amplification of anthrax spores by soil-dwelling amoebas. Appl Environ Microbiol. 2012; 78(22):8075–81.

34. Logares R, Haverkamp TH, Kumar S, Lanzén A, Nederbragt AJ, Quince C, Kauserud H. Environmental microbiology through the lens of high-throughput DNA sequencing: synopsis of current platforms and bioinformatics approaches. J Microbiol Methods. 2012;91(1):106–13.

35. Be NA, Thissen JB, Gardner SN, McLoughlin KS, Fofanov VY, Koshinsky H, Ellingson SR, Brettin TS, Jackson PJ, Jaing CJ. Detection of bacillus anthracis DNA in complex soil and air samples using next-generation sequencing. PLoS One. 2013;8(9):e73455.

36. Dragon DC, Rennie RP. Evaluation of spore extraction and purification methods for selective recovery of viable Bacillus anthracis spores. Lett Appl Microbiol. 2001;33(2):100–5.

37. Le Roux C, Grunow J, Morris J, Bredenkamp G, Scheepers J. A classification of the vegetation of the Etosha national park. S Afr J Bot. 1988;54(1):1–10.

38. Engert S. Spatial variability and temporal periodicity of rainfall in the Etosha National Park and surrounding areas in northern Namibia. Modoqua. 1997; 20(1):115–20.

39. Lauber CL, Metcalf JL, Keepers K, Ackermann G, Carter DO, Knight R. Vertebrate decomposition is accelerated by soil microbes. Appl Environ Microbiol. 2014;80(16):4920–9.

40. Bellan SE, Gimenez O, Choquet R, Getz WM. A hierarchical distance sampling approach to estimating mortality rates from opportunistic carcass surveillance data. Methods Ecol Evol. 2013;4(4):361–9.

41. Hyde ER, Haarmann DP, Petrosino JF, Lynne AM, Bucheli SR. Initial insights into bacterial succession during human decomposition. Int J Legal Med. 2015;129(3):661–71.

42. Lauber CL, Hamady M, Knight R, Fierer N. Pyrosequencing-based assessment of soil pH as a predictor of soil bacterial community structure at the continental scale. Appl Environ Microbiol. 2009;75(15):5111–20.

43. Albertsen M, Karst SM, Ziegler AS, Kirkegaard RH, Nielsen PH. Back to basics—the influence of DNA extraction and primer choice on phylogenetic analysis of activated sludge communities. PLoS One. 2015;10(7):e0132783.

44. Li H, Durbin R. Fast and accurate long-read alignment with burrows-wheeler transform. Bioinformatics. 2010;26(5):589–95.

45. Rasmussen S, Allentoft ME, Nielsen K, Orlando L, Sikora M, Sjögren K-G, Pedersen AG, Schubert M, Van Dam A, Kapel CMO. Early divergent strains of Yersinia pestis in Eurasia 5,000 years ago. Cell. 2015;163(3):571–82.

46. Valseth K, Nesbø CL, Easterday WR, Turner WC, Olsen JS, Stenseth NC, Haverkamp THA. Draft genome sequences of two bacillus anthracis strains from Etosha National Park, Namibia. Genome Announc. 2016;4(4):e00861-16.

47. Reichenberger ER, Rosen G, Hershberg U, Hershberg R. Prokaryotic nucleotide composition is shaped by both phylogeny and the environment. Genome Biol Evol. 2015;7(5):1380–9.

48. Nayfach S, Pollard KS. Average genome size estimation improves comparative metagenomics and sheds light on the functional ecology of the human microbiome. Genome Biol. 2015;16(1):51.

49. Raes J, Korbel JO, Lercher MJ, von Mering C, Bork P. Prediction of effective genome size in metagenomic samples. Genome Biol. 2007;8(1):R10.

50. Bengtsson-Palme J, Hartmann M, Eriksson KM, Pal C, Thorell K, Larsson DGJ, Nilsson RH. Metaxa2: improved identification and taxonomic classification of small and large subunit rRNA in metagenomic data. Mol Ecol Resour. 2015; 15(6):1403–14.

51. Louvel G, Der Sarkissian C, Hanghøj K, Orlando L. metaBIT, an integrative and automated metagenomic pipeline for analysing microbial profiles from high-throughput sequencing shotgun data. Mol Ecol Resour. 2016;16(6): 1415–27.

52. Truong DT, Franzosa EA, Tickle TL, Scholz M, Weingart G, Pasolli E, Tett A, Huttenhower C, Segata N. MetaPhlAn2 for enhanced metagenomic taxonomic profiling. Nat Methods. 2015;12(10):902–3.

53. Wood DE, Salzberg SL. Kraken: ultrafast metagenomic sequence classification using exact alignments. Genome Biol. 2014;15(3):R46.

54. Huson DH, Auch AF, Qi J, Schuster SC. MEGAN analysis of metagenomic data. Genome Res. 2007;17(3):377–86.

55. Lindgreen S, Adair KL, Gardner PP. An evaluation of the accuracy and speed of metagenome analysis tools. Sci Rep. 2016;6:19233.

56. Collins SL, Sinsabaugh RL, Crenshaw C, Green L, Porras-Alfaro A, Stursova M, Zeglin LH. Pulse dynamics and microbial processes in aridland ecosystems. J Ecol. 2008;96(3):413–20.

57. Dorsch M, Lovet D, Bailey GD. Fusobacterium equinum sp. nov., from the oral cavity of horses. Int J Syst Evol Microbiol. 2001;51(6):1959–63.

58. Schlater LK, Brenner D, Steigerwalt A, Moss CW, Lambert M, Packer R. Pasteurella caballi, a new species from equine clinical specimens. J Clin Microbiol. 1989;27(10):2169–74.

59. Konstantinidis KT, Tiedje JM. Trends between gene content and genome size in prokaryotic species with larger genomes. Proc Natl Acad Sci U S A. 2004;101(9):3160–5.

60. Prosser JI. Dispersing misconceptions and identifying opportunities for the use of 'omics' in soil microbial ecology. Nat Rev Microbiol. 2015; 13(7):439–46.

61. Teeling H, Glöckner FO. Current opportunities and challenges in microbial metagenome analysis–a bioinformatic perspective. Brief Bioinform. 2012; 13(6):728–42.

62. Fierer N, Bradford MA, Jackson RB. Towards an ecological classification of soil bacteria. Ecology. 2007;88(6):1354–64.

63. Tan J, Zuniga C, Zengler K. Unraveling interactions in microbial communities - from co-cultures to microbiomes. J Microbiol. 2015;53(5):295–305.

64. Staley JT, Konopka A. Measurement of in situ activities of nonphotosynthetic microorganisms in aquatic and terrestrial habitats. Annu Rev Microbiol. 1985;39(1):321–46.

65. Torsvik V, Øvreås L. Microbial diversity and function in soil: from genes to ecosystems. Curr Opin Microbiol. 2002;5(3):240–5.

66. Lindeque PM, Turnbull PCB. Ecology and epidemiology of anthrax in Etosha National Park, Namibia. Onderstepoort J Vet Res. 1994;61(1):71–83.

67. Sedlackova V, Dziedzinska R, Babak V, Kralik P. The detection and quantification of Bacillus thuringiensis spores from soil and swabs using quantitative PCR as a model system for routine diagnostics of Bacillus anthracis. J Appl Microbiol. 2017; https://doi.org/10.1111/jam.13445.

68. Ketola T, Mikonranta L, Laakso J, Mappes J. Different food sources elicit fast changes to bacterial virulence. Biol Lett. 2016;12(1):20150660.

69. Erlacher A, Cernava T, Cardinale M, Soh J, Sensen CW, Grube M, Berg G. Rhizobiales as functional and endosymbiontic members in the lichen symbiosis of Lobaria pulmonaria L. Front Microbiol. 2015;6:53.

70. Goodfellow M, Kämpfer P, Busse H-J, Trujillo ME, Suzuki K-I, Ludwig W, Whitman WB. Bergey's manual of systematic bacteriology : Volume 5: The Actinobacteria. In: vol. 5, 2 edn. New York: Springer Science & Business Media; 2012: 509–546.

71. Foesel B, Geppert A, Rohde M, Overmann J. Parviterribacter kavangonensis and Parviterribacter multiflagellatus a novel genus and two novel species within the order Solirubrobacterales and emended description of the classes Thermoleophilia and Rubrobacteria and its orders and families. Int J Syst Evol Microbiol. 2016;66:652–65.

72. Weber BS, Harding CM, Feldman MF. Pathogenic Acinetobacter: from the cell surface to infinity and beyond. J Bacteriol. 2015;198(6):880–7.

73. Berry C. The bacterium, Lysinibacillus sphaericus, as an insect pathogen. J Invertebr Pathol. 2012;109(1):1–10.

74. Vos P, Garrity G, Jones D, Krieg NR, Ludwig W, Rainey FA, Schleifer K-H, Whitman W. Bergey's manual of systematic bacteriology: Volume 3: The Firmicutes. In: vol. 3, 2 edn. New York: Springer Science & Business Media; 2011: 364–370.

75. Andrade LF, de Souza G, Nietsche S, Xavier AA, Costa MR, Cardoso AMS, Pereira MCT, Pereira D. Analysis of the abilities of endophytic bacteria associated with banana tree roots to promote plant growth. J Microbiol. 2014;52(1):27–34.

76. Batool F, Rehman Y, Hasnain S. Phylloplane associated plant bacteria of commercially superior wheat varieties exhibit superior plant growth promoting abilities. Front Life Sci. 2016;9(4):313–22.

77. Uroz S, Buée M, Murat C, Frey-Klett P, Martin F. Pyrosequencing reveals a contrasted bacterial diversity between oak rhizosphere and surrounding soil. Environ Microbiol Rep. 2010;2(2):281–8.

78. Vandenkoornhuyse P, Mahé S, Ineson P, Staddon P, Ostle N, Cliquet JB, Francez AJ, Fitter AH, Young JP. Active root-inhabiting microbes identified by rapid incorporation of plant-derived carbon into RNA. Proc Natl Acad Sci U S A. 2007;104(43):16970–5.

79. Bulgarelli D, Schlaeppi K, Spaepen S, Ver Loren van Themaat E, Schulze-Lefert P. Structure and functions of the bacterial microbiota of plants. Annu Rev Plant Biol. 2013;64:807–38.

80. Baldani JI, Rouws L, Cruz LM, Olivares FL, Schmid M, Hartmann A. The family Oxalobacteraceae. In: The prokaryotes: Alphaproteobacteria and Betaproteobacteria. Vol. 9783642301971; 2014. p. 919–74.

81. Pieterse CM, Zamioudis C, Berendsen RL, Weller DM, Van Wees SC, Bakker PA. Induced systemic resistance by beneficial microbes. Annu Rev Phytopathol. 2014;52:347–75.

82. Choudhary DK, Johri BN. Interactions of Bacillus spp. and plants–with special reference to induced systemic resistance (ISR). Microbiol Res. 2009;164(5): 493–513.

83. Potts M. Desiccation tolerance of prokaryotes. Microbiol Rev. 1994;58(4): 755–805.

84. Shuman S, Glickman MS. Bacterial DNA repair by non-homologous end joining. Nat Rev Microbiol. 2007;5(11):852–61.

85. Moeller R, Stackebrandt E, Reitz G, Berger T, Rettberg P, Doherty AJ, Horneck G, Nicholson WL. Role of DNA repair by nonhomologous-end joining in Bacillus subtilis spore resistance to extreme dryness, mono- and polychromatic UV, and ionizing radiation. J Bacteriol. 2007;189(8): 3306–11.

86. Garcia-Gonzalez A, Vicens L, Alicea M, Massey S. The distribution of recombination repair genes is linked to information content in bacteria. Gene. 2013;528(2):295–303.

87. Du Plessis W. Effective rainfall defined using measurements of grass growth in the Etosha National Park, Namibia. J Arid Environ. 2001;48(3): 397–417.

88. Clark JS, Campbell JH, Grizzle H, Acosta-Martìnez V, Zak JC. Soil microbial community response to drought and precipitation variability in the Chihuahuan Desert. Microb Ecol. 2009;57(2):248–60.

89. Martin M. Cutadapt removes adapter sequences from high-throughput sequencing reads. EMBnet journal. 2011;17(1):10–2.

90. Schmieder R, Edwards R. Quality control and preprocessing of metagenomic datasets. Bioinformatics. 2011;27(6):863–4.

91. Rice P, Longden I, Bleasby A. EMBOSS: the European molecular biology open software suite. Trends Genet. 2000;16(6):276–7.

92. Beszteri B, Temperton B, Frickenhaus S, Giovannoni SJ. Average genome size: a potential source of bias in comparative metagenomics. ISME J. 2010; 4(8):1075–7.

93. Buchfink B, Xie C, Huson DH. Fast and sensitive protein alignment using DIAMOND. Nat Methods. 2015;12(1):59–60.

94. Kanehisa M, Goto S. KEGG: kyoto encyclopedia of genes and genomes. Nucleic Acids Res. 2000;28(1):27–30.

95. Kanehisa M, Sato Y, Kawashima M, Furumichi M, Tanabe M. KEGG as a reference resource for gene and protein annotation. Nucleic Acids Res. 2016;44(D1):D457–62.

96. Love MI, Huber W, Anders S. Moderated estimation of fold change and dispersion for RNA-seq data with DESeq2. Genome Biol. 2014;15(12):550.

PERMISSIONS

The contributors of this book come from diverse backgrounds, making this book a truly international effort. This book will bring forth new frontiers with its revolutionizing research information and detailed analysis of the nascent developments around the world.

We would like to thank all the contributing authors for lending their expertise to make the book truly unique. They have played a crucial role in the development of this book. Without their invaluable contributions this book wouldn't have been possible. They have made vital efforts to compile up to date information on the varied aspects of this subject to make this book a valuable addition to the collection of many professionals and students.

This book was conceptualized with the vision of imparting up-to-date information and advanced data in this field. To ensure the same, a matchless editorial board was set up. Every individual on the board went through rigorous rounds of assessment to prove their worth. After which they invested a large part of their time researching and compiling the most relevant data for our readers.

The editorial board has been involved in producing this book since its inception. They have spent rigorous hours researching and exploring the diverse topics which have resulted in the successful publishing of this book. They have passed on their knowledge of decades through this book. To expedite this challenging task, the publisher supported the team at every step. A small team of assistant editors was also appointed to further simplify the editing procedure and attain best results for the readers.

Apart from the editorial board, the designing team has also invested a significant amount of their time in understanding the subject and creating the most relevant covers. They scrutinized every image to scout for the most suitable representation of the subject and create an appropriate cover for the book.

The publishing team has been an ardent support to the editorial, designing and production team. Their endless efforts to recruit the best for this project, has resulted in the accomplishment of this book. They are a veteran in the field of academics and their pool of knowledge is as vast as their experience in printing. Their expertise and guidance has proved useful at every step. Their uncompromising quality standards have made this book an exceptional effort. Their encouragement from time to time has been an inspiration for everyone.

The publisher and the editorial board hope that this book will prove to be a valuable piece of knowledge for researchers, students, practitioners and scholars across the globe.

LIST OF CONTRIBUTORS

Timothy J. Snelling and R. John Wallace
Rowett Institute, University of Aberdeen, Foresterhill, Aberdeen AB16 5BD, UK

Xihui Xu, Lu Li and Chen Chen
College of Life Sciences, Nanjing Agricultural University, Nanjing 210095, China

Guopeng Li
Agricultural Product Processing Research Institute, Chinese Academy of Tropical Agricultural Sciences, Zhanjiang 524001, China

Zhenzhu Su
State Key Laboratory of Rice Biology, Institute of Biotechnology, Zhejiang University, Hangzhou 310058, China

Chao Song, Hongdong Li, Yunhui Zhang and Jialin Yu
Department of Neonatology, Children's Hospital of Chongqing Medical University, Chongqing, China Ministry of Education Key Laboratory of Child Development and Disorders – Chongqing Key Laboratory of Pediatrics, China International Science and Technology Cooperation Base of Child Development and Critical Disorders, Chongqing, China

Nawel Jemil, Hanen Ben Ayed, Moncef Nasri and Noomen Hmidet
Laboratoire de Génie Enzymatique et de Microbiologie, Université de Sfax, Ecole Nationale d'Ingénieurs de Sfax, B.P, 1173-3038 Sfax, Tunisia

Angeles Manresa
Section of Microbiology, Department of Biology, Health and Environment, Faculty of Pharmacy, University of Barcelona, Joan XXIII s/n, 08028 Barcelona, Spain

Hironori Taniguchi and Volker F. Wendisch
Genetics of Prokaryotes, Faculty of Biology, Bielefeld University, Bielefeld, Germany Center for Biotechnology, Bielefeld University, Bielefeld, Germany

Tobias Busche and Jörn Kalinowski
Center for Biotechnology, Bielefeld University, Bielefeld, Germany.

Thomas Patschkowski and Karsten Niehaus
Center for Biotechnology, Bielefeld University, Bielefeld, Germany Proteome and Metabolome Research, Faculty of Biology, Bielefeld University, Bielefeld, Germany

Miroslav Pátek
Institute of Microbiology, Academy of Sciences of the Czech Republic, Prague, Czech Republic

Fernanda A. Dorella, Siomar C. Soares, Thiago L. P. Castro, Núbia Seyffert, Anderson Miyoshi, and Vasco Azevedo
Departamento de Biologia Geral, Instituto de Ciências Biológicas, Universidade Federal de Minas Gerais, Belo Horizonte, Minas Gerais, Brazil

Wanderson M. Silva
Departamento de Biologia Geral, Instituto de Ciências Biológicas, Universidade Federal de Minas Gerais, Belo Horizonte, Minas Gerais, Brazil INRA, UMR1253 STLO, 35042 Rennes, France Agrocampus Ouest, UMR1253 STLO, 35042 Rennes, France

Artur Silva
Instituto de Ciências Biológicas, Universidade Federal do Pará, Guamá, Belém, Pará, Brazil

Gustavo H. M. F. Souza
Waters Corporation, Waters Technologies Brazil, MS Applications Laboratory, Alphaville, São Paulo, Brazil

Yves Le Loir
INRA, UMR1253 STLO, 35042 Rennes, France Agrocampus Ouest, UMR1253 STLO, 35042 Rennes, France

Henrique Figueiredo
Aquacen, Escola de Veterinária, Universidade Federal de Minas Gerais, Belo Horizonte, Brazil

Laura E. Sellars, Jack A. Bryant and Stephen J. W. Busby
Institute of Microbiology and Infection, School of Biosciences, University of Birmingham, Edgbaston, Birmingham B15 2TT, UK

María-Antonia Sánchez-Romero
Departamento deGenética, Facultad de Biología, Universidad de Sevilla, 41080 Seville, Spain

Eugenio Sánchez-Morán
School of Biosciences, University of Birmingham, Edgbaston, Birmingham B15 2TT, UK

David J. Lee
Institute of Microbiology and Infection, School of Biosciences, University of Birmingham, Edgbaston, Birmingham B15 2TT, UK
Department of Life Sciences, Birmingham City University, Edgbaston, Birmingham B15 3TN, UK

Ana Flávia Canovas Martinez, Luís Gustavo de Almeida, and Fernando Luís Cônsoli
Laboratório de Interações em Insetos, Departamento de Entomologia e Acarologia, Escola Superior de Agricultura "Luiz de Queiroz", Universidade de São Paulo, Av Pádua Dias 11, 13418–900, Piracicaba, SP, Brazil

Luiz Alberto Beraldo Moraes
Laboratório de Espectrometria de Massas Aplicada a Produtos Naturais, Departamento de Química, Faculdade de Filosofia, Ciências e Letras de Ribeirão Preto, Universidade de São Paulo, Av Bandeirantes 3900, 14040–901, Ribeirão Preto, SP, Brazil

Chao Yu, Huamin Chen, Fang Tian, Fenghuan Yang and Chenyang He
State Key Laboratory for Biology of Plant Diseases and Insect Pests, Institute of Plant Protection, Chinese Academy of Agricultural Sciences, Beijing 100193, China

Sayaka Nakamura, Hiroaki Sato
Research Institute for Sustainable Chemistry, National Institute of Advanced Industrial Science and Technology (AIST), Higashi 1-1-1, Tsukuba, Ibaraki 305-8565, Japan

Reiko Tanaka, Yoko Kusuya, Hiroki Takahashi and Takashi Yaguchi
Medical Mycology Research Center, Chiba University, 1-8-1 Inohana, Chuo-ku, Chiba 260-8673, Japan

Ping Zhou, Jun-chang Guan, Zhi-ben Xu, Shuxian Gao and Qing-wei Zheng
Department of Microbiology and Anhui Key Laboratory of Infection and Immunity, Bengbu Medical College, 2600 Dong Hai Avenue, Bengbu, Anhui 233030, People's Republic of China

Xin-sheng Zhang
Editorial Board of Journal of Bengbu Medical College, Bengbu Medical College, Bengbu, Anhui 233030, People's Republic of China

Ming-zhu Xu
Department of Life Sciences, Bengbu Medical College, Bengbu, Anhui 233030, People's Republic of China

Lin Shen
Scientific Research Center, Bengbu Medical College, Bengbu, Anhui 233030, People's Republic of China

Feng Yu
Huzhou University Schools of Medicine and Nursing Science, Huzhou, Zhejiang 313000, People's Republic of China

Yasuo Igarashi and Feng Luo
Research Center of Bioenergy & Bioremediation, College of Resources and Environment, Southwest University, No.2 Tiansheng Road, BeiBei District, Chongqing 400715, China

Delyana Vasileva, Chiho Suzuki-Minakuchi and Kazunori Okada
Biotechnology Research Center, The University of Tokyo, 1-1-1 Yayoi, Bunkyo-ku, Tokyo 113-8657, Japan

Zongping Sun and Hideaki Nojiri
Research Center of Bioenergy & Bioremediation, College of Resources and Environment, Southwest University, No.2 Tiansheng Road, BeiBei District, Chongqing 400715, China
Biotechnology Research Center, The University of Tokyo, 1-1-1 Yayoi, Bunkyo-ku, Tokyo 113-8657, Japan

S. H. Flint
School of Food and Nutrition, Massey University, Palmerston North, NewZealand

D. Lindsay
Fonterra Research Institute, Palmerston North, New Zealand

M. P. Cox
Statistics and Bioinformatics Group, Institute of Fundamental Sciences, Massey University, Palmerston North, New Zealand.

S. A. Burgess
School of Food and Nutrition, Massey University, Palmerston North, NewZealand

Infectious Disease Research Centre, Institute of Veterinary, Animal and Biomedical Sciences, Massey University, Palmerston North, New Zealand.

P. J. Biggs
Statistics and Bioinformatics Group, Institute of Fundamental Sciences, Massey University, Palmerston North, New Zealand
Infectious Disease Research Centre, Institute of Veterinary, Animal and Biomedical Sciences, Massey University, Palmerston North, New Zealand

Suchismita Ghosh, Paul A. Ayayee, Laura G. Leff and Christopher B. Blackwood
Department of Biological Sciences, Kent State University, Kent, OH 44242, USA

Oscar J. Valverde-Barrantes
International Center for Tropical Botany (ICTB), Florida International University, Miami, FL 33199, USA

Todd V. Royer
School of Public and Environmental Affairs, Indiana University, Bloomington, Bloomington, IN 47405, USA

Robert Kasimir Kulis-Horn, Christian Rückert, Jörn Kalinowski and Marcus Persicke
Microbial Genomics and Biotechnology, Center for Biotechnology, Bielefeld University, Universitätsstraße 27, 33615 Bielefeld, Germany

Giantommaso Scarascia, Hong Cheng, Moustapha Harb and Pei-Ying Hong
Biological and Environmental Science & Engineering Division (BESE), King Abdullah University of Science and Technology (KAUST), Water Desalination and Reuse Center (WDRC), Thuwal 23955-6900, Saudi Arabia

Emina Ćudić, Kristin Surmann and Sabine Hunkel
FB 5 Microbiology, Department of Biology/Chemistry, University Osnabrück, Barbarastraße 11, 49076 Osnabrück, Germany

Gianna Panasia
FB 5 Microbiology, Department of Biology/Chemistry, University Osnabrück, Barbarastraße 11, 49076 Osnabrück, Germany
Department of Biology, Institute of Molecular Microbiology and Biotechnology, Universität Münster, Corrensstraße 3, 48149 Münster, Germany

Elke Hammer
Department of Functional Genomics, Interfaculty Institute of Genetics and Functional Genomics, University Medicine Greifswald, Friedrich-Ludwig-Jahn-Straße 15A, 17475 Greifswald, Germany

Nils Chr. Stenseth, Thomas H. A. Haverkamp and W. Ryan Easterday
Department of Biosciences, Centre for Ecological and Evolutionary Synthesis (CEES), University of Oslo, The Kristine Bonnevie Building, UiO, campus Blindern, Blindern, Oslo, Norway

Jaran S. Olsen
Norwegian Defence Research Establishment, Kjeller, Norway

Karoline Valseth
Department of Biosciences, Centre for Ecological and Evolutionary Synthesis (CEES), University of Oslo, The Kristine Bonnevie Building, UiO, campus Blindern, Blindern, Oslo, Norway
Norwegian Defence Research Establishment, Kjeller, Norway

Camilla L. Nesbø
Department of Biosciences, Centre for Ecological and Evolutionary Synthesis (CEES), University of Oslo, The Kristine Bonnevie Building, UiO, campus Blindern, Blindern, Oslo, Norway
Department of Biological Sciences, University of Alberta, Edmonton, Alberta, Canada.

Wendy C. Turner
Department of Biological Sciences, University at Albany, State University of New York, Albany, New York, USA

Index